Concrete and Masonry Movements

T0348666

Concrete and Masonry Movements

J. J. Brooks

AMSTERDAM • BOSTON • HEIDELBERG • LONDON • NEW YORK
OXFORD • PARIS • SAN DIEGO • SAN FRANCISCO • SINGAPORE
SYDNEY • TOKYO
Butterworth Heinemann is an imprint of Elsevier

Butterworth Heinemann is an imprint of Elsevier
The Boulevard, Langford Lane, Kidlington, Oxford OX5 1GB, UK
225 Wyman Street, Waltham, MA 02451, USA

Notices
Knowledge and best practice in this field are constantly changing. As new research and experience
broaden our understanding, changes in research methods, professional practices, or medical treatment
may become necessary.

Practitioners and researchers must always rely on their own experience and knowledge in evaluating and
using any information, methods, compounds, or experiments described herein. In using such information
or methods they should be mindful of their own safety and the safety of others, including parties for
whom they have a professional responsibility.

To the fullest extent of the law, neither the Publisher nor the authors, contributors, or editors, assume any
liability for any injury and/or damage to persons or property as a matter of products liability, negligence
or otherwise, or from any use or operation of any methods, products, instructions, or ideas contained in
the material herein.

Library of Congress Cataloging-in-Publication Data
Brooks, J. J.
  Concrete and masonry movements / Jeffrey Brooks.
    pages cm
  Includes bibliographical references and indexes.
  ISBN 978-0-12-801525-4
  1. Masonry. 2. Concrete–Creep. 3. Earth movements and building. I. Title.
  TA670.B76 2015
  624.1'83–dc23
                          2014023532

British Library Cataloguing in Publication Data
A catalogue record for this book is available from the British Library

  ISBN: 978-0-12-801525-4

For information on all Butterworth Heinemann publications
visit our web site at store.elsevier.com

This book has been manufactured using Print On Demand technology.

Contents

Preface

In the past, movements of concrete and masonry buildings have been attributed to causes structural failure but these instances have been rare because of good design practice. The latter recognizes that precise knowledge of movements is important in order to achieve the desired design serviceability criteria and avoid costly repairs due to lack of durability. Excessive movements are undesirable and even acceptable movements when restrained can cause undue local material failures, which may be dangerous, unsightly, and expensive to remedy. In some cases such as bridges and high-rise buildings large deflections can cause general alarm even though they may be structurally safe.

"Concrete and masonry movements" is a compilation of knowledge of four basic categories of movement: elasticity, shrinkage, creep and thermal movement but within each category there are several different types. All are explained and discussed in detail from theoretical viewpoints as well as from experimental observations. For concrete, up-to-date literature particularly on the effects of new chemical and mineral admixtures, and recycled waste materials are added to existing knowledge while, for masonry, comprehensive literature reviews, models, and viewpoints are presented. The role played by transfer of moisture at the unit/mortar interface is investigated together with the causes of cryptoflorescence and its effect on creep of masonry. Although the two materials are the oldest construction materials and tend to be treated separately by their respective professional institutions, the approach in this book considers deformations of concrete and masonry side by side or even together since they have many common features that result in similar properties and behaviour. On the other hand, they also have dissimilar features mainly emanating from the use of fired clay units in masonry, which can result in totally different behaviour; those features are highlighted in separate chapters.

This book has been written for undergraduate, postgraduate students and practicing civil engineers who wish to understand why movements occur and how to take them into account when designing concrete and masonry structures. For undergraduate students, underlying principles responsible for each type of movement are given and illustrated by worked examples and problems at the end of each chapter. The postgraduate student requires background knowledge of previous research on appropriate topics before embarking on new approaches, methods, and construction materials, and such background literature is presented for each topic with comprehensive list of references at the end of each chapter. For practicing civil and structural engineers, latest research findings are given together with relevant Codes of Practice prescribed by British, European, and American standards, the application and comparison of which are demonstrated with worked examples.

Acknowledgments

The author is grateful for the permission to reproduce or incorporate extracts from publications, full details of which can be found at the end of each chapter by a reference number given with captions to the illustrations and tables. In cases of full reproduction, specific copyright acknowledgment is given with illustrations and tables.

Particular thanks is given to the University of Leeds for granting permission to reproduce information from several masters and doctorate theses to compile tables and figures: (Figure 5.2; Table 6.1; Figures 8.18, 9.2, 6.21, and 6.22; Tables 10.3 and 10.4; Figure 10.16; 10.21; 12.3; 15.22; 16.7, and 16.9). Similarly, special appreciation is forthcoming to the following organizations for granting permission to use extensive data from numerous publications by the American Concrete Institute (ACI), The International Masonry Society (Stoke-on-Trent, formerly the British Masonry), the Institution of Civil Engineers (ICE) including Thomas Telford Ltd. (London), RILEM (Reunion Internationale des Laboratoires et Experts de Materiaux, Systems de Construction et Ouvrages), Bagneux, France and the British Standards Institution. Details of each publisher and locations in the book are given below:

- The following have been compiled using ACI sources (including ACI Journal, ACI Materials Journal, Special Publications, Committees-Manual of Concrete Practice, Concrete International): Figures 4.3 and 4.4; Figures 4.13–4.16; Tables 5.15 and 5.16; Figure 6.8; Figures 6.8–6.12; Figures 6.16 and 6.17; Figure 10.17; Figure 14.13; Table 13.2; Tables 13.1 and 13.2; Table 11.7; Table 11.15; Table 11.1; Figure 11.4; Figures 10.8 and 10.9.
- In the case of The International Masonry Society (including, British Ceramic Society, Masonry International), the figures and tables compiled are: Figure 5.1; Figure 5.5; Tables 5.2 and 5.3; Table 5.8; Figure 7.13; Figure 7.16; Figure 7.18; Figure 8.1; Figure 8.3; Figures 8.10–8.12; Tables 8.1–8.3; Figure 8.4; Figure 8.7; Figures 8.16 and 8.17; Figures 9.3–9.5; Figure 12.1; Figure 8.8; Figure 12.2; Figures 12.5 and 12.6; Figures 12.9–12.11; Figure 12.12; Figures 12.14 and 16.2.
- Data from ICE Publishing including Magazine of Concrete Research, ICE Proceedings and Civil Engineering and Public Works Review have been used in: Figure 3.3; Table 4.3; Figure 6.2; Figure 7.13(b); Figure 7.16 (a); Table 10.1; Figure 10.4; Figure 10.6; Figure 10.15; Figures 10.23–10.25; Figure 10.29; Figure 11.5; Figures 11.7–11.10; Table 11.16; Table 15.3; Figure 16.5; Figure 16.12; Figure 16.13; Figure 15.17; Figure 15.20; Figure 16.11; Figures 16.23 and 16.24.
- Extracts from RILEM publications (RILEM Bulletin, Materials and Structures Journal, Symposium Proceedings) are incorporated in: Figures 7.2 and 7.3; Figure 7.19; Figure 9.7; Table 10.5; Table 10.7; Figure 10.19; Table 12.2; Figures 13.4 and 13.5; Figures 15.2 and 16.4.

- British and European Standards have been used to compile: Tables 4.1 and 4.2; Figure 4.5; Figure 4.12; Tables 5.9–5.11; Tables 5.13 and 5.14; Table 11.3; Figures 11.1–11.3; Figure 13.3; Table 13.5; Tables 14.4–14.6; Tables 14.8 and 14.9; Figures 14.10–14.12.

Other publishers which are acknowledged for their contributions are now listed. The American Society for Testing and Materials (ASTM): Figure 10.10; Figures 15.18 and 16.25. Institution of Structural Engineers (ISE) including The Structural Engineer: Figures 11.11–11.16; Tables 11.19–11.22 The Concrete Society, including Concrete Journal: Figure 10.3 and Table 13.1. Society of Petroleum Engineers: Figure 10.13. Elsevier including Cement and Concrete Composites, Prentice-Hall: Figures 7.9 and 4.9. Lucideon, formerly CERAM Building Technology, British Ceramic Research Ltd.: Figure 5.6; Table 7.3; Figure 8.2; Figure 8.9; Figure 8.18; and Table 13.7. US National Institute of Standards and Technology: Figure 6.1. Prestressed Concrete Institute: Table 10.6. DIN (German Institute for Standardisation): Figure 10.5. Brick Development Association (BDA Design Note): Figure 14.15. Swedish Cement and Concrete Institute: Table 10.2. Portland Cement Association: Figure 10.7; Figure 15.1; Figure 15.4. CIRIA (Construction Industry Research and Information Association): Table 14.2. US Bureau of Reclamation: Figure 16.9. IHS BRE (The Building Research Establishment): Table 13.6. Pearson Education Ltd (including Prentice-Hall): Figures 2.1–2.4, Figure 2.5; Figure 4.9; Table 8.4; Figure 9.2; Figure 9.6; Figure 12.1; Figure 12.2; and Figure 13.2. Van Nostrand Reinhold (including Chapman and Hall): Figures 14.9 and 16.26. Japan Society of Civil Engineers: Figure 16.10. Japan Cement Association: Figure 15.3. Structural Clay Products: Figure 12.2. John Wiley and Sons: Table 15.2. Hanley Wood Business Media (Concrete Construction Magazine): Figure 6.15.

The author would like to acknowledge contributions made by numerous students who undertook research under the author's supervision, many of which are referenced throughout the book. Also, there is special acknowledgment to Adam Neville with whom the author studied and coauthored research papers and books over many years; his industry and meticulous attention to detail have been an inspiration. The important contributions by laboratory technical staff in teaching practical skills and experimental techniques to research students are not forgotten and so they are also acknowledged, and in particular the author is grateful to Vince Lawton. Lastly, but not least, the unwavering support of my wife Cath, who suggested the writing of this book in the first place, is gratefully acknowledged.

1 Introduction

"Concrete and masonry movements" is a compilation of existing and up-to-date knowledge of movements of two traditional construction materials, based upon the author's research and teachingover a period of 30 years. It is a reference book that brings together theory and engineering practice with worked examples and, consequently, is suitable for the practising engineer, research student, and undergraduate student studying civil engineering.

The presentation is somewhat different because it considers deformation properties of plain concrete and plain masonry together. Structural concrete and masonry containing steel reinforcing bars or prestressing tendons are not included. Conventionally, properties of concrete and masonry have been treated as separate composite materials by their respective professional institutions in spite of having common constituents: cement, sand, and coarse aggregate (brick or block). The theme of the book is to consider each type of movement of concrete and masonry in separate chapters, but to emphasize common features, except where behaviour and features are so common that treatment in different chapters is not warranted. It is the author's belief that the mutual exchange of knowledge in this manner will lead to a greater understanding of the movement properties of both materials.

What is essentially different about the two materials is when the clay brick or block is used as the "coarse aggregate" constituent, because of its behaviour under normal ambient conditions and how it can react with mortar to influence the movement of masonry. The book emphasizes the property of clay brick units exhibiting irreversible moisture expansion, which, under some circumstances, when combined with mortar to build free-standing masonry, can manifest itself as an enlarged moisture expansion due to the occurrence of cryptoflorescence at the brick/bond interface. When occurring in a control wall, this feature appears to increase creep of masonry because of the way in which creep is defined but, in practice, the enlarged moisture expansion is suppressed in masonry provided there is sufficient dead load or external load.

Summaries of all topics discussed are now presented chapter by chapter.

After defining terms and types of movement in Chapter 2, composite models for concrete and masonry are presented for: elasticity, creep, shrinkage or moisture expansion, and thermal movement. A new composite model is developed for masonry. Composite models are useful in understanding how individual components having different properties and quantities interact when combined. The models are applied and verified in other chapters, particularly for masonry, which has the advantage that movement properties of units can be physically measured in the laboratory. With concrete, this approach is not practicable because of the much smaller size of the coarse aggregate, a feature that makes it difficult to measure representative movement characteristics.

Concrete and Masonry Movements. http://dx.doi.org/10.1016/B978-0-12-801525-4.00001-7

An example of the above-mentioned problem is in Chapter 4, which deals with elasticity of concrete. Modulus of elasticity is related to strength empirically because of the difficulty in measurement of aggregate modulus in order to apply theoretical composite models. Short-term stress–strain behaviour in compression leading to different definitions of modulus of elasticity is described together with Poisson's ratio. Main influencing factors are identified and effects of chemical and mineral admixtures are discussed in detail. Relations prescribed by U.S. and European standards are given for estimating modulus of elasticity from strength in tension as well as corresponding relations in compression, but there is a large scatter mainly because of the failure to quantify the influence of aggregate precisely.

Chapter 5 deals with elasticity of masonry and, besides presenting current empirical relations between modulus of elasticity and strength, composite models are tested and developed for practical application. In the first instance, it is demonstrated that modulus of elasticity of units and mortar may be expressed as functions of their respective strengths so that the composite model for modulus of elasticity of masonry can be expressed in terms of unit and mortar strengths. However, a limitation of the theoretical approach is demonstrated in the case of units laid dry, which causes moisture transfer at the unit/mortar bond during construction. This mainly affects the elastic properties of the bed joint mortar phase. However, this effect can be quantified in terms of the water absorption of the unit, which is thus an additional factor taken into account by the composite prediction model.

The different types of deformation arising from moisture movement that occur in concrete are described in Chapter 6. These range from plastic, autogenous, carbonation, swelling, and drying shrinkage, but emphasis is given to autogenous shrinkage and drying shrinkage especially, in view of the recent developments in the use of high strength concrete made with low water/binder ratios, very fine cementitious material, and chemical admixtures. Influencing factors are identified and quantified, such as the effects of mineral admixtures: fly ash, slag, microsilica, and metakaolin, and the effects of chemical admixtures: plasticizers, superplasticizers, and shrinkage-reducing agents. Methods of determining autogenous shrinkage are described and the latest methods of prediction are presented with worked examples.

The drying shrinkage behaviour of calcium silicate and concrete masonry, and their component units and mortar joints, are the subjects of Chapter 7. After considering influencing factors, the importance of the moisture state of the units at the time of laying is emphasized because of its effect on shrinkage of the bonded unit, mortar, and masonry. A mortar shrinkage-reducing factor is quantified in terms of water absorption and strength of the unit. The geometry of the cross section of masonry, quantified in terms of the ratio of its volume to the drying, exposed surface area, is also shown to be an important factor. The main influencing factors are accommodated in the composite models, which are developed for practical use to estimate shrinkage of calcium silicate and concrete masonry. Methods prescribed by Codes of Practice are also presented and their application is demonstrated with worked examples.

Moisture movement of masonry built from most types of clay units behaves in a different manner to other types of masonry and to concrete due to the property of irreversible expansion of clay units, which begins as soon as newly-made units have cooled after leaving the kiln. The effect is partially restrained when units are bonded with mortar since the mortar joints shrink, but the net effect in masonry depends on the type of clay used to manufacture the unit and the firing temperature. In fact, masonry shrinks in the long-term when constructed from a low, expanding clay brick. In Chapter 8, a detailed review of irreversible moisture expansion of clay units is undertaken before proposing a model to estimate ultimate values from knowing the type of clay and the firing temperature. Laboratory methods of measuring irreversible moisture expansion of clay units are given. It is then demonstrated that prediction of moisture movement of clay brick masonry can be achieved successfully by composite modeling.

The phenomenon of enlarged moisture expansion of clay brickwork is the subject of Chapter 9, which occurs in special circumstances when certain types of clay unit are bonded with mortar to create conditions for the development of cryptoflorescence at the interface of the brick/mortar bond. In many instances, the clay units responsible for the phenomena are of low strength, have high suction rate, and are laid dry. The degree of enlarged expansion also depends on in-plane restraint of the masonry and, hence, can be suppressed by wetting or docking units before laying, and ensuring there is sufficient dead load acting on the masonry. Enlarged moisture expansion is of particular relevance in measuring creep of clay brickwork by using laboratory-sized specimens, and recommended test procedures are suggested. The chapter examines the nature of efflorescence, the influencing factors, and the mechanisms involved.

Chapters 10 and 11, respectively, deal with creep of concrete and standard methods of prediction of creep. Two chapters are allocated because of the number of factors influencing creep, and the numerous methods available to the designer for estimating elasticity, shrinkage, and creep of concrete, especially with the advent of high performance concrete containing mineral and chemical admixtures. Besides creep in compression, Chapter 10 highlights creep under tensile loading and creep under cyclic compression; prediction of creep under both those types of loading is included. Standard methods of estimating creep of concrete from strength, mix composition, and physical conditions are presented in Chapter 11 and their application demonstrated by worked examples. For greater accuracy, estimates by short-term testing are recommended and, finally, a case study is given to illustrate the recommended approach when new or unknown ingredients are used to make concrete.

Creep of masonry is the topic of Chapter 12. Compared with concrete, there has been only a small amount of research, and therefore there are fewer publications dealing with the subject. A brief historical review is given and a data bank of published results is complied. The chapter draws on the experience of knowledge of creep of concrete to develop a practical prediction model for masonry by quantifying creep of mortar and creep of different types of unit in terms of their respective strengths, water absorption of unit, and geometry of masonry. Current European and American Code of Practice guidelines are presented with worked examples. The association of creep with the presence of cryptoflorescence in certain types of clay brick masonry is also investigated.

Thermal movement of both concrete and masonry is considered together in Chapter 13 in terms of practical guidance prescribed in design documents and by composite modeling using thermal expansion coefficients of constituents: aggregate or unit and mortar, and their volumetric proportions. In practical situations, thermal movement and all the other various deformations of concrete or of masonry occur together and are often partially restrained in a complicated manner. The resulting effects, which may result in loss of serviceability due to cracking, are discussed in Chapter 14, together with remedies adopted in structural design to accommodate movements and to avoid cracking. Types and design of movement joints are described in detail and their application is demonstrated with worked examples.

Existing theories of creep and shrinkage of cement-based materials are based on those proposed for concrete. However, since none explain all the experimentally observed behaviour, a different theory is proposed and developed in Chapter 15, which is based on the movement of absorbed and interlayer water within and through the C-S-H pore structure. A key assumption is that the adsorbed water is load-bearing in having a structure and modulus of elasticity greater than that of "free" or normal water. If adsorbed water is removed, stress is transferred from adsorbed water in the pores to the solid gel of the cement paste, thus increasing its deformation. Drying shrinkage may be regarded as an elastic-plus-creep strain due to capillary stress generated by the removal of water. The theory is applied to several test cases of creep previously unexplained by existing theories.

The final Chapter 16 deals with the important subject of testing and measurement of elasticity, creep, and shrinkage of concrete and masonry. Measurement of the other types of movement are discussed in relevant chapters, and Chapter 16 concentrates on uniaxial-compressive and tensile-loading techniques and types of strain measurement with practical guidance for good, experimental practice in the laboratory. Prescribed American and European methods of test for determining creep of concrete are included, there being no equivalent standards for determining creep of masonry. Other prescribed, standard test methods are included in this chapter, which use length comparators for determining, independently of creep, shrinkage of concrete, mortar, and masonry units.

2 Classification of Movements

The types of movement discussed in this book will be briefly explained and defined in this chapter. The types are grouped as follows:

- Shrinkage and swelling, which includes plastic shrinkage, autogenous shrinkage, carbonation shrinkage, and drying shrinkage.
- Irreversible moisture expansion.
- Thermal expansion and contraction.
- Elastic strain and creep.

Whereas the first three groups of movement are determined from measurements using control, load-free specimens, elastic strain and creep are determined from the measurements of the total strain resulting from identical specimens subjected to external load, which is generally compression. The specific types of movement within a group will be defined shortly, after presentation of a general overview.

Concrete and masonry exhibit changes in strain with time, when no external stress is acting, due to the movement of moisture from or to the ambient medium. In the latter case, these changes are mainly due to drying shrinkage (although other types of shrinkage contribute), while swelling or moisture expansion arises from movement of moisture from the ambient medium. Other types of shrinkage that are usually measured with drying shrinkage are autogenous and carbonation. The generic term "shrinkage" is used for normal-strength concrete but, in the case of high-performance (high-strength and low-permeability) concrete, autogenous shrinkage is more significant and is determined separately.

In the past, the phenomenon of creep of concrete has been variously termed flow, plastic flow, plastic yield, plastic deformation, time yield, and time deformation [1]. This arose partly from the concept of the mechanism of the deformation as seen at the time and partly from a lack of agreement on what was still a newly discovered phenomenon. Nowadays, the term "creep" is universally accepted for both concrete and masonry.

When shrinkage and creep occur simultaneously, the common practice is to consider the two phenomena to be additive. The overall increase of strain of a stressed and drying member is therefore assumed to consist of shrinkage (equal in magnitude to that of an unstressed member of the same size and shape) and a change in strain due to stress, i.e., creep. This approach has the merit of simplicity and is suitable for analyses in many practical applications where shrinkage and creep occur together, but the additive definition is not really correct since the effect of shrinkage appears to increase the magnitude of creep. Nevertheless, this approach is followed in this book, since it has been followed by previous investigators and is universally adopted. To understand the phenomena under drying conditions, the extra component of creep is distinguished

Concrete and Masonry Movements. http://dx.doi.org/10.1016/B978-0-12-801525-4.00002-9

from creep under conditions of no moisture movement to or from the ambient medium. The former extra component is referred to as drying creep while the latter component is referred to as basic creep, as first used by Neville [2] and by Ali and Kesler [3]; thus, under drying conditions, the total creep is the sum of basic and drying creep.

The additive approach also applies to creep and shrinkage of masonry, except in the case of masonry built with clay bricks exhibiting irreversible moisture expansion and prone to cryptoflorescence at the brick/mortar bond. Here, unstressed free-standing masonry undergoes an enlarged expansion, whereas in masonry under compression cryptoflorescence is suppressed. Thus, use of the additive approach to quantify creep leads to a large and false magnitude of creep.

Definition of Terms Used

Shrinkage and Swelling

Shrinkage of concrete and masonry is caused by loss of moisture by evaporation, hydration of cement, and carbonation. The resulting reduction of volume as a fraction of the original volume is the *volumetric strain*, which is equal to three times the linear strain, so that shrinkage can be measured as a linear strain in units of mm per mm, usually expressed as microstrain (10^{-6}). Conversely, *swelling* is an increase in volume when there is continuous storage in water during hydration due to absorption of water by the cement paste; swelling is much smaller than drying shrinkage.

When freshly laid and before setting, mortar and concrete can undergo *plastic shrinkage* due to loss of water from the exposed surface or from suction by the drier layers underneath or adjacent. Plastic shrinkage is minimized by prevention of evaporation immediately after casting. Figure 2.1 illustrates that, from the initial setting of cement, *autogenous shrinkage* takes place due to internal consumption of water by the hydrating cement and there is no external moisture exchange to or from the set mortar or concrete. Autogenous shrinkage is determined using sealed specimens and occurs rapidly during the initial stages of hydration; it is small in normal-strength concrete and masonry mortar but can be very large in very-high-performance concrete made with a low water/cementitious materials ratio, containing chemical and mineral admixtures. When exposed to a dry environment at a later age, t_o, *drying shrinkage* takes place as a result of loss of moisture from the set concrete or mortar, while swelling occurs due to water storage. When reckoned from age t_o, Figure 2.1 shows that *total shrinkage* consists of drying shrinkage plus some autogenous shrinkage and, although the latter's contribution is less, it may still be significant depending on the type of concrete and the age of exposure to drying t_o.

Carbonation shrinkage takes place in surface layers of concrete and masonry mortar due to the reaction of carbon dioxide in the atmosphere with calcium hydroxide of the hardened cement paste in the presence of moisture. It occurs together and is measured with drying shrinkage but generally is much smaller. In normal-strength concrete and mortar, drying shrinkage is between 40% and 70% reversible on immersion of specimens in water after a period of drying, but this applies to first

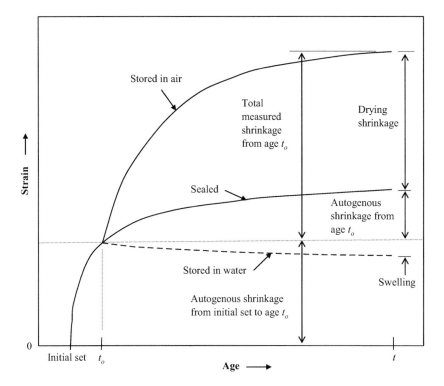

Figure 2.1 Types of shrinkage exhibited by concrete sealed and then stored in different conditions from age t_o.

drying, since subsequent cycles of drying and wetting are almost reversible. On the other hand, autogenous shrinkage and carbonation shrinkage are irreversible. Irreversibility of drying shrinkage is due to pore-blocking by-products of hydration of cement and carbonation during the process of drying.

Irreversible Moisture Expansion

This type of movement is a unique feature of fired clay bricks and blocks, and occurs after the units have cooled after leaving the kiln due to take-up of moisture from the atmosphere; the units expand rapidly at first, then slowly over a long period of time. The effect on clay masonry can also induce irreversible moisture expansion, but at a reduced level, and a net shrinkage is even possible because of opposing restraint by shrinking mortar joints. However, the process depends on type of clay unit and its age. Generally, to minimize irreversible moisture expansion, it is recommended that clay units not be used to construct masonry until they are at least 7 days old and that the design of clay masonry should include movement joints to allow for moisture expansion.

In some types of clay masonry, an *enlarged moisture expansion* may arise due to crystallization of salts at the clay unit/mortar interface, a process known as cryptoflorescence. Enlarged moisture expansion is defined as the expansion in excess of the irreversible moisture expansion of the clay unit used to construct the masonry. The phenomenon typically occurs in small, unrestrained masonry built with units of low strength, high water absorption, and high initial suction rate. On the other hand, enlarged moisture expansion is suppressed in masonry under compression, so that, as stated earlier, it poses a particular problem when determining creep in the laboratory in the traditional way, since the use of small load-free control specimens to allow for moisture movement yields unrealistic high values of creep.

Thermal Expansion and Contraction

Thermal movement arises from thermal expansion or contraction of concrete and masonry elements. Thermal movement is equal to the product of coefficient of thermal expansion and change in temperature and, for both concrete and masonry, the thermal coefficient is assumed to be independent of time.

Elastic Strain and Creep

In the most general form, the elastic strain plus creep–time curve for engineering materials exhibiting time-dependent failure is shown in Figure 2.2, creep being reckoned from the strain resulting from application of load. The strain at zero time is primarily elastic but may include a nonelastic component. Thereafter, there are three stages of creep. In the *primary creep* stage, the rate of creep is initially high and then decreases with time. If a minimum creep rate is exhibited, a *secondary creep* stage (sometimes called *stationary creep*) designates a stage of steady-state creep. The straight line relation of secondary creep may be a convenient approximation when the magnitude of this creep is large compared with primary creep. The *tertiary creep* stage may or may not exist, depending on the level of stress. For instance, in concrete, this may arise from an increase in creep due to growth of microcracks in the cement paste/mortar phase at stress greater than approximately 0.6–0.8 of the short-term strength in compression or in tension. Failure occurs when microcracks link and propagate in an unstable fashion through the whole material, which undergoes large strains prior to disintegration.

For normal levels of stress used in concrete and masonry elements, primary creep cannot be distinguished from secondary creep, and tertiary creep does not exist. The strain–time curve is of the form shown in Figure 2.3 and creep is simply defined as the gradual increase in strain with time for a constant applied stress after accounting for other time-dependent deformations not associated with stress, e.g., shrinkage and thermal movement. Creep may continue, although at a very low rate, for many years.

The strain at loading is classified as mainly an elastic strain and corresponds to the secant modulus of elasticity at the age when the load is applied. For the sake of accuracy, it should be noted that, since both concrete and masonry mature with age, the modulus of elasticity increases with time so that the elastic strain decreases with time under a sustained stress (see Figure 2.4(b)). Thus, strictly speaking, creep should be reckoned as strain in excess of the elastic strain at the time considered and not in

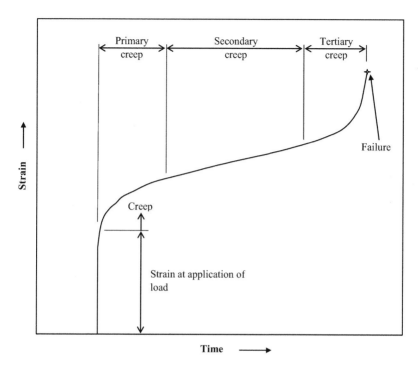

Figure 2.2 General form of the strain–time curve for material undergoing creep leading to failure [1].
Source: Creep of Plain and Structural Concrete, A. M. Neville, W. H. Dilger and J. J. Brooks, Pearson Education Ltd. © A. M. Neville 1983.

excess of the strain at the time of application of load. However, because the difference in the two methods is generally small and because of convenience, the change in elastic strain with age is ignored.

The strain at loading and the secant modulus of elasticity depend on the level of applied stress and its rate of application because the stress–strain curve is nonlinear, as shown in Figure 2.5. In fact, strictly speaking, concrete and masonry are classified as nonlinear and nonelastic materials, pure elasticity being defined as when strains appear and disappear on application and removal of load. For low stresses, the greater initial tangent modulus of elasticity is more appropriate to define the strain at loading of a creep test. If the load is applied extremely rapidly, the recorded strains and nonlinearity are reduced and, correspondingly, the secant modulus of elasticity becomes very similar to the initial tangent modulus. The dependency of instantaneous strain on rate of loading makes the demarcation between elastic and creep strains difficult so that reported test data should include the time taken to apply the load in a creep test, which is generally of the order of 1–2 min, depending on the type of loading apparatus.

The definition of terms is shown in Figure 2.4 using the additive definition of creep discussed earlier for the case when there is concomitant shrinkage. For concrete

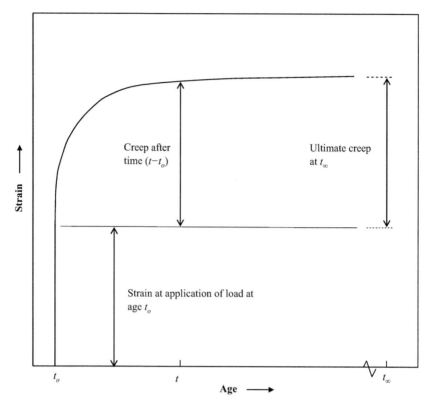

Figure 2.3 General form of the strain–time curve for concrete and masonry subjected to normal levels of sustained stress.

exposed to drying from age to when a compressive load is applied, the shrinkage as measured on a separate load-free specimen is given by Figure 2.4(a) and the total measured strain of the specimen under load is given by Figure 2.4(b) and consists of elastic strain, shrinkage, and *total creep*. For the case of sealed concrete or masonry, the total measured strain of the specimen under load is much less since there is no shrinkage component and creep is smaller, which is, in fact, termed *basic creep* (see Figure 2.4(c)). Thus, the total measured strain of a loaded and drying specimen consists of the components shown in Figure 2.4(d), where total creep comprises basic creep and *drying creep*. Drying creep is the extra creep induced even after allowing for free shrinkage as measured on an unstressed specimen.

It should be noted that basic creep is often used to describe creep of concrete stored in water. In such a case, swelling as measured on a control load-free specimen is usually small compared with creep under a compressive load, so that the conditions approximate to those of no moisture exchange or hygral equilibrium.

Compared with creep at normal temperature, creep is accelerated when heat is applied just before application of load. However, if heated to the same temperature

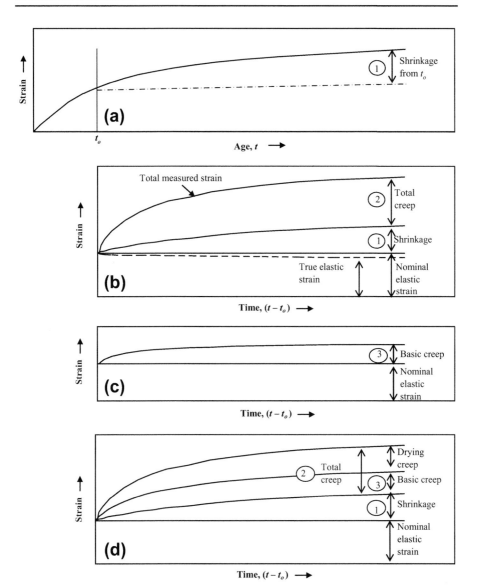

Figure 2.4 Definition of terms used for elastic strain and creep [1]. (a) Shrinkage of a load-free control specimen. (b) Total measured strain of a loaded and drying specimen. (c) Strain of a loaded and sealed specimen. (d) Components of strain of a loaded and drying specimen. *Source*: Creep of Plain and Structural Concrete, A. M. Neville, W. H. Dilger and J. J. Brooks, Pearson Education Ltd. © A. M. Neville 1983.

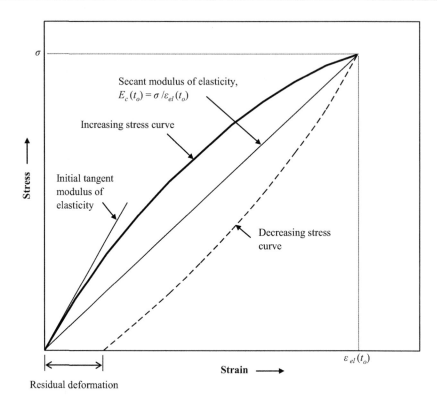

Figure 2.5 Generalized enlarged stress–strain curve for concrete and masonry [1].
Source: Creep of Plain and Structural Concrete, A. M. Neville, W. H. Dilger and J. J. Brooks, Pearson Education Ltd. © A. M. Neville 1983.

just after application of load, an additional component of creep occurs, which is termed *transitional thermal creep* or *transient creep*. At very high temperature, such as in fire, very high elastic and creep strains occur; collectively, these are termed *transient thermal strain*.

Shrinkage, elastic deformation, and creep are expressed as strain, i.e., as dimensionless quantities (mm per mm). However, sometimes it is convenient to give the magnitude of the elastic deformation and creep not for the actual stress applied (usually expressed as a proportion of the short-term strength) but per unit of stress. Such values are called *specific elastic strain* or *elastic compliance*, and *specific creep* or *creep compliance*, which are expressed in units of 10^{-6} per MPa. If $\sigma =$ stress applied, the specific elastic strain (ε_{sp}) is given by:

$$\varepsilon_{sp} = \frac{\varepsilon_{el}(t_o)}{\sigma} = \frac{1}{E(t_o)} \tag{2.1}$$

where $\varepsilon_{el} =$ elastic strain and $E(t_o) =$ modulus of elasticity at age t_o.

Specific creep, c_{sp} or $C(t, t_o)$, is given by:

$$c_{sp} = C(t, t_o) = \frac{c(t, t_o)}{\sigma} \tag{2.2}$$

where $c(t, t_o) =$ creep at age t due to a stress applied at age t_o.

The sum of the specific elastic strain at the time of application of load and the specific creep after time $(t - t_o)$ is termed the *compliance* or *creep function*, $\Phi(t, t_o)$, i.e.,

$$\Phi(t, t_o) = \frac{1}{\sigma}[\varepsilon_{el}(t_o) + c(t, t_o)] = \frac{1}{E(t_o)} + C(t, t_o) \tag{2.3}$$

The ratio of creep to the elastic strain is termed the *creep coefficient*, which is also known as the *creep factor*, viz.:

$$\phi(t, t_o) = \frac{c(t, t_o)}{\varepsilon_{el}(t_o)} = E(t_o) \times C(t, t_o) \tag{2.4}$$

The creep coefficient as defined in Equation (2.4) is the ratio of creep at age t to the elastic strain at the age of loading, t_o. An alternative term is the *28-day creep coefficient*, $\phi_{28}(t, t_o)$, which is defined as the ratio of creep age t measured from loading at age t_o, to the elastic strain at the age of 28 days. The two creep coefficients are related as follows:

$$\phi_{28}(t, t_o) = \phi(t, t_o) \frac{E_{28}}{E(t_o)} \tag{2.5}$$

where $E_{28} =$ modulus of elasticity at the age of 28 days.

A useful parameter in the analysis of modeling of creep effects is to quantify creep in terms of an *effective modulus* of elasticity, which decreases as the time under load increases. The effective modulus is equal to the stress divided by the sum of elastic strain and creep, i.e.,

$$E'(t, t_o) = \frac{\sigma}{\varepsilon_{el}(t_o) + c(t, t_o)} \tag{2.6}$$

where $E'(t, t_o) =$ effective modulus of elasticity at the age of t after load application at the age of t_o.

Hence, specific creep $C(t, t_o)$ is given by:

$$C(t, t_o) = \frac{1}{E'(t, t_o)} - \frac{1}{E(t_o)} \tag{2.7}$$

Creep Recovery

If a sustained load is removed, concrete and masonry undergo an instantaneous recovery followed by a slower time-dependent recovery, known as creep recovery.
Figure 2.6 illustrates this situation. Unlike the strain at application of load, the
instantaneous recovery is more elastic in nature and is lower in magnitude due to the
increase of modulus of elasticity with age. The instantaneous recovery strain is
determined by the unloading secant modulus of elasticity of the decreasing stress
curve in Figure 2.5 at the age of load removal. Creep recovery rapidly tends to a finite
value and is the reversible part of creep that is generally smaller than the preceding
creep; the remaining strain is the *residual deformation* or *permanent set* due to
irreversible creep. In young concrete, creep is large and is only approximately 20%
reversible, whereas in mature concrete creep is less but is more reversible.

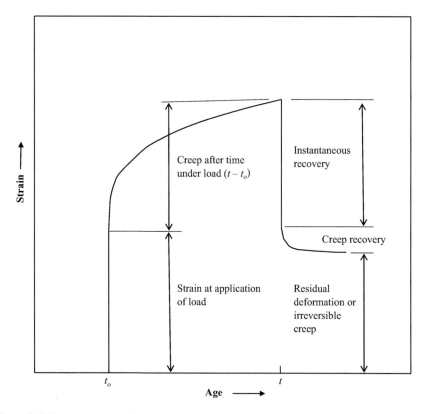

Figure 2.6 Instantaneous and creep recoveries after concrete or masonry has been subjected to
load from age t_o and unloaded at age t [1].
Source: Creep of Plain and Structural Concrete, A. M. Neville, W. H. Dilger and J. J. Brooks,
Pearson Education Ltd. © A. M. Neville 1983.

Relaxation

Under some circumstances, the deformations of concrete and masonry members are kept constant or vary in a predetermined manner. In the case of constant stress, strain steadily increases due to creep but, in the case of constant strain, the manifestation of creep is a lowering of the external stress, which is termed stress relaxation. Therefore,

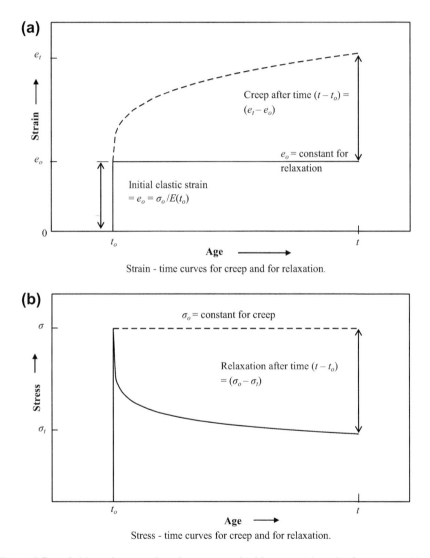

Figure 2.7 Definition of stress relaxation compared with creep: (a) strain-time curves (b) stress-time curves; initial stress σ_o, elastic strain e_o and $E(t_o)$ = modulus of elasticity at age t_o.

relaxation is a consequence of restrained creep. Figure 2.7 demonstrates the connection between the two phenomena. Relaxation is of considerable importance in the design of prestressed concrete and masonry structures. Since relaxation and creep are closely connected physically, they are affected by the same factors, though not to the same extent. For example, because the stress is continually decreasing, development of relaxation is faster than that of creep and, because of the relation between relaxation and creep, a separate discussion of relaxation as a property in its own right in this book is not thought to be warranted. Mathematical relations between relaxation and creep are presented by Neville et al. [1], such as that in Figure 2.7, where after time under load $(t-t_o)$ the ratio of stress σ_t, to the initial stress σ_o, which is known as the *relaxation ratio*, $R(t, t_o)$, is related to the creep coefficient of Equation (2.4) as follows:

$$R(t,t_o) = \frac{\sigma_t}{\sigma_o} = \frac{1}{1 + \phi(t,t_o)} \tag{2.8}$$

References

[1]   Neville AM, Dilger WH, Brooks JJ. Creep of plain and structural concrete. London and New York: Construction Press; 1983. 361pp.
[2]   Neville AM. Theories of creep in concrete. ACI J 1955;52:47–60.
[3]   Ali I, Kesler CE. Mechanisms of creep in concrete. In: Symposium on creep of concrete. ACI Special Publication No. 9; 1964. pp. 37–57.

3 Composite Models

The advantage of two-phase composite modeling of movement is that it facilitates a clear understanding of the interactive roles played by the two phases of the composite material subjected to external load or environmental changes in temperature and humidity. The phases generally have different properties, such as stress–strain behaviour, moisture movement, and thermal movement so that, to maintain compatibility of strain, intrinsic stresses are induced that, in the case of creep and moisture moment, are time dependent. Those effects can be quantified by composite models and therefore considered, and decisions made as to their significance.

In the case of masonry, a particular advantage is that movements of individual phases can be measured and studied at a practical level in the laboratory since, unlike concrete, the actual components are large enough to install normal laboratory instruments. Hence, measurements can be made on separate, not bonded components as well as on the components bonded within the masonry, which can then be used to verify the composite model theoretical solutions or study their limitations. An example of the latter is the study of the effect of moisture transfer across the brick/mortar interface, which is dealt with in Chapter 8. As will be shown later in Chapters 4, 7, 8, and 12, there are other significant effects arising from the interface that play important roles on the movement properties of masonry.

The chapter reviews several existing two-phase composite models originally developed for movements in concrete, from which a suitable model is selected to represent the arrangement of units and mortar in masonry. Theoretical analyses are undertaken for all types of movement in terms of the properties of brick or block and mortar and their relative proportions of the two phases. The detailed derivations are based on solid units with full-bedded mortar joints but, later, hollow and cellular units with full-bedded or faceshell-bedded mortar are also considered.

In the first instance, models for concrete are presented. Then models for masonry subjected to load are analyzed in order to determine the modulus of elasticity and creep of masonry, which is followed by the derivation of models for moisture movement, thermal movement, and Poisson's ratio. For each type of movement, anisotropy is considered by comparing horizontal and vertical movements. In all cases, the accuracy of simpler, more convenient models are then considered before recommending equations for general types of masonry that are suitable for practical application, many of which are developed for practical use in subsequent chapters.

Concrete and Masonry Movements. http://dx.doi.org/10.1016/B978-0-12-801525-4.00003-0

Concrete Models

In 1958, Hansen [1] suggested that any composite material can have two fundamentally different structures. The first of these is an ideal *composite hard material,* which has a continuous lattice of an elastic phase with a high modulus of elasticity with embedded particles of a lower modulus of elasticity. The second type of structure is that of an ideal *composite soft material,* which has elastic particles with a high modulus of elasticity embedded in a continuous matrix phase with a lower modulus of elasticity.

The differences between the two structures can be large when it comes to the calculation of the modulus of elasticity. In the case of a composite hard material, it is assumed that the strain is constant over any cross section, while the stresses in the phases are proportional to their respective moduli [2]. This is the situation in Figure 3.1(a), where the two phases are parallel to the external load and the stresses (σ_1 and σ_2) on each phase are different. The equation for the modulus of elasticity of the *parallel model* is:

$$E_c = (1 - g)E_m + gE_a \qquad (3.1)$$

where E_c = modulus of elasticity of the composite material, E_m = modulus of elasticity of the matrix phase, E_a = modulus of elasticity of the particle phase, and g = fraction volume of the particles.

On the other hand, the modulus of elasticity of a composite soft material is calculated from the assumption that the stress is constant over any cross section, while the strain in the phases is inversely proportional to their respective moduli. In this case, the phases are arranged in series and are perpendicular to the external load, as

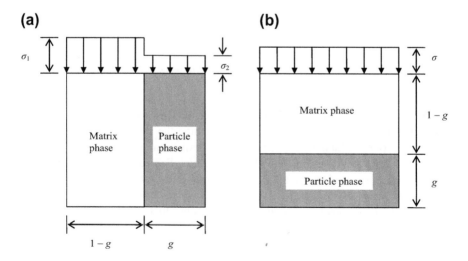

Figure 3.1 Parallel (a) and series (b) two-phase models for composite hard and soft materials subjected to external stress.

shown in Figure 3.1(b), and the stress on each phase is the same (σ). The equation for the modulus of elasticity of this *series model* is:

$$\frac{1}{E_c} = \frac{1-g}{E_m} + \frac{g}{E_a} \tag{3.2}$$

The above equations represent the boundaries of modulus of elasticity of composite two-phase materials. Neither boundary can be achieved in practice, as they do not satisfy the two requirements of equilibrium and compatibility. However, rather surprisingly, it has been found that application of the two-phase series model (Eq. (3.2)) to represent the aggregate (particle phase) and hardened cement paste (matrix phase) components of concrete works reasonably well when $E_a > E_m$. Conversely, when $E_a < E_m$, the parallel model (Eq. (3.1)) agrees reasonably well with experimental observations.

Hansen and Nielson [3] suggested a more sophisticated model in which a spherical particle, representing the aggregate, is embedded concentrically in a spherical mass of matrix representing the cement paste, the relative sizes of the two spheres being arranged in proportion to the fractional volumes of the two phases. Assuming that the Poisson's ratio of the phases are the same, equal to 0.2, Hansen [2] derived the modulus of elasticity of the concrete as:

$$E_c = \frac{(1-g)E_m + (1+g)E_a}{(1+g)E_m + (1-g)E_a} E_m \tag{3.3}$$

Equation (3.3) agreed well with experimental data, but even better results were yielded by the model shown in Figure 3.2, which was developed by Hirsch [4] and by Dougill [5]. The semi-empirical relationship developed for this model is:

$$\frac{1}{E_c} = 0.5\left[\frac{1-g}{E_m} + \frac{g}{E_a}\right] + 0.5\left[\frac{1}{(1-g)E_m + gE_a}\right] \tag{3.4}$$

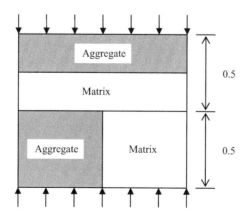

Figure 3.2 Composite model for Eq. (3.4).

According to Counto [6], the composite soft model and the above model are of limited validity in the case when E_a tends to zero, such as a porous cement paste, since E_c will also tend to zero. However, it is known that a porous material has a finite modulus. This limitation was demonstrated experimentally by Counto using a polythene aggregate concrete ($E_a = 0.29$ GPa) for which Eq. (3.4) predicted a much lower modulus for concrete than was measured. To overcome the limitation, Counto proposed the model shown in Figure 3.3. Here, the aggregate is in the form of a cylinder or prism placed at the center of a cylinder or prism of concrete, both of the cylinders or prisms having the same ratio of height to area of cross section. The model yields the following solution:

$$\frac{1}{E_c} = \frac{1 - g^{0.5}}{E_m} + \left[\left(\frac{1 - g^{0.5}}{g^{0.5}} \right) E_m + E_a \right]^{-1} \tag{3.5}$$

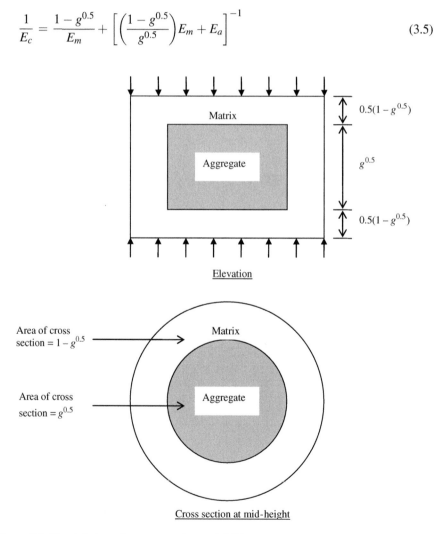

Figure 3.3 Counto's two-phase composite model [6].

Counto [6] verified the validity of Eq. (3.5) for a range of aggregate types and contents, and extended the applicability to creep of concrete. However, comparing the different models, in practice there is little difference in the predicted modulus of concrete for the normal range of aggregates. Figure 3.4 confirms that statement for a range of aggregate contents assuming $E_m = 25$ GPa and $E_a = 50$ GPa, where it can be seen that the estimates of concrete modulus of all three models (Eqs (3.3)–(3.5)) are virtually identical and lie between the parallel and series models.

By replacing the modulus of elasticity by an effective modulus of elasticity, the effects of creep can be taken into account by composite models. From the definition of creep in Chapter 2, creep is given as the total load strain under a constant stress after allowing for any shrinkage, swelling, or thermal strain, minus the elastic strain at loading. Thus, specific creep of concrete at age t due to load applied at age t_o, $C_c(t, t_o)$, is:

$$C_c(t, t_o) = \frac{1}{E'_c(t, t_o)} - \frac{1}{E_c(t_o)} \tag{3.6}$$

where $E'_c(t, t_o) =$ effective modulus of concrete at age t due to load applied at age t_o and $E_c(t_o) =$ modulus of elasticity at age t_o.

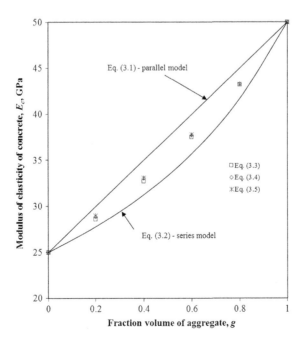

Figure 3.4 Modulus of elasticity of concrete as given by various two-phase composite models, assuming elastic modulus of aggregate $= 50$ GPa and elastic modulus of cement paste matrix $= 25$ GPa.

For example, in Counto's model [6], specific creep of the concrete is obtained from Eq. (3.5) as:

$$
C_c(t,t_o) = \left(1 - g^{0.5}\right)\left[\frac{1}{E'_m(t,t_o)} - \frac{1}{E_m(t_o)}\right]
$$
$$
+ \frac{\left(\frac{1-g^{0.5}}{g^{0.5}}\right)\left(E_m(t_o) - E'_m(t,t_o)\right)}{\left[E_a + \left(\frac{1-g^{0.5}}{g^{0.5}}\right)E'_m(t,t_o)\right]\left[E_a + \left(\frac{1-g^{0.5}}{g^{0.5}}\right)E_m(t_o)\right]} \tag{3.7}
$$

where E_a = modulus of elasticity of aggregate = constant and $E'_m(t,t_o)$ = effective modulus of matrix (hardened cement paste or mortar) at age t due to load applied at age t_o, i.e.,

$$
E'_m(t,t_o) = \frac{E_m(t_o)}{1 + C_m(t,t_o)E_m(t_o)} \tag{3.8}
$$

where $E_m(t_o)$ = modulus of elasticity of matrix at age t_o and $C_m(t,t_o)$ = specific creep of matrix at age t due to load applied at age t_o.

England [7] proposed a two-phase model for creep and shrinkage of concrete that consists of aggregate cubes surrounded symmetrically by a matrix of cement paste or mortar. The cubes are arranged in close-packed layers with cubes of adjacent layers staggered on both transverse directions. The matrix is thus made up of columns and slabs capable of carrying direct compressive or tensile stress. To obtain creep and shrinkage-time solutions, England carried out "step-by-step" analysis using computer software.

Other models exist for concrete [8,9], of which Hobbs [8] developed a composite model for the bulk modulus of concrete by analyzing the volumetric strain and stress of a two-phase material subjected to an applied hydrostatic stress. The bulk modulus of the composite, k_c, was determined as:

$$
k_c = k_m\left[1 + \frac{2g(k_a - k_m)}{(k_a + k_m) - g(k_a - k_m)}\right] \tag{3.9}
$$

where k_m = bulk modulus of the matrix and k_a = bulk modulus of the aggregate.
In terms of the modulus of elasticity, the bulk modulus is related as follows:

$$
k_m = \frac{E_m}{3(1 - 2\mu_m)}, \quad k_a = \frac{E_a}{3(1 - 2\mu_a)} \quad \text{and} \quad k_c = \frac{E_c}{3(1 - 2\mu_c)} \tag{3.10}
$$

In Eq. (3.10), μ_m, μ_a, and μ_c = Poisson's ratios for the matrix, aggregate, and composite, respectively, and if those ratios are assumed to be equal, then from Eq. (3.9) the modulus of elasticity of the composite becomes:

$$
E_c = E_m\left[1 + \frac{2g(E_a - E_m)}{(E_a + E_m) - g(E_a - E_m)}\right] \tag{3.11}
$$

In fact, Eq. (3.11) is identical to that derived by Hansen and Neilson (Eq. (3.3)).

Hobbs [8] also derived models for creep, shrinkage, and thermal movement and showed they were applicable to concrete. For specific creep:

$$C_c = C_m\left(\frac{1-g}{1+g}\right) \tag{3.12}$$

where C_c = specific creep of the composite and C_m = specific creep of matrix phase.

For linear shrinkage of a two-phase material having equality of Poisson's ratio for each phase:

$$S_c = S_m - \frac{(S_m - S_a)g2E_a}{(E_m + E_a) + g(E_a - E_m)} \tag{3.13}$$

where S_c, S_m, and S_a = shrinkage of the composite, matrix, and aggregate, respectively.

When $S_a = 0$ and $E_a \to \infty$, Eq. (3.13) reduces to the same form as that for creep (Eq. (3.12)), viz.:

$$S_c = S_m\left[\frac{1-g}{1+g}\right] \tag{3.14}$$

Similar expressions apply to thermal movement of a two-phase composite:

$$\alpha_c = \alpha_m - \frac{(\alpha_m - \alpha_a)g2E_a}{E_m + E_a + g(E_a - E_m)} \tag{3.15}$$

and, when $\alpha_a = 0$ and $E_a \to \infty$:

$$\alpha_c = \alpha_m\left[\frac{1-g}{1+g}\right] \tag{3.16}$$

where α_c, α_m, and α_a = coefficient of thermal expansion of composite, matrix, and aggregate, respectively.

Masonry Models

The first attempt to model the modulus of elasticity of masonry appears to be that by Sahlin [10], who used the series model of Eq. (3.2) and reported reasonable agreement between estimated and measured values. Base and Baker [11] used the same model and found estimates to within 1–14%. Jessop et al. [12] combined the parallel and series composite models to represent the repetitive elements in a wall. Small and large elements were considered as indicated in Figure 3.5. The small element is divided into

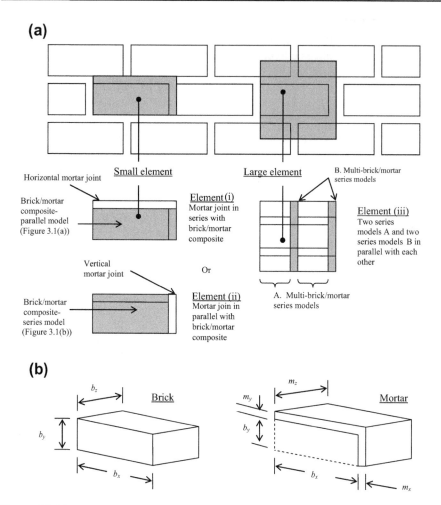

Figure 3.5 Identification and modeling of repeating elements in a masonry wall [12]. (a) Repeating elements in a wall subjected to vertical load and composite model arrangements [13]. (b) Dimensions of brick and mortar components for modeling repeating elements.

brick and mortar subelements arranged in parallel or in series, while the large element is divided into series subelements acting in parallel with each other. Using conditions of equilibrium and compatibility, three equations representing the two small elements and one large element were developed by Jessop et al., the equation for Element (i) of Figure 3.5 being:

$$E_c = \frac{(b_y + m_y)(b_x + m_x)m_z E_m (b_x b_z E_b + m_x m_z E_m)}{A_w \left[((b_x + m_x)m_z b_y E_m + b_x b_z m_y E_b + m_x m_z m_y E_m) \right]} \tag{3.17}$$

where:

$$A_w = \frac{(b_y + m_y)(b_x + m_x)(b_x b_z + m_x m_y)m_z}{[(b_x + m_x)m_z b_y + b_x b_z m_y + m_x m_z m_y]}$$

and E_c, E_m, and E_b = moduli of elasticity of concrete, mortar, and brick, respectively; other symbols are shown in Figure 3.5(b).

Shrive et al. [13] slightly simplified the foregoing expressions by substituting some linear dimensions by areas, but Eq. (3.17) is simplified considerably when actual dimensions for brick or block and mortar are used. For example, for a single-leaf wall built with a standard brick: $b_x = 215$ mm, $b_y = 65$ mm, $b_z = 102.5$ mm, mortar joint thickness: $m_x = m_y = 10$ mm, and $m_z = b_z$, Eq. (3.17) becomes:

$$E_c = E_m \frac{(21.5E_b + E_m)}{(18.85E_m + 2.75E_b)} \tag{3.18}$$

Jessop et al. [12] found that there was little difference between modulus of elasticity as predicted by the equations representing all three elements shown in Figure 3.5(a) and, since it was the simplest to use, they recommended the use of Eq. (3.17).

For a five-stack bonded (no vertical mortar joints) concrete hollow block prism, Ameny et al. [14] amended the expression by Jessop et al. [12], which turned out to be the same as the series composite model (Eq. (3.2)). Later, Ameny et al. [15] proposed a slightly different arrangement of the block and mortar phases in the small repetitive elements for full-bedded hollow stack-bonded prisms, and developed expressions for different arrangements of mortar on the bed face of hollow block units. Based on the analysis of Jessop el al. [12] and Shrive and England [16] for the large repetitive element of Figure 3.5(a), Ameny [17] developed an expression for a single-leaf wall in stretcher bond with full-bedded mortar joints.

Attempts have been made to model creep and shrinkage by Jessop et al. [12] and by Shrive and England [16] using the large repeating element in Figure 3.5(a). Shrinkage was taken to be a multiple of creep, which was assumed to take place in the mortar only. To account for varying stress and strain with time, analysis was carried out using a "step-by-step" approach and an effective modulus of elasticity. Ameny et al. [18] extended their own elastic analysis of different combinations of full-bedded solid and face-shell-bedded hollow concrete masonry to derive expressions involving creep; they also considered methods for dealing with time dependency, other than the effective modulus. They claimed experimental verification, but it was necessary to empirically adjust creep of unbonded and mortar specimens in order for the composite model prediction to agree with measured creep of masonry.

After reviewing existing models for masonry movements, Brooks [19] derived composite models for elasticity, creep, and moisture movement of single-leaf brickwork The theoretical approach was based on Counto's model for concrete [6] but using rectangular prisms instead of cylinders to represent the two phases, the arrangement of which seemed particularly appropriate for mortar-bonded units in masonry. Movements in the lateral or horizontal direction were considered as well as the axial or vertical direction. Subsequently, the same approach was extended to cover any size and type of masonry and thermal movement of masonry [20,21]. The full derivation of models for elasticity, creep, Poisson's ratio, and moisture and thermal movements of masonry is now given together with consideration of approximate, more practical, expressions.

The original objective was to derive full theoretical composite model expressions for all types of movements of masonry in the vertical and in the horizontal directions and then to assess their validity using strain measurements of unbonded brick or block and mortar specimens. However, the theoretical analysis revealed that some of the developed expressions were rather complex and unwieldy for the models to be considered for practical application. In this section, the full solutions are derived before simplifying into approximate solutions by neglecting the contribution of vertical mortar joints to the overall movement, and then their accuracy is compared with the full solutions. The main purpose of this approach was to develop the simplest expressions for use in practice in order to estimate any type of movement in any direction and for any type and size of masonry. The approximate solutions are, in fact, directly applicable to those types of masonry that do not have vertical mortar joints, such as *full- and face-shell-bedded, interlocking hollow concrete blockwork*.

Modulus of Elasticity

In the derivation of modulus of elasticity, the basic assumptions used are as follows:

- No effects of bond between brick or block and mortar.
- Poisson's effect is neglected.
- Strain is proportional to stress.
- There is no external restraint.
- Deformations of mortar and block are isotropic, but deformation of brick is anisotropic.

Vertical Loading

Consider masonry subjected to a vertical loading, as shown in Figure 3.6(a). The overall change in height of the masonry under a stress, σ_{wy}, is equal to the sum of the changes in height of the brick/vertical mortar joint composite and the horizontal mortar joint, shown in Figure 3.6(b). Hence:

$$H\varepsilon_{wy} = b_y C\varepsilon_{bmy} + m_y(C+1)\varepsilon_{hmy} \tag{3.19}$$

where H = height of masonry, C = number of brick courses, $C+1$ = number of mortar courses, b_y = depth of brick or block, m_y = depth of horizontal mortar joint, and ε_{wy}, ε_{bmy}, and ε_{hmy} = vertical strains in masonry, brick/mortar composite, and horizontal mortar joint, respectively.

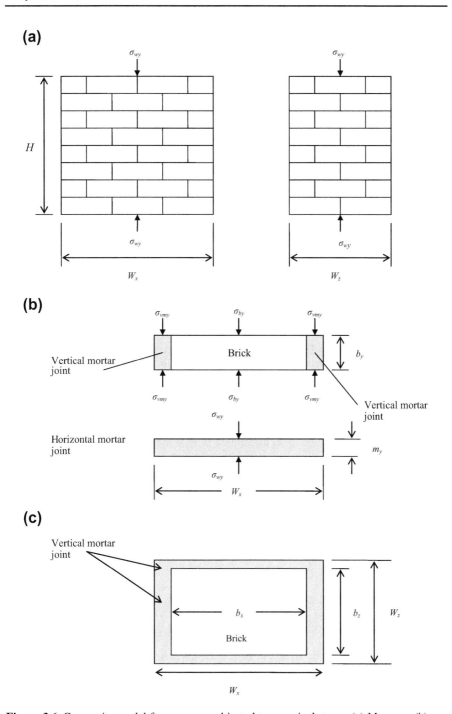

Figure 3.6 Composite model for masonry subjected to a vertical stress. (a) Masonry. (b) Section through brick/mortar composite and horizontal mortar joint. (c) Plan of brick/mortar composite.

From the stress–strain relations:

$$\varepsilon_{wy} = \frac{\sigma_{wy}}{E_{wy}}, \quad \varepsilon_{bmy} = \frac{\sigma_{wy}}{E_{bmy}} \quad \text{and} \quad \varepsilon_{hmy} = \frac{\sigma_{wy}}{E_m} \tag{3.20}$$

where E = appropriate modulus of elasticity: E_{wy} for masonry; E_{bmy} for brick/mortar composite; E_m for mortar.

Substitution of Eq. (3.20) in Eq. (3.19) yields:

$$\frac{1}{E_{wy}} = \frac{b_y C}{H} \left(\frac{1}{E_{bmy}} \right) + \frac{m_y(C+1)}{H} \left(\frac{1}{E_m} \right) \tag{3.21}$$

To obtain the brick/mortar modulus, E_{bmy}, in terms of the moduli of brick and of mortar, consider the compatibility of strain in the brick/mortar composite, i.e., $\varepsilon_{bmy} = \varepsilon_{by} = \varepsilon_{vmy}$, so that

$$\frac{\sigma_{wy}}{E_{bmy}} = \frac{\sigma_{by}}{E_{by}} = \frac{\sigma_{vmy}}{E_m} \tag{3.22}$$

where ε_{vmy} = strain in vertical mortar joint, and σ_{by} and σ_{vmy} are stresses in the brick and vertical mortar joint, respectively.

The relation between σ_{by} and σ_{vmy} is found by equating the force on the masonry to the sum of forces acting on the composite, viz.:

$$A_w = A_b \sigma_{by} + A_{vm} \sigma_{vmy} \tag{3.23}$$

where A_w, A_b, and A_{vm} = cross-sectional area of masonry, bricks, and vertical mortar joints, respectively.

Now since $A_{vm} = A_w - A_b$, and from Eq. (3.4) $\sigma_{vmy} = \sigma_{by} \frac{E_m}{E_{by}}$, so that from Eq. (3.23):

$$\sigma_{by} = \frac{A_w \sigma_{wy}}{A_b + A_{vm} \frac{E_m}{E_{by}}}$$

substitution of σ_{by} in Eq. (3.22) gives:

$$\frac{1}{E_{bmy}} = \frac{A_w}{E_{by} A_b + E_m A_{vm}}$$

and further substitution for $1/E_{bmy}$ in Eq. (3.21) give the modulus of elasticity of the masonry in terms of the moduli of brick and mortar, viz.:

$$\frac{1}{E_{wy}} = \frac{b_y C}{H} \left[\frac{A_w}{E_{by} A_b + E_m A_{vm}} \right] + \frac{m_y(C+1)}{H} \frac{1}{E_m} \tag{3.24}$$

If the vertical mortar joints are ignored, then $A_{vm} = 0$ and $A_b = A_w$ so that Eq. (3.21) becomes:

$$\frac{1}{E_{wy}} = \frac{b_y C}{H} \frac{1}{E_{by}} + \frac{m_y(C+1)}{H} \frac{1}{E_m} \tag{3.25}$$

Horizontal Loading

Now, considering masonry subjected to horizontal loading, stresses are induced in the phases as shown in Figure 3.7. For compatibility of lateral movement, the lateral strains in the masonry, horizontal mortar joint, and brick/mortar composite are equal (Figure 3.7(b)), viz:

$$\varepsilon_{wx} = \varepsilon_{bmx} = \varepsilon_{hmx} \tag{3.26}$$

From the stress–strain relations:

$$\frac{\sigma_{wx}}{E_{wx}} = \frac{\sigma_{bmx}}{E_{bmx}} = \frac{\sigma_{hmx}}{E_m} \tag{3.27}$$

where σ_{wx}, σ_{bmx}, and σ_{hmx} are the horizontal stresses on the masonry, brick/vertical mortar composite, and horizontal mortar joint, respectively.

Since the change in length of the brick/mortar composite is equal to the sum of the change in lengths of the mortar and brick (Figure 3.7(c)):

$$W_x \frac{\sigma_{bmx}}{E_{bmx}} = m_x \frac{\sigma_{bmx}}{E_m} + b_x \frac{\sigma_{bx}}{E_{bx}} \tag{3.28}$$

To solve Eq. (3.27), σ_{bx} is required as a function of σ_{bmx}, which is obtained by considering the inner brick/mortar composite as shown in Figure 3.7(d). For compatibility of strain:

$$\sigma'_{hmx} \frac{1}{E_{bx}} = \sigma_{bx} \frac{1}{E_m} \tag{3.29}$$

and, for equilibrium of forces:

$$\sigma'_{hmx} m_z b_y + \sigma_{bx} b_z b_y = \sigma_{bmx} W_z b_y \tag{3.30}$$

It should be noted that vertical mortar in Figure 3.7(c) is arranged in proportion to its actual thickness over the cross section of the masonry; in other words, m_x = total thickness in the x-direction and m_z = total thickness in the z-direction. Substitution of Eq. (3.29) in Eq. (3.30) leads to:

$$\sigma_{bx} = \frac{\sigma_{bmx} W_z}{\frac{E_m}{E_{bx}} m_z + b_z} \tag{3.31}$$

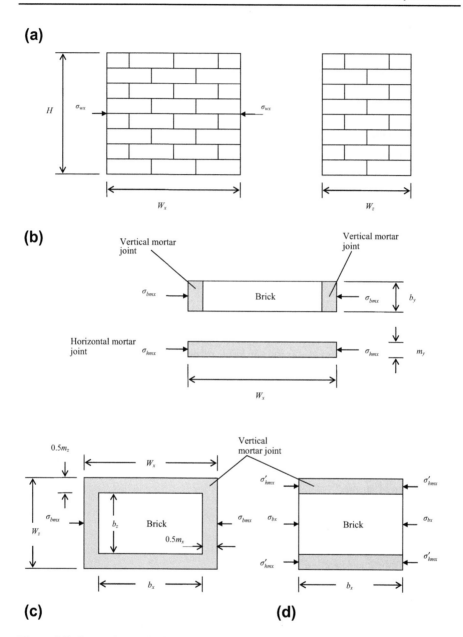

Figure 3.7 Composite model for masonry subjected to a horizontal stress. (a) Masonry. (b) Section through brick/mortar composite and horizontal mortar joint. (c) Plan of brick or mortar composite. (d) Plan of inner brick or mortar composite.

Hence, substitution of Eq. (3.31) in Eq. (3.28) gives:

$$\frac{1}{E_{bmx}} = \frac{m_x}{W_x}\frac{1}{E_m} + \frac{b_x}{W_x}\frac{W_z}{(E_m m_z + E_{bx} b_z)} \tag{3.32}$$

Before Eq. (3.27) can be solved to obtain the modulus of elasticity of masonry, the relation between σ_{bmx} and σ_{wx} is required. Since the lateral force on the masonry is equal to the sum of forces acting on the brick/mortar composite and the horizontal mortar joint:

$$\sigma_{wx} H W_z = m_y(C+1)\sigma_{hmx} W_z + C\sigma_{bmx} b_y W_z$$

Hence,

$$\sigma_{bmx} = \frac{\sigma_{wx} H - m_y(C+1)\sigma_{hmx}}{Cb_y} \tag{3.33}$$

Now, from Eq. (3.27) $\sigma_{hmx} = \frac{E_m}{E_{bmx}}\sigma_{bmx}$ so that substitution in Eq. (3.33) gives:

$$\sigma_{bmx} = \frac{\sigma_{wx} H}{Cb_y + m_y(C+1)\frac{E_m}{E_{bmx}}} \tag{3.34}$$

and, therefore, from Eq. (3.27):

$$E_{wx} = \frac{Cb_y}{H}E_{bmx} + \frac{m_y(C+1)}{H}E_m$$

Now E_{bmx} can be obtained from Eq. (3.32) and substitution in the above equation gives:

$$E_{wx} = \frac{Cb_y}{H}\left[\frac{W_x E_m(E_m m_z + b_z E_{bx})}{m_x(E_m m_z + b_z E_{bx}) + E_m b_x W_z}\right] + \frac{m_y(C+1)}{H}E_m \tag{3.35}$$

Neglecting the contribution made by the vertical mortar joints implies $m_x = m_z = 0$, $b_x = W_x$ and $b_z = W_z$, so that Eq. (3.35) reduces to:

$$E_{wx} = \frac{Cb_y}{H}E_{bx} + \frac{m_y(C+1)}{H}E_m \tag{3.36}$$

For masonry constructed from standard bricks or blocks, the expressions for vertical and horizontal modulus of elasticity are not affected to a significant extent by height, width, and cross-sectional area so that, for example, there is little difference between walls and piers, an observation that concurs with observations reported in practice (Chapter 5). Taking the case of a 4-brick-wide × 26-course-high single-leaf wall built with 215 × 102.5 × 65 mm standard bricks, Eq. (3.24) gives the vertical modulus of elasticity as:

$$\frac{1}{E_{wy}} = 0.862\left[\frac{1}{0.961E_{by} + 0.043E_m}\right] + 0.138\frac{1}{E_m} \tag{3.37}$$

and Eq. (3.20) gives the horizontal modulus of the wall as:

$$E_{wx} = 0.862\left[\frac{E_m E_{bx}}{0.039E_{bx} + 0.961E_m}\right] + 0.138E_m \tag{3.38}$$

Figure 3.8 shows the influence of the modulus of brick and of mortar on the comparison between the vertical and horizontal moduli of brickwork according to Eqs (3.37) and (3.38), respectively; in the comparison, it is assumed that the bricks are isotropic, i.e., $E_{bx} = E_{by}$. It appears that the brickwork moduli are approximately equal for a strong or high-modulus mortar, but the effect of a weaker or low-modulus mortar results in significant anisotropy of brickwork modulus when the brick modulus exceeds 15–20 GPa.

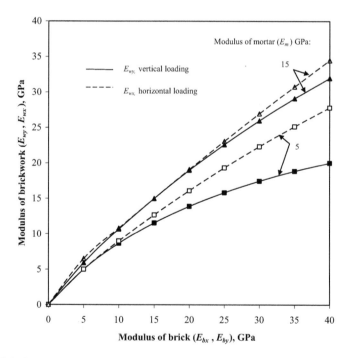

Figure 3.8 Influence of elastic moduli of brick and mortar on elastic modulus of brickwork subjected to vertical and horizontal loading, according to Eqs (3.37) and (3.38), respectively; single-leaf wall: 4 bricks wide × 26 courses.

It is of interest to observe that E_{wy} as given by the transposition of Eq. (3.37) is the same as that given by Jessop et al. [12] for a single-leaf wall (Eq. (3.18)) due to the fact that the composite model arrangement for element (i) in Figure 3.5(a) is the same as that in Figure 3.7(a).

When the contribution made by the vertical mortar joints is neglected for brick-work built from standard bricks, Eqs (3.25) and (3.36), respectively, become:

$$\frac{1}{E_{wy}} = \frac{0.86}{E_{by}} + \frac{0.14}{E_m} \tag{3.39}$$

and

$$E_{wx} = 0.86E_{bx} + 0.14E_m \tag{3.40}$$

In fact, for a wide range of brick and mortar moduli of elasticity, the approximate solutions are acceptable for estimating the modulus of elasticity of brickwork, except in the case of horizontal loading when the mortar modulus is low. Figure 3.9 confirms this statement by plotting the approximate solutions for brickwork modulus without vertical mortar joints against the full solutions for brickwork modulus including the vertical mortar joints. In the case of horizontal loading, the approximate solution

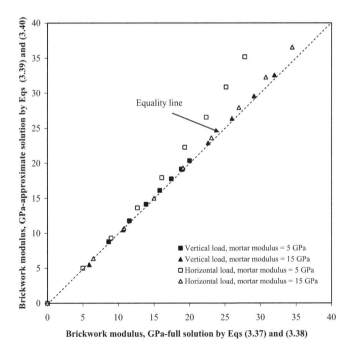

Figure 3.9 Accuracy of approximate composite model solutions for modulus of elasticity of brickwork.

clearly overestimates the brickwork modulus when $E_m = 5$ GPa, but for a higher value of, say, 10 GPa the estimates would probably be acceptable.

Considering single-leaf blockwork masonry, four blocks wide and 12 courses high, built with standard blocks of size $440 \times 215 \times 100$ mm, Eqs (3.24) and (3.35) give the vertical and horizontal moduli, respectively, as:

$$\frac{1}{E_{wy}} = 0.952 \left[\frac{1}{0.987E_{by} + 0.022E_m} \right] + 0.048 \frac{1}{E_m} \tag{3.41}$$

and

$$E_{wx} = 0.952 \left[\frac{E_m E_{bx}}{0.02E_{bx} + 0.98E_m} \right] + 0.048E_m \tag{3.42}$$

Compared with brickwork with $E_{bx} = E_{by}$, there is less anisotropy of modulus of elasticity of blockwork for a weak mortar, so that equality of moduli may be assumed for both vertical and horizontal loadings [20].

When the contribution of the vertical mortar joints to the elastic modulus of blockwork built with standard blocks is neglected, then Eqs (3.25) and (3.36) yield the approximate relationships:

$$\frac{1}{E_{wy}} = \frac{0.952}{E_{by}} + \frac{0.048}{E_m} \tag{3.43}$$

and

$$E_{wx} = 0.952E_{bx} + 0.048E_m \tag{3.44}$$

For a wide range of block and mortar moduli, it has been shown [20] that the accuracy of the approximate solutions for vertical and horizontal modulus of elasticity of blockwork is better than those for brickwork. Furthermore, it is of interest to note that the foregoing approximate solutions for vertical and horizontal modulus of elasticity of masonry are the same expressions presented earlier for the composite series model (Eq. (3.2)) and composite parallel model (Eq. (3.1)), respectively.

In the previous analysis and discussion, equality of unit modulus in the vertical and horizontal directions was assumed or, in other words, it was assumed that bricks or blocks were isotropic. However, for clay bricks this may not be the situation in practice because, for example, extruded or perforated bricks generally have a greater elastic modulus between bed faces than between header faces (see Chapter 5). Therefore, according to the composite model, the effect of anisotropy of brick modulus would be to decrease the horizontal modulus, E_{wx}. On the other hand, the assumption of isotropy for concrete blocks is likely to be correct, and therefore anisotropy is not an additional factor to be considered in the comparison of vertical and horizontal modulus of elasticity of concrete blockwork.

Creep

As in the case of concrete, modeling of creep is accomplished by treating the elastic modulus as an effective modulus to allow for time-dependent increase in strain under sustained loading. Hence, creep is given as the total strain under a unit stress minus the elastic strain at loading. Thus, the vertical creep of brickwork (C_w) is:

$$C_{wy} = \frac{1}{E'_{wy}} - \frac{1}{E_{wy}} \tag{3.45}$$

and for horizontal loading:

$$C_{wx} = \frac{1}{E'_{wx}} - \frac{1}{E_{wx}} \tag{3.46}$$

where E'_{wy} and E'_{wx} are the respective effective moduli, in the vertical and horizontal directions as adapted from Eqs (3.37) and (3.38):

$$E'_{wy} = 0.862 \left[\frac{1}{0.961E'_{by} + 0.043E'_m} \right] + 0.138\frac{1}{E'_m} \tag{3.47}$$

and

$$E'_{wx} = 0.862 \left[\frac{E'_m E'_{bx}}{0.039E'_{bx} + 0.961E'_m} \right] + 0.138E'_m \tag{3.48}$$

where the effective moduli of brick and mortar $(E'_{by}, E'_{bx}$ and $E'_m)$ are given by:

$$\frac{1}{E'_{by}} = \frac{1}{E_{by}} + C_{by}; \quad \frac{1}{E'_{bx}} = \frac{1}{E_{bx}} + C_{bx}; \quad \frac{1}{E'_m} = \frac{1}{E_m} + C_m \tag{3.49}$$

and C_{by}, C_{bx}, and C_m are the vertical creep of brick, horizontal creep of brick, and creep of mortar, respectively.

Figure 3.10 illustrates the influence of creep of mortar and creep of brick on creep of brickwork according to Eqs (3.47) and (3.48). It is assumed that the mortar has an initial modulus of elasticity of 10 GPa and a final effective modulus of 2 GPa so that the specific creep ranges from 0 to 400×10^{-6} per MPa. It is also assumed that the brick has an initial modulus of elasticity of: (a) 20 GPa with a creep range of $0–10 \times 10^{-6}$ per MPa, and (b) 5 GPa with a creep range of $0–100 \times 10^{-6}$ per MPa. The two conditions are representative of, firstly, a strong brick exhibiting little creep and, secondly, a weak brick exhibiting significant creep. The analysis shows that, for weaker bricks, creep of brickwork in the vertical and horizontal directions is similar but, for strong bricks, creep of brickwork under vertical loading is greater than under

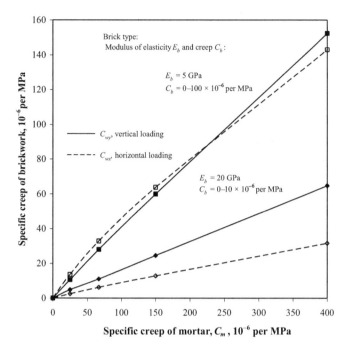

Figure 3.10 Influence of creep of mortar and brick on creep of brickwork subjected to vertical and horizontal loading, according to Eqs (3.47) and (3.48); single-leaf wall: 4 bricks wide × 26 courses high; modulus of elasticity of mortar, $E_m = 10$ GPa.

horizontal loading. Similar trends are apparent for blockwork, where overall creep is less than for brickwork and differences in creep between loading directions are less noticeable [20].

According to the composite model, when E_b is less than 5 GPa, creep of solid brickwork pier is appreciably greater than for a single-leaf wall; otherwise, creep is similar for the two types of masonry [20]. However, it should be emphasized that the comparison of walls and piers on the basis of equality of creep of mortar is somewhat invalid when masonry is allowed to lose moisture to the environment. In reality, the effect of drying increases creep of mortar but, in a solid pier, moisture migration is slower than in a single-leaf wall so that creep is actually less in piers than in walls (see Chapter 12).

The expression for creep of masonry in the vertical direction is simplified when the contribution of the vertical mortar joints is neglected. For example, in the case of brickwork, by analogy with the expression for elastic modulus, from Eq. (3.39) the effective modulus is given by:

$$\frac{1}{E'_{lwy}} = \frac{0.86}{E'_{lby}} + \frac{0.14}{E'_m} \tag{3.50}$$

After substituting Eqs (3.39) and (3.50) in Eq. (3.45), creep of brickwork becomes:

$$C_{wy} = 0.86C_{by} + 0.14C_m \tag{3.51}$$

The equivalent relationship for creep in the horizontal direction is not so readily expressed, and creep has to be obtained from Eq. (3.46) after calculating the effective modulus of brickwork:

$$E'_{wx} = 0.86E'_{bx} + 0.14E'_m \tag{3.52}$$

The assessment of accuracy of the approximate solutions for creep of brickwork is shown in Figure 3.11 and the solutions tend to underestimate creep slightly, especially for brickwork built from a strong brick under horizontal load. However, in that case, the level of creep is so small that the underestimate may be ignored, so that the approximate expressions for estimating creep can be recommended for all types of brickwork.

When the vertical mortar joints are neglected, the equivalent expressions to obtain creep of blockwork subjected to load in the vertical and horizontal directions are, respectively:

$$C_{wy} = 0.952C_{by} + 0.048C_m \tag{3.53}$$

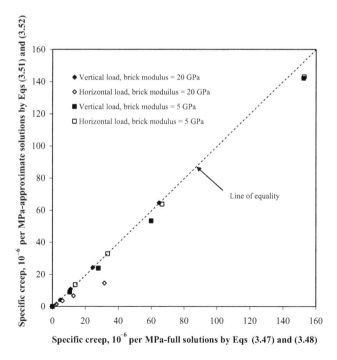

Figure 3.11 Accuracy of approximate composite model expressions for creep of brickwork subjected to vertical and horizontal loading.

and

$$E'_{wx} = 0.952E'_{bx} + 0.048E'_m \tag{3.54}$$

When compared with the full solutions allowing for vertical mortar joints, the accuracy of the above approximate expressions for estimating creep of blockwork is excellent and better than for brickwork. Consequently, the approximate expressions can be recommended for practical use.

Shrinkage

Moisture movement of masonry refers to movements caused by transfer of moisture to or from the component phases of masonry. Typically, mortar loses moisture to the atmosphere mainly through drying and mortar undergoes drying shrinkage, while clay bricks absorb moisture from the atmosphere and undergo moisture expansion. For convenience, shrinkage is assumed to be positive in the following analysis and moisture expansion is regarded as negative shrinkage.

Vertical Shrinkage

Figure 3.12 shows the arrangement of the brick or block and mortar phases representing masonry undergoing vertical shrinkage. Since the brick and vertical mortar joint phases in the brick/mortar composite (Figure 3.12(a)) have different shrinkage characteristics, in order to maintain compatibility of strain, stresses are required to be induced in the brick (σ_{by}) and in the vertical mortar joint (σ_{vmy}). The overall vertical shrinkage of the masonry (S_{wy}) is obtained by summing the changes in lengths of the brick/mortar composite and the horizontal mortar joint, viz.:

$$HS_{wy} = b_y C\varepsilon_{bmy} + m_y(C + 1)S_m$$

and, therefore:

$$S_{wy} = \frac{b_y C}{H}\varepsilon_{by} + \frac{m_y(C + 1)}{H}S_m \tag{3.55}$$

where S_m = shrinkage of mortar and the other terms are as defined previously and in Figure 3.12.

For compatibility of strain in the brick/mortar composite:

$$\varepsilon_{bmy} = \varepsilon_{by} = \varepsilon_{vmy} \tag{3.56}$$

where:

$$\varepsilon_{by} = S_{by} + \frac{\sigma_{by}}{E_{by}} \tag{3.57}$$

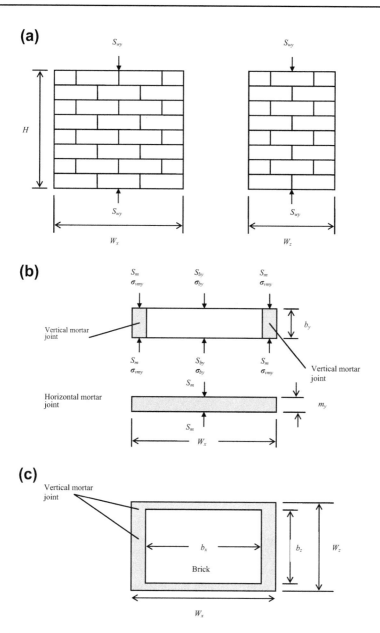

Figure 3.12 Composite model for masonry undergoing vertical shrinkage. (a) Masonry.
(b) Section through brick/mortar composite and horizontal mortar joint. (c) Plan of brick/mortar composite.

$$\varepsilon_{vmy} = S_m + \frac{\sigma_{vmy}}{E_m} \tag{3.58}$$

and S_{by} = vertical shrinkage of brick or block.

Since there is no net vertical force on the masonry:

$$A_b \sigma_{by} + A_m \sigma_{vmy} = 0 \tag{3.59}$$

where A_b and A_m are cross-sectional areas of bricks and vertical mortar joints, respectively.

Hence, from Eqs (3.57)–(3.59):

$$\sigma_{by} = \frac{E_m \left(S_m - S_{by} \right)}{\left[\frac{A_b}{A_m} + \frac{E_m}{E_{by}} \right]} \tag{3.60}$$

Substitution of σ_{by} in Eqs (3.57) and (3.58) yields:

$$\varepsilon_{bmy} = S_{by} + \frac{\left(S_m - S_{by} \right)}{\left[1 + \frac{A_b}{A_m} \frac{E_{by}}{E_m} \right]} \tag{3.61}$$

and, consequently, substitution of ε_{bmy} in Eq. (3.55) gives the vertical shrinkage of masonry:

$$S_{wy} = \frac{b_y C}{H} S_{by} + \frac{m_y(C+1)}{H} S_m + \frac{b_y C}{H} \frac{\left(S_m - S_{by} \right)}{\left[1 + \frac{A_b}{A_m} \frac{E_{by}}{E_m} \right]} \tag{3.62}$$

Horizontal Shrinkage

Figure 3.13 shows the composite model arrangement for masonry exhibiting horizontal moisture movement. In this case, for compatibility of horizontal strain due to differential shrinkage, the brick and brick/mortar composite are required to be subjected to an induced stress (σ_{bmx}) and the horizontal mortar bed joint to a stress (σ_{hmx}) (see Figure 3.13(b)). Consequently, shrinkage of the masonry (S_{wx}) can be equated to the strains in the brick/mortar composite (ε_{bmx}) and the horizontal mortar joint (ε_{hmx}) as follows:

$$S_{wx} = \varepsilon_{bmx} = \varepsilon_{hmx} \tag{3.63}$$

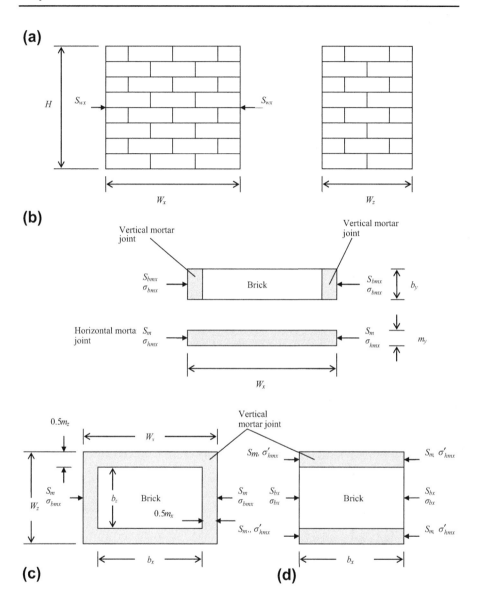

Figure 3.13 Composite model for masonry undergoing horizontal shrinkage. (a) Masonry. (b) Section through brick/vertical mortar composite and horizontal mortar joint. (c) Plan of brick/mortar composite. (d) Plan of inner brick/mortar composite.

where

$$\varepsilon_{bmx} = S_{bmx} + \frac{\sigma_{bmx}}{E_{bmx}} \tag{3.64}$$

and

$$\varepsilon_{hmx} = S_m + \frac{\sigma_{hmx}}{E_m} \tag{3.65}$$

In Eqs (3.64) and (3.65), S_{bmx} and S_m are the horizontal shrinkage of the brick/mortar composite and horizontal shrinkage of the mortar bed joint, respectively.

Now consider the change in horizontal length of the brick/mortar composite, the plan of which is shown in Figure 3.13(c) and (d):

$$W_x \left[S_{bmx} + \frac{\sigma_{bmx}}{E_{bmx}} \right] = 2m_x \left[S_m + \frac{\sigma_{bmx}}{E_m} \right] + b_x \left[S_{bx} + \frac{\sigma_{bx}}{E_{bx}} \right] \tag{3.66}$$

The stress on the brick (σ_{bx}) can be expressed as a function of the stress on the brick/mortar composite (σ_{bmx}) by considering the strain of the inner brick/mortar composite (Figure 3.13(d)).

For compatibility of strain:

$$S_{bx} + \frac{\sigma_{bx}}{E_{bx}} = S_m + \frac{\sigma'_{hmx}}{E_m} \tag{3.67}$$

and, for the equilibrium of forces:

$$2\sigma'_{hmx}\frac{m_z}{2}b_y + \sigma_{bx}b_zb_y = \sigma_{bmx}W_zb_y \tag{3.68}$$

In the case of single-leaf walls, $m_z = 0$ and $b_z = W_z$.
Substitution of Eq. (3.67) in Eq. (3.68) leads to:

$$\sigma_{bx} = \frac{\sigma_{bmx}W_z - (S_{bx} - S_m)E_mm_z}{\left[\frac{E_m}{E_{bx}}m_z + b_z \right]} \tag{3.69}$$

Hence, substitution of σ_{bx} in Eq. (3.66) yields:

$$S_{bmx} + \frac{\sigma_{bmx}}{E_{bmx}} = \frac{\sigma_{bmx}}{W_x}\left[\frac{m_x}{E_m} + \frac{b_x}{(E_mm_z + b_zE_{bx})} \right] + \frac{m_z}{W_x}S_m + \frac{b_x}{W_x}S_{bx}$$
$$- \frac{(S_{bx} - S_m)E_mm_zb_x}{W_x(E_mm_z + b_zE_{bx})} \tag{3.70}$$

Since the sum of forces acting on the brick/mortar composite and the horizontal mortar joint is zero (Figure 3.13(b)):

$$\sigma_{bmx} = -\frac{m_y(C+1)}{b_yC}\sigma_{hmx} \tag{3.71}$$

From Eqs (3.63) and (3.65):

$$\sigma_{hmx} = (S_{wx} - S_m)E_m \tag{3.72}$$

so that substitution in Eq. (3.71) gives:

$$\sigma_{bmx} = -\frac{m_y(C+1)}{b_yC}(S_{wx} - S_m)E_m \tag{3.73}$$

Also, from Eqs (3.63) and (3.64):

$$S_{wx} = S_{bmx} + \frac{\sigma_{bmx}}{E_{bmx}} \tag{3.74}$$

Therefore, equating Eq. (3.70) with Eq. (3.74) and substituting for σ_{bmx} from Eq. (3.73) yields the expression for horizontal shrinkage of masonry in terms of the brick or block and mortar proportions, and their moisture movement properties, viz.:

$$\left.\begin{array}{l} S_{wx}\left[1 + \dfrac{m_y(C+1)}{b_yCW_x}m_x + \dfrac{m_y(C+1)}{b_yCW_x}\dfrac{E_mW_zb_z}{E_mm_z + b_zE_{bx}}\right] = S_{bx}\left[\dfrac{b_x}{W_x} - \dfrac{E_mm_zb_x}{W_x(E_mm_z + b_zE_{bx})}\right] \\[4mm] +S_m\left[\dfrac{m_x}{W_x} + \dfrac{m_y(C+1)}{b_yCW_x}m_x + \dfrac{m_y(C+1)}{b_yCW_x}\dfrac{E_mW_zb_x}{E_mm_z + b_zE_{bx}} + \dfrac{E_mm_zb_x}{W_x(E_mm_z + b_zE_{bx})}\right] \end{array}\right\} \tag{3.75}$$

The expressions for vertical and horizontal moisture movement of masonry are simplified considerably when specific sizes are considered. In the case of a brickwork wall, 4 bricks wide × 26 courses high, the vertical moisture movement given by Eq. (3.62) becomes:

$$S_{wy} = 0.862S_{by} + 0.138S_m + \frac{0.862(S_m - S_{by})}{\left[1 + 24.43\frac{E_{by}}{E_m}\right]} \tag{3.76}$$

and the horizontal moisture movement given by Eq. (3.75) reduces to:

$$S_{wx} = \frac{0.955S_{bx} + \left[0.046 + 0.152\frac{E_m}{E_{bx}}\right]S_m}{\left[1 + 0.152\frac{E_m}{E_{bx}}\right]} \tag{3.77}$$

The equivalent vertical and horizontal expressions for a blockwork wall, 4 blocks wide × 12 courses high are, respectively:

$$S_{wy} = 0.952S_{by} + 0.048S_m + \frac{0.952\left(S_m - S_{by}\right)}{\left[1 + 43.75\frac{E_{by}}{E_m}\right]} \tag{3.78}$$

and

$$S_{wx} = \frac{0.979S_{bx} + \left(0.006 + 0.049\frac{E_m}{E_{bx}}\right)S_m}{\left(1 + 0.049\frac{E_m}{E_{bx}}\right)} \tag{3.79}$$

Some examples of moisture movements of the brickwork single-leaf wall estimated by Eqs (3.76) and (3.77) can be compared for different levels of mortar shrinkage and three levels of brick moisture movement, viz.: a clay brick having an irreversible expansion of -200×10^{-6}, a brick with zero moisture movement, and a concrete or calcium silicate brick having a shrinkage of 200×10^{-6}. In the analysis, it is assumed that $E_{by} = E_{bx}$ and $E_{by}/E_m = 2$. Figure 3.14 shows the comparisons, and it is apparent that as mortar shrinkage increases, horizontal moisture movement becomes

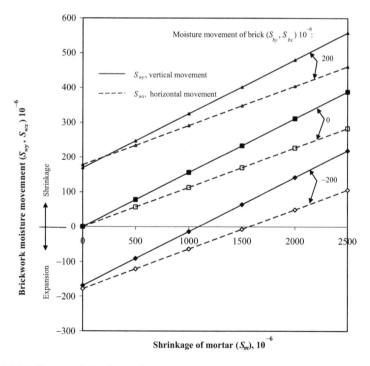

Figure 3.14 Influence of shrinkage of mortar and moisture movement of brick on vertical and horizontal moisture movement of a single-leaf wall, 4 bricks wide × 26 courses high, according to Eqs (3.76) and (3.77), assuming $E_b/E_m = 2$.

appreciably less than vertical moisture movement. Also, for the clay brick undergoing a moisture expansion, higher levels of mortar shrinkage cause the moisture movement of the wall to change from moisture expansion to shrinkage.

Application of Eqs (3.76) and (3.77) assumes that the *brick/mortar modulus ratio*, $E_b/E_m = 2$. In fact, because the moisture movement process is time dependent, the ratio is really an effective modulus ratio that usually increases with time due to creep of mortar, which causes E_m to decrease at a faster rate than E_b. The effect of a higher creep of mortar is to exaggerate the difference between horizontal and vertical moisture movements but, on the other hand, a lower creep and a lower brick/mortar modulus ratio has the effect of reducing the difference between horizontal and vertical moisture movements [21]. However, for convenience and the reasons stated in the section entitled: Mortar/brick modulus ratio'. of Chapter 8, in this analysis the effective modulus ratio is assumed to be equal to the elastic modulus ratio.

A further assumption is that the bricks are isotropic with respect to moisture movement, i.e., $S_{bx} = S_{by}$, which may not be true in practice, especially for clay bricks (see Chapter 8). If the moisture expansion between bed faces is greater than between header faces, again this would have the effect of exaggerating the difference between vertical and horizontal moisture movement of masonry, thus possibly causing significant anisotropic behaviour of moisture movement in masonry.

In contrast to modulus of elasticity, there are appreciable differences in moisture movement between walls and piers. For the same level of brick and mortar moisture movements, composite model analysis reveals that moisture movement of the pier is slightly greater than that of the wall [21]. However, in practice, like creep, the rate of drying is an overriding factor; so under drying conditions moisture loss is slower in piers and, consequently, shrinkage is less in piers than in single-leaf walls. The drying effect is quantified by the volume/exposed surface area ratio (see Chapters 6 and 7).

When the influence of the vertical mortar or header face joints is ignored by putting $m_x = m_z = 0$, $b_x = W_x$, and $b_z = W_z$, the expressions for vertical and horizontal moisture movements of single-leaf brickwork (Eqs (3.76) and (3.77)), respectively, become:

$$S_{wy} = 0.86S_{by} + 0.14S_m \tag{3.80}$$

and

$$S_{wx} = S_{bx} + \frac{S_m - S_{bx}}{\left[1 + 6.26\frac{E_{bx}}{E_m}\right]} \tag{3.81}$$

The corresponding relationship for vertical moisture movement of blockwork is:

$$S_{wy} = 0.952S_{by} + 0.048S_m \tag{3.82}$$

and the relationship for horizontal moisture movement is:

$$S_{wx} = S_{bx} + \frac{S_m - S_{bx}}{\left[1 + 19.85\frac{E_{bx}}{E_m}\right]} \tag{3.83}$$

The accuracy of the approximate expressions for moisture movement of brickwork is compared with the full solutions in Figure 3.15. There is a general tendency for the approximate expressions to underestimate moisture movement and especially for horizontal movement with high mortar shrinkage. While the approximate expression for vertical movement is acceptable (Eq. (3.80)), it is recommended that the full solution (Eq. (3.77)) be used for estimating horizontal movement.

Compared with brickwork having the same mortar shrinkage and equal shrinkage of block and brick, the moisture movement of blockwork is less [21]. Furthermore, for blockwork, there is a smaller difference between vertical and horizontal shrinkage, and the accuracy of the approximate solutions (Eqs (3.82) and (3.83)) is acceptable for vertical and horizontal moisture movements.

Composite model analysis also suggests that, for the same moisture movement of brick or block and the same shrinkage of mortar, the moisture movement of piers is similar to that of walls [21]. However, as is the situation for brickwork, the condition of equality of mortar shrinkage does not occur in practice. As stated earlier, under drying conditions, the loss of moisture from the mortar joints is slower in piers due to a longer drying path length from inside the pier to the outside environment, which results in less shrinkage in piers than in walls.

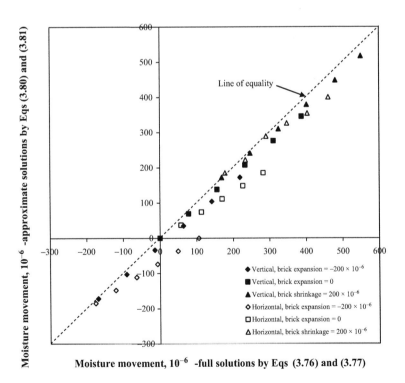

Figure 3.15 Accuracy of approximate composite model expressions for moisture movement of brickwork in the vertical and horizontal directions.

Thermal Movement

The expressions derived for moisture movement in the previous section are also applicable to thermal movement of masonry. Thus, from knowing the coefficients of thermal expansion of mortar, brick, and block, the models give solutions for thermal movement in the vertical and horizontal directions. Like moisture movement, thermal movements are virtually independent of height and geometry, and taking the example of single-leaf brickwork, 4 bricks wide × 26 courses high, the equivalent full solutions of the moisture movement expressions of Eqs (3.76) and (3.77) for thermal movement are as follows:

$$\alpha_{wy} = 0.862\alpha_b + 0.138\alpha_m + \frac{0.862(\alpha_m - \alpha_b)}{\left[1 + 24.43\frac{E_{by}}{E_m}\right]} \tag{3.84}$$

in the vertical direction and, in the horizontal direction:

$$\alpha_{wx} = \frac{0.955\alpha_b + \left(0.046 + 0.152\frac{E_m}{E_{bx}}\right)\alpha_m}{\left[1 + 0.152\frac{E_m}{E_{bx}}\right]} \tag{3.85}$$

where α_{wy} = vertical coefficient of thermal expansion of brickwork; α_{wx} = horizontal coefficient of thermal expansion of brickwork; α_b = coefficient of thermal expansion of brick, which is assumed to be isotropic; and α_m = coefficient of thermal expansion of mortar.

The corresponding expressions for a blockwork wall, 4 blocks wide × 12 courses high, are:

$$\alpha_{wy} = 0.952\alpha_b + 0.048\alpha_m + \frac{0.952(\alpha_m - \alpha_b)}{\left[1 + 43.75\frac{E_{by}}{E_m}\right]} \tag{3.86}$$

vertically, and horizontally:

$$\alpha_{wx} = \frac{0.979\alpha_b + \left[0.006 + 0.049\frac{E_m}{E_{bx}}\right]\alpha_m}{\left[1 + 0.049\frac{E_m}{E_{bx}}\right]} \tag{3.87}$$

In Eqs (3.86) and (3.87), α_b = coefficient of thermal expansion of block, which is assumed to be isotropic.

Figure 3.16 demonstrates that the vertical and horizontal thermal movements for brickwork are similar and are mainly dependent on the thermal coefficient of the brick since there is only a small influence when the mortar thermal coefficient

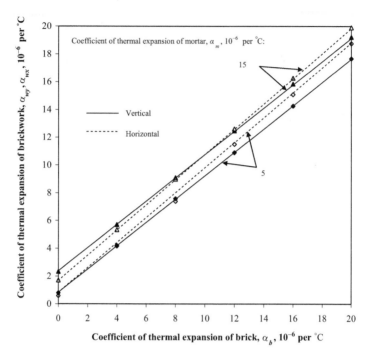

Figure 3.16 Influence of thermal expansions of brick and mortar on coefficient of thermal expansion of brickwork in the vertical and horizontal directions, according to Eqs (3.84) and (3.85); single-leaf wall, 4 bricks wide × 26 courses high, assuming $E_b/E_m = 2$.

changes from 5 to 15×10^{-6} per °C. There is a negligible influence of the modulus ratio E_b/E_m and, the trends for blockwork are almost identical to those of brickwork [21].

When the contribution of the vertical mortar joints is ignored, the vertical and horizontal expressions for brickwork become, respectively:

$$\alpha_{wy} = 0.86\alpha_b + 0.14\alpha_m \tag{3.88}$$

and

$$\alpha_{wx} = \alpha_b + \frac{(\alpha_m - \alpha_b)}{\left[1 + 6.26\frac{E_{bx}}{E_m}\right]} \tag{3.89}$$

For blockwork, the approximate expressions for coefficient of thermal expansion in the vertical direction are:

$$\alpha_{wy} = 0.952\alpha_b + 0.048\alpha_m \tag{3.90}$$

and the approximate expression for the horizontal direction is:

$$\alpha_{wx} = \alpha_b + \frac{(\alpha_m - \alpha_b)}{\left[1 + 19.85\frac{E_{bx}}{E_m}\right]} \tag{3.91}$$

Neglecting the contribution of vertical mortar joints to the thermal movement of masonry does not cause any significant loss of accuracy, as demonstrated for the case of brickwork shown in Figure 3.17.

Poisson's Ratio

It has been shown earlier that the composite modeling of masonry can be represented by combinations of the series model and the parallel models shown in Figure 3.1. That approach is now used for the analysis of Poisson's ratio of masonry subjected to vertical load and horizontal load.

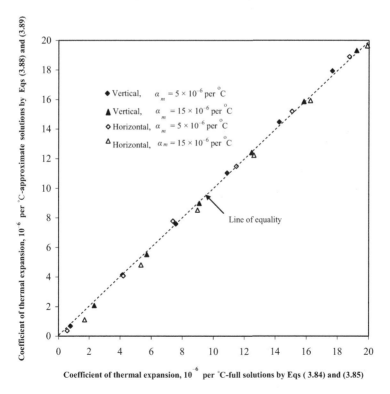

Figure 3.17 Accuracy of approximate composite model expressions for thermal movement of brickwork in the vertical and horizontal directions assuming $E_b/E_m = 2$.

Vertical Loading

In this case, the simplified approach is used by first analyzing the brick or block/ vertical mortar composite (see Figure 3.18(a)) as a two-phase composite parallel model shown in Figure 3.1(a).

Under a vertical stress, σ_{wy}, acting on the masonry, i.e., on the unit/mortar composite, the stresses on the respective unit and mortar are σ_{by} and σ_{vmy}, as shown. The

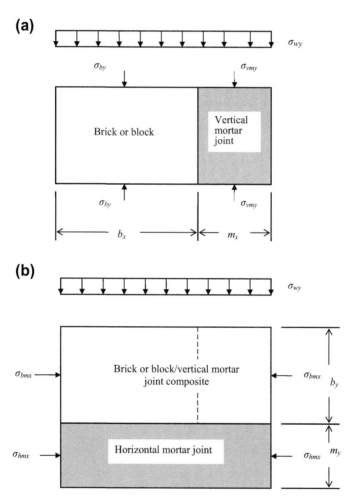

Figure 3.18 Composite model for Poisson's ratio for masonry subjected to vertical load. (a) Parallel model for brick or block/vertical mortar joint composite. (b) Series model for brick or block/vertical mortar joint composite and horizontal mortar joint.

lateral movement of the unit/vertical mortar joint composite is equal to the sum of the lateral movements of the brick or block and mortar:

$$-\mu_{bmy}\frac{\sigma_{wy}}{E_{wy}}(b_x + m_x) = -\mu_{by}\frac{\sigma_{by}}{E_{by}}(b_x) - \mu_m\frac{\sigma_{vmy}}{E_m}(m_x) \tag{3.92}$$

where μ_{bmy}, μ_{bx}, and μ_m are the Poisson's ratios of unit/mortar composite, unit, and mortar, respectively, under vertical loading.

Since the vertical strains are equal:

$$\frac{\sigma_{wy}}{E_{wy}} = \frac{\sigma_{by}}{E_{by}} = \frac{\sigma_{vmy}}{E_m}$$

so that σ_{by} and σ_{vmy} can be expressed as a function of σ_{wy} and substituted in Eq. (3.92) to yield:

$$\mu_{bmy} = \mu_{by}\left[\frac{b_x}{b_x + m_x}\right] + \mu_m\left[\frac{m_x}{b_x + m_x}\right] \tag{3.93}$$

Now, considering the standard dimensions of brick, block, and mortar, Eq. (3.93) becomes:

$$\left.\begin{array}{l}\mu_{bmy} = 0.956\mu_{by} + 0.044\mu_m \text{ (brick/mortar composite)} \\[2mm] \text{and} \\[2mm] \mu_{bmy} = 0.978\mu_{by} + 0.022\mu_m \text{(block/mortar composite)}\end{array}\right\} \tag{3.94}$$

The parallel two-phase composite model, shown in Figure 3.18(b), now represents the brick/vertical mortar composite and the horizontal (bed) mortar joint. Here, under vertical loading, the different lateral strains of the brick/mortar composite and bed mortar joint induce horizontal stresses σ_{bmx} and σ_{mx} so that for equality of horizontal strain in the masonry, brick/mortar composite, and bed joint mortar:

$$-\mu_{wy}\frac{\sigma_{wy}}{E_{wy}} = -\mu_{bmy}\frac{\sigma_{wy}}{E_{bmy}} + \frac{\sigma_{bx}}{E_{bmx}} = -\mu_m\frac{\sigma_{wy}}{E_m} + \frac{\sigma_{hmx}}{E_m} \tag{3.95}$$

By equating the forces acting horizontally, σ_{hmx} can be expressed in terms of σ_{bx}:

$$\sigma_{hmx} = -\sigma_{bx}\frac{b_y}{m_y}$$

and then substituted in Eq. (3.95) to express σ_{bx} in terms of σ_{wy}:

$$\sigma_{bx} = \sigma_{wy}\frac{\left[\dfrac{\mu_{bmy}}{E_{bmy}} - \dfrac{\mu_m}{E_m}\right]}{\left[\dfrac{1}{E_{bmx}} + \dfrac{1}{E_m}\dfrac{b_y}{m_y}\right]}$$

The Poisson's ratio of masonry μ_{wy} can now be obtained from Eq. (3.95):

$$\mu_{wy} = E_{wy} \frac{\left[\mu_m + \mu_{bmy}\frac{E_{bmx}}{E_{bmy}}\frac{b_y}{m_y}\right]}{\left[E_m + E_{bmx}\frac{b_y}{m_y}\right]} \tag{3.96}$$

Now substituting μ_{bmy} from Eq. (3.94), neglecting the contribution of the vertical mortar to the modulus of elasticity of the brick/mortar composite and assuming that $E_{bmx} \approx E_{bx}$ and $E_{bmy} \approx E_{by}$, Eq. (3.96) becomes, respectively, for standard-sized brickwork and blockwork:

$$\left.
\begin{aligned}
\mu_{wy} &= E_{wy}\left[\frac{\mu_m + 6.5(0.956\mu_{by} + 0.044\mu_m)\frac{E_{bx}}{E_{by}}}{E_m + 6.5E_{bx}}\right] \\[2em]
\mu_{wy} &= E_{wy}\left[\frac{\mu_m + 21.5(0.978\mu_{by} + 0.022\mu_m)\frac{E_{bx}}{E_{by}}}{E_m + 21.5E_{bx}}\right]
\end{aligned}
\right\} \tag{3.97}$$

The approximate solution for masonry is obtained by neglecting the contribution of the vertical mortar joints to the Poisson's ratio, which implies $\mu_{bmy} \approx \mu_{by}$ and, consequently, for brickwork:

$$\left.
\begin{aligned}
\mu_{wy} &= E_{wy}\left[\frac{\mu_m + 6.5\mu_{by}\frac{E_{bx}}{E_{by}}}{E_m + 6.5E_{bx}}\right] \\[2em]
\text{and, for blockwork:}& \\[2em]
\mu_{wy} &= E_{wy}\left[\frac{\mu_m + 21.5\mu_{by}\frac{E_{bx}}{E_{by}}}{E_m + 21.5E_{bx}}\right]
\end{aligned}
\right\} \tag{3.98}$$

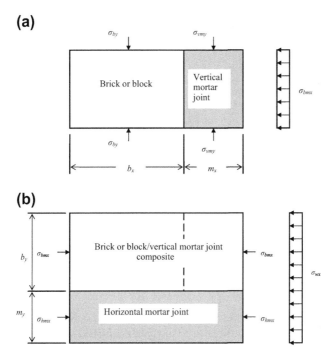

Figure 3.19 Composite model for Poisson's ratio for masonry subjected to horizontal load. (a) Series model for brick or block/vertical mortar joint composite. (b) Parallel model for brick or block/mortar composite and horizontal mortar joint.

Horizontal Loading

Under horizontal loading, σ_{wx}, the representation of masonry by the series and parallel two-phase models is reversed so that the brick/vertical mortar composite becomes the series model (see Figure 3.19(a)); the solution for Poisson's ratio of the composite is:

$$\mu_{bmx} = E_{bmx}\frac{\left[\mu_m + \mu_{bx}\frac{E_{by}}{E_{bx}}\frac{b_x}{m_x}\right]}{\left[E_m + E_{by}\frac{b_x}{m_x}\right]} \tag{3.99}$$

Considering now the parallel model as representing the brick/vertical mortar composite and the horizontal mortar bed joint (Figure 3.19(b)), the general solution for Poisson's ratio for masonry under horizontal loading is:

$$\mu_{wx} = \mu_{bmx}\left(\frac{b_y}{b_y + m_y}\right) + \mu_m\left(\frac{m_y}{b_y + m_y}\right)$$

and, after substitution for μ_{bmx} from Eq. (3.99):

$$\mu_{wx} = E_{bmx} \frac{\left[\mu_m + \mu_{bx}\frac{E_{by}}{E_{bx}}\frac{b_x}{m_x}\right]}{\left[E_m + E_{by}\frac{b_x}{m_x}\right]} \left(\frac{b_y}{b_y + m_y}\right) + \mu_m \left(\frac{m_y}{b_y + m_y}\right) \tag{3.100}$$

For standard-sized masonry, Eq. (3.100) yields:

$$\left.\begin{array}{l}\text{Brickwork}: \quad \mu_{wx} = 0.867E_{bmx}\dfrac{\left[\mu_m + 21.5\dfrac{E_{by}}{E_{bx}}\mu_{bx}\right]}{\left[E_m + 21.5E_{by}\right]} + 0.133\mu_m \\[4ex] \text{Blockwork}: \quad \mu_{wx} = 0.956E_{bmx}\dfrac{\left[\mu_m + 44\dfrac{E_{by}}{E_{bx}}\mu_{bx}\right]}{\left[E_m + 44E_{by}\right]} + 0.044\mu_m\end{array}\right\} \tag{3.101}$$

If the influence of vertical mortar to Poisson's ratio of the unit/mortar composite is neglected and the unit is assumed to be isotropic ($E_{bx} = E_{by}$), the approximate solutions for Poisson's ratios of brickwork and blockwork under horizontal loading are now, respectively:

$$\left.\begin{array}{l}\mu_{wx} = 0.867\mu_{bx} + 0.133\mu_m \\[1ex] \mu_{wx} = 0.956\mu_{bx} + 0.044\mu_m\end{array}\right\} \tag{3.102}$$

The extent of influencing factors on Poisson's ratio of masonry is illustrated by the following example, in which E_b ranges from 5 to 40 GPa and $E_m = 5$ and 15 GPa. It is assumed that $\mu_m = 0.20$, $\mu_{bx} = \mu_{by} = 0.12$, and $E_{bx} = E_{by} = E_b$.

Figure 3.20(a) shows the outcome of applying Eqs (3.97) and (3.98) to obtain the full and approximate solutions for Poisson's ratio of brickwork subjected to vertical load; E_{wy} was calculated using Eq. (3.37). It is apparent that there is little effect of neglecting the vertical mortar joints, but Poisson's ratio of brickwork is less for a low-modulus mortar, especially with a high brick modulus.

In contrast, Figure 3.20(b) indicates hardly any significant change to Poisson's ratio of brickwork subjected to horizontal loading either by neglecting the vertical mortar joints or by changing the type of mortar and brick. In fact, it may be taken as equal to that given by the approximate solution of Eq. (3.101).

The corresponding results for blockwork are shown in Figure 3.20(c). Again, there is little effect of neglecting the vertical mortar joints on Poisson's ratio under vertical loading (Eqs (3.97) and (3.98)) but, under horizontal loading with a high modulus block, the full solution for Poisson's ratio of blockwork (Eq. (3.101)) with a low modulus mortar is somewhat less than that given by the approximate solution (Eq. (3.102)).

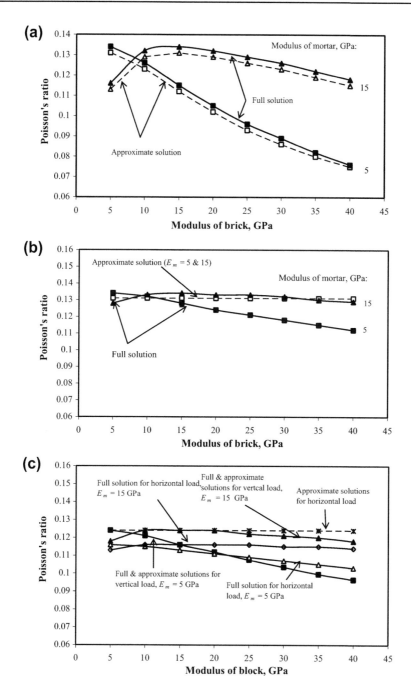

Figure 3.20 Poisson's ratio of masonry subjected to vertical and horizontal loadings. (a) Brickwork loaded in the vertical direction. (b) Brickwork loaded in the horizontal direction. (c) Blockwork loaded in vertical and horizontal directions.

Summary

Based on the theoretical analysis by two-phase composite modeling of brickwork and blockwork, full solutions for elastic, creep, moisture movement, and thermal movement have been developed. In addition, after consideration of the contribution made by the vertical mortar joints, in most cases approximate and simpler solutions can be justified after comparing their accuracy with the full solutions. The recommended composite model equations for estimating movements of masonry built from standard-size units are summarized in Tables 3.1 and 3.2.

The analysis in this chapter is applicable for solid units bonded with full-bedded mortar joints, a solid unit being defined as when the net area is greater than 75% of the gross cross-sectional area, or Group 1 according to BS EN 1996-1-1: 2005 [22]. However, the equations can also be applied to units with net areas less than 75% of the gross area, such as hollow units and shell-bedded units, but with modified coefficients

Table 3.1 Recommended Formulae for Estimating Movements of Brickwork Built with Standard Solid Bricks

Type of Movement	Direction	Formula (Brick Size: 215 × 102.5 × 65 mm)		Applicability
Elasticity	Vertical	$\frac{1}{E_{wy}} = \frac{0.86}{E_{by}} + \frac{0.14}{E_m}$	(3.39)	Any height and geometry
	Horizontal	$E_{wx} = 0.86E_{bx} + 0.14E_m$	(3.40)	
		$(E_m \geq 10 \text{ GPa})$		
		$E_{wx} = 0.86\left[\frac{E_m E_{bx}}{0.04E_{bx}+0.96E_m}\right] + 0.14E_m$		
		$(E_m \leq 10 \text{ GPa})$	(3.38)	
Creep	Vertical	$C_{wy} = 0.86C_{by} + 0.14C_m$	(3.51)	Single-leaf walls.
	Horizontal	$C_{wx} = \frac{1}{E'_{wx}} - \frac{1}{E_{wx}}$	(3.46)	For other
		$E'_{wx} = 0.86E'_{bx} + 0.14E'_m$	(3.52)	geometries, convert by V/S factor (see Figure 7.15)
Moisture movement	Vertical	$S_{wy} = 0.86S_{by} + 0.14S_m$	(3.80)	Single-leaf walls. For other
	Horizontal[a]	$S_{wx} = \frac{0.955S_{bx}+\left[0.046+0.152\frac{E_m}{E_{bx}}\right]S_m}{\left[1+0.152\frac{E_m}{E_{bx}}\right]}$	(3.77)	geometries, convert by V/S factor (see Figure 7.15)
Thermal movement	Vertical	$\alpha_{wy} = 0.86\alpha_b + 0.14\alpha_m$	(3.88)	Any height and geometry
	Horizontal[a]	$\alpha_{wx} = \alpha_b + \frac{(\alpha_m-\alpha_b)}{\left[1+6.26\frac{E_{bx}}{E_m}\right]}$	(3.89)	
Poisson's ratio	Vertical	$\mu_{wy} = E_{wy}\left[\frac{\mu_m+6.5\mu_{by}\frac{E_{bx}}{E_{by}}}{E_m+6.5E_{bx}}\right]$	(3.98)	Any height and geometry
	Horizontal	$\mu_{wx} = 0.867\mu_{bx} + 0.133\mu_m$	(3.102)	

[a]When estimating moisture movement and thermal movement in the horizontal direction, the modulus ratio should be assumed to be the elastic modulus ratio, the moduli of unit and mortar being obtained from their respective strengths, as detailed in Chapter 4.

Table 3.2 Recommended Formulae for Estimating Movements of Blockwork Built with Standard Blocks; for Solid Blocks, A'_m(Net Area) $= A_m$(Gross Area)

Type of Movement	Direction	Formula (Block Size: 440 × 215 × S100 mm)		Applicability
Elasticity	Vertical	$\frac{1}{E_{wy}} = \frac{0.952}{E_{by}} + 0.048\frac{A_m}{A'_m}\frac{1}{E_m}$	(3.43)	Any height and geometry
	Horizontal	$E_{wx} = 0.952E_{by} + 0.048\frac{A'_m}{A_m}E_m$	(3.44)	
Creep	Vertical	$C_{wy} = 0.952C_{by} + 0.048\frac{A_m}{A'_m}C_m$	(3.53)	Single-leaf walls. For other geometries, convert by V/S factor (see Figure 7.15)
	Horizontal	$C_{wx} = \frac{1}{E'_{wx}} - \frac{1}{E_{wx}}$	(3.46)	
		$E'_{wx} = 0.952E'_{bx} + 0.048\frac{A'_m}{A_m}E'_m$	(3.54)	
Moisture movement	Vertical	$S_{wy} = 0.952S_{by} + 0.048S_m$	(3.82)	Single-leaf walls. For other geometries, convert by V/S factor (see Figure 7.15)
	Horizontal[a]	$S_{wx} = S_b + \frac{(S_m - S_{bx})}{\left[1 + 19.85\frac{A_m}{A'_m}\frac{E_{bx}}{E_m}\right]}$	(3.83)	
Thermal movement	Vertical	$\alpha_{wy} = 0.952\alpha_b + 0.048\alpha_m$	(3.90)	Any height and geometry
	Horizontal[a]	$\alpha_{wx} = \alpha_b + \frac{(\alpha_m - \alpha_b)}{\left[1 + 19.85\frac{A_m}{A'_m}\frac{E_{bx}}{E_m}\right]}$	(3.91)	
Poisson's ratio	Vertical			Any height and geometry
		$\mu_{wy} = E_{wy}\left[\frac{\mu_m + 21.5\mu_{by}\frac{A_m}{A'_m}\frac{E_{bx}}{E_{by}}}{E_m + 21.5\frac{A_m}{A'_m}E_{bx}}\right]$	(3.97)	
	Horizontal	$\mu_{wx} = 0.867\mu_{bx} + 0.133\mu_m$	(3.102)	

[a]When estimating moisture movement and thermal movement in the horizontal direction, the modulus ratio should be assumed to be the elastic modulus ratio, the moduli of unit and mortar being obtained from their respective strengths, as detailed in Chapter 4.

to allow for the reduced cross-sectional area of the horizontal mortar bed joints. In the analysis, the movement properties of hollow units, for example E_{by}, are based on gross cross-sectional area, but the net area or, strictly speaking, the unit/bond contact area is required to be known for amending the movement properties of the bed joint mortar. Where required, the composite model formula for hollow blockwork is amended by the gross/net area ratio, A_m/A'_m, or vice versa, as indicated in Table 3.2. The major effect of the smaller volume of bed joint mortar is to reduce the modulus of elasticity and increase creep of masonry.

Problems

3.1 Define two ideal two-phase composite models.

3.2 What is the basic requirement for the analysis of the series model subjected to external load?

3.3 What is the basic requirement for the analysis of the parallel model subjected to external load?

3.4 In the analysis of any composite model, what are the assumptions used?

3.5 In your opinion, in Question 3.4 what is the weakest assumption and why?

3.6 What advantage is there with composite modeling of masonry compared with concrete?

3.7 It is required to produce concrete with an elastic modulus of 35 GPa. Calculate the minimum elastic modulus of the aggregate when the hydrated cement paste occupies 30% by volume and has an elastic modulus of 25 GPa.
Answer = 41.4 GPa.

3.8 Explain how creep can be taken into account in composite models.

3.9 For brickwork, the elastic modulus of the solid clay brick between bed faces is 25 GPa and the elastic modulus of mortar is 5 GPa. Use Table 3.1 to estimate the elastic modulus of the brick between header faces when there is no anisotropy of brickwork elastic modulus.
Answer = 20.0 GPa.

3.10 In what manner can creep of mortar influence moisture movement of masonry?

3.11 Is moisture movement in single-leaf walls different from moisture movement in solid piers? Explain your answer.

3.12 What has the greater influence on thermal movement of masonry; (a) the unit or (b) the mortar type? Use Eq. (3.84) to illustrate your answer when α_b changes from 5 to 15×10^{-6} per °C, and then α_m changes from 5 to 15×10^{-6} per °C.

3.13 How would you model masonry built from hollow blocks with face-shell-bedded mortar?

3.14 Use the series two-phase composite model to derive the equation for the modulus of elasticity of concrete in terms of the moduli and volumes of aggregate and cement paste.

3.15 Use the parallel two-phase composite model to derive the equation for the modulus of elasticity of concrete in terms of the moduli and volumes of aggregate and cement paste.

3.16 Define the bulk modulus of elasticity and express it in terms of modulus of elasticity and Poisson's ratio.

3.17 Discuss the influence of perforations, frogs, and voids on the isotropy of bricks and blocks.

References

[1] Hansen TC. Creep of concrete, bulletin no. 33. Stockholm: Swedish Cement and Concrete Research Institute; 1958. pp. 48.

[2] Hansen TC. Theories of multi-phase materials applied to concrete, cement mortar and cement paste. In: Proceedings of an international conference on the structure of concrete. London: Cement and Concrete Association; 1968. pp. 16–23.

[3] Hansen TC, Neilson KEC. Influence of aggregate properties on concrete shrinkage. ACL J 1965;62:783–94.

[4] Hirsch TJ. Modulus of elasticity of concrete as affected by elastic moduli of cement paste and aggregate. ACI J 1962;59:427–51.

[5] Dougill JW. Discussion on reference 7. ACI J 1962;59:1362–5.

[6] Counto UJ. The effect of the elastic modulus of the aggregate on the elastic modulus, creep and creep recovery of concrete. Mag Concr Res 1964;16(48):129–38.

[7] England GL. Method of estimating creep and shrinkage strains of concrete from properties of constituent materials. ACI J 1965;62:1411–20.

[8] Hobbs DW. The dependence of the bulk modulus, Young's modulus, creep, shrinkage and thermal expansion upon aggregate volume concentration. Mater Struct 1971;4(20):107–14.

[9] Kameswara Rao CVS, Swamy RN, Mangat PS. Mechanical behaviour of concrete as a composite material. Mater Struct 1974;7(40):265–70.

[10] Sahlin S. Structural masonry. New Jersey: Prentice-Hall Inc.; 1971. 290 pp.

[11] Base GD, Baker LR. Fundamental properties of structural brickwork. J Aust Ceram Soc 1973;9:1–6.

[12] Jessop EL, Shrive NG, England GL. Elastic and creep properties of masonry. In: Proceedings of the North American conference, Colorado; 1978. 12.1–12.17.

[13] Shrive NG, Jessop EL, Khalil MR. Stress-strain behaviour of masonry walls. In: Proceedings of 5th international brick masonry conference. Washington (DC): Brick Institute of America; 1979. pp. 453–8.

[14] Ameny P, Loov RE, Jessop EL. Strength, elastic and creep properties of concrete masonry. Int J Mason Constr 1980;1(Part. 1):33–9.

[15] Ameny P, Loov RE, Shrive NG. Prediction of elastic behaviour of masonry. Int J Mason Constr 1983;3(1):1–9.

[16] Shrive NG, England GL. Elastic, creep and shrinkage behaviour of masonry. Int J Mason Constr 1981;1(Part. 3).

[17] Ameny P. Modelling the deformation of masonry [Ph.D. thesis]. The University of Calgary; 1982.

[18] Ameny P, Loov RE, Shrive NG. Models for long-term deformation of brickwork. Mason Int 1984;1:27–8.

[19] Brooks JJ. Composite models for predicting elastic and long-term movements in brickwork walls. Proc Br Mason Soc 1986;1:20–3.

[20] Brooks JJ. Composite modelling of elasticity and creep of masonry. Dept. Civil Engineering Report. University of Leeds; 1987. 23 pp.

[21] Brooks JJ. Composite modelling of moisture movement and thermal movement of masonry. Dept. Civil Engineering Report. University of Leeds; 1987. 21 pp.

[22] BS EN 1996-1-1, 2005, Eurocode no. 6: design of masonry structures. General rules for reinforced and unreinforced masonry structures, British Standards Institution. See also: UK National Annex to BS EN 1996-1-1: 2005.

4 Elasticity of Concrete

To comply with service design requirements for deflection and deformation of concrete structures, the relation between stress and strain is required. Like other engineering materials, concrete behaves almost elastically when subjected to short-term stress, but in a slightly more complex manner when first loaded and especially at early ages, so that the modulus of elasticity has to be carefully defined. In fact, there are different types of modulus of elasticity that may be classified as either static or dynamic. As the load increases, the stress–strain becomes more nonlinear because of microcracking at the aggregate/hardened cement paste interfaces, until microcracks eventually link and form larger cracks that lead to failure at the ultimate load or strength. This chapter describes the full stress–strain behaviour of concrete. including the descending branch when determined under constant rate of strain, and Poisson's ratio, which is of interest in calculating multiaxial and volumetric strains. Relationships between modulus of elasticity and strength of concrete containing admixtures and corresponding behaviour of concrete subjected to tension are highlighted.

Stress–Strain Behaviour

Application of uniaxial compression to a concrete specimen at a constant rate of stress up to failure produces the stress–strain behaviour, as shown in Figure 4.1. Although the longitudinal or axial strain, ε_a, is a contraction, the lateral or radial strain is an extension, the ratio of lateral strain to longitudinal strain being termed the *Poisson's ratio*, which is discussed in a later section. Below approximately 30% of the ultimate stress, strains are approximately proportional to stress but, beyond this point, they start to become nonlinear due to the formation of vertical cracks, which gradually become unstable as the stress increases so that the concrete specimen eventually is no longer a continuous body. The latter is indicated by the volumetric strain, ε_v, of Figure 4.1, which changes from a slow contraction to a rapid extension due to additional cracking strain just prior to failure.

The point at which the axial stress–strain curve begins to exhibit significant nonlinear behaviour is termed the *limit of proportionality* and is attributed to very fine bond cracks or microcracks at the interface of the aggregate and hardened cement paste, which may exist prior to application of load. Microcracks occur as a result of differential volume changes between the aggregate and cement paste due to thermal and moisture movements, and stress–strain behaviour under subsequent loading. Figure 4.2 shows that the stress–strain relations are linear for separate specimens of hardened cement paste and aggregate, but the stress–strain curve for concrete becomes significantly curvilinear at higher stresses, mainly due to the micocracks

Concrete and Masonry Movements. http://dx.doi.org/10.1016/B978-0-12-801525-4.00004-2

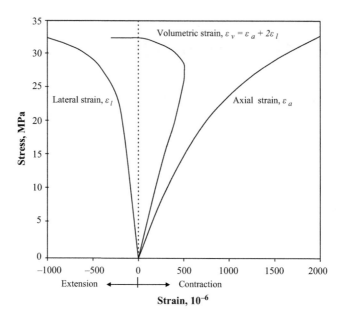

Figure 4.1 Strains in a concrete cylinder subjected to uniaxial compressive stress up to failure [1].
Source: Properties of Concrete, Fourth Edition, A. M. Neville, Pearson Education Ltd. ©
A. M. Neville 1963, 1973, 1975, 1977, 1981, 1995.

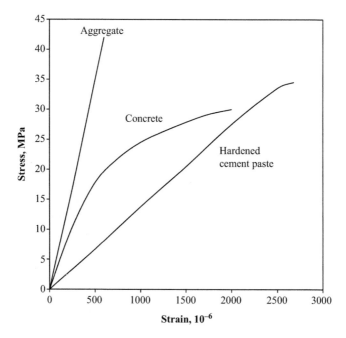

Figure 4.2 Stress–strain characteristics for hardened cement paste, aggregate, and concrete [1].
Source: Properties of Concrete, Fourth Edition, A. M. Neville, Pearson Education Ltd. ©
A. M. Neville 1963, 1973, 1975, 1977, 1981, 1995.

formed at the interfaces of the two component phases; creep of the cement paste may also contribute. Above 30% of the ultimate stress, microcracks begin to increase in length, width, and number, and collectively cause the strain to increase at a faster rate than the stress so that stress–strain curve becomes convex to the stress axis. This is the stage of slow propagation of microcracks but, at higher stresses, from 70 to 90% of the ultimate stress, cracks open up through the mortar matrix (cement paste and fine aggregate), and thus bridge the bond cracks so that a continuous crack pattern is formed. This is the stage of fast propagation of cracks and, if the stress is sustained, failure will occur with the passage of time due to *static fatigue or creep rupture* [2] (see also Chapter 14). Of course, if the load continues to increase, rapid failure at the nominal ultimate strength will take place.

As already stated, the foregoing is a description of the stress–strain behaviour when the concrete specimen is loaded at a *constant rate of stress* when the stress–strain curve terminates at the peak stress or ultimate strength. However, if the test machine frame is very stiff relative to that of the specimen, a postpeak descending curve may be monitored and, in fact, this is always achieved if the test is carried out at a *constant rate of strain* or *rate of deformation*. In this case, beyond the peak stress, the load has to be gradually decreased as the strain increases until failure occurs. This is the procedure adopted in the flexural tensile test used to determine *fracture energy* where the midpoint deflection of a simply supported notched prism is controlled at a constant rate until the specimen breaks in two halves [3]. Typical results for concrete loaded in compression at a constant rate of strain are shown in Figure 4.3. The existence of the descending branch implies that concrete has a capacity to withstand

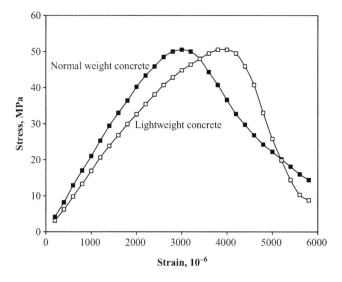

Figure 4.3 Stress–strain relations for normal-weight concrete and lightweight concrete tested at a constant rate of strain [1,4].

some load beyond the maximum load because of the controlled linking of micro-cracks, thus delaying sudden failure. Compared with normal weight aggregate concrete, lightweight concrete has a lower ascending stress–strain gradient and therefore a smaller modulus of elasticity, but has a steeper descending curve and is therefore more brittle in nature.

According to Neville [1], if the stress–strain curve ends abruptly at the peak stress, the material would be classified as *brittle*, and the less steep the descending branch of the stress–strain curve, the more *ductile* the behaviour. If the slope beyond the peak is zero, the material is said to be *perfectly plastic*. Generally, the area enclosed by the full stress–strain curve represents the *fracture toughness*, i.e., the work necessary to cause the failure [2].

Knowledge of the full stress–strain curve is required for the design of structural reinforced concrete, especially when very high strength or high performance concrete is used. Besides having a higher modulus of elasticity, high strength concrete is more brittle than normal strength concrete and has a steeper ascending branch, as well as a steeper descending branch of its stress–strain curve. High-strength concrete also develops a smaller amount of cracking than normal concrete throughout all stages of loading [1].

The effect of high temperature on stress–strain behaviour of normal- and high-strength concretes was studied by Castillo and Durrani [5]. Figure 4.4 shows

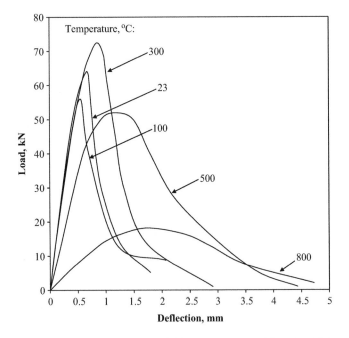

Figure 4.4 Load-deflection characteristics of 30 MPa concrete subjected to high temperature just before loading at a constant rate of strain [5].

examples for normal strength concrete and, for both normal and high strength concretes, the increase of temperature caused concrete to become more ductile. The strain at peak load did not vary significantly between room temperature and 200 °C, although strength decreased by some 6–20%. Between 300 and 400 °C, strain at peak load increased slightly, whereas strength increased above room temperature strength by 8–14%. Between 500 and 800 °C, peak load strain increased significantly with the 800 °C peak load strain being four times that at room temperature; over the same temperature range, there was a progressive loss of strength.

For design purposes, BS EN 1992-1-1:2004 [6] prescribes the following analytical expressions in order to calculate stress–strain curves of normal weight aggregate and lightweight aggregate concretes at normal temperature. Tables 4.1 and 4.2 give values of the salient features of the curves for different strength classes of concrete, as identified by the 28-day characteristic cylinder compressive strength, f_{ck}. The complete analytical expression for the ascending and descending stress (σ_c)-strain (ε_c) curves of normal weight concrete under short-term loading up to nominal failure [6] is:

$$\sigma_c = f_{cm} \frac{\varepsilon_c}{\varepsilon_{c1}} \left[\frac{\varepsilon_{c1}k - \varepsilon_c}{\varepsilon_{c1} + (k-2)\varepsilon_c} \right] \tag{4.1}$$

where $f_{cm} = $ 28-day mean cylinder strength $= f_{ck} + 8$

$$k = 11.58(f_{cm})^{-0.7}\varepsilon_{c1}$$

$$\varepsilon_{c1} = 700(f_{cm})^{0.31} \times 10^{-6} \leq 2800 \times 10^{-6} = \text{strain at peak stress}$$

Equation (4.1) is valid for $0 < \varepsilon_c < \varepsilon_{cu1}$, where $\varepsilon_{cu1} = $ nominal ultimate strain, viz:

$$\left. \begin{array}{ll} \varepsilon_{cu1} = 3500 \times 10^{-6} & \text{when } f_{cm} \leq 50 \text{ MPa} \\[2mm] \varepsilon_{cu1} = \left\{ 2800 + 27 \left[\frac{98 - f_{cm}}{100} \right]^4 \right\} 10^{-6} & \text{when } f_{cm} \geq 50 \text{ MPa} \end{array} \right\} \tag{4.2}$$

In Table 4.1, for each strength class of normal weight concrete, the characteristic cube strength, $f_{cck} \approx 1.2\,f_{ck}$.

In the case of lightweight aggregate concrete (Table 4.2), the 28-day mean cylinder strength $f_{lcm} = f_{lck} + 8$ when $f_{lck} \geq 20$ MPa, and the characteristic cube strength $f_{cck} \approx 1.1\,f_{ck}$. The data compiled in Table 4.2 apply to lightweight aggregate concrete having a closed structure and a density of not more than 2200 kg/m^3 and not less than 800 kg/m^3, consisting of or containing a proportion of artificial or natural lightweight aggregate having a particle density less than 2000 kg/m^3 [6]. Furthermore, the ascending branch of the stress–strain curve is assumed to be linear up to the ultimate strength, and there is no descending

Table 4.1 Strength and Deformation Data for Normal Weight Aggregate Concrete According to BS EN 1992-1-1: 2004 [6]

Property	Strength Class (Characteristic Cylinder Strength, f_{ck}, MPa)													
	12	16	20	25	30	35	40	45	50	55	60	70	80	90
Characteristic cube strength, f_{cck}, MPa	15	20	25	30	37	45	50	55	60	67	75	85	95	105
28-day mean cylinder strength, f_{cm}, MPa	20	24	28	33	38	43	48	53	58	63	68	78	88	98
28-day mean tensile strength, f_{tcm}, MPa	1.6	1.9	2.2	2.6	2.9	3.2	3.5	3.8	4.1	4.2	4.4	4.6	4.8	5.0
28-day mean modulus of elasticity, E_{cm}, GPa	27	29	30	31	33	34	35	36	37	38	39	41	42	44
Peak stress compressive strain, ε_{c1}, 10^{-6}	1800	1900	2000	2100	2200	2250	2300	2400	2450	2500	2600	2700	2800	2800
Ultimate compressive strain, ε_{cu1}, 10^{-6}	3500									3200	3000	2800	2800	2800

Table 4.2 Strength and Deformation Data for Lightweight Aggregate Concrete According to BS EN 1992-1-1: 2004 [6]

Property	Strength Class (Characteristic Cylinder Strength, f_{ck}, MPa)												
	12	**16**	**20**	**25**	**30**	**35**	**40**	**45**	**50**	**55**	**60**	**70**	**80**
Characteristic cube strength, f_{cck}, MPa	13	18	22	28	33	38	44	50	55	60	66	77	88
28-day mean cylinder strength, f_{lcm}, MPa	17	22	28	33	38	43	48	53	58	63	68	78	88
28-day mean tensile strength, f_{lctm}, MPa	$f_{lctm} = f_{ctm} \times [0.4 + (\rho/2200)]$												
28-day mean modulus of elasticity, E_{lcm}, GPa	$E_{lcm} = E_{cm} \times (\rho/2200)^2$												
Peak stress compressive strain, ε_{lc1}, 10^{-6}	$\varepsilon_{lc1} = k \times f_{lcm}/(E_{lcm})$, where $k = 1.1$ for sanded lightweight aggregate concrete and $k = 1.0$ for all lightweight aggregate concrete												
Ultimate compressive strain, ε_{lcu1}, 10^{-6}	$\varepsilon_{lcu1} = \varepsilon_{lc1}$												

branch of the stress–strain curve so that strain at ultimate strength is also the ultimate strain, i.e.,

$$\varepsilon_{lc1} = \frac{f_{lcm}}{E_{lcm}} = \varepsilon_{lcu1} \qquad (4.3)$$

where $E_{lcm} = $ 28-day mean modulus of elasticity of lightweight concrete, given by Eq. (4.8).

Using the above expressions, examples of stress–strain curves estimated for lightweight aggregate concrete are compared with curves estimated for normal weight aggregate concrete in Figure 4.5.

It is of interest to compare stress–strain behaviour of concrete subjected to uniaxial tension with that for compression, the former being required for the study of crack formation or fracture mechanics. An example of the complete stress–strain curve for concrete loaded in direct tension under constant rate of strain is shown in Figure 4.6 Compared with compressive stress–strain behaviour, the limit of proportionality on the ascending curve where microcracks begin to develop is lower in tension (40–50% of peak stress), whereas the shape of the descending branch of the curve is steeper, which indicates rapid development of microcracks in a localized zone where the final crack and fracture will occur. The presence of the localized zone of microcracks implies that the postpeak stress–strain behaviour is dependent on the strain gauge length and location on the specimen. Within the fracture zone, the deformation of

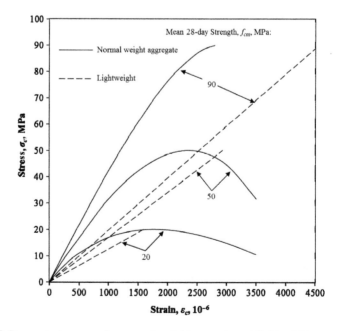

Figure 4.5 Stress–strain curves for normal-weight concrete and lightweight aggregate concrete according to Eq. (4.1) given by BS EN 1992-1-1:2004 [6].

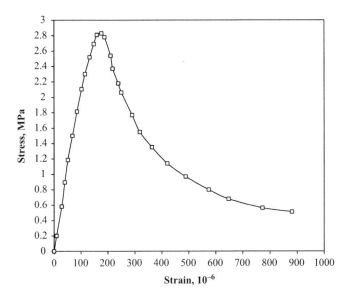

Figure 4.6 Example of stress–strain curve in direct tension loaded at constant rate of strain [1,7].

concrete corresponds to the development of crack width and, consequently, it is preferable to describe the postpeak behaviour by a stress-deformation (crack width) rather than by a stress–strain relation [8].

The factors affecting the tensile stress–strain curve are the same as those affecting concrete loaded in compression. Komlos [9] found that a decrease in the aggregate/cement ratio slightly increased the strain throughout the complete range of the stress–strain relationship, but the effect of aggregate grading was insignificant when the maximum particle size remained constant. For a given stress, concrete having a higher water/cement ratio had a slightly higher level of strain, but no difference could be found in the strain at failure. However, other tests indicated that a lower water/cement ratio increased the strain at failure even though the tensile strength was less [10]. Adding fly ash without changing the water/cement ratio also increased the strain at failure without affecting strength of concrete [10]. For a given stress, water curing causes an increase in the strain, whereas dry curing causes a decrease in strain [9].

Hughes and Ash [11] found that the limit of proportionality in tensile testing increased with a decrease in aggregate size or, in other words, smaller sized aggregate required a larger stress to induce microcracking. The strain at peak stress increased as the tensile strength increased [7], and a review of many test data indicated the general trend was that ultimate strain (*strain capacity*) increases as the tensile strength increases [12].

Under compressive loading or tensile loading, strength, modulus of elasticity, and the overall stress–strain relationship is affected by the rate of application of load. Increasing the rate of loading has the effect of increasing the modulus of elasticity and strength, together with generally causing the stress–strain curve to become less curvilinear. According to Oh [13], the effect of strain rate is more sensitive under tension than under compression.

Static Modulus of Elasticity

For most structural concrete design applications, a value for the modulus of elasticity is required to estimate elastic strain, deflection, and deformation of structural elements in order to comply with serviceability requirements. Although the full stress–strain behaviour of concrete up to failure has already been described, it is relevant to consider how to define and determine the modulus of elasticity of concrete, and to consider it within the context of other engineering materials subjected to a short-term cycle of load up to a maximum of typical working load that is approximately between 0.2 and 0.4 of the failure load.

The definition of *pure elasticity* is that strains appear and disappear immediately on application and removal of stress. Figure 4.7 illustrates two categories of pure elasticity: (a) linear and (b) nonlinear, and two cases of nonelasticity: (c) linear and (d) nonlinear. Whereas in the pure elastic categories, the loading and unloading curves coincide, in the nonelastic instances, there are separate loading and unloading

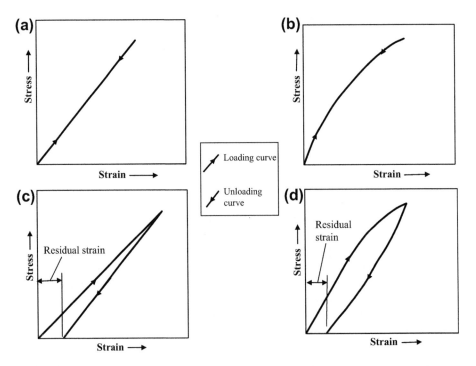

Figure 4.7 (a) Linear and elastic (b) Non-linear and elastic, (c) Linear and non-elastic (d) Non-linear and non-elastic.
Source: Concrete Technology, Second Edition, A. M. Neville and J. J. Brooks, Pearson Education Ltd. © Longman Group UK Ltd. 1987.

curves, and a permanent residual strain of deformation remains on the complete removal of load. Of the common engineering materials, steel exhibits near pure linear and elastic behaviour (Figure 4.7(a)), some plastics and timber exhibit pure nonlinear and elastic behaviour (Figure 4.7(b)), and the stress–strain behaviour of brittle-type materials (glass and rocks) is described as linear and nonelastic (Figure 4.7(c)). When first subject to a cycle of compressive load, concrete exhibits nonlinear and nonelastic stress–strain behaviour, as typified by Figure 4.7(d). The area between the loading and unloading curves is known as *hysteresis* and represents the irreversible energy of deformation (e.g., creep, microcracking); subsequent cycles of loading decrease the hysteresis and nonlinearity. The stress–strain behaviour is similar for concrete loaded to moderate and high stresses in tension.

The slope of the stress–strain curve gives the modulus of elasticity, but the term *Young's modulus* can only be applied to the linear categories of Figure 4.7. In the case of concrete, the modulus of elasticity of concrete can be determined in several ways, as indicated in Figure 4.8, which is an enlarged version of Figure 4.7(d). In the absence of a straight portion of the loading stress–strain curve, the *initial tangent modulus*, E_{to}, is given by a tangent to the curve at the origin where the stress is near zero. The initial tangent modulus is approximately equal to the *dynamic modulus of elasticity*, E_d, which is determined by vibrating a concrete specimen at high frequency

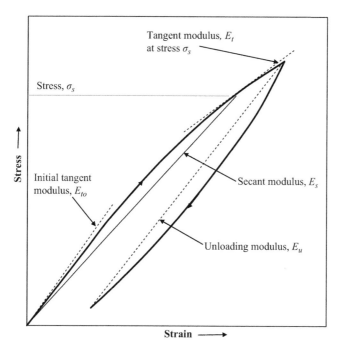

Figure 4.8 Generalized magnified stress–strain curve for concrete [2].
Source: Concrete Technology, Second Edition, A. M. Neville and J. J. Brooks, Pearson Education Ltd. © Longman Group UK Ltd. 1987.

(see p. 72). This modulus represents just pure elastic effects, since there is only a small stress involved and no microcracking or creep.

When the strain response is required after a particular application of stress σ_s, the appropriate strain is given by the *secant modulus*, E_s; for example, in creep tests; the secant modulus is also known as the *chord modulus*. The strain for small changes in stress around the nominal stress of σ_s, is determined by the *tangent modulus*, E_t. There is also an *unloading modulus*, E_u, which quantifies stress and strain in the case of decreasing load. Since, generally, the unloading modulus is parallel to the initial tangent modulus, E_u is often used to determine E_{to} since the tangent to the loading curve at near zero stress is difficult to judge; as stated earlier, the initial tangent modulus is also approximately equal to the dynamic modulus.

Composite models for representing the modulus of elasticity of concrete have been presented in Chapter 3, and all models reveal the importance of influencing factors: moduli of aggregate and hardened cement paste phases as well as their proportions by volume. Table 4.3 compares predictions with measured values, and demonstrates the effects of those influencing factors. When modulus of the mortar or cement paste are known, Counto's composite model (Eq. (3.5)) is superior for predicting the modulus of elasticity of concrete followed by the Hirsch–Dougill model (Eq. (3.4)); the limitations of the parallel model (Eq. (3.1)) and series model (Eq. (3.2)) are discussed in Chapter 3. The pattern of behaviour of aggregate type and fraction volume on the modulus of elasticity of concrete is also illustrated in Figure 4.9 made with different materials and a cement paste matrix phase having a modulus of elasticity of 19 GPa. For any type of aggregate, the modulus of elasticity of concrete increases as the fractional volume of aggregate increases, and the modulus of elasticity of concrete increases when the modulus of aggregate exceeds 19 GPa, but decreases when the modulus of aggregate is less than 19 GPa.

In practice, the range of modulus of elasticity of normal weight aggregate used in concrete is 50–120 Gpa, whereas the range for lightweight aggregate is much lower, viz. 12–40 GPa. The range of modulus of elasticity for hardened cement paste varies from 7 to 28 GPa [14]. By contrast, the range of total aggregate content or cement paste content by volume in concrete mixes only changes by a small amount. For example, the total volume of concrete having mix proportions by mass (g) of 1 (cement): 6 (total aggregate) with a water/cement ratio $= 0.55$ is $(1/3.15 + 6/2.65 + 0.55) = 3.131$ cm^3, assuming specific gravities of cement and aggregate are, respectively, 3.15 and 2.65 g/cm^3. Therefore, the percentage volume of total aggregate is $[(6/2.65)/3.131] \times 100 \approx 72\%$. Comparing this with two other concrete mixes having aggregate/cement ratios of 9 and 4.5, and corresponding water/cement ratios of 0.75 and 0.40, their percentage aggregate contents are 76% and 70%, respectively, so that there is only little variation in volumetric content. In terms of cement paste content (volume of cement plus water), the corresponding values of the three mixes having aggregate/cement ratios of 9, 6, and 4.5 by mass are, respectively, 24, 28, and 30%.

Although occupying a smaller volume than the aggregate phase, the hardened cement paste phase influences the modulus of elasticity of concrete through the water/cement ratio or, more fundamentally, the capillary porosity (P_c). Verbeck and

Table 4.3 Influence of Aggregate Type and Volume Fraction on Predicted and Measured Modulus of Elasticity of Concrete [15]

	Coarse Aggregate		Predicted Modulus of Elasticity of Concrete by Model				Measured Modulus
Type	Modulus of Elasticity, GPa	Fraction Volume	Parallel (Eq. (3.1))	Series (Eq. (3.2))	Hirsch/Dougill (Eq. (3.4))	Counto (Eq. (3.7))	of Elasticity, GPa
Series 1: Modulus of Elasticity of Mortar = 40.5 GPa							
Cast iron	104.8	0.50	72.7	58.4	64.8	64.7	71.7
		0.25	51.3	47.0	49.1	52.0	54.3
Flint gravel	74.5	0.50	57.5	52.5	54.9	55.0	55.4
		0.25	49.0	45.7	47.3	47.5	47.2
Glass	72.4	0.50	56.5	51.9	54.1	54.2	54.2
		0.25	48.5	45.5	50.4	47.2	46.9
Polythene	0.29	0.50	20.4	0.58	1.13	15.2	14.5
		0.25	30.5	1.14	2.20	27.1	22.9
Series 2: Modulus of Elasticity of Hardened Cement Paste = 10.6 GPa							
Steel	206.8	0.55	118.5	22.2	37.4	34.4	33.4
Flint gravel	72.5	0.55	37.4	20.0	26.1	26.6	27.4
Glass	72.4	0.55	37.4	20.0	26.0	26.6	26.5

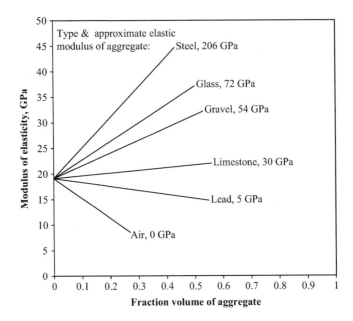

Figure 4.9 Effect of type of aggregate on the modulus of elasticity of concrete [14].

Helmuth [16] demonstrated (see Figure 15.3) the proportionality between modulus of elasticity and the term $(1 - P_c)^3$, so that as the water/cement ratio increases, the porosity increases and the modulus decreases. Since strength is similarly dependent on capillary porosity [2], the modulus of elasticity and strength of cement paste are closely related and, correspondingly, it can be expected that modulus of elasticity and strength of concrete are related. In fact, when required for design of concrete structures, modulus of elasticity is usually estimated from strength by the expressions presented shortly.

When designing concrete elements, composite models are not used for estimating modulus of elasticity of concrete because of the difficulty of measuring strain of small aggregate particles. Research investigations have involved determining modulus on larger specimens of aggregate prepared from parent rock in order to attach the strain measuring device. However, the properties of the larger test specimen are unlikely to be the same as for smaller coarse and fine aggregate particles, because the crushing process reduces the number of natural flaws. Furthermore, two-phase modeling of concrete does not account for *transition zone* effects [1,2] at the aggregate–cement paste interface where properties of cement paste can differ from properties of the bulk of cement paste, either due to "incomplete packing" of cement particle during mixing, or due to crystalline $Ca(OH)_2$ formed on the surface of aggregate during hydration of cement [1]. The zone can be regarded as a third phase that, although has a small fractional volume, can have a significantly greater strength and modulus of elasticity than hydrated cement paste. For example, very fine cementitious materials, such as microsilica, react with $Ca(OH)_2$ to yield a greater modulus of elasticity of concrete compared with that expected by estimation by a two-phase composite model [2].

If the temperature is raised some time before testing concrete, hydration of cement paste will be accelerated resulting in a greater stiffness and, hence, modulus of elasticity of concrete [17]. On the other hand, increasing the temperature of mature concrete just before testing is to decrease the modulus of elasticity [5]. For example, compared with that of concrete stored at 23 °C, the modulus of elasticity has been reported to reduce by almost threefold when stored at a temperature of 650 °C [18,19] (see also Table 10.4). Over a temperature range of 23–800 °C, the reduction of modulus of elasticity relative to that at room temperature is similar for both normal and high-strength concretes, as shown in Figure 4.10. Also shown in the same figure are results several investigators, the overall trends suggesting an influence of moisture state of the test specimens prior to loading. It seems that the rate of loss of modulus of elasticity is greater for wet and saturated concrete, when more evaporable moisture can be expelled from the concrete, than for the loss of modulus for dry concrete, which has already lost some evaporable moisture prior to testing. CEB Model Code 90 [25,26], recommends that the effect of elevated temperature at the time of testing on the modulus of elasticity of concrete (E_{cT}), at temperature (T), at any age t without moisture exchange for a temperature range of 5–80 °C, may be estimated from:

$$E_{cT} = E_{c20}(1.06 - 0.003T) \qquad (4.4)$$

where E_{c20} = modulus of elasticity at normal temperature (≈ 20 °C).

As has already been mentioned, a main factor affecting the degree of curvature of the stress–strain curve and, therefore, the modulus of elasticity is the rate of

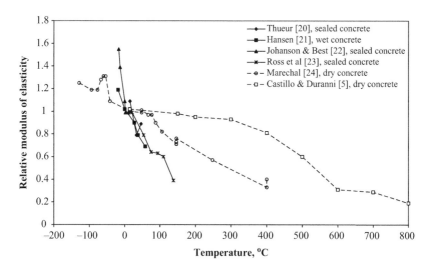

Figure 4.10 Influence of temperature on modulus of elasticity of concrete heated shortly before testing; values expressed relative to modulus of elasticity at temperature ≈ 20 °C.

application of stress. When the load is applied rapidly, the curvature becomes small and strains are reduced, the more so the higher the rate of loading, so that extremely rapid loading (≤ 0.01 s) results in near linear stress–strain behaviour. On the other hand, an increases in loading time (5 s to about 2 min) can increase the curvature and strain, but within the range of 2 min to about 10 min—a time normally required to test a specimen, the change of curvature and increase in strain and are small [2].

Since the nonlinear behaviour of the stress–strain curve at normal levels of stress is mainly due to and creep (including any microcracking), the demarcation between elastic and creep strains is difficult. For practical purposes, an arbitrary distinction is made: the deformation resulting from application of the design stress is considered elastic (*initial elastic strain*), and the subsequent increase in strain under sustained stress is regarded as creep. Thus, as stated earlier, the modulus of elasticity defined in this way is the secant modulus of Figure 4.8. There is no standard method of determining the secant modulus, but it is usually measured at stresses ranging from 15 to 50% of the short-term strength. Since the secant modulus is dependent on level of stress and its rate of application, the stress and time should always be stated [2].

Several cycles of loading and unloading reduce creep so that the subsequent experimental stress–strain curve tends to become linear. This procedure is prescribed by ASTM C 469-02 [27] and BS 1881-121: 1983 [28] for the determination of the *static modulus of elasticity*, using a test cylinder or prism. In the BS test, the preferred specimen is a 150-dia. × 300-mm cylinder loaded and unloaded at a rate of 0.6 MPa/s between a minimum stress of 0.5 MPa and a maximum stress of 0.3 of the mean cylinder strength determined on companion cylinders. For two cycles of loading and unloading, the load is sustained for short periods (0.5–1 min) at the maximum and minimum stresses, then afterwards, a third loading static modulus of elasticity is determined from the quotient of maximum stress minus minimum stress and corresponding strains.

Dynamic Modulus of Elasticity

So far, the various terms used to describe modulus of elasticity refer to the static modulus of elasticity since they are determined from the strain response under static loads. On the other hand, the test procedure for the determination of dynamic modulus of elasticity, as prescribed by BS 1881-209: 1990 [29] and ASTM C 215-02 [30], is to determine the resonant or lowest fundamental frequency of a concrete specimen by subjecting it to high frequency vibrations. Figure 4.11(a) shows the test setup for a 100- × 100- × 500-mm prismatic specimen clamped at its midpoint with an electro-magnet exciter unit at one end face and a pick-up transducer at the other end. The exciter is driven by a variable frequency oscillator with a range of 100–10,000 Hz so that longitudinal vibrations can be propagated through the specimen and received by the pick-up, amplified, and measured. Resonance is indicated when the amplitude is a maximum, and if the frequency when this occurs is n (Hz), the length of the specimen

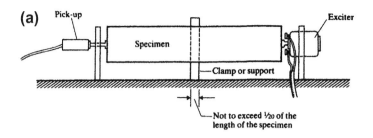

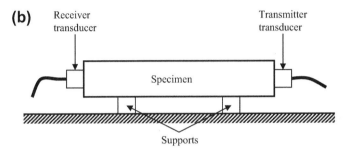

Figure 4.11 Measurement of dynamic modulus of elasticity (a) Dynamic modulus test and (b) ultrasonic pulse velocity test setups.

is L (m), ρ (kg/m^3) is the density of concrete and g (m/s^2) = acceleration due to gravity, then the dynamic modulus, E_d (kg/m^2), is:

$$E_d = 4n^2L^2\frac{\rho}{g} \tag{4.5}$$

To yield the dynamic modulus in GPa with L in mm, Eq. (4.5) becomes:

$$E_d = 4n^2L^2\rho \times 10^{-15} \tag{4.6}$$

In addition to the test based on the longitudinal resonance frequency, tests based on the transverse (flexural) frequency and the torsional frequency are possible [14,30]. The fundamental frequency in the torsional mode of vibration gives the modulus of rigidity or modulus of elasticity in shear.

Since it is approximately equal to the initial tangent modulus, the dynamic modulus is always greater than the secant or static modulus of elasticity because of the shape of the stress–strain curve. However, the ratio of static modulus, E_c, to dynamic modulus is not constant since, for example, an increase in compressive strength or in age results in a higher ratio due to a reduction in the curvature of the stress–strain curve. A general relation between E_c and E_d was given in the superseded BS 8110-2: 1985 [31]:

$$E_c = 1.25E_d - 19 \tag{4.7}$$

Equation (4.7) does not apply to concrete containing a cement content of more than 500 kg/m^3 or to lightweight aggregate concrete. For the latter, the following has been proposed [1,32]:

$$E_c = 1.04E_d - 4.1 \qquad (4.8)$$

The dynamic modulus of elasticity test apparatus is limited to laboratory testing where it is extremely useful for assessing durability of concrete, such as deterioration from chemical attack.

It is also possible to determine the dynamic modulus of elasticity of concrete using the ultrasonic pulse velocity apparatus [1,33]. Figure 4.11(b) shows the test set-up in which the normal apparatus generates a pulse of compressive wave vibrations at ultrasonic frequency (10–150 Hz), which are transmitted by an electro-acoustic transducer held in contact with the surface of the end of the concrete beam. After passing through the concrete, the vibrations are received by a second electro-acoustic transducer held at the other end of the beam, and the time taken is amplified before being digitally displayed. The time to pass through the beam is measured to with ± 0.1 μs and, knowing the length of path, the pulse velocity can be calculated. Unlike the dynamic modulus apparatus, the ultrasonic pulse velocity apparatus is portable and therefore a useful, nondestructive test method that can be used on site to assess quality control, frost or chemical damage, and crack detection of concrete structures [1,2]. However, to estimate dynamic modulus of elasticity, prior knowledge of Poisson's ratio is required (see Eq. (4.22)) and, moreover, access to opposite faces of the concrete member is normally needed to transmit compression ultrasonic waves (direct method), which is not always possible for in-situ construction concrete. However, the technique proposed by Qixian and Bungey [34] purports to overcome this problem by measuring and relating velocities of shear and surface ultrasonic wave propagations (semidirect and indirect transmissions) to compression ultrasonic wave propagation (direct transmission), thus enabling both Poisson's ratio and dynamic modulus of elasticity to be calculated.

Relationship of Modulus of Elasticity with Strength

In general, the modulus of elasticity increases with an increase of compressive strength of concrete. However, some factors that influence strength of concrete affect modulus of elasticity in a different manner. These can be summarized as follows:

- Although the type of aggregate as a factor is acknowledged indirectly in modulus–strength expressions in terms of density of concrete, the modulus of elasticity of concrete is strongly dependent on the modulus of elasticity of aggregate and its proportion by volume, whereas strength is mainly dependent on the water/binder ratio.
- Modulus of elasticity depends on the level of stress due to the nonlinearity of the stress–strain curve and on rate of loading, whereas strength depends only on the rate of loading.

- Moisture condition of the concrete affects the modulus of elasticity and strength differently. A wet specimen has a lower strength than a dry specimen, but a wet specimen has higher modulus than a dry specimen (see Chapter 15).
- Size of the concrete test specimen affects strength, but not its modulus of elasticity.

The effect of differences in influencing factors is to contribute to scatter in the modulus–strength relationship but, for normal-weight aggregate and lightweight concrete, relationships are sufficiently accurate to be used confidently for estimating modulus of elasticity for most design applications, typically being within $\pm30\%$. For example, BS EN 1992-1-1: 2004 [6] tabulates values of the mean 28-day modulus of elasticity (E_{cm}, GPa) for different characteristic strengths or strength classes (f_{ck}) as shown in Table 4.1, the values being analytically related by:

$$E_{cm} = 11.03 (f_{cm})^{0.3} \qquad (4.9)$$

where $f_{cm} = (f_{ck} + 8)$, and $f_{cm} = 28$-day mean cylinder compressive strength (MPa).

The above expression is for the secant modulus of elasticity determined between 0 and $0.4 f_{cm}$ and applies to normal weight aggregate concretes made with quartzite aggregates. For concrete made with limestone and sandstone aggregates, the modulus of elasticity should be reduced by 10% and 30%, respectively. On the other hand, for basalt aggregates, the value should be increased by 20% [6]. The values of modulus of elasticity given in Table 4.1 should be regarded as indicative for general applications and, if necessary, values should be specifically assessed when structures are likely to be sensitive to deviations from the general values [6].

For lightweight aggregate concretes with a closed structure and density of not more than 2200 kg/m^3, containing a proportion of artificial or natural lightweight aggregate having a particle density less than 2000 kg/m^3, the mean secant modulus of elasticity of lightweight aggregate concrete, E_{lcm}, is dependent on the density (oven-dry) of concrete, and is estimated by BS EN 1992-1-1: 2004 [6], as follows:

$$E_{lcm} = E_{cm} \left(\frac{\rho}{2200} \right)^2 \qquad (4.10)$$

where the density, ρ, is between 800 and 2200 kg/m^3.

It should be noted that Eq. (4.10) does not apply to aerated concrete either autoclaved or normally cured, nor lightweight aggregate concrete with an open structure. When more accurate data are need, e.g., where deflections are of importance, tests should be carried out to determine the modulus of elasticity of lightweight aggregate concrete.

The ACI Building Code 318-05 [35] gives the following expression for normal-weight concrete:

$$E_c = 4.7 (f_c)^{0.5} \qquad (4.11)$$

When the density of concrete is between 500 and 2500 kg/m³, the ACI 318-05 expression for modulus of elasticity becomes:

$$E_c = 43\rho^{1.5}f_c^{0.5} \times 10^{-6} \tag{4.12}$$

Other modulus–strength expressions exist, the details of which are summarized by Neville [1]; they are essentially similar in form to Eqs (4.11) and (4.12) with slight variations in coefficients and power indices.

Age

The modulus of elasticity increases as the age of concrete increases but at a lower rate than strength. According to BS EN 1992-1-1: 2004 [6], the increase in the modulus of elasticity with time may be estimated from the 28-day strength, f_{cm}, and modulus, E_{cm}, by:

$$E_c(t) = E_{cm}\left[\frac{f_c(t)}{f_{cm}}\right]^{0.3} \tag{4.13}$$

where $E_c(t)$ and $f_c(t)$ are modulus of elasticity and strength, respectively, at the age of t (days).

Now the development of strength with time is expressed as follows:

$$f_c(t) = \exp\left\{S\left[1 - \left(\frac{28}{t}\right)^{0.5}\right]\right\}f_{cm} \tag{4.14}$$

so that Eq. (4.13) becomes

$$E_c(t) = E_{c28}\left(\exp\left\{S\left[1 - \left(\frac{28}{t}\right)^{0.5}\right]\right\}\right)^{0.3} \tag{4.15}$$

where S = coefficient to allow for type of cement, viz:

0.20 for cement classes CEM 42.5R, CEM 52.5N and CEM 52.5R (Class R).
0.25 for cement of classes CEM 32.5R, CEM 42.5N (Class N).
0.38 for cement of strength classes CEM 32.5N (Class S).

The cement classes (CEM) are those specified according to BS EN 197-1: 2000 [36] and are grouped according to their rate of hardening: Class R = rapid, Class N = normal, and Class S = slow.

Figure 4.12 illustrates the development of strength and modulus of elasticity with age according to Eqs (4.14) and (4.15), respectively, for Class 50 concrete. Also shown is the corresponding development of elastic strain for a unit stress, i.e., $(E_c(t))^{-1}$, which is of interest in defining the starting point for creep. It is apparent that

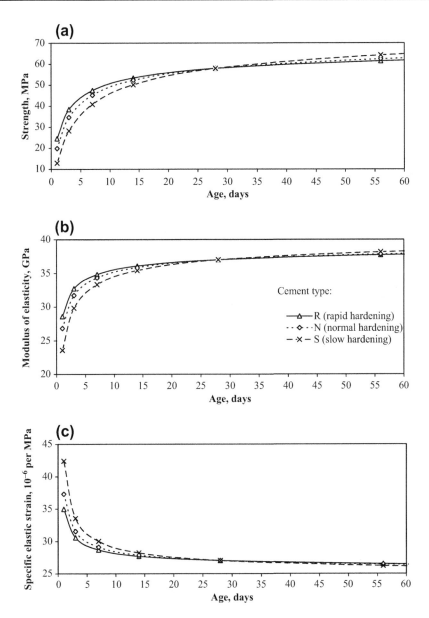

Figure 4.12 Effect of age on (a) strength, (b) modulus of elasticity, and (c) elastic strain at loading of class 50 concrete made with different cement types, according to BS EN 1992-1-1:2004 [6].

the highest rate of change occurs prior to the age of 28 days, but the rate of change of modulus of elasticity and elastic strain is much lower than strength. For example, for concrete made with normal cement, strength increases by a factor of almost two from 1 day to 28 days, but there is only a 40% increase in modulus of elasticity and 30% decrease in elastic strain over the same period. In Chapter 10, the definition of creep assumes that the elastic stain remains constant throughout the creep process and is equal to that on first loading, but Figure 4.12(c) clearly this is not the case when concrete is subjected to load at very early ages, say 1–3 days, because elastic strain decreases with time under sustained load.

Admixtures

In 2000, a review of previous research revealed that the effect of admixtures on modulus of elasticity of concrete is essentially the same as the effect of admixtures on strength, so that the relation between modulus and strength is unaffected [37]. Concretes containing water-reducing admixtures (plasticizers and superplasticizers) and cement replacement materials (fly ash, microsilica, metakaolin, and ground granulated blast-furnace slag), are compared with their respective control (admixture-free) concretes in Figures 4.13–4.16. In previous publications where different specimen sizes were used for strength, values have been adjusted to those of an equivalent cylinder with a height/diameter ratio of two. The modulus–strength data cover

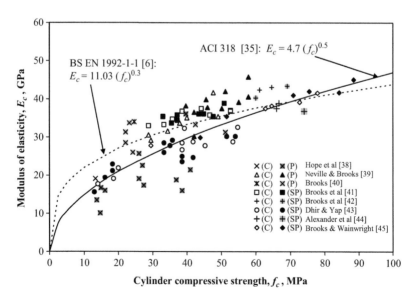

Figure 4.13 Modulus of elasticity as a function of compressive strength for concretes with and without plasticizing and superplasticizing admixtures. (C) = control concrete, (P) = concrete with plasticizer and (SP) = concrete with superplasticizer [37].

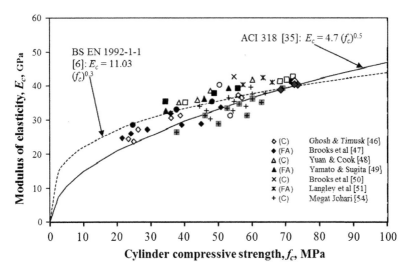

Figure 4.14 Modulus of elasticity as a function of compressive strength for concrete with and without fly ash. (C) = control concrete, (FA) = fly ash.

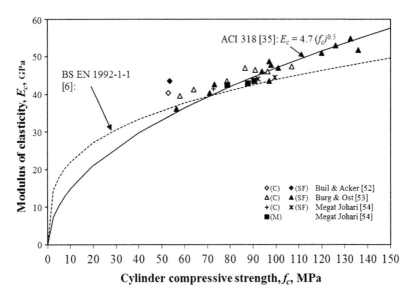

Figure 4.15 Modulus of elasticity as a function of compressive strength for concrete with and without microsilica (SF) and metakaolin (M). (C) = control.

different ages and environmental storage conditions and, in the majority of cases, the modulus reported is the secant modulus as measured at the start of creep tests.

Although Figure 4.13 indicates considerable scatter, both Code of Practice expressions represent the overall trend of modulus of elasticity with strength, and there

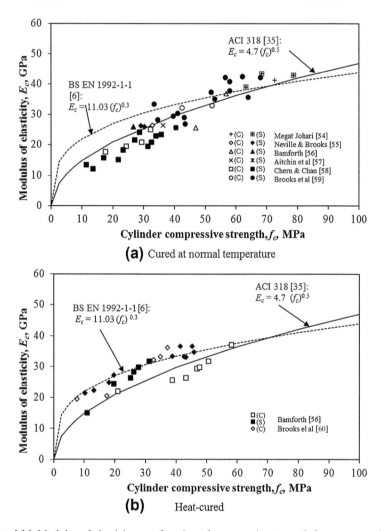

Figure 4.16 Modulus of elasticity as a function of compressive strength for concrete with and without blast-furnace slag. (a) Cured at normal temperature and (b) temperature-matched curing of insulated concrete [37].

is no consistent, discernible difference between admixture and control concretes. The plasticizing admixtures were used to make flowing concretes as well as water-reduced concretes, and covered a wide range of types: lignosulfonate plasticizers [38], carboxylic acid plasticizers [38–40], lignosufonated naphthalene condensate superplasticizers [41–44], and sulfonated formaldehyde condensate superplasticizers [45].

There is no obvious influence of the mineral admixture fly ash on the modulus of elasticity–strength relationship, as can be seen in Figure 4.14 and, again, for the range of plotted strength data, the Code of Practice expressions represent trends satisfactorily. Similarly, for very high-strength concrete, replacement of cement by finer cementitious materials (microsilica and metakaolin) does not affect the modulus–strength relationship (Figure 4.15). However, in the case of replacement of cement by blast-furnace slag, the modulus–strength overall trend of concrete cured at normal temperature appears to be best described by the ACI 318 expression (Figure 4.16(a)). On the other hand, the BS EN 1992-1-1 expression appears to more accurately represent the modulus–strength trend for heat-cured concrete (Figure 4.16(b)). In the latter tests, adiabatic curing of mass concrete was simulated by storing test specimens in water storage tanks heated according to the heat of hydration temperature of a large cube of concrete cast in an insulated mold [60].

Tension

Neville [1] refers to tests that broadly showed that the modulus of elasticity in tension is equal to the modulus of elasticity in compression. BS EN 1992-1-1: 2004 [6], also assumes equality of moduli in tension and in compression, with the modulus for different strength classes of concrete is shown as a function of direct tensile strength in Figure 4.17; the modulus and strength values are taken from Table 4.1 and are represented by the dashed curve. It is apparent that the BS EN 1992-1-1 estimates represent the general trend of experiment data. An expression for relating tensile modulus (E_t) as a function of direct tensile strength (f_t) is also indicated in Figure 4.17 and is:

$$E_t = 20f_t^{0.4} \qquad\qquad (4.16)$$

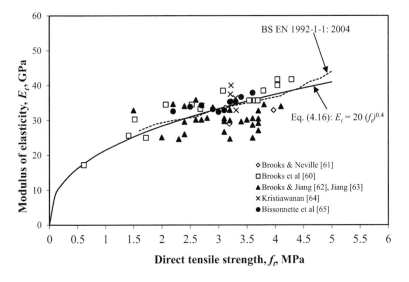

Figure 4.17 Tensile modulus of elasticity as a function of direct tensile.

In this connexion, the relationship of direct tensile strength and compressive strength is of interest. Using the notation of BS EN 1992-1-1: 2004 [6], the mean 28-day tensile strength (f_{ctm}) is related to characteristic compressive strength (f_{ck}) or strength class (C) as follows:

$$f_{ctm} = 0.3f_{ck}^{0.67} \qquad \text{when } f_{ck} \leq C50/60$$

and

$$f_{ctm} = 2.12 \ln[1 + 0.1f_{cm}] \quad \text{when } f_{ck} > C50/60$$

$$(4.17)$$

Values of tensile strength for different classes on concrete are given in Table 4.1. Since it's in terms of characteristic strength, the mean compressive strength, $f_{cm} = f_{ck} + 8$, Eq. (4.17) can be written in simplified notation:

$$f_t = 0.3[f_c - 8]^{0.67} \qquad \text{when } f_c \leq 60 \text{ MPa}$$

and

$$f_t = 2.12 \ln[1 + 0.1f_c] \quad \text{when } f_c > 60 \text{ MPa}$$

$$(4.18)$$

Figure 4.18 shows that Eq. (4.46) is generally satisfactory for expressing the overall trend of experimental measurements of direct tensile strength plotted as a function of compressive strength. An alternative analytical expression shown in Figure 4.18 is:

$$f_t = 0.11f_c^{0.89} \tag{4.19}$$

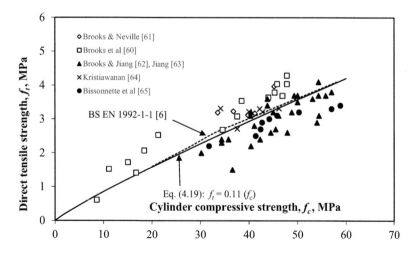

Figure 4.18 Direct tensile strength as a function of cylinder compressive.

Neville [1] reviews several expressions for relating tensile and compressive strengths, which are essentially similar in form to Eq. (4.47) with different coefficients and power indices, depending on the type of test specimen used to determine strength, e.g., tensile splitting strength and cube compressive strength. Experimental scatter is a feature of all expressions is a result of factors that affect tensile and compressive strengths differently, as well as inherent variability [1].

Poisson's Ratio

The term Poisson's ratio, μ, has been defined on p. 61 as the ratio of lateral strain to axial strain. Below approximately 30–40% of the ultimate stress, Poisson's ratio is approximately constant but, at higher stresses, it begins to increase due to the formation of vertical cracks, which become unstable and extensive when it becomes an *apparent Poisson's ratio* (see Figure 4.1). Under a further increase of stress, the volumetric strain becomes tensile and Poisson's ratio exceeds 0.5. At this stage, the concrete is no longer a continuous body and failure is imminent [1]. The design and analysis of some types of structures require the knowledge of Poisson's ratio in connection with volumetric movement and deformation under multiaxial stress. Under uniaxial compression, the axial strain is contraction and the lateral strain is extension, but the sign of the strain is ignored when quoting a value for Poisson's ratio.

Generally, the range of Poisson's ratio (μ) is 0.15–0.20 for lightweight and normal-weight aggregate concretes when determined from strain measurements taken during the test to determine static modulus of elasticity by the method of ASTM C469-10 [27] or BS 1881-121: 183 [28]. Those values apply to the elastic phase, i.e., below the limit of proportionality of stress and strain where Poisson's ratio is constant. By analogy with the composite model developed for masonry (p. 49), Poisson's ratio of concrete depends on the complex interaction of elastic modulus of concrete, and Poisson's ratios, volume fractions, and moduli of elasticity of aggregate and hardened cement paste. Theoretical considerations infer that Poisson's ratio of concrete increases as the modulus of the cement paste decreases, and decreases as the modulus of the aggregate increases. Experimental evidence reveals that Poisson's ratio for hardened cement paste (0.22) is greater than that of concrete (0.19) [66], and Poisson's ratios of wet- and dry-stored concrete are similar [42]. When made from a large range of aggregates, Poisson's ratio of normal-weight aggregate concrete ranged from 0.18 (marble aggregate) to 0.36 (rounded quartz aggregate) [67]. For the design of structural concrete, BS EN 1992-1-1:2004 [6] recommends Poisson's ratio for concrete to be 0.2 for uncracked concrete and zero for cracked concrete.

An alternative method of determining Poisson's ratio to the static tests is by dynamic means where the fundamental resonant frequency of longitudinal vibration of a concrete beam specimen is measured, and also the velocity of a pulse of ultrasonic waves is measured using the method of ASTM C597-02 [68] or BS EN

12,504-4: 2004 [69]. Both methods have been described earlier in this chapter. The principle of the ultrasonic test is that the velocity of compression sound waves transmitted by the direct method in a solid material, V (m/s), is related to its dynamic modulus of elasticity, E_d (kg/m^2), density, ρ (kg/m^3), g = acceleration due to gravity (9.81 m/s^2), and *dynamic Poisson's ratio*, μ_d, as follows [2]:

$$V^2 = \frac{E_d g(1 - \mu_d)}{\rho(1 + \mu_d)(1 - 2\mu_d)} \tag{4.20}$$

Expressing V in units of mm/s and E_d in GPa, Eq. (4.20) becomes:

$$V^2 = \frac{E_d(1 - \mu_d)10^{15}}{\rho(1 + \mu_d)(1 - 2\mu_d)} \tag{4.21}$$

Substitution of E_d from Eq. (4.6) in Eq. (4.21) permits dynamic Poisson's ratio to be calculated from the expression:

$$\left(\frac{V}{2nL}\right)^2 = \frac{1 - \mu_d}{(1 + \mu_d)(1 - 2\mu_d)} = F(\mu_d) \tag{4.22}$$

where V = pulse velocity (mm/s), n = resonant frequency (Hz), and L = length of the concrete beam (mm).

Typically, for a range of types of concrete, the pulse velocity varies from 2.0 to 5.0 km/s and the resonant frequency varies from 3000 to 8000 Hz. Thus, if V, n, and L are known from tests, Eq. (4.22) provides an analytical solution for dynamic Poisson's ratio. Alternatively, Figure 4.19 shows dynamic Poisson's ratio plotted as a function of $F(\mu_d)$, which allows a graphical estimate of μ_d. For example, suppose $V = 4.5$ km/s (4.5×10^6 mm/s), $n = 4200$ Hz and $L = 500$ mm, then $F(\mu) = 1.148$ and, from Figure 4.19, $\mu_d = 0.225$.

In addition to estimating dynamic modulus of elasticity, it is also possible to calculate dynamic Poisson' ratio from the relationships developed by Qixian and Bingey [34], which express the velocities of wave propagations transmitted through concrete in different directions.

Poisson's ratio determined dynamically is somewhat greater than that determined from static tests, typically ranging from 0.2 to 0.24 [1]. If the dynamic modulus is determined from the longitudinal or transverse mode of vibration, and the *dynamic modulus of rigidity* (G) from the torsional mode of vibration, the dynamic Poisson's ratio (μ_d) can be found as follows:

$$\mu_d = \frac{E_d}{2G} - 1 \tag{4.23}$$

According to Neville [1], values of dynamic Poisson's ratio obtained by Eq. (4.18) lie between Poisson's ratio obtained from direct measurements and dynamic Poisson's ratio calculated from dynamic modulus/ultrasonic tests (Eq. (4.22)).

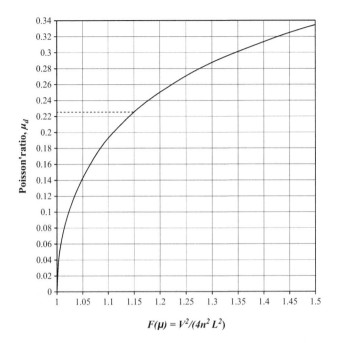

$$F(\mu) = V^2/(4n^2 L^2)$$

Figure 4.19 Values of Poisson's ratio given by solution of Eq. (4.22). Dashed line given an estimate of $\mu_d = 0.225$ when $F(\mu) = 1.15$.

Problems

1. Define pure elasticity.
2. What is the difference between static and dynamic moduli of elasticity?
3. What are the influences of aggregate properties on the modulus of elasticity of concrete?
4. What is hysteresis?
5. Explain (a) initial tangent modulus of elasticity and (b) tangent modulus of elasticity.
6. Sketch the stress–strain curves for normal-weight and lightweight concretes loaded at (a) constant rate of stress and (b) constant rate of strain.
7. What is the significance of the descending branch of the stress–strain curve for concrete?
8. Quote typical values of Poisson's ratio as determined by (a) static testing and (b) dynamic testing.
9. Is Poisson's ratio constant for all stresses up to failure? If not, explain why Poisson's ratio changes.
10. Why is two-phase composite modeling of elasticity unsuitable for concrete?
11. What is the transition zone?
12. Discuss the effect of mineral admixtures on modulus of elasticity of concrete.
13. Discuss the effect of chemical admixtures on modulus of elasticity of concrete.
14. How does rate of loading affect modulus of elasticity of concrete?

15. Describe a standard method of test to determine the modulus of elasticity of concrete.
16. Using the BS EN 1992-1-1: 2004 method, calculate the 90-day modulus of elasticity of concrete made with Class N cement, given that the characteristic cube strength is 60 MPa.
Answer: 38.6 GPa.

References

[1] Neville AM. Properties of concrete. 4th Ed. Pearson Prentice Hall; 2006. 844 pp.

[2] Neville AM, Brooks JJ. Concrete technology. 2nd Ed. Pearson Prentice Hall; 2010. 422 pp.

[3] RILEM. Determination of the fracture energy of mortar and concrete by means of three-point bending on notched beams, Draft Recommendations. Mater Struct 1985; 18(106):285–96.

[4] Wang PT, Shah SP, Naaman AE. Stress-strain curves of normal weight and lightweight concrete in compression. ACI J November 1978;75:603–11.

[5] Castillo C, Durrani AJ. Effect of transient high temperature on high strength concrete. ACI Mater J 1990;87(1):47–53.

[6] BS EN 1992-1-1. Eurocode 2, design of concrete structures. General rules and rules for buildings (see also UK National Annex to Eurocode 2). British Standards Institution; 2004.

[7] Zhen-Hai G, Xiu-Qin Z. Investigation of complete stress-deformation curves for concrete in tension. ACI Mater J 1987;84(4):278–85.

[8] Li Z, Shah SP. Localisation of microcracking in concrete under uniaxial tension. ACI Mater J 1994;91(4):372–81. ASTM C469/C469M-10. Standard test method for static modulus of elasticity and Poisson's ratio of concrete in compression. American Society for Testing and Materials, 2010.

[9] Komlos K. Factors affecting the stress-strain relation of concrete in uniaxial tension. ACI J 1969;66(2):111–4.

[10] Brooks JJ, Kristiawan SA. The effect of loading rate on tensile stress-strain behaviour of concrete. In: Proceedings of the 11th annual BCA/concrete society conference on higher education and the concrete industry, UMIST, Manchester; July 2001.

[11] Hughes BP, Ash JF. Short-term loading and deformation of concrete in uniaxial tension and pure tension. Mag Concr Res 1968;20(64):145–54.

[12] Tasdemir MA, Lydon FD, Barr BIG. The tensile strain capacity of concrete. Mag Concr Res 1996;48(176):211–8.

[13] Oh BH. Behaviour of concrete under dynamic tensile loads. ACI Mater J 1987;84(1):8–13.

[14] Mindess S, Young JF. Concrete. Prentice-Hall; 1981. 671 pp.

[15] Counto UJ. The effect of the elastic modulus of the aggregate on the elastic modulus, creep and creep recovery of concrete. Mag Concr Res 1964;16(48):129–38.

[16] Verbeck GJ, Helmuth FH. Structure and physical properties of cement paste. In: Proceedings 5th international symposium on the chemistry of cement, Tokyo, 1968, vol. 3. Cement Association of Japan; 1969. pp. 1–32.

[17] Illston JM, Dinwoodie JM, Smith AA. Concrete, timber and metals. The nature and behaviour of structural materials. Von Rostrand Reinhold; 1979. 663 pp.

[18] Cruz CR. Apparatus for measuring creep of concrete at elevated temperature. PCA Research and Development Bulletin 225, Portland Cement Association, Stokie, Ill. 1968;10(3):36–42.

[19] ACI Committee 209.1R-05. Report on factors affecting shrinkage and creep of hardened concrete. ACI Manual of Concrete Practice, Part1; 2012.

[20] Theuer AU. Effect of temperature on the stress deformation of concrete, National Bureau of Standards, Washington. D.C. Journal of Research 1937;18(2):195–204.

[21] Hansen TC. Creep and stress relaxation of concrete. Stockholm: Proceedings No. 31, Swedish Cement and Concrete Research Institute; 1960. 112 pp.

[22] Johansen R, Best CH. Creep of concrete with and without ice in the system. RILEM Bulletin, Paris 1962;(16):47–57.

[23] Ross AD, Illston JM, England GL. Short and long-term deformations of concrete as influenced by physical structure and state. London: Proceedings International Conference on the Structure of Concrete, Cement and Concrete Association; 1968. pp. 407–422.

[24] Marechal JC. Variations in the modulus elasticity and Poisson's ratio with temperature, Concrete for Nuclear Reactors. ACI Special Publication 1972;1(34):495–503.

[25] CEB, CEB-FIP Model Code 1990. CEB bulletin d'Information No. 213/214. Lausanne, Switzerland: Comite Euro-International du Beton; 1993. pp. 33–41.

[26] ACI Committee 209.2R-08. Guide for modelling and calculating shrinkage and creep in hardened concrete. American Concrete Institute; 2012. ACI Manual of Concrete Practice, Part 1.

[27] ASTM C469/C469M-10. Standard test method for static modulus of elasticity and Poisson's ratio of concrete in compression. American Society for Testing and Materials; 2010.

[28] BS 1881-121. Testing concrete. Method for determination of static modulus of elasticity in compression. British Standards Institution; 1983.

[29] BS 1881-209. Testing concrete. Recommendations for the measurement of dynamic modulus of elasticity. British Standards Institution; 1990.

[30] ASTM C215-08. Standard test method for fundamental transverse, longitudinal, and torsional frequencies of concrete specimens. American Society for Testing and Materials; 2008.

[31] BS 8110-2. Structural use of concrete. Code of practice for special circumstances. British Standards Institution; 1985. Withdrawn in 2010.

[32] Swamy RN, Bandoppadhyay AK. The elastic properties of structural lightweight concrete. In: Proceedings of the Institution of Civil Engineers, Part 2, vol. 59; 1975. pp. 381–94.

[33] BS 1881-203. Testing concrete. Recommendations for measurement of velocity of ultrasonic pulses in concrete. British Standards Institution; 1990.

[34] Quixian L, Bungey JH. Using compression wave ultrasonic transducers to measure velocity of surface waves and hence determine dynamic modulus of elasticity for concrete. Constr Build Mater 1996;10(4):237–42.

[35] ACI Committee 318-05. Building code requirements for structural concrete and commentary, Part 3. ACI Manual of Concrete Practice; 2007.

[36] BS EN 197-1. Cement. Composition, specifications and conformity criteria for common cements. British Standards Institution; 2011.

[37] Brooks JJ. Elasticity, creep and shrinkage of concretes containing admixtures. In: Al Manaseer A, editor. Proceedings Adam Neville symposium: creep and shrinkage-structural design effects. ACI Publication SP 194; 2000. pp. 283–360.

[38] Hope BB, Neville AM, Guruswami A. Influence of admixtures on creep of concrete containing normal weight aggregate. In: Proceedings RILEM symposium on admixtures for mortar and concrete, Brussels; 1967. pp. 17–32.

[39]   Neville AM, Brooks JJ. Time-dependent behaviour of concrete containing a plasticizer. Concrete 1975;9(10):33–5.

[40]   Brooks JJ. Influence of plasticizing admixtures Cormix P7 and Cormix 2000 on time-dependent properties of flowing concrete. Research Report. Dept. of Civil Engineering, University of Leeds; 1984. 12 pp.

[41]   Brooks JJ, Wainwright PJ, Neville AM. Time-dependent properties of concrete containing superplasticizing admixtures. In: Superplasticizers in concrete. ACI Publication SP 62; 1979. pp. 293–314.

[42]   Brooks JJ, Wainwright PJ, Neville AM. Time-dependent behaviour of high-early strength concrete containing a superplasticizer. In: Developments in the use of superplasticizers. ACI Publication SP 68; 1981. pp. 81–100.

[43]   Dhir RK, Yap AWF. Superplasticized flowing concrete: strength and deformation properties. Mag Concr Res 1984;36(129):203–15.

[44]   Alexander KM, Bruere GM, Ivansec I. The creep and related properties of very high-strength superplasticized concrete. Cem Concr Res 1980;10(2):131–7.

[45]   Brooks JJ, Wainwright PJ. Properties of ultra high strength concrete containing a superplasticizer. Mag Concr Res 1983;35(125):205–14.

[46]   Ghosh RS, Timusk J. Creep of fly ash concrete. ACI J Proc 1981;vol. 78. Title No. 78-30.

[47]   Brooks JJ, Wainwright PJ, Cripwell JB. Time-dependent properties of concrete containing pulverised fuel ash and a superplasticizer. In: Cabrera JG, Cusens AR, editors. Proceedings international conference on the use of PFA in concrete, vol. 1. Dept. Civil Engineering, University of Leeds; 1982. pp. 221–9.

[48]   Yuan RL, Cook JE. Time-dependent deformations of high strength fly ash concrete. In: Cabrera JG, Cusens AR, editors. Proceedings international conference on the use of PFA in concrete, vol. 1. Dept. Civil Engineering, University of Leeds; 1982. pp. 255–60.

[49]   Yamato T, Sugita H. Shrinkage and creep of mass concrete containing fly ash. In: Proceedings 1st international conference on the use of fly ash, silica fume, slag, and other mineral by-products in concrete. ACI Publication, SP 79; 1983. pp. 87–102.

[50]   Brooks JJ, Gamble AE, Al-Khaja WA. Influence of pulverised fuel ash and a super-plasticizer on time-dependent performance of prestressed concxrete beams. In: Proceedings symposium on utilisation of high strength concrete, Stavanger, Norway; 1987. pp. 205–14.

[51]   Langley WS, Carette GG, Malhotra VM. Structural concrete incorporating high volumes of ASTM Class F fly ash. ACI Mater J Proc 1989;86(5):507–14.

[52]   Buil M, Acker P. Creep of a silica fume concrete. Cem Concr Res 1985;15:463–6.

[53]   Burg RG, Ost BW. Engineering properties of commercially available high strength concretes. PCA Research and Development Bulletin RD 104T. Skokie, Illinois: Portland Cement Association; 1992. 55 pp.

[54]   Megat Johari MA. Deformation of high strength concrete [Ph.D. thesis]. School of Civil Engineering, University of Leeds; 2000. 295 pp.

[55]   Neville AM, Brooks JJ. Time-dependent behaviour of Cemsave concrete. Concrete 1975;9(3):36–9.

[56]   Bamforth PB. An investigation into the influence of partial Portland cement replacement using either fly ash or ground granulated blast furnace slag on the early age and long-term behaviour of concrete. Research Report 014J/78/2067. Southall, UK: Taywood Engineering; 1978.

[57]   Aitchin PC, Laplante R. Volume changes and creep measurements of slag cement concrete. Research Report. Universite de Sherbrooke; 1986. 16 pp.

[58] Chern JC, Chan YW. Deformations of concrete made with blast-furnace slag cement and ordinary Portland cement. ACI Mater J Proc 1989;86(4):372–82.

[59] Brooks JJ, Wainwright PJ, Boukendakji M. Influences of slag type and replacement level on strength, elasticity, shrinkage and creep of concrete. In: Proceedings 4th international conference on fly ash, silica fume, slag and natural pozzolans in concrete, vol. 2. ACI Publication SP 132; 1992. pp. 1325–42.

[60] Brooks JJ, Wainwright PJ, Al-Kaisi AF. Compressive and tensile creep of heat-cured ordinary Portland and slag cement concretes. Mag Concr Res 1991;43(154):1–12.

[61] Brooks JJ, Neville AM. A comparison of creep, elasticity and strength of concrete in tension an in compression. Mag Concr Res 1977;29(100):131–41.

[62] Brooks JJ, Jiang X. Cracking resistance of plasticized fly ash concrete. In: Cabrera JG, Rivera-Villarreal R, editors. Proceedings of international RILEM conference on the role of admixtures in concrete; 1999. pp. 493–506. Monterrey.

[63] Jiang, X., The effect of creep in tension on cracking resistance of concrete, [MSc (Eng) thesis], School of Civil Engineering, University of Leeds, 140 pp.

[64] Kristiawan SA. Restrained shrinkage cracking of concrete [Ph.D. Thesis]. School of Civil Engineering, University of Leeds; 2002. 217 pp.

[65] Bissonnette B, Pigeon M, Vaysburd AM. Tensile creep of concrete: study of its sensitivity to basic parameters. ACI Mater J 2007;104(4):360–8.

[66] ASTM C597-02. Test for pulse velocity through concrete. American Society for Testing and Materials; 2002.

[67] BS EN 12504-4. Testing concrete. Determination of ultrasonic pulse velocity. British Standards Institution; 2004.

[68] Gopalakrishnan KS, Neville AM, Ghali A. A hypothesis on the mechanism of creep of concrete with reference to multiaxial compression. ACI J 1970;67:29–35.

[69] Kordina K. Experiments on the influence of the mineralogical character of aggregates on the creep of concrete. RILEM Bull, Paris 1960;(6):7–22.

5 Elasticity of Masonry

To comply with serviceability specifications, the designer of masonry structures needs to know the modulus of elasticity, creep, shrinkage (or moisture expansion), and thermal movement so that he or she can estimate the total time-dependent movement of masonry. This chapter deals with the first of those movements, namely, elasticity. Besides being required for calculating the elastic deformation due to first application of load, modulus of elasticity is needed for estimating creep arising from sustained load, since it is expressed in design guides in terms of the creep coefficient (ratio of creep to elastic strain) or an effective (reduced) modulus. Codes of Practice [1−3] relate modulus of elasticity to characteristic strength or specified compressive strength of masonry, which is dependent on unit strength and mortar strength. However, there is little information and guidance as to the accuracy of modulus of elasticity estimated by design guides. In addition to Codes of Practice, there are other methods [4−6] for estimating modulus of masonry from strength of the unit. All methods have the merit of simplicity in that they express the modulus in terms of properties known to the designer, but they do not appear to have been reviewed with regard to their applicability and accuracy for published test results.

The aims of this chapter are to review the stress−strain behaviour of masonry and Poisson's ratio. The factors influencing the modulus of elasticity are identified. Relationships are developed for unit and mortar in terms of known properties: strength of mortar and strength of unit before incorporating them in the composite models for masonry developed in Chapter 3. The important influences of curing conditions before application of load and unit water absorption are included in composite models for improved estimates of modulus of elasticity of masonry. Methods of prediction are then compared and their accuracies assessed.

Stress−Strain Behaviour

When calculating the modulus of elasticity of masonry, it is tacitly assumed that the relationship between stress and strain is linear. In Chapter 4, it was shown that this is strictly not the case for concrete when it is first subjected to a cycle of load, because it falls into the nonlinear and nonelastic category (see page 70). In the case of masonry, Allen [7,8] obtained stress−strain relationships for walls, prisms, and piers, and reported linearity up to 17% of the ultimate stress for prisms, but up to 50% for other types of masonry. There was no difference between the modulus of elasticity of prisms and piers, but walls had a slightly higher value that, according to Sahlin [9], only occurred when high-strength units were used. Lenczner [10] obtained load-contraction curves for brickwork cubes built with different mortars, and the curves

Concrete and Masonry Movements. http://dx.doi.org/10.1016/B978-0-12-801525-4.00005-4

showed nonlinear behaviour initially followed by linear behaviour up to approximately 50—80% of the failure load (Figure 5.1). He reported that part of the strain was truly elastic in that it was recoverable, the rest being nonrecoverable due to viscous flow of the mortar and "bedding-down" effects. Fattal and Cattaneo [11] also reported linearity up to 50% of the ultimate load for piers and walls. Bradshaw and Hendry [12] reported parabolic stress—strain curves for story-high walls, the secant modulus decreasing as the stress increased. On the other hand, Yokel et al. [13] found that hollow concrete masonry walls exhibited linear behaviour up to 80% of the failure stress. After undertaking a comprehensive review of masonry with regard to stocky and slender specimens, and from laboratory tests, Edgell [14] concluded that all masonry exhibited a continuously decreasing modulus of elasticity up to the ultimate stress. Several other researchers have reported linear-parabolic shapes of the stress—strain curve while investigating the influence of various factors on elasticity of masonry, such as size of unit [15], strength of unit [16], biaxial loading [17], type of mortar [18], and type of unit [19]. Clearly, the consensus view is that the stress—strain curve up to or near to failure for masonry is nonlinear.

Figure 5.2 shows typical, full stress—strain behaviour up to failure of the three types of masonry, together with the corresponding behaviour of unbonded samples of unit and mortar from which the masonry was constructed [20]. In all cases, the stress—strain curves of masonry are nonlinear even for clay brickwork built from a clay brick, which itself exhibits linearity of stress—strain up until failure (Figure 5.2(a)); it is apparent that the nonlinear stress—strain behaviour of brickwork arises from the corresponding behaviour of the mortar phase. In the cases of calcium silicate brickwork (Figure 5.2(b)) and concrete blockwork (Figure 5.2(c)), the nonlinear stress—strain

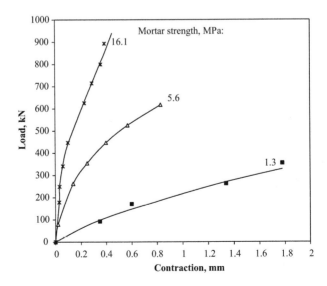

Figure 5.1 Load-contraction curves for clay brickwork cubes [10].

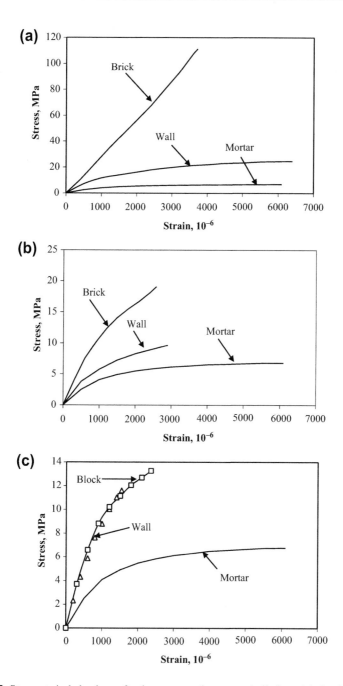

Figure 5.2 Stress-strain behaviour of unit, mortar and masonry built from (a) clay brick, (b) calcium silicate brick, and (c) concrete block [20].

characteristics of both the units and mortar contribute to the nonlinearity of stress and strain of masonry. Moreover, it is possible that another source of nonlinearity is the presence of interfaces or transition zone at the unit/mortar bond, where weak material and microcracks exist as a result of moisture transfer from mortar to unit and vice-versa (see page 112). Under compression, weak material and microcracks contract more than the surrounding stiffer material, thus causing strain to increase more than stress. Whereas the stress–strain behaviour of clay and calcium silicate brickwork lies between the curves for unit and mortar, the stress-strain curve the of the masonry block wall in Figure 5.2(c) virtually coincides with that of the concrete block, because some 95% of the wall consists of block and there is a only a small contribution from the mortar bed joint.

From the above discussion, it seems that masonry can be classified in the same category as concrete with regard to short-term stress–strain behaviour; namely, the *nonlinear and nonelastic* category. Therefore, the appropriate parameter to express the quotient of stress and strain is the secant modulus of elasticity (see p. 72), which varies according to the level of stress. However, in the case of low levels of working stress, for example 1–3 MPa, these apply to the initial part of the full stress–strain curves where nonlinearity is small, and there is little difference between the initial tangent and secant moduli.

Poisson's Ratio

The discussion in this section refers to modulus of elasticity of masonry subjected to loading in the vertical or axial direction. Under the same loading, lateral strain occurs, the ratio of lateral to axial strain being defined as the Poisson's ratio. Dhanasekar et al. [17] reported a value of 0.19 for solid pressed clay brickwork having a modulus of 5.7 GPa. The degree of lateral restraint from the test machine platens was shown to be a factor by Thomas and O'Leary [21], who found Poisson's ratio of calcium silicate brickwork piers to be 0.14 with plywood packing placed between the steel platens and brickwork, but 0.11 when brush packing was used; brush packing is capable of transmitting vertical load but permits unrestrained lateral movement of the specimen. A similar value for calcium silicate (0.13) was reported by Brooks [22], and this compared with a much lower value of 0.05 for Fletton clay brickwork. Somewhat higher values of Poisson's ratio for calcium silicate brickwork (0.17–0.23) were reported by Meyer and Schubert [23] when determined using masonry specimens as specified by the German Standard DIN 18,554: Part 1, 1985.

More detailed tests were carried out by Amjad [20], the results of which are given in Table 5.1. In those tests, clay brickwork, calcium silicate brickwork, and concrete blockwork walls and piers were built with three types of mortar and loaded between two types of platen: steel and concrete. Although Poisson's ratio of mortar (μ_m) increased as its strength (f_m) decreased, there was no consistent influence of type of mortar on Poisson's ratio of masonry (μ_{wy}). Similarly, there was no consistent influence of platen type and geometry on Poisson's ratio of masonry, and mean values

Table 5.1 Poisson's Ratio of Units, Mortar, and Masonry

	Property								
	Unit			Mortar			Masonry		
Type of Masonry	f_{by} MPa	E_{bx} GPa	μ_{by}	f_m, MPa	E_m GPa	μ_m	f_{wy} MPa	E_{wy} GPa	μ_{wy}
Clay brickwork	105.0	17.8	0.09	13.0	9.3	0.15	31.1	20.5	0.12
				8.9	6.5	0.17	23.3	16.5	0.10
				3.0	3.1	0.19	21.7	10.6	0.08
Calcium silicate brickwork	26.1	10.9	0.12	10.9	8.1	0.16	13.3	10.5	0.12
				5.6	5.2	0.17	10.3	8.1	0.12
				3.0	2.9	0.19	10.6	7.2	0.13
Concrete blockwork	13.8	12.4	0.16	10.0	7.5	0.16	11.6	12.1	0.19
				5.0	4.9	0.18	11.9	11.4	0.15
				3.2	3.5	0.19	10.6	11.3	0.18

of walls, piers, and different platens are given in Table 5.1. It should be added that any trends would likely be masked by a high variability of $\pm20\%$. The overall average values of Poisson's ratio for masonry for the three mortars in Table 5.1 are: 0.10 for clay brickwork (Engineering Class B), 0.12 for calcium silicate, and 0.17 for solid concrete blockwork. In fact, the average values suggest that Poisson's ratio for masonry may be assumed to be equal to the Poisson's ratio of the unit (μ_{by}) from which it is constructed. The theoretical Poisson's ratio of masonry built with standard brick or solid blocks, neglecting the contribution of the vertical mortar and assuming isotropy of unit modulus ($E_{bx} = E_{by}$), is given by Eq. (3.98):

$$\mu_{wy} = E_{wy}\left[\frac{\mu_m + k\mu_{by}}{E_m + kE_{bx}}\right] \tag{5.1}$$

where E_{wy} = vertical modulus of elasticity of masonry, E_m = modulus of elasticity of mortar, E_{bx} = horizontal modulus of elasticity of brick or block, and $k = 6.5$ for brickwork and 21.5 for blockwork.

Using the parameters listed in Table 5.1, the Poisson's ratio calculated using Eq. (5.1) for each type of masonry and mortar is in good agreement with the measured values given in the last column of Table 5.1.

Geometry of Cross-Section

Lenczner [5,6] found that there was no appreciable difference between the elastic modulus of story-high walls, piers, and cavity walls. Abdullah [24] came to the same conclusion after testing single-leaf walls, cavity walls, hollow piers, and solid piers. It may be recalled that in Chapter 3, the theoretical analysis by composite modeling also suggested no appreciable effect of cross-section shape and size on the modulus of elasticity.

Amjad's tests [20] on clay, calcium silicate, and concrete masonry walls and piers involved various heights and cross sections, and he found no significant influence of geometry or *platen restraint* (assessed by height/width ratio); the results are shown in Figure 5.3. Interestingly, the modulus of calcium silicate brickwork built with "frog-down" units (not recommended practice) was appreciably less than with units laid "frog-up", probably because of entrapped air inside the frog of the "frog-down" units and a different stress distribution in the mortar bed joints [25].

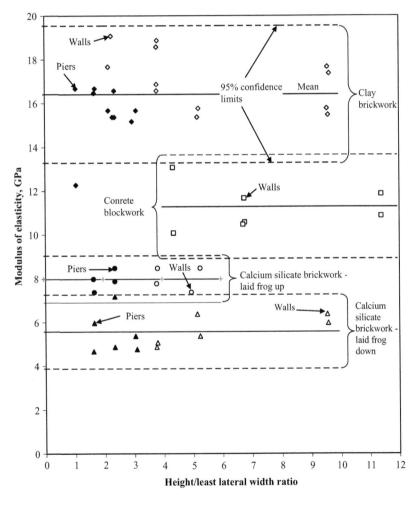

Figure 5.3 Modulus of elasticity of various sizes of clay, calcium silicate, and concrete walls and piers [20].

Type of Mortar

In an early investigation into the effect of mortar on modulus of elasticity of masonry, Bragg [26] found little difference in modulus of elasticity of masonry piers built with cement−sand mortar and cement−lime−sand mortar. The modulus of elasticity of piers built with pure lime was much lower, but this only occurred when all three types of mortar were compared using masonry built with low-strength units. Lenczner's results [10], shown in Figure 5.1, clearly demonstrate that the strength of mortar−mortar has a strong influence on the modulus of elasticity of masonry. Sahlin [9] confirmed Bragg's observations [26] when he referred to the relationships developed by Hilsdorf [27] from tests on piers, which showed that the effect of using cement−sand mortar significantly increased the modulus of elasticity compared with lime−sand mortar.

Forth and Brooks [28] measured the modulus of elasticity of clay brickwork built with different mortars and found that an overall change in strength of mortar of 65% produced a change in brickwork modulus of 20% (Table 5.2). Brickwork modulus was generally greater for a higher mortar strength, but a common relationship was not so apparent for all types of mortar.

The theoretical influence of mortar type or, specifically, its modulus of elasticity on the modulus of elasticity of masonry, has been quantified and shown to be significant by composite model analysis in Chapter 3. Since the type of mortar is usually identified by strength, either by design or by quality control testing, it is pertinent to investigate the relationship of modulus of elasticity and compressive strength of mortar. Glanville and Barnett [29] appear to be the first to report modulus of elasticity and strength data for one type of mortar at the age of 28 days; the modulus was determined at a stress of 4.5 MPa, but the storage conditions and specimen size details were not given. Lenczner [10] carried out tests at the age of 28 days on a range of mortar types by using standard wet cubes for strength and cylinders stored under polythene sheet for modulus, which was determined at a stress of 50% of the strength. Other researchers [15,27,30,31] report results for a variety of specimen

Table 5.2 Effect of Different Types of Mortar on Modulus of Elasticity of Clay Brickwork 13-Course-High × 2-Brick-Wide Single-Leaf Walls Built with Class B Engineering Brick

Mortar		Brickwork Modulus of Elasticity, GPa
Type	14-day Strength, MPa	
1:3, cement−sand	22.4	25.0
1:$\frac{1}{4}$:3, cement−lime−sand	22.7	25.0
1$\frac{1}{2}$:4$\frac{1}{2}$, cement−lime−sand	10.3	23.8
1:3$\frac{1}{2}$, masonry cement−sand	13.7	23.4
1:4, cement/plasticizer−sand	9.0	20.8
1:1:6, cement−lime−sand	6.5	20.3
1:5, cement/ggbs−sand	8.7	19.5

ggbs = Ground granulated blast-furnace slag.

sizes and storage conditions, but in many cases test details are unknown. Later data [28,32−38] are for strength determined using 75-mm wet cubes and secant modulus of elasticity by using wet- or dry-stored 200- × 75- × 75-mm prisms subjected to a stress of 1.5 MPa at various ages. The foregoing published test results have been collated to produce 184 sets of modulus-strength data, and they are plotted in Figure 5.4.

In 1971, Sahlin [9] simply equated the mortar modulus of elasticity, E_m (GPa), to strength, f_m (MPa):

$$E_m = f_m \tag{5.2}$$

The above equation tends to overestimate the modulus for high strength mortars, as can be seen in Figure 5.4. A more representative relationship is:

$$E_m = \frac{f_m}{0.975 + 0.0125 f_m} \tag{5.3}$$

However, for a given compressive strength, it is apparent that there is a large variation in modulus, which can be quantified by an error coefficient, M (%), defined as:

$$M = \frac{1}{E_m} \sqrt{\frac{\sum (E_p - E_a)^2}{n - 1}} \times 100 \tag{5.4}$$

where E_m = actual mean modulus for n number of observations, E_p = predicted modulus, and E_a = actual modulus.

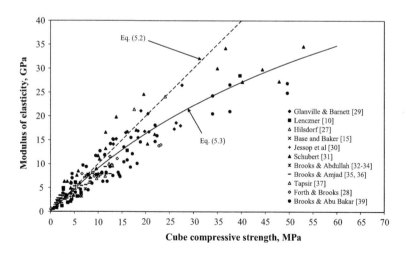

Figure 5.4 Modulus of elasticity of mortar as a function of compressive strength.

The error coefficient is analogous to the coefficient of variation, but deviation is measured from the actual value of modulus. The error coefficient of Eq. (5.3) is 27.2% for a mean modulus of elasticity = 9.2 GPa.

For lower strength mortars, say $f_c < 25$ MPa, a linear relationship may be assumed:

$$E_m = 0.8f_c \tag{5.5}$$

In this case, the error coefficient is 30.9% for a mean modulus = 7.3 GPa.

Besides the natural variability of the material, the scatter of data points in Figure 5.4 can be attributed to several factors that, in the case of concrete, are known to affect modulus of elasticity in a different manner to strength [41]. By analogy, it may be surmised that for mortar:

- Modulus of elasticity is strongly dependent on the type of sand (modulus of elasticity) and its proportion by volume in the mortar, whereas strength is not so much affected (for a constant water content).
- Modulus of elasticity depends on the level of stress due to the nonlinearity of the stress-strain curve and on rate of loading, whereas strength depends only on the rate of loading.
- Moisture condition of the mortar affects the modulus of elasticity and strength differently. A wet specimen has a lower strength than a dry specimen, but a wet specimen has higher modulus than a dry specimen.
- Size of the specimen affects strength of mortar, but not its modulus of elasticity.

Type of Unit and Anisotropy

The traditional method of quantifying the modulus of elasticity of masonry is to relate it to strength of masonry. This section reviews those relations for the separate categories of clay brickwork, calcium silicate brickwork, and concrete blockwork. Then, corresponding elastic behaviour of unbonded units is discussed.

With regard to anisotropic behaviour of masonry, very little testing seems to have been reported for masonry subjected to load applied in the horizontal direction or parallel to the mortar bed joints. According to composite model theory (see Figure 3.3), for units having isotropic elastic modulus, there is only significant anisotropy of masonry modulus of elasticity when there is a combination of weak mortar and high strength unit, otherwise moduli in the vertical and horizontal directions are similar. However, in reality, units often have different bed face and header face moduli because of the presence of perforations and frogs so that corresponding masonry behaviour is affected. Moreover, moisture absorption and suction rate of the unit when the latter is laid dry can affect the properties of the mortar at the interface; this latter influence is discussed in the next section.

Results obtained by Brooks and Abu Bakar [41] for anisotropic tests on clay brickwork, calcium silicate brickwork, and concrete blockwork are shown in Table 5.3. To ensure that the height/width ratio was the same for vertical and horizontal loading, the single-leaf walls were constructed as shown in Figure 5.5, the horizontally loaded walls actually being tested in the vertical direction by

Table 5.3 Modulus of Elasticity of Walls Subjected to Vertical and Horizontal Loading [41]

	Unit Strength, MPa		Mortar Strength, MPa	Masonry Modulus, GPa	
Masonry Type	Bed Face	Header Face		Bed Face	Header Face
Clay, perforated brick	95.7	8.7	14.7	13.6	9.0
Calcium silicate, solid brick	29.1	22.2	14.3	10.6	12.7
Concrete, solid dense aggregate block	11.6	11.6	16.1	13.4	13.4

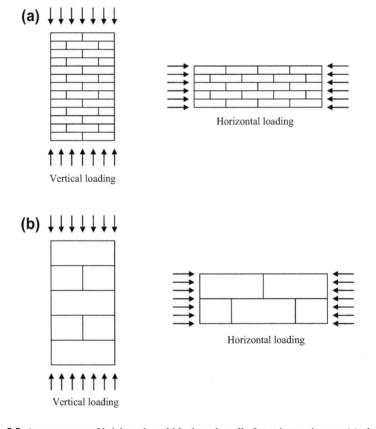

Figure 5.5 Arrangement of brickwork and blockwork walls for anisotropic tests (a) clay and calcium silicate brickwork (b) concrete blockwork [41].

rotating them through 90° in order to fit in the similar test rigs used for the vertically loaded walls. Units were laid dry in $1:\frac{1}{2}:4\frac{1}{2}$ cement−lime−sand mortar loads being applied at the age of 14 days after curing under polythene sheet. Except for the clay brickwork, Table 5.3 indicates that the elastic moduli of calcium silicate brickwork and concrete blockwork were similar for both vertical and horizontal loadings. In the case of clay brickwork, the degree of anisotropy as given by the horizontal/vertical modulus ratio was 0.66, which apparently reflected the degree of anisotropy of the unit strength and reduced stiffness in the horizontal or header face direction due to the presence of perforations.

Clay Brickwork

Figure 5.6 shows the secant modulus of elasticity plotted as a function of ultimate strength of brickwork for all sources of clay masonry, as reported by CERAM Building Technology [44]. It is apparent that there is a large scatter of data, as mostly enclosed by the dashed lines, which has a mean modulus/strength ratio of 0.75. Brooks and Amjad [35] reported elastic moduli and strength data for small walls and piers of various height/width ratios built with a Class B Engineering clay brick and a $1:\frac{1}{2}:4\frac{1}{2}$ cement−lime−sand mortar; they found the average ratio of modulus of elasticity to compressive strength of brickwork to be 0.72.

Although the foregoing results apply to the ultimate strength of brickwork, BS 5628-1: 2005 specifies that modulus of elasticity is related to a minimum value of strength; namely, the characteristic strength, f_k, which is dependent on the type of

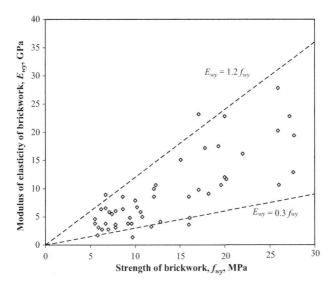

Figure 5.6 Secant modulus of elasticity as a function of strength of clay brickwork [42].

mortar as well as the strength of unit. The modulus of clay masonry is simply related to the characteristic strength of masonry as follows:

$$E_{wy} = 0.9f_k \tag{5.6}$$

Calcium Silicate Brickwork

CERAM Building Technology's review [42] expresses modulus of elasticity of calcium silicate masonry as a function of strength of masonry as in Figure 5.7, which includes the experimental data from different sources. The relationship is:

$$E_{wy} = 0.65f_{wy} \tag{5.7}$$

Earlier, Meyer and Schubert [23] had found that the modulus of masonry built from perforated calcium silicate bricks was slightly greater than the modulus of masonry built from solid bricks. Moreover, using a calcium silicate brick used of strength 26.1 MPa, Brooks and Amjad [36] found that when built (incorrectly) with "frog-down bricks,", masonry had an 11% less strength and 30% less masonry modulus than masonry built (correctly) with "frog-up bricks", probably because of entrapped air and a different stress distribution in the "frog-down" case [25].

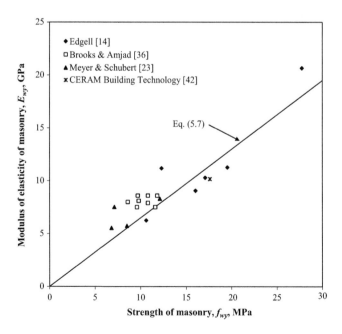

Figure 5.7 Modulus of elasticity of masonry as a function of strength of masonry for calcium silicate brickwork; Edgell's data are for failure stress and initial tangent modulus of elasticity.

Like concrete, the compressive strength of calcium silicate brickwork is much greater than its *tensile strength*, the latter being approximately 0.33 MPa [42]. However, unlike concrete, the *tensile modulus of elasticity* of calcium silicate brickwork is also much lower than in compression and, moreover, is anisotropic. CERAM Building Technology [42] carried out direct tensile tests on calcium silicate brickwork and demonstrated the higher stiffness in the horizontal direction than in the vertical direction. Although the variability was high (coefficient of variation $\approx 80\%$), the mean modulus of elasticity was 3.34 GPa when the load was applied parallel to the bed face, compared with a value of 0.90 GPa when the load was applied perpendicular to the bed face.

Concrete Blockwork

Test results of modulus of elasticity for *dense aggregate concrete* and *lightweight aggregate concrete* blockwork are scant and somewhat inconsistent, as shown in Table 5.4.

It is apparent that the ratio of masonry modulus/block strength ratio varies considerably. The recorded secant modulus of elasticity is mostly that corresponding to strain at the first application of stress in creep tests so that likely sources of variability are rate of loading or time to apply the load, the level of stress (high stress/strength ratios above 0.3 can cause nonlinear strain), type of mortar, and age at test.

In Ameny et al.'s tests [45], the average ratio of concrete blockwork modulus to concrete blockwork strength was 0.70, whereas in Brooks and Amjad's tests [36], the average concrete blockwork modulus of elasticity (GPa) was numerically equal to the average concrete blockwork strength (MPa); like clay and calcium silicate brickwork, both modulus and strength were independent of height/least lateral width ratio.

In the report by CERAM Building Technology [42], tests of *autoclaved aerated concrete (AAC)* or *aircrete* masonry indicated that moduli of elasticity given by the BRE digest [44] were higher than measured ones. Table 5.5 shows that, for a range

Table 5.4 Modulus of elasticity of aggregate concrete block masonry

Data Source	Aggregate Concrete Type	Block Strength, MPa	Masonry Secant Elastic Modulus, GPa
Lenczner [43]	Dense	5.6	8.1
Lenczner [44]	Lightweight	3.3	6.0
Ameny et al. [45]	Lightweight [a]	12.3	6.4
Brooks & Amjad [36]	Dense	12.9	11.9
Tapsir [37]	Dense	14.9	13.6
CERAM Building Technology [42]	Lightweight	16.8	7.3
CERAM Building Technology [42]	Dense	20.3	6.3

[a]hollow block.

Table 5.5 Modulus of Elasticity of AAC Masonry

Data Source	Dry Density, kg/m³	Unit Compressive Strength, f_{by}, MPa	Elastic Modulus of Masonry, E_{wy}, GPa
CERAM Building	–	2.75	1.48
Technology [42]	–	2.95	0.79
BRE Digest 342 [46]	450	3.2	1.60
	525	4.0	2.00[a]
	600	4.5	2.40
	675	6.3	2.55
	750	7.5	2.70

[a]Interpolated value [42].

of AAC block strengths of 3.2–7.5 MPa, the modulus of elasticity of masonry was 1.6–2.7 GPa and, generally, BRE suggest that the modulus (GPa) can be taken as 0.6 × block strength (MPa). However, for low-strength AAC blocks, CERAM Building Technology found a lower average masonry modulus of 0.4 × block strength.

When the modulus of block is low, the modulus of masonry is approximately equal to the block modulus because the influence of the mortar is negligible, as indicated by Eq. (3.40), since the modulus of blockwork is dominated by the block modulus term. Hence, for AAC blockwork, $E_{wy} = E_{by}$ and from the average results of the BRE Digest and CERAM Building Technology shown in Table 5.5:

$$E_{wy} = 0.5f_{by} \tag{5.8}$$

Clay Units

With regard to the elastic properties of unbonded clay bricks, Glanville and Barnett's results [29] were analyzed by Sahlin [9], who reported a linear relationship between modulus of elasticity and compressive strength for *extruded* clay bricks. The modulus was measured between header faces (E_{bx}) and the strength was measured in the usual way, i.e., between bed faces (f_{by}). Figure 5.8(a) shows the original data [29] together with other published data of investigators [32–39, 48]. The general trend is expressed as follows:

$$E_{bx} = 0.25f_{by} \tag{5.9}$$

Equation (5.9) has an error coefficient of 37.0% for an average modulus of 15.3 GPa, the coefficient of 0.25 being slightly less than the equation proposed by Sahlin, who also observed that the *pressed* clay bricks of Glanville and Barnett's tests had a slightly greater header face modulus than extruded bricks, a trend that is confirmed by other results, as shown in Figure 5.8(b). Although there is a large degree of scatter, the average header face modulus-strength trend is:

$$E_{bx} = 0.5f_{by} \tag{5.10}$$

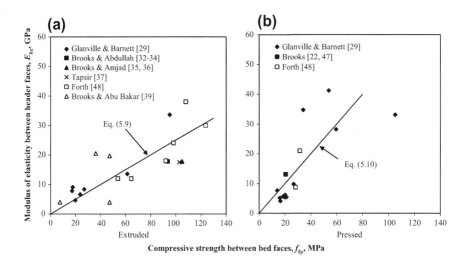

Figure 5.8 Modulus of elasticity of extruded and pressed clay bricks between header faces as a function of their compressive strength between bed faces.

Since the bed face modulus of elasticity of bricks, E_{by}, determines the vertical modulus of elasticity of brickwork, its relationship with compressive strength is of special interest. For extruded clay bricks, Figure 5.9 again shows appreciable scatter, but the general trend is linear, viz.

$$E_{by} = 0.4f_{by} \qquad (5.11)$$

The error coefficient of Eq. (5.11) is 31.9% for a mean bed-face modulus of 21.9 GPa. When Eqs (5.9) and (5.11) are compared, *anisotropic behaviour* is clearly demonstrated with respect to the modulus of elasticity of extruded clay bricks, since $E_{by} > E_{bx}$. For pressed clay bricks, the only modulus-strength data available in the literature are for low-strength bricks and these are bracketed independently in Figure 5.9; in general, the modulus of pressed bricks between bed faces is less than between header faces (Figure 5.8(b)). Thus, the anisotropic behaviour of pressed clay bricks appears to be opposite to that of extruded clay bricks.

Shrive and Jessop [49] reported anisotropic behaviour of modulus of elasticity of Canadian extruded clay solid units after testing specimens cut from full-size units. Bed-face modulus was measured in compression using 12- × 12- and 20- × 20-mm section prisms, and header face and stretcher-face moduli were measured in tension by using 16- × 6-mm section specimens. For a unit strength of 87.0 MPa, the average values of modulus of elasticity were: 30.8 GPa (bed face), 21.4 GPa (header face) and 16.4 GPa (stretcher-face). For the given strength, the values of elasticity between header and bed faces are in close agreement with those of Figures 5.8(a) and 5.9.

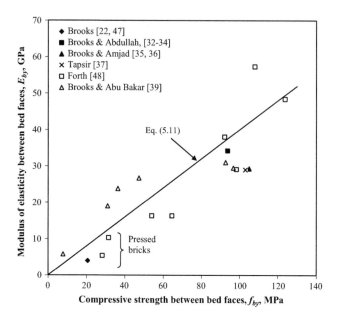

Figure 5.9 Modulus of elasticity of extruded clay bricks between bed faces as a function of compressive strength between bed faces.

Calcium Silicate Units

Compared with clay bricks, there are only a few published test results for modulus of elasticity of calcium silicate bricks [22,29,33,36,37], for which the average trend of bed-face modulus as a function of strength is:

$$E_{by} = 0.46f_{by} \tag{5.12}$$

and the corresponding average header-face modulus vs strength is:

$$E_{bx} = 0.41f_{by} \tag{5.13}$$

Concrete Blocks

Modulus of elasticity data for concrete blocks made from different aggregates were reported by Sahlin [9] and these are shown in Figure 5.10, together with some other data. Both modulus and strength values are based on *gross area*, since solid, cellular, and hollow concrete blocks are included. Although there is a high degree of scatter, the average trend of Figure 5.10 is linear, viz:

$$E_{by} = 0.9f_{by} \tag{5.14}$$

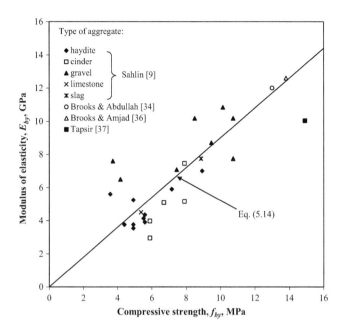

Figure 5.10 Modulus of elasticity of concrete blocks as a function of compressive strength.

Equation (5.14) has an error coefficient of 25% for a mean modulus of elasticity of 6.7 MPa. There is little published information on degree of anisotropy of cellular and hollow blocks, and limited test data suggests solid concrete blocks are isotropic [34,36,37].

All of the above simple linear modulus-strength relationships and degree of anisotropy of modulus for different types of unit are summarized in Table 5.6. In the case of

Table 5.6 General Elastic Modulus-Strength Equations and Anisotropy of Elasticity for Different Types of Unit

| Type of Unit | Modulus (GPa)-Strength (MPa) Equation | | Elastic Anisotropy, E_{by}/E_{bx} |
	Vertical (Between Bed Faces)	Horizontal (Between Header Faces)	
Clay brick: extruded	$E_{by} = 0.40\, f_{by}$	$E_{bx} = 0.25\, f_{by}$	1.6
pressed	$E_{by} = 0.25\, f_{by}$	$E_{bx} = 0.50\, f_{by}$	0.5
Calcium silicate brick	$E_{by} = 0.46\, f_{by}$	$E_{bx} = 0.41\, f_{by}$	1.12
Concrete block	$E_{by} = 0.90\, f_{by}$ (All types)	$E_{bx} = 0.90\, f_{by}$ (Solid only)	1.0 (Solid only)

extruded clay units, the significant anisotropy of elasticity is due to the presence of perforations, method of manufacture, and type of clay, whereas for calcium silicate bricks, there is only a small degree of anisotropy and solid concrete blocks are isotropic. In the majority of cases, bed-face modulus is equal to or greater than header-face modulus but, for pressed clay bricks, Table 5.6 shows that bed-face modulus is only about 50% of the header-face modulus.

The average values of Table 5.6 are based on test results of different types of full-size units given in Table 5.7. In the same table, corresponding strength and elasticity results are given for 25-mm-diameter cylinders cored from the solid parts of the unit. The latter tests were performed in order to isolate the influence of frog and perforations from other factors. A comparison of the properties of full-size units and core samples of Table 5.7 reveals the following:

- For pressed clay, there appears to be no effect of the presence of a frog on anisotropy, since strength ratios are similar (≈ 0.78) and modulus ratios are similar (≈ 0.55).
- For extruded clay bricks, perforations markedly reduce the header-face strength of the full-size unit so that its strength ratio is much greater than that for the core sample. The same influence is apparent for elasticity, but less so.
- For the calcium silicate brick, the strength and modulus ratios of the full-size unit slightly exceed those of the core sample ratio, possibly due to platen restraint (see Table 5.7).
- The core sample ratios indicate that clay bricks exhibit anisotropy of strength and elasticity due to the type of clay and manufacturing process. Thus, the extrusion process itself is responsible for basic anisotropy, which is then magnified by the number and size of perforations. On the other hand, the calcium silicate and concrete block properties are isotropic.

Another factor that affects compressive strength is platen restraint of the testing machine, which may impose a state of triaxial stress on the unit, depending on the type of platen and the height/width ratio of the unit [40,52]. Thus, for a standard size of brick, strength between bed faces will be affected more than strength between header faces. With extruded bricks, the number and size of perforations will clearly reduce strength because of the lower cross-sectional area, but *stress concentration* at the edge of perforations is also a factor [52]. In that case, strength between header faces is more strongly influenced than strength between bed faces. Modulus of elasticity may also be affected by the same factors, but to a much lesser extent than in the case of strength.

Unit/Mortar Interaction

Application of composite models derived in Chapter 3 to estimate the modulus of elasticity of masonry revealed that, under certain conditions, the assumption that elastic properties of mortar bed joints were equal to those of unbonded mortar specimens was invalid. Analysis showed that predictions were good for brickwork built from low water absorption units and sealed under polythene sheet until application of load [20,24,47]. On the other hand, for brickwork built with low absorption units and stored in a drying environment before application of load, and for brickwork built

Table 5.7 Anisotropic Properties of Units [24,39,50,51]

| | Unit Details | | Strength | | | | Elasticity | | | |
| | | | Full Unit | | 25-mm-Dia. Core | | Full Unit | | 25-mm-Dia. Core | |
Name	Manufacture	Type	f_{by} (MPa)	f_{by}/f_{bx}	f_{by} (MPa)	f_{by}/f_{bx}	E_{by} (GPa)	E_{by}/E_{bx}	E_{by} (GPa)	E_{by}/E_{bx}
UK Clay Brick										
Fletton clay	Pressed	Frogged	12.5[a]	0.78	20.7	0.79	4.7	0.57	5.6	0.53
Birtley old english	Slop moulded	Solid	31.5	—	—	—	10.3	0.49	—	—
Marshall's class B	Extruded	3-hole perforated	85.2	10.27	120.3	1.09	28.0	1.89	33.6	1.30
Marley Dorket Honeygold	Extruded	3-hole perforated	60.0	—	—	—	16.5	1.34	—	—
Steetley smooth red	Extruded	14-hole perforated	92.2	—	—	—	38.0	2.1	—	—
Marshall's nori	Extruded	3-hole perforated	108.0	—	—	—	57.3	1.51	—	—
Butterly waingrove red rust	Extruded	10-hole perforated	123.7	—	—	—	48.4	1.61	—	—
Malaysia Clay Brick										
Tajo	Extruded	5-hole perforated	47.3	7.89	53.4	1.62	26.7	1.36	29.9	1.23
Kim ma	—	Solid	36.3	2.30	38.2	1.58	323.8	1.17	25.1	1.19
Butterworth	Extruded	3-hole Perforated	7.6	2.09	8.9	1.60	5.8	1.45	5.8	1.17
Calcium Silicate Brick										
Esk, UK	Pressed	Frogged	25.4[b]	1.26	21.2	1.01	11.7	1.13	11.9	0.93
Batamas, Malaysia	Pressed	Solid	16.4	1.17	20.2	1.07	8.9	1.11	1.02	1/02
Concrete Block										
Tilcon	Pressed	Solid	13.0[c]	—	13.8	1.05	12.0[c]	—	12.6	1.04

[a]Frog not filled with mortar.
[b]Frog filled with mortar.
[c]Using 100- × 100- × 200-mm prisms.

from high absorption units, but not wetted or *docked* before being laid and sealed until application of load, the modulus of elasticity was overestimated by composite model equations [32−37,48]. Therefore, in those instances where moisture transfer occurred, either internally from mortar to unit or externally from brickwork to the surrounding environment, it was apparent that the important assumption of the composite model that the properties of unbonded specimens were the same as the properties of bonded phases in the masonry was incorrect. The overestimation of modulus of elasticity is thought to be caused by the presence of "softer material" due to the formation of weak zones of plastic shrinkage microcracks at the mortar/unit interfaces due to early rapid loss of moisture from the freshly-laid plastic mortar to the unit. In addition, shrinkage microcracks and voids can occur within the main body of mortar bed joints due to loss of moisture to the surrounding drying environment. Under a compressive load, any microcracks would tend to close, thus contributing to additional strain over and above the "true" elastic strain and, consequently, result in a lower-than-expected modulus of elasticity.

The sensitivity of the modulus of elasticity to transfer of moisture from the mortar bed joint is demonstrated in Table 5.8, which summarizes results from 5-stack brickwork built with: (1) low water absorption clay brick, and (2) high-absorption calcium silicate brick. In the case of clay brickwork sealed until loading, there was no effect of docking the bricks prior to laying on modulus of elasticity of brickwork since the bricks had low water absorption and so absorbed little or no moisture from the mortar. On the other hand, for sealed calcium silicate brickwork, compared with docked bricks, the brickwork built with undocked bricks resulted in a much lower modulus of elasticity because moisture transferred from the mortar to the part-filled or empty pores of the brick. Table 5.8 reveals that the influence of external loss of moisture to the environment also results in a lower modulus of elasticity for both clay and calcium silicate brickwork regardless of whether the units were docked or undocked.

Table 5.8 Influence of Storage Condition and Unit Moisture State on Modulus of Elasticity of Brickwork [39]

Brick			14−Day Modulus of Elasticity of Brickwork, GPa	
Type	Water Absorption, %	Storage Before Load	Undocked (Dry)	Docked (Wetted)
Clay	4.5	Sealed (under polythene)	14.2	14.1
Clay	4.5	1 day sealed, 13 days drying	12.9	13.8
Calcium silicate	14.0	Sealed	6.7	10.6
Calcium silicate	14.0	1 day sealed, 13 days drying	5.7	8.7

Mortar Elasticity Reduction Factor

The influence of loss of moisture on the reduction of modulus of elasticity of the mortar bed joint due to water absorption of unit and due to drying by environment has been quantified for units having a wide range of water absorption [39]. By reviewing previous research test data [32−37,48], it was possible to compare the modulus of elasticity of the mortar bed joint (E_{mc}), calculated from the composite model, with that measured on unbonded mortar prismatic specimens (E_m). From the measured elastic moduli of brick (E_{by}) and brickwork (E_{wy}), E_{mc} was calculated by rearranging Eqs (3.39) or (3.43). For example, in the case of brickwork:

$$E_{mc} = 0.14 \left[\frac{1}{E_{wy}} - \frac{0.86}{E_{by}} \right]^{-1} \tag{5.15}$$

For various types of brick and curing or storage condition, the calculated modulus was compared with the measured modulus in terms of the relative modulus ratio, E_{mc}/E_m, which is defined as the mortar elasticity reduction factor (γ_e). Two distinct trends were apparent as shown in Figure 5.11, one trend being for brickwork sealed under polythene sheet, which represents the influence of just the unit absorption (W_a), whereas the second trend is for brickwork, initially sealed for 1 day and then exposed to drying in air, which represents the combined action of unit absorption and environmental drying. It can be seen that the modulus of elasticity of the mortar bed joint is

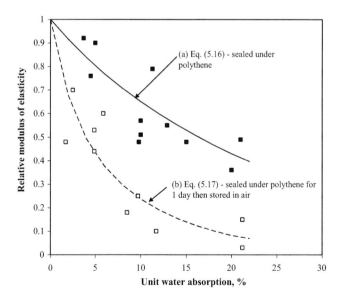

Figure 5.11 Reduction factor γ_e for modulus of elasticity of mortar bed joint due to water absorption of unit for brickwork (a) sealed and (b) sealed then exposed to drying.

quite sensitive to loss of moisture, especially in the second case, when brickwork is exposed to external drying. The general relationships for the two storage conditions are:

1. Sealed under polythene:

$$\gamma_e = \frac{1 - 0.016W_a}{1 + 0.029W_a} \tag{5.16}$$

2. Sealed under polythene for 1 day and then stored in air:

$$\gamma_e = \frac{1 - 0.030W_a}{1 + 0.195W_a} \tag{5.17}$$

Equations (5.16) and (5.17) apply to masonry built with dry units, i.e., undocked. In practice, if masonry is assumed to be protected from excessive drying for the first day, then Eq. (5.17) is appropriate, provided that the units are laid in a dry state. However, it is the recommended practice that high absorption clay bricks, i.e., those with suction rates exceeding 1.5 kg/m^2/min, should be wetted or docked before laying, and in that case $\gamma_e = 1$ may be assumed. It should be noted that it is *not* the recommended practice to wet or dock concrete and calcium silicate units because subsequent shrinkage could cause cracking problems in the masonry (see Chapter 7).

Age

Most of the methods of predicting modulus of elasticity of masonry given in the next section imply that there is no effect of age on modulus of elasticity of masonry. Assuming that units are mature and their strength is unchanged by age, any effect would arise from a gain in mortar strength as the masonry matures, and yet most methods relate the modulus to the design strength, which is based on 28-day strength of mortar. Since masonry structures may be subjected to load at different ages, it is of interest to comment on how the modulus of elasticity is affected as mortar strength increases due to hydration of cement. Figure 5.12 shows test results for various type of masonry. When masonry is sealed there is a progressive increase in modulus due to the gain in strength of the mortar due to availability of moisture for hydration, but when masonry is allowed to dry from the first day after construction, the modulus tends to decrease. At early ages, therefore, there is significant influence of age on the modulus of elasticity of masonry; however, beyond the age of 14 days the changes in modulus are small, so that for that mature masonry, the general assumption is acceptable.

Prediction of Modulus of Elasticity of Masonry

From what has been discussed in the previous sections, it is apparent that the factors affecting the modulus of elasticity of masonry are the same as those affecting the unit and mortar, although to a different extent, in addition to the factor arising from the

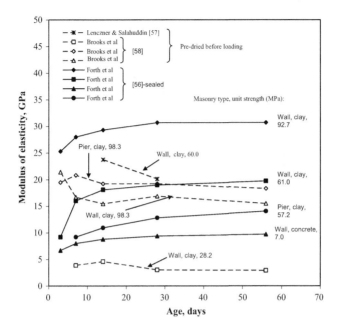

Figure 5.12 Modulus of elasticity of masonry as affected by age and curing condition.

unit/mortar interface or bond. That latter feature is included with composite models developed for practical prediction, which are given later in this section after current design methods of estimating modulus of elasticity of masonry are presented.

Current Methods

There are several expressions available for estimating the modulus of elasticity of masonry. Those that simply depend on unit strength were proposed by Plowman [4] and by Lenczner [6], their respective equations being:

$$E_{wy} = 0.2f_{by} + 4.1 \tag{5.18}$$

and:

$$E_{wy} = 3.75\sqrt{f_{by}} - 10.0 \tag{5.19}$$

where E_{wy} = vertical modulus of elasticity of masonry (GPa) and f_{by} = strength of unit, (MPa).

The modulus of elasticity of masonry, as given by Codes of Practice: BS 5628-2: 2005 [1] and BS EN 1996-1-1: 2005 (Eurocode 6) [2], is based on *characteristic strength* (f_k). BS EN 5628-2: 2005 expresses f_k in terms mortar strength or designation, and strength and type of unit as shown in Tables 5.9 and 5.10. The various types of

Table 5.9 Characteristic strength of masonry constructed with clay and calcium silicate bricks having a standard format of height to least lateral dimension of 0.6, and having no more than 25% of formed voids or 20% of frogs according to BS 5628-1: 2005 [53].

Unit compressive strength, MPa	Characteristic strength of masonry (MPa) for mortar type:			
	M12 (i)	M6 (ii)	M4 (iii)	M2 (iv)
5	2.5	2.5	2.5	2.5
10	4.0	3.8	3.4	2.8
15	5.3	4.8	4.3	3.6
20	6.4	5.6	5.0	4.1
30	8.3	7.1	6.3	5.1
40	10.0	8.4	7.4	6.1
50	11.6	9.5	8.4	7.1
75	15.2	12.0	10.5	9.0
100	18.3	14.2	12.3	10.5
125	21.2	16.1	14.0	11.6
150	23.9	17.9	15.4	12.7

mortar mixes prescribed by BS 5628-1: 2005 are given in Table 5.11, together with strength class and designation. The strength class is based on the 28-day minimum strength and, if required, 28-day strength can be estimated from strength at earlier ages by Table 5.12. According to BS 5628-1: 2005 [53], the modulus of elasticity of masonry is simply related to characteristic strength as follows:

$$E_{wy} = 0.9 f_k \tag{5.20}$$

The approach of BS EN 1996-1-1:2005 (Eurocode six) is similar, but a formula is preferred for relating characteristic strength (f_k) to unit strength (f_{by}) and mean mortar strength (f_m):

$$f_k = K f_{by}^{\alpha} f_m^{\beta} \tag{5.21}$$

where K is dependent on unit type and geometry, and mortar type; values of K are given in Table 5.13.

Equation (5.21) applies for the unit strengths $f_{by} \leq 110$ MPa, when units are laid in general purpose mortar, and for unit strengths $f_{by} \leq 50$ MPa, when laid in thin layer mortar, as defined in Table 5.13. The mean mortar strength, f_m is limited to:

- 12 MPa when units are laid in general purpose mortar.
- 10 MPa when units are laid in lightweight mortar.
- 2 × unit strength.
- Coefficient of variation of unit strength of 25%.

The values of K apply for single leaf masonry and are dependent on type of mortar, type of unit, and its group classification. The latter depends on the unit's geometrical requirements for volume of holes or perforations, as defined in Table 5.14. For

Table 5.10 Characteristic strength of masonry constructed with concrete blocks according to
BS 5628-1: 2005 [53]

Unit compressive strength, MPa	Characteristic strength of masonry (MPa) for mortar type:			
	M12 (i)	M6 (ii)	M4 (iii)	M2 (iv)
Autoclave aerated aggregate block having a ratio of height to least horizontal dimension of 0.6				
2.9	1.4	1.4	1.4	1.4
3.6	1.7	1.7	1.7	1.7
5.2	2.5	2.5	2.5	2.5
7.3	3.4	3.2	3.2	21.8
10.4	4.4	4.2	4.1	3.5
17.5	-	-	-	-
22.5	-	-	-	-
30.0	-	-	-	-
≥40	-	-	-	-
Aggregate concrete block having a ratio of height to least horizontal dimension of 0.6				
2.9	1.4	1.4	1.4	1.4
3.6	1.7	1.7	1.7	1.7
5.2	2.5	2.5	2.5	2.5
7.3	3.4	3.2	3.2	2.8
10.4	4.4	4.2	4.1	3.5
17.5	6.3	5.5	5.1	4.6
22.5	7.5	6.5	6.0	5.3
30.0	9.5	7.9	7.2	6.2
≥40	11.2	9.3	8.2	7.1
Aggregate concrete block having not more than 25 % of formed voids and a ratio of height to least horizontal direction between 2.0 and 4.5				
2.9	2.8	2.8	2.8	2.8
3.6	3.5	3.5	3.5	3.5
5.2	5.0	5.0	5.0	4.4
7.3	6.8	6.4	6.4	5.6
10.4	8.8	8.4	8.2	7.0
17.5	12.5	11.1	10.1	9.1
22.5	15.0	13.0	12.0	10.5
30.0	18.7	15.9	14.5	12.5
≥40	22.1	18.7	16.8	14.5
Autoclaved concrete block having a ratio of height to least horizontal dimension between 2.0 and 4.5				
2.9	2.8	2.8	2.8	2.8
3.6	3.5	3.5	3.5	3.5
5.2	5.0	5.0	5.0	4.4
7.3	6.8	6.4	6.4	5.6
10.4	8.8	8.4	8.2	7.0
17.5	-	-	-	-
22.5	-	-	-	-
30.0	-	-	-	-
≥40	-	-	-	-

Table 5.11 Prescribed mortar designation according to BS 5628-1: 2005 [53]

28-day strength, MPa	Strength class	Mortar designation	Type of mortar (mix proportions by volume)*				Increasing capacity to accommodate movement, e.g. due to settlement, temperature and moisture changes
			Cement:lime: sand, with or without air entrainment	Cement: sand, with or without air entrainment	Masonry cement: sand, with inorganic filler other than lime	Masonry cement: sand, with lime	
12	M12	(i)	$1: (0 \text{ to } \frac{1}{4}) : 3$	-	-	-	
6	M6	(ii)	$1: \frac{1}{2} : (4 \text{ to } 4.5)$	$1 : (3 \text{ to } 4)$	$1 : 2\frac{1}{2} : 3$	$1 : 3$	
4	M4	(iii)	$1 : 1 : (5 \text{ to } 6)$	$1 : (5 \text{ to } 6)$	$1 : (4 \text{ to } 5)$	$1 : (3\frac{1}{2} \text{ to } 4)$	
2	M2	(iv)	$1 : 2 : (8 \text{ to } 9)$	$1 : (7 \text{ to } 8)$	$1 : (5\frac{1}{2} \text{ to } 6\frac{1}{2})$	$1 : 4\frac{1}{2}$	

* Proportioning by mass will give more accurate batching provided the bulk densities are checked on site. Also, where a range of sand proportion is given, the lower figure should be used with sands containing a low fraction of fines.

Table 5.12 Strength of Mortar Relative to 28-Day Strength, f_t/f_{28}

Age, Days	1	3	7	14	28
Strength ratio, f_t/f_{28}	0.4	0.59	0.72	0.83	1

example, the original definitions of solid and perforated clay bricks of BS 3921 are now classified by BS EN 1996-1-1: 2005 as Group 1 and Group 2, respectively, whereas clay units of Groups 3 and 4 have not traditionally been used in the United Kingdom. Likewise, combinations of lightweight mortar and calcium silicate units (Groups 1 and 2) have not traditionally been used in the United Kingdom.

In Eq. (5.21), α and β depend on type and thickness of the mortar joints. For general purpose mortar and lightweight mortar, the characteristic strength is:

$$f_k = K f_{by}^{0.7} f_m^{0.3} \tag{5.22}$$

For thin layer mortar (bed joint thickness 0.5−3 mm) using clay units of Group 1, calcium silicate units of Groups 1 and 2, and autoclaved concrete units of Group 1:

$$f_k = K f_{by}^{0.85} f_m^0 = K f_{by}^{0.85} \tag{5.23}$$

Also for thin layer mortar, but using clay units of Group 2:

$$f_k = K f_{by}^{0.7} \tag{5.24}$$

In the case of BS EN 1996-1-1: 2005, UK National Annex [2], the modulus of elasticity (GPa) is numerically equal to characteristic strength (MPa):

$$E_{wy} = f_k \tag{5.25}$$

In the United States, ACI 530-05 [53] expresses modulus of elasticity of masonry as a function of specified compressive strength, f_m', as below:

$$\left. \begin{array}{l} \text{For clay masonry :} \quad E_{wy} = 0.7 f_m' \\ \text{For concrete masonry :} \quad E_{wy} = 0.9 f_m' \end{array} \right\} \tag{5.26}$$

In the case of autoclaved aerated concrete (AAC) masonry of specified strength f_{aac}', the ACI expression is:

$$E_{wy} = 6.5 \left(f_{aac}' \right)^{0.6} \tag{5.27}$$

Although the specified strength of AAC masonry is dependent only on the strength of the unit, the specified compressive strengths of clay and concrete masonry are dependent on both the strength of unit and type of mortar, as shown in Table 5.15. The ACI method

Table 5.13 Values of K to be used in Eq. (5.21) [2]

| | Type of unit and group number (see Table 5.14) | | | | | | | | | | | | | |
| Type of mortar | Clay | | | | Calcium silicate | | Aggregate concrete | | | | | Autoclaved aerated concrete | Manufactured stone | Cut natural stone |
	1	2	3	4	1	2	1* (laid flat)	1	2	3	4	1	1	1
General purpose (joint thickness ≥6 ≤15 mm)	0.55 (0.5)	0.45 (0.4)	0.35 (-)	0.35 (-)	0.55 (.5)	0.45 (0.4)	- (0.50)	0.55 (0.45)	0.45 (0.52)	0.40 (-)	0.35 (-)	0.55	0.45	0.45
Thin layer mortar (joint thickness ≥0.5 ≤3 mm)	0.75	0.70	0.50 (-)	0.35 (-)	0.80	0.65 (0.70)	- (0.70)	0.80	0.65 (0.76)	0.50	-	0.80	0.75	-
Lightweight mortar: (Density ≥ 600 ≤ 800 kg/m³)	0.30	0.25	0.20 (-)	0.20 (-)	-	-	- (0.40)	0.45	-	-	-	0.45	-	-
(Density ≥ 800 ≤ 1300 kg/m³)	0.40	0.30	0.25 (-)	0.25 (-)	-	-	- (0.40)	0.45	-	-	-	0.45	-	-

* If units contain formed vertical voids, multiply K by (100−n)/100 where n = % voids (max. 25%). The numbers in parenthesis are those listed by the UK National Annex of BS EN 1996-1-1: 2005.

Table 5.14 Geometrical requirements for grouping of masonry units according to BS EN 1996-1-1: 2005

Limits of requirement for unit type

Group number	Total volume of voids, % gross vol.			Volume of any hole, % gross vol.			Thickness of web and shell, mm						Combined thickness[a] of web and shell, % overall width		
	Clay	Calcium silicate	Concrete[b]	Clay	Calcium silicate	Concrete[b]	Clay		Calcium silicate		Concrete[b]		Clay	Calcium silicate	Concrete[b]
							Web	Shell	Web	Shell	Web	Shell			
1	≤ 25	≤ 25		≤ 12.5	≤ 12.5										
2 Vertical holes	> 25 ≤ 55	> 25 ≤ 55	> 25 ≤ 60	Each of multiple holes ≤ 2. Grip holes up to a total of 12.5	Each of multiple holes ≤ 15. Grip holes up to a total of 30	Each of multiple holes ≤ 30. Grip holes up to a total of 30	≥ 5	≥ 8	≥ 5	≥ 10	≥ 15	≥ 15	≥ 16	≥ 20	≥ 18
3 Vertical holes	> 25 ≤ 70	-	> 25 ≤ 70	Each of multiple holes ≤ 2. Grip holes up to a total of 12.5	-	Each of multiple holes ≤ 30. Grip holes up to a total of 30	≥ 3	≥ 6	-	-	≥ 15	≥ 15	≥ 12	-	≥ 15
4 Horizontal holes	> 25 ≤ 70	-	> 25 ≤ 50	Each of multiple holes ≤ 30.	-	Each of multiple holes ≤ 25.	≥ 5	≥ 6	-	-	≥ 20	≥ 20	≥ 12	-	≥ 45

[a]Combined thickness of web and shell measured horizontally.
[b]For conical or cellular holes, use mean value of webs and shells

Table 5.15 Specified Compressive Strength of Clay and Concrete Masonry
According to ACI 530.1R-05 [54]

Net Area Compressive Strength of Unit, MPa		Net Area Compressive Strength of Masonry, MPa
Type M or S Mortar	Type N Mortar	
Clay Masonry		
11.7	14.5	6.9
23.1	28.6	10.3
34.1	42.7	13.8
45.5	56.9	17.2
56.9	71.0	20.7
68.3	–	24.1
91.0	–	27.6
Concrete Masonry		
8.6	9.0	6.9[a]
13.1	14.8	10.3[a]
19.3	21.0	13.8[a]
25.9	27.9	17.2[a]
33.1	36.2	20.7[a]

[a]For units less than 100 mm in height, use 85% of values listed.

differs from the British and European methods in that strength is based on *net area*, which allows for any perforations and voids in the units. The different types of mortar, as specified by ACI 530.1R-05, are given in Table 5.16, where it is emphasized that the specification is based on either volume proportions or properties, but not both.

Composite Model

In the composite model analysis, the effect of any water absorbed by the units from the freshly-laid mortar is assumed to affect the properties of the whole mortar joint. In reality, the absorption process is likely to cause an interfacial or transition zone effect at the unit/mortar bond, because of transfer of soluble salts, such as that discussed on page 241. However, to simplify the analysis, it is convenient to assume the change in mortar properties occurs in the whole mortar joint, otherwise for modeling purposes, a third phase would have to be considered in addition to the unit and mortar. Another assumption is that the unit properties are unchanged by an increase in moisture content brought about by the absorption process. The modulus of elasticity of some materials is sometimes increased slightly by water saturation, but in the present situation, any effect due to partial absorption of the unit is assumed to be small.

To allow for the effect of water absorption from the mortar by the bricks on the modulus of elasticity of brickwork, Eq. (3.39) can be modified as follows:

$$\frac{1}{E_{wy}} = \frac{0.86}{E_{by}} + \frac{0.14}{\gamma_e E_m} \tag{5.28}$$

Table 5.16 ACI 530.1R-05 Specification Requirements for Mortar by Proportions or by Laboratory-Based Properties: Strength, Air Content, and Minimum Retentivity of 75% [54]

Mortar	Type	Proportions by vol. of Cementitious Materials								Aggregate Ratio, Measured in Damp Loose Conditions	Average Compressive Strength at 28 days, MPa	Max. Air Content, %
		Portland Cement or Blended Cement	Mortar Cement M	Mortar Cement S	Mortar Cement N	Masonry Cement M	Masonry Cement S	Masonry Cement N	Hydrated Lime or Lime Putty			
Cement–lime	M	1	—	—	—	—	—	—	$\frac{1}{4}$	Not less than $2\frac{1}{4}$ and not more than 3 times the sum of the separate volumes of cementitious materials	17.2	12
	S	1	—	—	—	—	—	—	$>\frac{1}{4}-\frac{1}{2}$		12.4	12
	N	1	—	—	—	—	—	—	$>\frac{1}{2}-1\frac{1}{4}$		5.2	14[a]
	O	1	—	—	—	—	—	—	$>1\frac{1}{4}-2\frac{1}{2}$		2.4	14[a]
Mortar cement	M	1	—	—	1	—	—	—	—		17.2	12
	M	—	1	—	—	—	—	—	—		17.2	12
	S	$\frac{1}{2}$	—	—	1	—	—	—	—		12.4	12
	S	—	—	1	—	—	—	—	—		12.4	12
	N	—	—	—	1	—	—	—	—		5.2	14
	O	—	—	—	1	—	—	—	—		2.4	14
Masonry cement	M	1	—	—	—	—	—	1	—		17.2	18
	M	—	—	—	—	1	—	—	—		17.2	18
	S	$\frac{1}{2}$	—	—	—	—	—	1	—		12.4	18
	S	—	—	—	—	—	1	—	—		12.4	18
	N	—	—	—	—	—	—	1	—		5.2	20[b]
	O	—	—	—	—	—	—	1	—		2.4	20[b]

[a]when reinforcement is incorporated, maximum air content is 12%.
[b]when reinforcement is incorporated, maximum air content is 18%.

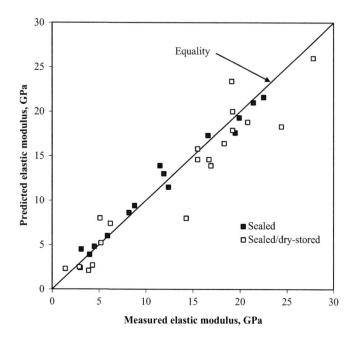

Figure 5.13 Comparison of predicted and measured modulus of elasticity of brickwork allowing for water absorption factor (Eq. (5.28)).

Using the same data as those used to derive Eqs (5.16) and (5.17), Figure 5.13 demonstrates a satisfactory correlation between the predicted modulus of elasticity using Eq. (5.28) and the measured modulus, the error coefficient being 17.6% for a mean measured modulus of 12.7 GPa.

The influence of unit water absorption factor on modulus of masonry also applies to concrete blocks, however, because the modulus of blockwork is less affected by the mortar joint than in the case of brickwork, it follows that the water absorption influence is smaller. For blockwork, Eq. (3.43) now becomes:

$$\frac{1}{E_{wy}} = \frac{0.952}{E_{by}} + \frac{A_m}{A'_m}\frac{0.048}{\gamma_e E_m} \tag{5.29}$$

As stated earlier, the mortar elasticity reduction factor expresses the ratio of deduced measured strain in the mortar bed joint[a] and strain as measured on a separate mortar specimen subjected to the same stress. Direct experimental measurement of strain in the mortar bed joint in concrete blockwork was carried out by Khalaf et al. [55], using 10-mm electrical-resistance strain gauges, and they reported appreciably

---

[a] as deduced from measured masonry and unit strain.

less strain in the mortar joint than was measured in separate mortar cylinders. In fact, the average ratio of bed joint strain to cylinder strain was 0.45, which is of a similar order to values of γ_e, shown in Figure 5.11.

Since the moduli of unit and mortar are likely to be unknown without special tests, Eqs (5.28) and (5.29) are not really suitable for use in design. However, earlier in this chapter, E_m and E_{by} were shown to be functions of strength of mortar (Eq. (5.6)) and strength of unit (Table 5.6), respectively, so that substitution of those functions in Eqs (5.28) and (5.29) now allows E_{wy} to be expressed in terms of properties known to the designer. For example, using the relations given in Table 5.6, the vertical modulus of elasticity of different types of brickwork is as follows:

Extruded Wire-Cut Perforated Clay Brickwork

$$\frac{1}{E_{wy}} = \frac{2.15}{f_{by}} + \frac{0.175}{\gamma_e f_m} \tag{5.30}$$

Pressed and Slop Moulded Clay Brickwork

$$\frac{1}{E_{wy}} = \frac{3.44}{f_{by}} + \frac{0.175}{\gamma_e f_m} \tag{5.31}$$

Calcium Silicate Brickwork

$$\frac{1}{E_{wy}} = \frac{1.87}{f_{by}} + \frac{0.175}{\gamma_e f_m} \tag{5.32}$$

Concrete Blockwork

$$\frac{1}{E_{wy}} = \frac{1.058}{f_{by}} + \frac{A_m}{A'_m} \frac{0.060}{\gamma_e f_m} \tag{5.33}$$

Accuracy of Prediction

The accuracy of Eqs (5.30)–(5.33) has been assessed by comparing their predictions with measured moduli of elasticity of masonry, as reported in previous publications. In addition, the accuracy of estimates by all other methods has been compared: Plowman [4], Lenczner [6], and Codes of Practice [1–3]. The accuracy is quantified in terms of the error coefficient, M, (Eq. (5.4)) and P, i.e., the percentage of results falling within ±30% of the measured values. For clay brickwork, the error coefficient is based on 74 sets of data with an average measured modulus of elasticity of 15.8 GPa. In most cases of extruded or pressed clay bricks, the precise volumes of perforations or frogs was unknown and so, for the BS EN 1996-one to one method, it was assumed that perforated clay bricks conformed to Group two category of Table 5.14; also, some of the mortar strengths [29] exceeded the maximum allowable limit of Eq. (5.22);

however, analysis of those particular data sets revealed no perceptible difference in error coefficient compared with that of the total data sets. For the same reason of lack of details of perforations and frogs, analysis by the ACI 530.1R method of clay bricks was based on gross area rather than net area.

With all methods, Figures 5.14—5.19 show that there is a large variability of prediction. Error coefficients (M) lie between 27 and 53%, with percentage of values (P) lying between 36 and 76%. The composite model is the most accurate, with the lowest error coefficient of 27% and 49% of estimated values being within 20% of the actual values. It is thought that some of the variability is due to the different test conditions used by the various investigators, such as rate of loading, which is hardly ever quoted in publications. Also, the relatively high stresses used by Glanville and Barnett [29] and Lenczner [10] could have encroached into the region of nonlinear stress—strain behaviour. Moreover, the same authors used high strength mortars that, strictly, are outside the recommended range stipulated by BS EN 1996-1-1:2005 (see p. 121). However, analysis of those data did not reveal any difference in accuracy compared with the general accuracy using all data. The worst estimates were by BS 5625-1: 2005 (Figure 5.16), this outcome being attributed to the lower characteristic strengths adopted in the current standard than in the superseded BS 5628-2:1995; it was demonstrated previously that the latter gave similar accuracies to other methods [46]. Of the other standards, BS EN 1996-1-1 (Figure 5.17) had the lowest error coefficient, followed by the ACI

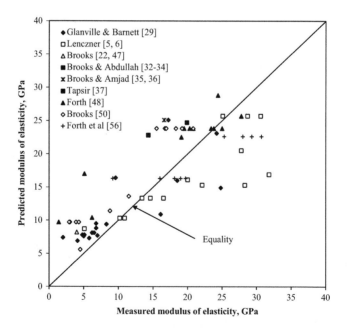

Figure 5.14 Comparison of predicted elastic modulus by Plowman (Eq. (5.18)) and measured elastic modulus of clay brickwork; $M = 34\%$, $P = 60\%$.

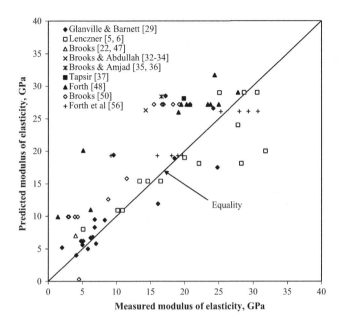

Figure 5.15 Comparison of predicted elastic modulus by Lenczner (Eq. (5.19)) with measured elastic modulus of clay brickwork; $M = 39\%$, $P = 53\%$.

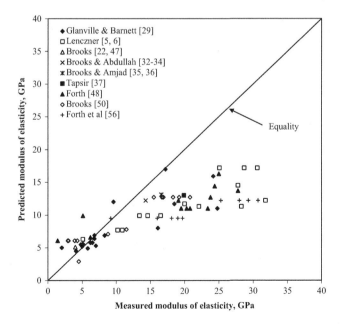

Figure 5.16 Comparison of predicted elastic modulus by BS 5628-1: 2005 (Eq. (5.20)) with measured elastic modulus of clay brickwork; $M = 53\%$, $P = 36\%$.

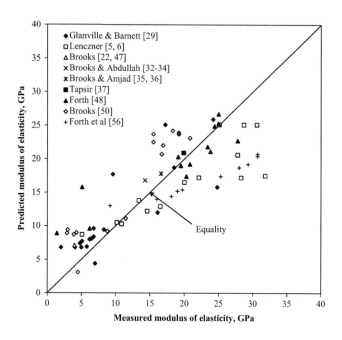

Figure 5.17 Comparison of predicted elastic modulus by BS EN 1996-1-1: 2005 (Eq. (5.25)) with measured elastic modulus of clay brickwork; $M = 32\%$, $P = 55\%$.

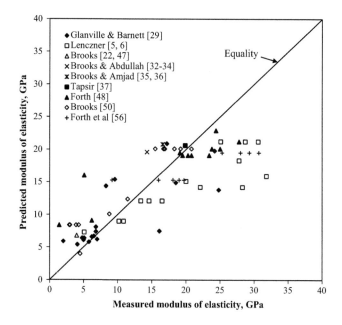

Figure 5.18 Comparison of predicted elastic modulus by ACI 530 method (Eq. (5.26)) with measured elastic modulus of clay brickwork; $M = 35\%$, $P = 66\%$.

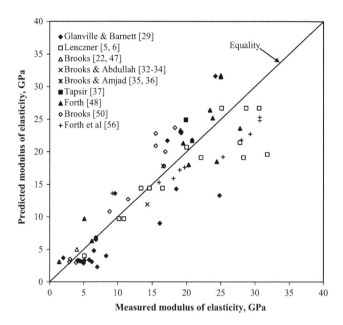

Figure 5.19 Comparison of predicted elastic modulus by composite model (Eqs (5.30) and (5.31)) with measured elastic modulus of clay brickwork; $M = 27\%$, $P = 76\%$.

530 method (Figure 5.18), although the latter had a greater number of estimates falling within 30% of actual values.

Despite the fact that the properties mortar is not taken into account, the accuracy of the Plowman (Figure 5.14) and Lenzcner (Figure 5.15) methods is not much different from that of the standards methods, which implies that the unit is far more influential than the mortar on the modulus of elasticity of clay brickwork. Although the standard methods do account for the type of mortar, they tend to underestimate the modulus of brickwork built with high-strength units. The improvement in accuracy of prediction of modulus of elasticity by the composite model method (Figure 5.19) is mainly achieved by accounting for the additional influences of water absorption of unit and curing condition prior to application of load.

For other types of masonry other than clay brickwork, the accuracy of estimating modulus of elasticity has been assessed by analyzing 22 sets of data for concrete blockwork built with solid blocks [34,36,37] and nine sets analyzed for calcium silicate brickwork [22,29,36,37,50]. All Codes of Practice underestimated modulus of blockwork, particularly that of BS EN 1996-1-1: 2005 with ACI 530.1R-05 being the best with an average underestimate (M) of 21%. For calcium silicate brickwork, estimates by BS 5628: 2005 and BS EN 1996-1-1: 2005 gave similar error coefficients (44%). The composite model prediction errors were much lower, being 9 and 18% for concrete blockwork and calcium silicate brickwork, respectively, but the improved accuracy could be expected since the same data sets were used to derive the modulus-strength prediction expressions (Eqs (5.32) and (5.33)).

Example

It is required to estimate the modulus of elasticity of masonry by using the six methods described earlier in this chapter. The unit is an extruded/perforated clay type of standard size, manufactured in the United Kingdom, has a strength, f_{by}, = 60 MPa and water absorption, γ_{wa}, = 6%; the perforations amount to 10% of the gross unit volume. The type of mortar is $1:\frac{1}{2}:4\frac{1}{2}$ (designation ii), or M10 (28-day strength, f_m, = 10 MPa). The solutions are as follows:

1. Plowman's method

The modulus of elasticity is simply given by Eq. (5.18), i.e., 16.1 GPa.

2. Lenczner's method

In this case, the modulus of elasticity is 19.0 GPa as given by Eq. (5.19).

3. BS 5628-2: 2005 method

From Table 5.9, the characteristic strength for M10 mortar and a unit strength of 60 MPa = 12.2 MPa (by interpolation). Therefore, from Eq. (5.20), the modulus of elasticity of brickwork = 11.0 GPa.

4. BS EN 1996-1-1: 2005 method

The unit is categorized as Group 1 according to Table 5.14. The characteristic strength is 17.5 MPa as given by Eq. (5.22), with $K = 0.5$ (Table 5.13-UK value). Consequently, the modulus of elasticity of brickwork = 17.5 GPa.

5. ACI 530.1R-05 method

The mortar is categorized by mix proportions as Type S according to Table 5.16. Since the volume of perforations in the brick is small, the net area is assumed to be equal to the gross area and, from Table 5.15, the design strength of masonry = 21.6 MPa (by interpolation). Hence, from Eq. (5.26), the modulus of elasticity = 15.1 GPa.

6. Composite model method

Assuming the brickwork to be covered for 1 day and then exposed to drying, the water absorption factor is given by Eq. (5.17), viz. 0.38. The modulus of elasticity is given by Eq. (5.30) as 12.2 GPa.

Summary

In this chapter, it has been demonstrated that the stress–strain behaviour of masonry is classified as nonelastic and nonlinear. Also, there is no significant influence of geometry or height of masonry on the modulus of elasticity. Two of the main factors influencing the modulus of elasticity of masonry are type of mortar and type of unit for which linear modulus-strength relations have been developed, although there is appreciable scatter of test data. Whereas the modulus of elasticity of calcium silicate bricks and solid concrete blocks are virtually isotropic, the modulus of clay bricks can be significantly anisotropic due to the manufacturing process and the presence of frogs and perforations. At early ages, there is significant influence of age on the modulus of elasticity of masonry. However, beyond the age of 14 days, the changes in modulus are small, so that for mature masonry, the general assumption of little influence of age is acceptable. All current Codes of Practice recommend the modulus of elasticity of masonry be estimated from characteristic strength or specified design strength of masonry and give accuracies ranging from 32 to 50%. An improvement in accuracy of prediction is achieved by the composite model method (27%) by allowing for the additional influences of water absorption of unit and curing condition of masonry prior to application of load.

Problems

1. What are the causes of nonlinear stress—strain behaviour of masonry?
2. Quote typical values of Poisson's ratio for concrete, calcium silicate, and clay masonry.
3. How does geometry of masonry affect modulus of elasticity and Poisson's ratio?
4. What are the reasons for the high variability of modulus of elasticity of mortar?
5. Do the same factors affect strength of mortar and elastic modulus of mortar?
6. Discuss the effect of type of material and manufacturing process on modulus of elasticity of units.
7. Why are the anisotropic elastic properties different for pressed and extruded clay bricks?
8. Discuss the factors affecting the relationships between modulus and strength of units?
9. What is the effect of frogged bricks laid frog-down on the modulus of elasticity of brickwork?
10. Does age affect the modulus of elasticity of masonry?
11. What is the effect of water absorption of the unit on modulus of elasticity of masonry?
12. Would you expect the modulus of elasticity of masonry to be the same in the horizontal and vertical directions?
13. Estimate the modulus of elasticity by the BS EN 1996-one to one method for masonry constructed with hollow aggregate concrete blocks with 10-mm mortar joints, and a $1:\frac{1}{2}:4\frac{1}{2}$ cement—lime—sand mortar having a 28-day strength $= 6$ MPa. The standard-sized blocks have 50% cavity volume and have a strength of 8 MPa based on gross area.
    Answer: 3.8 GPa (based on gross area).
14. For Question 5.13, calculate the modulus of elasticity by the ACI 530.1R method.
    Answer: 9.9 GPa (based on net area and Type N mortar).
15. For Question 5.13, calculate the modulus of elasticity by the composite model method, assuming the blocks have a water absorption $= 10\%$.
    Answer: 4.6 GPa.

References

[1]   BS 5628−2: Code of Practice for Use of Masonry, Part 2: structural use of reinforced and prestressed masonry. British Standards Institution; 2005.

[2]   BS EN 1996-1-1, 2005, Eurocode No. 6: Design of Masonry Structures. General rules for reinforced and unreinforced masonry structures, British Standards Institution. See also: UK Nationall Annex to BS EN 1996-1-1; 2005.

[3]   ACI Committee 530-05. Building Code Requirements for Masonry Structures. ACI Manual of Concrete Practice, Part 6. American Concrete Institute; 2007.

[4]   Plowman JM. The modulus of elasticity of brickwork. Trans. Br. Ceram. Soc. 1965;(4): 37−64.

[5]   Lenczner D. The effect of strength and geometry on the elastic and creep properties of masonry members. In: Proceedings 1st North American masonry conference, Boulder; 1978. 23.1−23.15.

[6]   Lenczner D. Brickwork: Guide to creep. SCP 17. Newcastle, Staffs: Structural Clay Products Ltd.; 1980. 26 pp.

[7]   Allen MH. Compressive, transverse and racking strength tests of 4-in brick walls. Research Report No. 9. Geneva, Illinois: Structural Clay Products; 1965.

[8]   Allen MH. Compressive and transverse strengths of 8-in brick walls. Research Report No. 10. Geneva, Illinois: Structural Clay Products; 1966.

[9]   Sahlin S. Structural masonry. New Jersey: Prentice-Hall Inc.; 1971. 290 pp.

[10] Lenczner D. Strength and elastic properties of a 9 inch brickwork cube. Trans. Br Ceram Soc 1966;65(6). 363−282.

[11] Fattal SG, Cattaneo LE. The structural performance of masonry walls under compression and flexure, NBS (National Bureau of Standards). In Building science series, vol. 73; 1976.

[12] Bradshaw RE, Hendry AW. Further crushing tests on storey-high walls 4.5-in thick. In: Proceedings of British ceramic society, 11; 1968. pp. 25−53.

[13] Yokel FY, Mathey RG, Dikkers RD. Compressive strength of slender concrete masonry walls, NBS (National Bureau of Standards). In Building science series, vol. 73; 1978.

[14] Edgell GJ. Stress-strain relationships for brickwork-their application in the theory of unreinforced slender members. Technical Note No. 313. British Ceramic Research Association; 1980.

[15] Base GD, Baker LR. Fundamental properties of structural brickwork. J Aust Ceram Soc 1973;19:16−23.

[16] Fisher K. The effect of low strength bricks in high strength brickwork. In: Proceedings of British ceramic society, 21; 1973. pp. 79−97.

[17] Dhanasekar M, Kleeman PW, Page AW. Biaxial stress-strain relations for brick masonry. J Struct Eng 1985;111(5):324−33.

[18] Ameny P, Loov RE, Jessop EL. Strength, elastic, and creep properties of concrete masonry. Int J Mason Constr 1980;1(1):33−9.

[19] Ameny P, Loov RE, Shrive NG. Prediction of elastic behaviour of masonry. Int J Mason Constr 1983;3(1):1−9.

[20] Amjad MA. Elasticity and strength of masonry, units and mortar [Ph.D. thesis]. Department of Civil Engineering: University of Leeds; 1990. p. 253.

[21] Thomas K, O'Leary DC. Tensile testing of bricks. Constr Res Dev J 1970;2(3):141−8.

[22] Brooks JJ. Time-dependent behaviour of calcium silicate and fletton clay brickwork walls. In: Proceedings of British masonry society, 1; 1986. pp. 17−9.

[23] Meyer V, Schubert P. Stress-strain relationship for masonry. In: Proceedings of the 9th international brick and block masonry conference, vol. 3; 1991. pp. 1306−13. Berlin.

[24] Abdullah CS. Influence of geometry on creep and moisture movement of clay, calcium silicate and concrete masonry [Ph.D. thesis]. Department of Civil Engineering: University of Leeds; 1989. p. 90.

[25] Sinha MH, Hendry AW. The effect of brickwork bond on the load-bearing capacity of model brick walls. In: Proceedings of the British ceramic society, 11; 1968. pp. 55−72.

[26] Bragg JC. Compressive strength of large brick piers. Technical Paper No. 11. Bureau of Standards, U. S.; 1919.

[27] Hilsdorf HK. Investigation into the failure mechanism of brick masonry loaded in axial compression. In: Johnson RB, editor. Designing, engineering and constructing with masonry products; 1969. pp. 34−41.

[28] Forth JP, Brooks JJ. Influence of mortar type on long-term deformation of single-leaf masonry. In: Proceedings of British masonry society, 6; 1995. pp. 157−61.

[29] Glanville WH, Barnett PW. Mechanical properties of bricks and brickwork masonry. DSIR, Building Research. London: HMSO; 1934. Special Report No. 22.

[30] Jessop EL, Shrive NG, England GL. Elastic and creep properties of masonry. In: Proceedings of 1st North American masonry conference, Boulder; 1978. pp. 12.1−12.17.

[31] Schubert P. Modulus of elasticity of masonry. In: Wintz JA, Yorkdale AH, editors. Proceedings of 5th international brick/block masonry association conference. Mclean, VA: BIA; 1982. p. 139.

[32] Brooks JJ, Abdullah CS. Composite model prediction of the geometry effect on creep and shrinkage of clay brickwork. In: de Courcy JW, editor. Proceedings of 8th international brick/block masonry conference, vol. 1. Dublin: Elsevier Applied Science; 1988. pp. 316−23.

[33] Brooks JJ, Abdullah CS. Composite modelling of the geometry influence on creep and shrinkage of calcium silicate brickwork. In: Proceedings of British masonry society, 4; 1990. pp. 36−8.

[34] Brooks JJ, Abdullah CS. Creep and drying shrinkage of concrete blockwork. Mag Concr Res 1990;42(150):15−22.

[35] Brooks JJ, Amjad MA. Elasticity and strength of clay brickwork test units. In: deCourcy JW, editor. Proceedings of 8th international brick/block masonry conference, vol. 1. Dunlin: Elsevier Applied Science; 1988. pp. 342−9.

[36] Brooks JJ, Amjad MA. Strength and elasticity of calcium silicate and concrete masonry. In: Proceedings of British masonry society, 4; 1990. pp. 30−2.

[37] Tapsir SH. Time-dependent loss of post-tensioned diaphragm and fin masonry walls [Ph.D. Thesis]. Department of Civil Engineering: University of Leeds; 1994. p. 272.

[38] Brooks JJ, Forth JP. Influence of clay unit on the moisture expansion of masonry. In: Proceedings of British masonry society, 6; 1995. pp. 80−4.

[39] Brooks JJ, Abu Baker BH. The modulus of elasticity of masonry. Mason Int 1998;12(2): 58−63.

[40] Neville AM. Properties of concrete. fourth ed. London: Pearson Prentice-Hall; 2006. 844pp.

[41] Brooks JJ, Abu Baker BH. Anisotropy of elasticity and time-dependent movements of masonry. In: Proceedings of the fifth international masonry conference, masonry, vol. 8. British Masonry Society; 1998. pp. 44−7.

[42] CERAM Building Technology. Movement in masonry. Partners in Technology Report; 1996. 119 pp.

[43] Lenczner D. Creep in concrete blockwork piers. Struct Eng 1974;52(3):97−101.

[44]   Lenczner D. Creep in brickwork and blockwork in cavity walls and piers. In: Proceedings fifth international symposium on loadbearing brickwork, 27. London: British Ceramic Society; 1979. pp. 53–66.

[45]   Ameny P, Loov RE, Jessop EL. Strength, elastic and creep properties of concrete masonry. Int J Mason Constr 1984;1(Part 1):33–9.

[46]   BRE Digest 342. Autoclaved aerated concrete. Building Research Establishment; March 1989.

[47]   Brooks JJ. Composite models for predicting elastic and long-term movements in brickwork walls. In: Proceedings of the British masonry society, 1; 1986. pp. 20–3.

[48]   Forth JP. Influence of mortar and brick on long-term movement of clay masonry [Ph.D. Thesis]. Department of Civil Engineering: University of Leeds; 1995. 300 pp.

[49]   Shrive NG, Jessop EL. Anisotropy in extruded clay units and its effects on masonry behaviour. In: Proceedings of the second Canadian masonry symposium. Carlton University; 1980. pp. 39–55.

[50]   Brooks JJ. Time-dependent behaviour of masonry and its component phases. Brussels: EC Science and Technology Programme, Contract No. C11-0925; 1996. 106 pp.

[51]   Ekolu SO. Structural differences between methods of brickmaking and how those differences influence the properties of fired clay bricks [MSc Thesis]. Department of Civil Engineering: University of Leeds; 1997. 100 pp.

[52]   Neville AM, Brooks JJ. Concrete technology. Second Edition. Pearson Prentice-Hall; 2010. 438 pp.

[53]   BS 5628–1: 2005, Code of Practice for Use of Masonry: Part 1: structural use of unreinforced masonry, British Standards Institution.

[54]   ACI Committee 530.1R-05. Commentary on specifications for masonry structures. ACI Manual of Concrete Practice, Part 6. American Concrete Institute; 2007.

[55]   Khalaf FM, Hendry AW, Fairbairn DR. Elastic modulus and strength of hollow concrete block masonry with reference to the effect of lateral ties. Mag Concr Res 1992;44(160): 185–94.

[56]   Forth JP, Bingel PR, Brooks JJ. Effect of loading age on creep of sealed clay and concrete masonry. In: Proceedings of the British masonry society, 8; 1998. pp. 52–5.

[57]   Lenczner D, Salahuddin J. Creep and moisture movements in brickwork walls. In: Proceedings of fourth IBMAC, Bruges, section 2; 1976. pp. 2a. 4.0 − 2a. 4.5.

[58]   Brooks JJ, Abdullah CS, Forth JP, Bingel PR. The effect of age on deformation of masonry. Mason Int 1997;11(2):51–5.

6 Shrinkage of Concrete

Besides deformations due to externally applied stress, volume changes due to moisture migration in and out of hardened concrete are of importance. A contraction or shrinkage results from loss of moisture and an expansion or swelling results from ingress of moisture. Even in mass concrete or when concrete is sealed, hydration of cement leads to internal consumption of moisture, resulting in autogenous shrinkage and, prior to setting while still in a plastic state, plastic shrinkage of cement paste will occur if loss of moisture is permitted. Under normal drying environmental conditions, moisture diffuses from the hardened concrete, resulting in drying shrinkage, which includes carbonation shrinkage due to the reaction with carbon dioxide in the environmental air. Influencing factors in all those different types of shrinkage are discussed, together with measurement and prediction of autogenous shrinkage; standard methods of predicting of drying shrinkage are considered, together with creep, in Chapter 11.

For normal-strength concrete, autogenous shrinkage is historically assumed to be small and is included with drying shrinkage. However, with the advent of high-strength concrete made with low water/binder ratios, very fine cementitious mineral admixtures, and superplasticizing chemical admixtures, autogenous shrinkage is much more significant, especially at very early ages. Influencing factors are discussed in detail in this chapter, together with methods of measurement and prediction.

Whatever the cause or type of shrinkage, whether by loss of water to the environment, hydration of cement, or carbonation, the resulting reduction in volume as a proportion of the original volume, or volumetric strain, is equal to three times the linear contraction. In practice, therefore, shrinkage can be measured as a linear strain so that its units are mm per mm, which are usually expressed as microstrain (10^{-6}). For example, typical values of ultimate drying shrinkage range from 100 to 400×10^{-6}.

Plastic Shrinkage

While cement paste is fresh, it undergoes a volumetric contraction whose magnitude is of the order of 1% of the absolute volume of dry cement. This contraction is known as plastic shrinkage, and it occurs if water is allowed to escape from the surface of concrete or by suction of dry concrete below. The contraction induces tensile stress in the surface layers because they are restrained by the nonshrinking inner concrete and, since concrete is very weak in its plastic state, plastic cracking at the surface can readily occur [1]. Plastic shrinkage is greater the greater the rate of evaporation of water, which in turn depends on the air temperature, the concrete temperature, the relative humidity of the air, and wind speed. Prevention of evaporation immediately after

Concrete and Masonry Movements. http://dx.doi.org/10.1016/B978-0-12-801525-4.00006-6

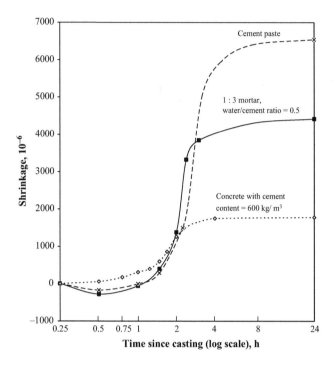

Figure 6.1 Early-shrinkage cement paste, mortar, and concrete stored in air at 20 °C and 50% relative humidity with a wind velocity of 1.0 m/s [3].

casting reduces plastic shrinkage and, according to ACI 305R-99 [2], evaporation rates greater than 0.5 kg/h/m^2 of the exposed concrete surface have to be avoided in order to reduce plastic shrinkage and prevent plastic cracking.

Plastic shrinkage is greater the larger the cement content of the mix, or lower the larger the volumetric aggregate content. The effect is demonstrated in Figure 6.1, where the plastic shrinkage of cement paste is over three times that of concrete. The relation between plastic shrinkage cracking and bleeding is complex, since retardation of setting allows more bleeding and leads to more plastic shrinkage. On the other hand, greater bleeding capacity reduces evaporation from the surface of the concrete and this reduces plastic shrinkage cracking [1]. Bleeding is caused by mix water being forced upwards when heavier solid particles settle downwards. A large amount of bleeding and settlement may lead to plastic settlement cracking if there is some form of restraint or obstruction to free settlement [4].

Swelling/Expansion

If there is a continuous supply of water during hydration, concrete expands due to absorption of water by the cement gel, this process being known as swelling (Figure 6.2).

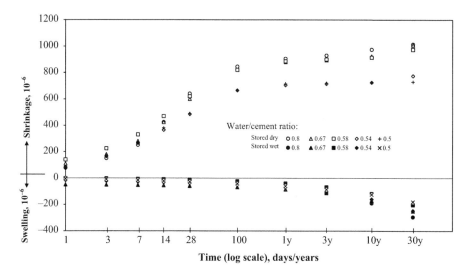

Figure 6.2 Long-term shrinkage and swelling of normal-weight aggregate concrete stored in air and in water from the age of 14 days [5]; concrete was made with a rapid-hardening Portland cement, quartzitic sand, and gravel in the proportions 1:1.71:3.04.

In neat cement paste, swelling is very high and, although lower in concrete, it is dependent on the cement content. In fact, the influencing intrinsic factors in swelling are the same as for drying shrinkage, which is discussed later. For example, in concrete made with normal-weight aggregate, both drying shrinkage and swelling increase as the water/cement ratio increases, but swelling is much smaller than drying shrinkage, especially after short periods of exposure, as can be seen in Figure 6.2. At later periods of exposure, the ratio of drying shrinkage to swelling decreases, viz. the average ratios of Figure 6.2 are 17.0, 8.7, and 3.8 after 100 days, 3 years, and 30 years, respectively. Thus, swelling occurs slowly at a steady rate, whereas drying shrinkage develops rapidly in the first 100 days and then, subsequently, the opposite trend occurs, namely, rate of swelling increases and the rate of drying shrinkage decreases. Similar trends have been reported for semi- or sand-lightweight aggregate concrete [5].

Neville [6] reports an increase in mass of around 1% as well as an increase in volume due to the swelling process. The swelling is caused by absorption of water by the cement gel, the water molecules acting against the cohesive forces and tending to force the gel particles apart with a resultant swelling pressure. In addition, the ingress of water decreases the surface tension of the gel so that further expansion occurs [1,3]. Swelling is larger in deep sea water and can lead to increased swelling due to increased external water pressure; this has implications for potential corrosion of steel reinforcement due to ingress of chlorides into concrete [1].

Swelling can arise from adverse chemical reactions between the aggregate and hydrated cement paste, the most common being between the active silica constituents of the aggregate and the alkalis in cement [6]. Commonly known as *alkali–aggregate reaction*, the alkali-silica gel that is formed absorbs water and hence has a tendency to swell, but since the gel is confined by the surrounding cement paste, internal pressure occurs, which may lead to expansion and cracking. Another type of harmful aggregate reaction is *alkali—carbonate reaction*, which occurs between some dolomitic limestones and alkalis in cement in humid conditions; this reaction can also result in swelling of concrete [6].

Carbonation Shrinkage

At the same time as contracting due to drying shrinkage, concrete probably undergoes carbonation shrinkage. Many experimental data labeled drying shrinkage include both types of shrinkage, but their mechanisms are different. *Carbonation* arises from the reaction of carbon dioxide (CO_2) with the hydrated cement. The gas CO_2 is present in the atmosphere: about 0.03% by volume in rural air, 0.1% or even more in an unventilated laboratory, and generally up to 0.3% in large cities. In the presence of moisture, CO_2 forms carbonic acid, which reacts with crystals of calcium hydroxide ($Ca(OH)_2$) to form calcium carbonate ($CaCO_3$), which is deposited in empty pores; other cement compounds are also decomposed. A concomitant of the process of carbonation is a contraction of concrete, i.e., carbonation shrinkage, which is irreversible in nature.

Carbonation proceeds from the surface of concrete inwards but does so extremely slowly. The actual rate of carbonation depends on the permeability of concrete, its moisture content, and the CO_2 content and relative humidity of the ambient medium. Since the permeability is determined by the water/cement ratio and the effectiveness of curing, concrete with a high water/cement ratio and inadequately cured will be more prone to carbonation, i.e., there will be a greater depth of carbonation. The extent of carbonation may easily be determined by treating a freshly broken surface with phenolphthalein—the free $Ca(OH)_2$ is colored pink while the carbonated portion is uncolored. The contribution of carbonation shrinkage to drying shrinkage is more extensive in smaller specimens or in specimens having a smaller surface area [7]. The latter effect plus the dependency on CO_2 level in the atmosphere should be borne in mind when extrapolating shrinkage data determined by small specimens in an unventilated laboratory to estimate long-term shrinkage of full-size concrete structural members.

Figure 6.3 shows the contribution of carbonation shrinkage to drying shrinkage of mortar specimens stored in air at different relative humidity (RH): (a) dried first in CO_2-free air (drying shrinkage) and then subjected to carbonation to yield total shrinkage; (b) subjected to simultaneous drying and carbonation to yield total shrinkage. In the case of drying followed by carbonation, Figure 6.3(a) shows that at intermediate humidity, carbonation increases the total shrinkage, but not at humidity of 100% or 25%. In the latter case, there is insufficient water in the pores within the paste for CO_2 to form carbonic acid. On the other hand, when the pores are full of water (100% RH) the diffusion of CO_2 into the paste is very slow. In the case of simultaneous drying and carbonation (Figure 6.3(b)), total shrinkage and carbonation shrinkage are greater at lower humidity than in Figure 6.3(a) because water is available to form carbonic acid but, at higher humidity, total shrinkage and carbonation

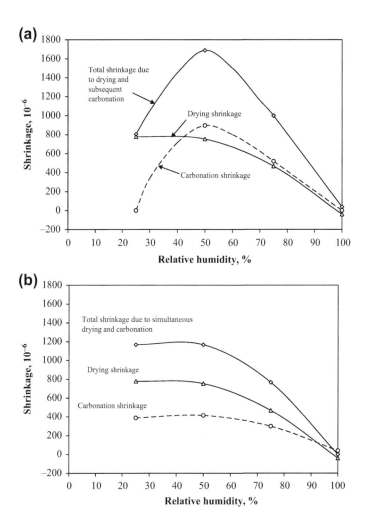

Figure 6.3 Carbonation shrinkage and drying shrinkage of mortar stored at different relative humidities [8]. (a) Carbonation shrinkage after first drying followed by carbonation. (b) Carbonation shrinkage due to simultaneous drying and carbonation.

shrinkage are slightly less than in Figure 6.3(a) because presence of pore water slows diffusion of CO_2. A practical consequence of the foregoing is that carbonation is greater in concrete protected from direct rain but exposed to moist air than in concrete periodically washed down by rain [6].

Besides causing shrinkage, carbonation of concrete made with ordinary Portland cement results in a slightly increased strength and reduced permeability, possibly because water released by the decomposition of $Ca(OH)_2$ on carbonation aids the process of hydration and $CaCO_3$ is deposited in the empty pores and voids of the cement paste. However, importantly, carbonation neutralizes the alkaline nature of the hydrated cement paste, and thus the protection of steel from corrosion is impaired. Consequently, in structural concrete members, if the full depth of cover to reinforcement is carbonated and moisture and oxygen can ingress, corrosion of steel and possibly cracking will result [1,6].

Total shrinkage of concrete subjected to wetting and drying cycles is greater when carbonation occurs in the drying cycle and, in fact, may lead to crazing (shallow cracking) of exposed surfaces due to restraint of surface layers by nonshrinking inside layers [6]. However, carbonation of concrete prior to exposure to cycles of wetting and drying reduces reversible shrinkage (see Figure 6.4) and renders precast products more dimensionally stable. Such a procedure can be carried out in practice by exposing

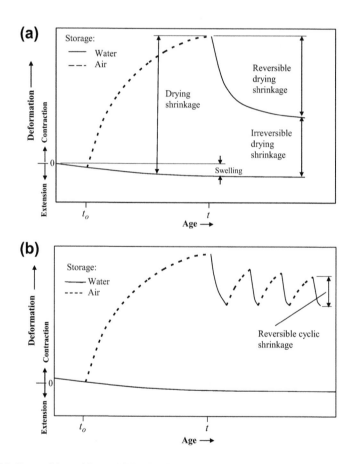

Figure 6.4 Reversible and irreversible shrinkage of concrete stored in water until exposure to drying compared with swelling of concrete stored continuously in water [1]. (a) Deformation of concrete stored continuously in water and concrete stored in air from age t_o until age t and then resaturated. (b) Deformation of concrete stored dry from age t_o and subjected to cycles of wetting and drying from age t.

Source: Line Concrete Technology, Second Edition, A. M. Neville and J. J. Brooks, Pearson Education Ltd. © Longman Group UK Ltd. 1987.

precast products to flue gases. Neville [6] refers to ACI 517.2R-92 [9], which prescribes various techniques of carbonating concrete products.

Drying Shrinkage

This section deals first with factors affecting drying shrinkage of normal-strength concrete before discussing drying shrinkage of high-strength concrete and the influence of
chemical and mineral admixtures.

The *relative humidity* of the air surrounding the concrete greatly affects the magnitude of drying shrinkage, as shown in Figure 6.5. In the shrinkage test prescribed by
BS 1881-5: 1970 [11], the specimens are dried for a specified period under specified
conditions of temperature and humidity. The drying shrinkage occurring under those
accelerated conditions is of the same order as that after a long exposure to air at a relative humidity of 65%, the latter being representative of the average of indoor (45%)
and outdoor (85%) conditions of the United Kingdom. In the United States, ASTM
C157−08 [12] specifies a temperature of 23 °C and relative humidity of 50% for
the determination of drying shrinkage. Methods of determining drying shrinkage are
given in Chapter 16.

Type and fineness of Portland cement have little influence on drying shrinkage of
concrete [6], but Roper [13,14] found that cements with low quantities of sulfate may

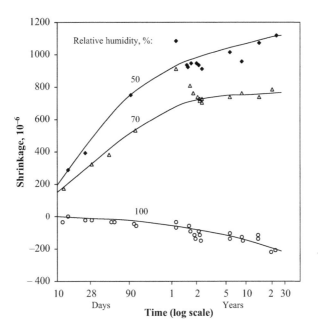

Figure 6.5 Drying shrinkage-time curves of concrete stored at different relative humidities,
after wet curing for 28 days [10].

exhibit more shrinkage when determined on mortar specimens. Drying shrinkage of concrete made with high-alumina cement is of the same magnitude as when Portland cement is used but occurs more rapidly, and finer-ground cement results in greater drying shrinkage of concrete than coarser-ground cement [6,14]. Expansive cements used to make shrinkage-compensating concrete are discussed later in this chapter.

Reversibility

To some extent drying shrinkage is reversible, i.e., the reabsorption of water will cause expansion of concrete but not to its original volume, as illustrated in Figure 6.4(a). In normal-strength concrete, reversible shrinkage is between 40% and 70% of the preceding drying shrinkage, depending on the age when first drying occurs. If fully hydrated at the time of exposure, drying shrinkage is more reversible, but if drying is accompanied by hydration and carbonation then additional bonding occurs, and porosity of the hardened cement paste is less and strength greater, thus preventing ingress of water on resaturation; this results in more irreversible drying shrinkage [1].

After commencement of drying from age t_o, the pattern of concrete subjected to cycles of wetting and drying from a later age t is shown in Figure 6.4(b). Conditions giving rise to reversible *cyclic shrinkage* in practice are daily climatic changes. The magnitude of the cyclic change depends on the duration of wetting and drying periods, the ambient humidity, and the composition of the concrete, but it is important to note that drying is slower than wetting. Consequently, shrinkage resulting from a prolonged period of dry weather can be reversed by a short period of rain. In general, compared with normal-weight aggregate concrete, drying shrinkage of lightweight aggregate concrete is more reversible.

Aggregate and Water/Cement Ratio

Clearly, the intrinsic drying shrinkage of the hardened cement paste is the source of the drying shrinkage of concrete, but it is reduced by the amount and quality of the fine and coarse aggregate. In consequence, the presence of the aggregate should not simply be regarded as a filler to increase bulk volume of concrete, because of its important role in resisting the drying shrinkage of cement paste and thus giving concrete dimensional stability. The effect is demonstrated in Figure 6.6, where, for a constant water/cement ratio, the greater the volume of aggregate the less the drying shrinkage of concrete. Also indicated in the same figure is the increase of drying shrinkage as the water/cement ratio increases when the volume content of aggregate remains unchanged.

Pickett [16] developed a theoretical relationship between drying shrinkage and cement paste content by considering a small particle of elastic aggregate at the center of a large sphere of shrinking and elastic material like concrete. For a constant water/cement ratio, he showed that:

$$\frac{dS_c}{S_c} = \frac{3(1 - \mu_c)}{1 + \mu_c + 2(1 - 2\mu_a)E_c/E_a} \times \frac{dg}{1 - g} \tag{6.1}$$

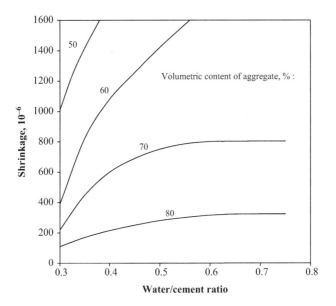

Figure 6.6 Influence of water/cement ratio and aggregate content on drying shrinkage of concrete [15].

where S_c = shrinkage of concrete, g = fractional volume of aggregate, E_c = modulus of elasticity of concrete, E_a = modulus of elasticity of aggregate, μ_c = Poisson's ratio of concrete, and μ_a = Poisson's ratio of aggregate.

By putting

$$\alpha = \frac{3(1 - \mu_c)}{1 + \mu_c + 2(1 - 2\mu_a)E_c/E_a} \tag{6.2}$$

and assuming α is constant for mixes of the same proportions, Eq. (6.1) becomes

$$\frac{dS_c}{S_c} = \alpha \frac{dg}{1 - g}$$

and then integration leads to:

$$S_c = S_p(1 - g)^\alpha \tag{6.3}$$

where S_p = shrinkage of neat cement paste for which $g = 0$.

For different types of sand and two water/cement ratios, Pickett [16] verified the relation between the shrinkage ratio (S_c/S_p) and $(1 - g)$ with $\alpha = 1.7$, as shown in Figure 6.7, which confirmed the importance of volumetric aggregate content (g) or, conversely, the importance of hardened cement paste volumetric content ($1 - g$), as

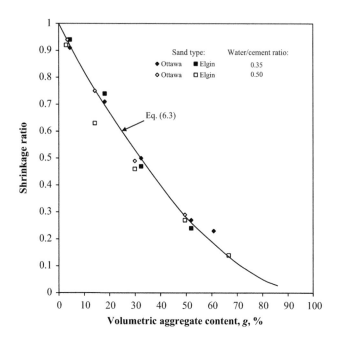

Figure 6.7 Influence of volumetric content of aggregate on relative shrinkage (the ratio of shrinkage of concrete to shrinkage of neat cement paste) [16].

a factor in drying shrinkage of concrete. In practice, for concrete of equal workability but having a wide range of mix proportions, the range of cement paste content is quite small, viz. 24–30%, or the corresponding range of volumetric aggregate content is 76–70%. However, although those changes appear small, they have the effect of increasing shrinkage of concrete by about 45% according to Eq. (6.3) with $\alpha = 1.7$. It should be noted that, besides volumes of hydrated cement and water, calculations involving the cement paste volumetric content should include entrapped air and air entrainment, while those involving aggregate volumetric content should include fine and coarse aggregate as well as unhydrated cement.

For a constant water/cement ratio and volume of aggregate, the *maximum size and grading of aggregate* do not influence drying shrinkage of concrete. However, the use of larger aggregate permits the use of a leaner mix, so that larger aggregate leads to lower drying shrinkage. A leaner mix needs less water for wetting the surface of aggregate and, hence, has a lower volume of cement paste [1].

The expression for α (Eq. (6.2)) implies that the ratio of elastic modulus of concrete to elastic modulus of aggregate (E_c/E_a) is also a factor in shrinkage of concrete. If it is assumed that $\mu_c = \mu_a = 0.16$, then

$$\alpha = \frac{2.52}{1.16 + 1.36 E_c/E_a} \tag{6.4}$$

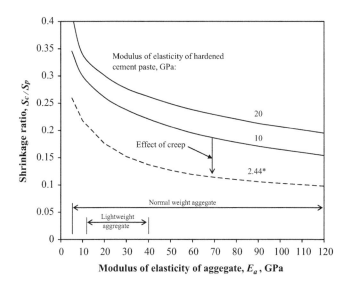

Figure 6.8 Influence of modulus of elasticity of aggregate on shrinkage of concrete estimated by Pickett's model [16]; * indicates an effective modulus of elasticity with an elastic modulus = 10 GPa and a cement paste creep of 315×10^{-6}/MPa.

The assumption of equality of Poisson's ratio can be justified by considering the effect of significant changes of μ_a. For example, a 30% variation causes the shrinkage ratio S_c/S_p (estimated by the procedure described below) to change by less than 10%.

Values of the concrete/aggregate modulus ratio in Eq. (6.4) may be obtained using the two-phase composite model presented in Chapter 3 (Eq. (3.5)). By assuming $g = 0.7$, the expression for modulus of elasticity of concrete becomes:

$$\frac{1}{E_c} = \frac{0.163}{E_m} + [0.195 + E_a]^{-1} \tag{6.5}$$

From the above expression, E_c/E_a can be obtained for different levels of cement paste modulus (E_m) and substituted in Eq. (6.4) to obtain α and then S_c/S_p from Eq. (6.3). The results are plotted in Figure 6.8. It can be seen that the effect of a higher modulus of elasticity of aggregate is to lower the shrinkage ratio or, in other words, to reduce the shrinkage of concrete by restraining the shrinkage of hardened cement paste. It follows that, in general, concrete made with lightweight aggregate exhibits more shrinkage than concrete made with normal-weight aggregate, although in practice there is a considerable variation in shrinkage of the latter due to the available range of modulus of elasticity of aggregate (Figure 6.8). The influence of type of nonshrinking normal-weight aggregate, as quantified by its modulus of elasticity, is illustrated by the well-known long-term experimental drying shrinkage results obtained by Troxell et al. [10], shown in Figure 6.9.

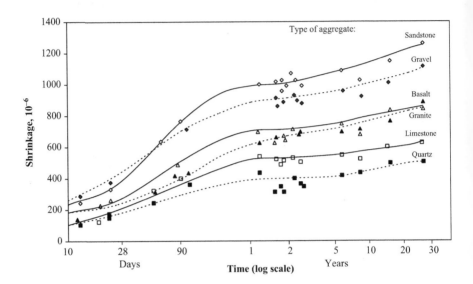

Figure 6.9 Shrinkage of concrete made with different types of aggregate, stored wet for 28 days and then exposed to air at 21 °C and 50% RH [10].

Returning to Figure 6.8, for a given aggregate modulus, the shrinkage ratio becomes less as the modulus of elasticity of cement paste decreases, a feature that suggests that creep is a factor in drying shrinkage of concrete. This can be demonstrated by using an effective modulus in the above calculations to allow for creep of hardened cement paste to estimate creep of concrete using Eq. (6.5). In the case of Pickett's model, it is thought that creep of concrete has the effect of relieving internal stresses induced by the shrinking concrete on aggregate particles, with the consequent reduction of shrinkage of concrete.

According to Neville [6], natural aggregate, such as granite, limestone, and quartzite, is not normally subjected to shrinkage, but there exist rocks that shrink on drying by up to 900×10^{-6}; they are some dolerites and basalts, and also some sedimentary rocks such as greywache and mudstone. Those aggregates usually have high water absorption, and this property can be treated as a warning sign that shrinkage may be a problem. The presence of clay in aggregate, such as breccia, lowers its restraining effect on shrinkage and, because clay itself is subjected to shrinkage, clay coatings on aggregate can increase shrinkage by up to 70% [6,14,17]. The uses of *recycled concrete aggregate* (up to 5% of masonry), and *recycled aggregate* (up to 100% of masonry) to make new concrete have to be evaluated carefully [1].

It may be noted that an important factor in drying shrinkage is water/cement ratio, an increase of which causes an increase in drying shrinkage, which, as will be seen later, is the opposite effect to that reported for autogenous shrinkage (see p. 169). The usual explanation given is that high water/cement ratio increases drying shrinkage because the volumetric cement paste content is greater, or there is less restraint to movement from a lower volumetric content of aggregate. However, Figure 6.6 indicates that drying shrinkage still increases with an increase in water/cement for a constant aggregate content. This explanation should also lead to an increase in autogenous shrinkage, which is not the case. Capillary tension theory is attributed to be the cause of both drying shrinkage and autogenous shrinkage. The theory postulates that removal of moisture from pores induces surface tension in pore water menisci and compression in the calcium silicate hydrates (C-S-H), the effect being greater for smaller pores in cement paste having a low water/cement ratio. The apparent discrepancy is possibly explained by the different modes of moisture transport. In the case of drying shrinkage, loss of moisture takes place at the outer surface layers of the concrete, whereas, in the case of autogenous shrinkage, water is consumed uniformly in the sealed or mass concrete during hydration of cement by internal self-desiccation. When the water/cement ratio is high, concrete is more porous, the pores being larger with more connectivity—in other words concrete with a high water/cement ratio is more permeable than in the case of low water/cement ratio [1,6]. This means that moisture can move more readily to the surface and escape to the environment in concrete made with a high water/cement ratio than in the case of concrete made with a low water/cement ratio. Thus, the permeability factor in the process of drying shrinkage is important since larger pores will empty, menisci form, and capillary stress be induced at a faster rate for concrete made with a higher water/cement ratio than for concrete having smaller pores produced by a lower water/cement ratio.

Size and Shape of Member

The process of drying shrinkage arises from *diffusion* of moisture to the outer surface of concrete exposed to a drying environment. Diffusion can be defined as the flow of fluid through partial dry concrete caused by a chemical potential or moisture potential, which includes combined effects of gradients of (1) solute concentration giving osmotic pressure, (2) temperature and (3) moisture content, the latter giving rise to surface forces and vapor/water interface forces in capillaries [18]. Diffusion differs from *permeability*, in which fluid is forced through concrete by absolute pressure, such as water through the concrete of a dam due to pressure of the water on the upstream side. Carlson [19] first introduced *linear diffusion theory* to predict drying shrinkage of concrete, in which it is assumed that shrinkage is proportional to the moisture loss. However, proportionality has not been found over long periods of time and drying in concrete is really a nonlinear diffusion process [20]. In fact, as drying progresses the remaining moisture is lost with ever increasing difficulty, so that diffusivity decreases with decreasing water content.

The actual drying shrinkage of a concrete member is affected by its size and shape, and application of *nonlinear diffusion theory* permits mathematical solutions for moisture movement and consequent drying shrinkage for different geometries [21]. Before the advent of those solutions, however, and after experiencing difficulty with linear diffusion theory for irregular-shaped members, Ross [22] introduced the surface/volume ratio as an indicator of size and shape with the concept that members having the same surface/volume ratio undergo the same drying shrinkage. Nowadays, the

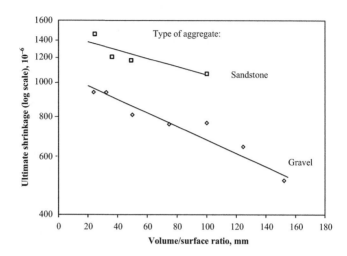

Figure 6.10 Relation between ultimate shrinkage and volume/surface ratio [23].

inverse of that ratio, i.e., volume/surface ratio (*V/S*), is more commonly used, together with other terms, which are defined later. Figure 6.10 shows that there is a linear relationship between the logarithm of ultimate drying shrinkage and volume/surface ratio.

Keeton [24] found higher and faster drying shrinkage for cylinders than for square prisms, with shrinkage of I- and T-sections in between. Similar findings were reported by Kesler et al. [25]. After comparing cylindrical specimens and I-shaped specimens, Hanson and Mattock [23] found that both the rate and ultimate drying shrinkage were affected by size of specimen. Their results, shown in Figure 6.11, indicated that I-shaped specimens generally exhibited 14% less drying shrinkage than cylindrical specimens having the same volume/surface ratio. The shape of the specimen affects the moisture distribution within it. For instance, in a square prism the variation in relative humidity along a diagonal is different than along a normal to the surface, as demonstrated in Figure 6.12. Hence, for the same volume/surface ratio, the shape causes drying shrinkage to be slower for square prisms than for cylinders. However, for design purposes, Neville [6] suggests that the effect of shape is secondary and can be neglected.

Hobbs [27] carried out tests on small slabs and prisms, the latter being partly sealed to simulate drying of slabs. It was reported that rate of drying shrinkage depends on specimen size but becomes approximately the same after 2−3 years of exposure. Hobbs also suggested that theoretical ultimate drying shrinkage is independent of size but, in practice, this may not occur due to the contribution of carbonation shrinkage to drying shrinkage, which is greater in smaller specimens. The technique of part-sealing of laboratory-size specimens to simulate drying of thin mortar joints in masonry is used to determine their drying shrinkage and creep properties in Chapters 7, 8, and 12.

Bryant and Vadhanavikkit [28] quantified the effect of both size and shape on drying shrinkage in terms of *equivalent thickness*. The difference between size and shape

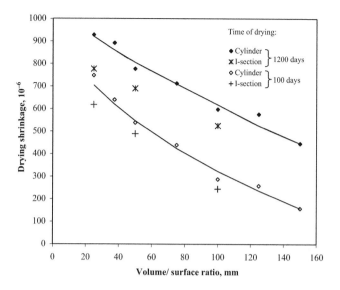

Figure 6.11 Influence of volume/surface ratio and section shape of member on drying shrinkage of concrete [23].

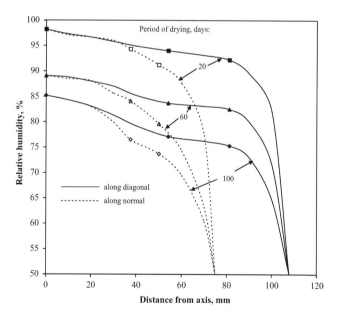

Figure 6.12 Distribution of relative humidity along diagonal and normal to the surface in a prismatic concrete specimen stored at a relative humidity of 50% [26].

can be explained in terms of variation in mean distance that the water has to travel to the surface, and is known as the *drying path length*. The equivalent thickness (T_d), together with other terms *effective thickness* (T_e) or *theoretical thickness* (h_o) and *average thickness* (T_a), are defined as follows:

$$\frac{V}{S} = \frac{\text{Total volume of member}}{\text{Total drying surface area of member}} \tag{6.6}$$

$$T_e = h_o = \frac{\text{Area of cross section}}{\text{Exposed semi-perimeter}} 2\frac{A_c}{u} = 2\frac{V}{S} \tag{6.7}$$

$$T_a = 2T_e \tag{6.8}$$

$$T_d = 4\bar{d}_p \tag{6.9}$$

$$\bar{d}_p = \frac{\sum V_d d_p}{V} \tag{6.10}$$

where A_c = cross-sectional area, u = perimeter exposed to drying, d_p = drying path length = distance between the centroid of the section and drying surface of element of volume V_d, and $\bar{d}_p$ = average drying path length of all elements in the total volume, V.

It is of interest to note that the factor 2 of Eq. (6.7) is included to make T_e equal to the thickness of an infinite extended slab. This is shown in Figure 6.13, along with other examples of cross sections of concrete members and values of the various terms used to quantify the influence of size and shape on drying shrinkage.

The lower drying shrinkage of large concrete members may be regarded as being due to restraint of the outer part of drying concrete by the nonshrinking core. In practice, therefore, it is a case of *differential* or *restrained shrinkage*, the latter being classified in Chapter 14 as internal restraint, which may lead to shrinkage cracking. In practice, it is not possible to measure unrestrained shrinkage or "true" shrinkage as an intrinsic property of concrete, so that the specimen size used to determine shrinkage should always be quoted. In theory, true drying shrinkage only occurs when there is little or no moisture gradient, such as in a thin concrete specimen, which, of course, has to have a minimum thickness because of the presence of coarse aggregate. It is possible to cut thin concrete specimens and make thin cement paste or mortar specimens, but to prepare them without introducing microcracks is difficult. It is important, therefore, to realize that when referring to drying shrinkage of concrete elements what is measured is restrained shrinkage, which is a combination of true shrinkage and strain due to internal restraint. This topic is explained more fully in Chapter 14.

Period of Moist Curing

Hobbs [27] found no effect of length of moist curing on drying shrinkage of concrete having a water/cement ratio of 0.47. However, according to ACI

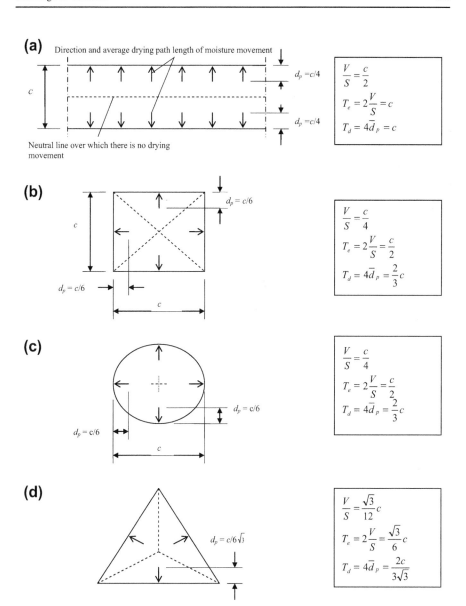

Figure 6.13 Examples of drying path length (d_P), volume/surface ratio (V/S), effective thickness (T_e), and equivalent thickness (T_d) of different concrete member cross-section shapes drying from the outer surfaces. (a) Slab of thickness c. (b) Square prism of side c. (c) Cylinder of diameter c. (d) Equilateral triangular prism of side c. (e) Rectangular prism b. (f) I-section beam.

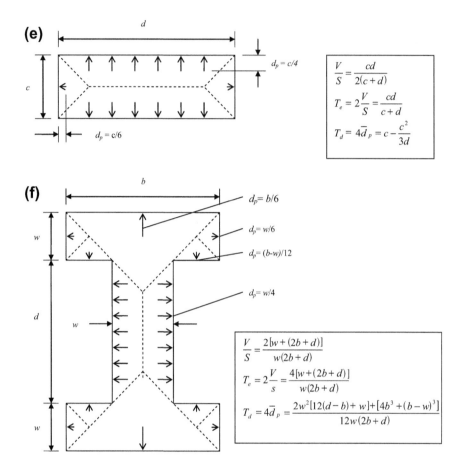

Figure 6.13 *(continued).*

209.1R-05 [14], extended periods of moist curing will usually reduce the amount of drying shrinkage by 10−20%, but the effect depends on the water/cement ratio. The effect of curing period on concrete made with different water/cement ratios was investigated by Perenchio [29]. The results, shown in Figure 6.14, indicate that 1-year drying shrinkage is higher for curing periods between 3 and 7 days, but is less for extended periods of moist curing; a similar trend has been observed for concrete containing chemical and mineral admixtures (see Figure 11.15). It is apparent that the pattern of behaviour of drying shrinkage with water/cement ratio is complex, being affected by interaction of several factors: degree of hydration, pore size and distribution (which affect strength and permeability and thus diffusion of moisture through concrete), and a possible contribution from autogenous shrinkage, especially at low water/cement ratios. Neville [6] states that, although

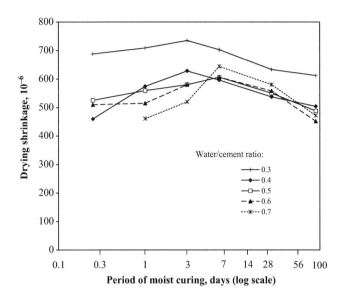

Figure 6.14 Influence of period of moist curing on 1-year drying shrinkage of concrete having different water/cement ratios [29].

complex, the effect of moist curing period on the magnitude of drying shrinkage of normal-strength concrete can be regarded as small.

For high-strength concrete, autogenous shrinkage, drying shrinkage, and total (autogenous-plus-drying) shrinkage have been compared after moist curing periods of 1 day and 28 days; the types of concrete investigated are detailed in Table 6.1. In general, although there were appreciable variations, it was found that increasing the length of moist curing reduced average total shrinkage by 22% and average autogenous shrinkage by 41%, while the average drying shrinkage was unchanged.

Development with Time

The loss of moisture from concrete in a drying environment is a gradual process, so an important factor in drying shrinkage is *duration of exposure*. Figures 6.2 and 6.9 indicate that drying shrinkage takes place over long periods and, as already mentioned, a part of long-term drying shrinkage may be due to carbonation. In any case, the rate of drying shrinkage decreases rapidly with time so that generally:

- 20–40% of 30-year drying shrinkage occurs in the first 2 weeks of exposure to drying.
- 50–85% of 30-year drying shrinkage occurs in 3 months.
- 80–90% of 30-year drying shrinkage occurs in 1 year.

Table 6.1 Effect of Mineral Admixtures on 200-Day Drying Shrinkage of High-Strength Concrete Exposed to Drying at Different Ages [30]

Concrete Type: Admixture-Replacement, %	28-Day Cube Strength, MPa	Exposed to Drying from 1 Day				Exposed to Drying from 28 Days			
		Total Shrinkage, 10^{-6}	Autogenous Shrinkage, %	Drying Shrinkage %	Drying Shrinkage Relative	Total Shrinkage, 10^{-6}	Autogenous Shrinkage, %	Drying Shrinkage %	Drying Shrinkage Relative
OPC control	86.7	558	25.4	74.6	1	415	23.6	76.4	1
MS-5	105.7	458	34.1	65.9	0.73	325	22.5	77.5	0.79
MS-10	113.9	486	51.2	48.8	0.57	327	40.1	59.9	0.62
MS-15	117.5	554	50.0	50.0	0.57	338	45.9	54.1	0.58
MK-5	91.5	499	54.3	45.7	0.55	377	38.7	61.3	0.73
MK-10	103.7	455	56.3	43.7	0.48	340	31.5	68.5	0.74
MK-15	103.4	410	53.9	46.1	0.45	296	38.9	61.1	0.57
FA-10	86.7	478	32.6	67.4	0.77	430	32.3	67.7	0.92
FA-20	84.3	513	32.7	67.3	0.83	420	26.2	73.8	0.98
FA-30	82.1	482	33.6	66.4	0.77	427	17.1	82.9	1.12
GGBS-20	95.3	520	41.0	59.0	0.74	405	33.6	66.4	0.93
GGBS-40	87.6	398	56.8	43.2	0.41	402	27.1	72.9	0.71
GGBS-60	86.7	469	57.8	42.2	0.48	410	15.9	84.1	0.84

OPC control = ordinary Portland cement concrete plus superplasticizer; all other concretes had the same mix proportions plus the mineral admixture.
Types of mineral admixture: MS = microsilica; MK = metakaolin; FA = fly ash; GGBS = ground granulated blast-furnace slag.
Total shrinkage = autogenous shrinkage + drying shrinkage.
Relative drying shrinkage = ratio of drying shrinkage of admixture concrete to drying shrinkage of OPC concrete.

The above ranges of shrinkage reflect the influence of factors in drying shrinkage of concrete described earlier. In Chapter 11, methods of prediction generally quantify the development of drying shrinkage, $S_c(t, t_o)$, with time by some form of hyperbolic expression, such as:

$$S_c(t, t_o) = S_{c\infty} \left[\frac{(t - t_o)}{a_s S_{c\infty} + (t - t)} \right]^{0.5}$$
(6.11)

where t = age of concrete, t_o = age at exposure to drying after moist curing, $S_{c\infty}$ = ultimate drying shrinkage, and a_s = a coefficient related to the initial rate of shrinkage.

Depending on the method of prediction, some of the influencing factors discussed earlier are taken into account by the term $S_{c\infty}$, such as type of concrete, size and shape of the concrete member (expressed in terms of V/S), and ambient relative humidity of storage. The coefficient a_s is also a function of size and shape of concrete member.

Temperature

According to Mindess and Young [7], elevated temperature during moist curing reduces the irreversible drying shrinkage of cement paste but reversible shrinkage is unaffected, the decrease depending on the maximum temperature. At 65 °C, irreversible shrinkage can be reduced by about two-thirds and total shrinkage by about one-third. The effect depends on the length of time the paste is maintained at higher temperature. However, the time of exposure at high temperature that is required to reduce shrinkage can be relatively short and may be less than the total specified curing time.

ACI 209.1R-05 [14] refers to Klieger [31], who found that heat and steam curing can significantly reduce drying shrinkage of normal-strength concrete by as much as 30%. Hanson [32] investigated the effect of *low-pressure steam curing* and *autoclave curing* on drying shrinkage of various types of concrete made with different aggregates and cements. It was found that, compared with moist-cured concrete of the same compressive strength, steam curing reduced drying shrinkage by 16−30% for type I Portland cement concrete and by 26−39% for type III Portland cement concrete. Corresponding reductions of drying shrinkage of autoclave-cured concrete were much greater, viz. 73% and 79%, respectively.

When concrete is subjected to elevated temperature while drying, drying shrinkage is accelerated, and this effect is taken into account by the prediction method of CEB MC90 [33]. However, the effect of elevated temperature during the period of curing on subsequent drying shrinkage is not taken into account. Likewise, the duration of moist curing less than 14 days is assumed not to significantly affect drying shrinkage. To allow for the effect of elevated temperature during drying in air, CEB MC90-99 [34] proposes the following relationship for drying shrinkage of concrete subjected to a constant temperature above 30 °C:

$$S_{cT}(t, t_c) = K \left(\beta'_{RH,T} \right) \beta_{sT}(t - t_c)$$
(6.12)

where K = a constant depending on strength, type of cement, and ambient humidity at normal temperature (20 °C),

$$\beta'_{RH,T} = \left[1 + \left(\frac{0.08}{1.03 - h}\right)\left(\frac{T - 20}{40}\right)\right]$$

and

$$\beta_{sT}(t - t_c) = \left[\frac{(t - t_c)}{0.14\left(\frac{V}{S}\right)^2 \exp[-0.06(T - 20)] + (t - t_c)}\right]^{0.5}$$

In the above expressions, S_{cT} = drying shrinkage at temperature T (°C), t = age of concrete, t_c = age at end of moist curing, $(t - t_c)$ = duration of drying (days), h = ambient relative humidity = RH (%)/100, and V/S = volume/surface ratio (mm).

Considering a concrete member having a V/S = 50 mm and stored at 65% RH so that h = 0.65, then Eq. (6.12) becomes

$$S_{cT}(t, t_c) = K\left[1 + (5.264 \times 10^{-3})(T - 20)\right]$$
$$\times \left[\frac{(t - t_c)}{350 \exp[-0.06(T - 20)] + (t - t_c)}\right]^{0.5} \tag{6.13}$$

When T = 20 °C, Eq. (6.13) reduces to

$$S_{c20}(t, t_c) = K\left[\frac{(t - t_c)}{350 + (t - t_c)}\right]^{0.5} \tag{6.14}$$

The effect of temperature is conveniently assessed by the relative drying shrinkage, $R_{sT}(t, t_c)$, viz. the ratio of drying shrinkage at temperature T to drying shrinkage at 20 °C, i.e., from the above expressions:

$$R_{ST}(t, t_c) = \frac{S_{cT}(t, t)}{S_{c20}((t, t_c))} \tag{6.15}$$

Figure 6.15 shows relative drying shrinkage as a function of temperature for different drying times. It is apparent that drying shrinkage is very sensitive to elevated temperature and, at early stages of drying, can be much greater than drying shrinkage at normal temperature. However, sensitivity reduces with age and, for this particular example, long-term 50-year drying shrinkage at 80 °C is around 30% greater than drying shrinkage at 20 °C.

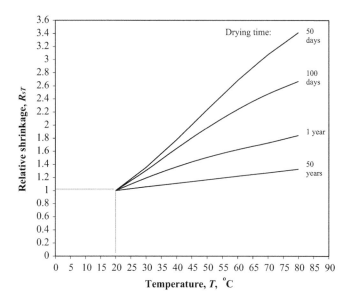

Figure 6.15 Influence of elevated temperature during drying on drying shrinkage of concrete after moist-curing at 20 °C according to Eq. (6.15); relative shrinkage is ratio of drying shrinkage at elevated temperature to drying shrinkage at 20 °C.

Admixtures

In many instances, the incorporation of an admixture is accompanied by a change in mix proportions in order to achieve a desirable property, e.g., use of a plasticizer to achieve a reduction in water/cement ratio to enhance strength. In such cases, if there is a change in drying shrinkage it is not known whether the change is due to the admixture as well as to the reduction in water/cement ratio. In order to isolate any effect of admixtures on drying shrinkage, the *relative deformation method* has been developed to adjust the drying shrinkage for any change in mix proportions of the admixture concrete compared with drying shrinkage of the control plain concrete [35]. The relative deformation method is applicable for concretes having the same constituents (apart from the admixture) and the same operating conditions (curing, age at exposure, storage environment, and time of drying). Specifically, the approach is used to adjust reported drying shrinkage data when there are accompanying changes in water/cementitious materials ratio (or water/binder ratio) and the volumetric content of cementitious materials of the admixture concrete. The method is also applicable for adjusting creep to allow for changes in mix proportions.

Relative Deformation Method

As has already been stated, for a constant water/cement ratio, Pickett [16] showed that drying shrinkage is related to cement paste content, as given by Eq. (6.3). If now drying shrinkage of concrete having a water/cement ratio of w/c and cement paste content of $(1 - g)$ is expressed relative to drying shrinkage of concrete having "standard" mix proportions of water/cement ratio $= 0.5$ and cement paste content $= 0.3$, Eq. (6.3) becomes:

$$R_{cs} = R_{ps}\left(\frac{1-g}{0.3}\right)^n = R_{ps}R_g \tag{6.16}$$

where $R_{cs} =$ relative drying shrinkage of concrete and $R_{ps} =$ drying shrinkage of cement paste of w/c relative to drying shrinkage of cement paste with $w/c = 0.5$, and $R_g =$ relative volume fraction of cement paste.

Analysis of many sets of data reported for shrinkage and creep of plain (admixture-free) concrete [35] revealed average values of n of 1.8 for normal-weight aggregates and 1.0 for lightweight aggregates. Hence, from Eq. (6.16), for normal-weight aggregate concrete:

$$R_{gs} = 8.73(1 - g)^{1.8} \tag{6.17}$$

and, for lightweight aggregate concrete:

$$R_{gs} = 3.33(1 - g) \tag{6.18}$$

Now analysis also showed that the average relative shrinkage of cement paste could be expressed as an exponential function of water/cement ratio, w/c, as follows:

$$R_{ps} = 1.1 \exp\left[-13(0.8 - w/c)^4\right] \tag{6.19}$$

The above expressions were obtained by analyzing suitable reported shrinkage data having large ranges of water/cement ratio and cement paste content [35]. For each data source, it was possible to select drying shrinkage for mixes with approximate average $w/c = 0.5$ and $(1 - g) = 0.3$, and then to calculate relative values of drying shrinkage for other values of w/c and $(1 - g)$ to obtain the shrinkage coefficients of Eqs (6.17)–(6.19); general trends are shown graphically in Figure 6.16. Interestingly, coefficient R_{ps} indicates a nondependence of relative drying shrinkage on water/cement ratios greater than 0.8, implying that, at that level, water takes the form of "free" water in the hardened cement paste. Unlike physically and chemically bound water, removal of "free" water does not contribute to drying shrinkage of concrete.

For relative drying shrinkage of normal-weight aggregate concrete, combining Eq. (6.17) with Eq. (6. 19) leads to:

$$R_{cs} = 9.61(1 - g)^{1.8} \exp\left[-13(0.8 - w/c)^4\right] \tag{6.20}$$

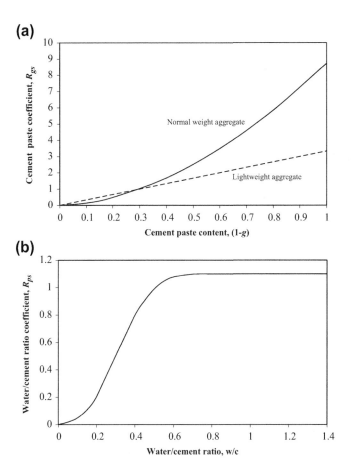

Figure 6.16 Drying shrinkage coefficients for Eqs (6.17)−(6.19) [35]. (a) Cement paste coefficient. (b) Water/cement ratio coefficient.

Correspondingly, the relative drying shrinkage of lightweight aggregate concrete is:

$$R_{cs} = 3.67(1 - g)\exp\left[- 13(0.8 - w/c)^4\right] \tag{6.21}$$

It should be emphasized that R_{gs} and R_{ps} are average functions (Figure 6.16), and there are large variations, particularly for R_{ps} at high water/cement ratios. Nevertheless, their use to estimate drying shrinkage of plain concrete having a wide range of mix composition yields a reasonable accuracy, with an average error coefficient (see Eq. (5.4)) of 14% [34]. Provided drying shrinkage of a particular type of concrete is known in terms of its cement paste volumetric content and water/cement ratio, the relative deformation method allows drying shrinkage to be estimated for another type of concrete having different mix proportions. The cement paste fractional volume content is calculated as follows. When the mix proportions by mass of concrete, $1:a:w$, are given the volume fraction of cement paste is:

$$(1-g) = \frac{A}{100} + \frac{d \times 10^{-3}}{m}\left(\frac{1}{3.15}+w\right) \qquad (6.22)$$

where A = air content (%), d = density of concrete (kg/m3), m = mass of concrete = $(1+a+w)$ (kg), a = total aggregate/cement ratio, w = free water/cement ratio, and 3.15 = specific gravity of cement. It should be noted that the free water/cement ratio is that after deducting any water absorbed by aggregate. When the mix contains mineral admixtures, appropriate values of density, *water/cementitious materials ratio*, and specific gravity are used in Eq. (6.22) (see also p. 302).

The following example illustrates the use of the relative deformation method.

Example

Suppose the ultimate drying shrinkage is 500×10^{-6} for concrete with normal-weight concrete having a total aggregate/cement ratio = 5.5 and a water/cement ratio = 0.55. It is required to estimate the ultimate drying shrinkage for concrete having a total aggregate cement ratio = 4.0 and water/cement ratio = 0.45. Assume the air content is zero and the density of concrete for both mixes is 2400 kg/m^3. The solution is detailed below.

Let parameters for the two concrete mixes be denoted by suffixes 1 and 2. For 1 kg of cement, their respective total masses are:

$$m_1 = (1+5.5+0.55) = 7.05\,kg; \quad m_2 = (1+4+0.4) = 5.4\,kg$$

The respective cement paste contents are obtained by Eq. (6.22), viz:

$$(1-g_1) = \frac{2400 \times 10^{-3}}{7.05}\left(\frac{1}{3.15}+0.55\right) = 0.295;$$

$$(1-g_2) = \frac{2400 \times 10^{-3}}{5.4}\left(\frac{1}{3.15}+0.4\right) = 0.319$$

Hence, from Eq. (6.20):

$$R_{cs1} = 9.61(0.295)^{1.8}e^{[-13(0.8-0.55)^4]} = 1.015;$$

$$R_{cs2} = 9.61(0.319)^{1.8}e^{[-13(0.8-0.4)^4]} = 0.881$$

Example—cont'd

Recalling that R_{sc} = ratio of drying shrinkage of concrete with any $(1 - g)$ and w/c to drying shrinkage of a reference concrete with $(1 - g) = 0.3$ and $w/c = 0.5$, and, given the drying shrinkage of mix $1 = 500 \times 10^{-6}$, the drying shrinkage of the reference mix is $(500 \times 10^{-6}) \div R_{sc1} = (500 \times 10^{-6}) \div 1.015 = 492.6 \times 10^{-6}$. Hence, the drying shrinkage of mix 2 is $R_{sc2} \times 492.6 \times 10^{-6} = 0.881 \times 492.6 \times 10^{-6} = 434 \times 10^{-6}$.

The relative deformation method has been used to analyze the influence of *chemical admixtures* on drying shrinkage of concrete in cases where the mix composition of the admixture concrete differed from that of the control plain concrete [35]. The admixtures investigated were: *plasticizers (water reducers)* and *superplasticizers (high-range water reducers)*.The procedure was to adjust reported drying shrinkage of the plain (control) concrete in proportion to R_{cs}, calculated for the admixture concrete mix composition, so that any difference between the observed shrinkage of the admixture concrete and the adjusted drying shrinkage of the control concrete could be attributed to the admixture, per se. Of course, in the case where admixtures are used to make *high-workability* or *flowing concrete* without any change in mix composition from that of the control concrete, no adjustment to the drying shrinkage of the control concrete is necessary [35].

The outcome of the analysis and review of the effect of chemical admixtures of drying shrinkage of concrete is shown in Figure 6.17. The review consisted of 14 sets of reported data involving plasticizers and 49 sets of reported data involving superplasticizers. The specific types of plasticizer were lignosulfonate [35–37] and carboxylic acid [35–38], and the types of superplasticizer were sulfonated naphthalene formaldehyde condensate [39–46], sulfonated melamine formaldehyde condensate [43,44,47], and copolymer [38,43–45]. It is apparent from Figure 6.17 that no individual type of admixture behaves differently and there is an overall trend that indicates general increase of drying shrinkage by approximately 20% when plasticizers and superplasticizers are used to make flowing concrete, i.e., the increase in drying shrinkage is due solely to the presence of the admixtures. Clearly, the scatter of points and standard deviation indicate that there is a large variation of the effect, but it is thought that there is a general likelihood of an increase in drying shrinkage that is probably associated with the ability of the admixture to entrain air; such air may be considered as aggregate with zero elastic modulus, which lowers the resistance to drying shrinkage of the cement paste. Consequently, it is quite simple to allow for the presence of chemical admixtures on drying shrinkage of concrete. For instance, in the example shown on p. 159, if mix 2 concrete contains a plasticizer, it may be assumed that there is an increase in drying shrinkage of 20%, Hence, the estimated ultimate drying shrinkage of the admixture concrete is 521×10^{-6}.

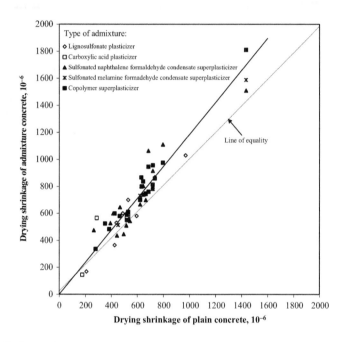

Figure 6.17 Comparison of drying shrinkage with and without plasticizing and superplasticizing admixtures; mean percentage increase and standard deviation of drying shrinkage of admixture concrete = 119 ± 20 [35].

Regarding the subject of *air entrainment* used to protect concrete from damage due to alternating freezing and thawing, *air-entraining agents* in the form of additives or admixtures are used to introduce discrete bubbles of air [1]. Although, as implied in the previous paragraph, entrained air would be expected to increase drying shrinkage compared with non-air-entrained concrete, in practice air entrainment generally improves workability and therefore permits the use of a lower water/cement ratio or a leaner mix. Those factors lead to a lower drying shrinkage and, thus, the net effect is probably not significant. Indeed, tests have shown that air entrainment has little effect on drying shrinkage [6], provided the total *air content* of concrete is less than 8% [14].

Ai and Young [48] reported a 30% reduction in drying shrinkage of cement paste using a *shrinkage-reducing admixture*. Also, short-term tests on concrete containing a shrinkage-reducing admixture without adjusting mix proportions revealed significant reductions in drying shrinkage compared with the control admixture-free concrete [49,50]. It is believed that a reduction in drying shrinkage of concrete is achieved by the shrinkage-reducing admixture lowering the surface tension of pore water, thus reducing capillary pore stress. However, when used with plain concrete without changing mix proportions, concrete containing the shrinkage-reducing admixture appears to suffer a small loss of strength.

The effectiveness of shrinkage-reducing admixtures is not impaired by the combined use of mineral and chemical admixtures in the same concrete mix. Al-Manaseer and Ristanovic [51] reported test data for different dosage levels of shrinkage-reducing admixture added to mixes made with type II Portland cement and 5% microsilica or metakaolin, the mixes having various amounts of fly ash with either a superplasticizer or superplasticizer/plasticizer. The cementitious material content, water/cementitious material ratio, and coarse/fine aggregate ratio were held constant at 350 kg/m³, 0.23, and 1.12, respectively. The results are plotted in Figure 6.18 in terms of ratio of drying shrinkage of the concrete containing the admixture to the drying shrinkage of the admixture-free concrete after 100 days of exposure in a drying environment at 50% RH. Although there is appreciable scatter, the general trend appears to suggest that drying shrinkage reduces with increasing dosage of shrinkage-reducing admixture. Also shown is the expression proposed by Al-Manaseer and Ristanovic [51] for a factor to modify prediction of drying shrinkage of concrete to allow for dosage of the shrinkage-reducing admixture (SRA), viz.

$$\beta(SRA) = \frac{2}{(2 + SRA^{0.7})} \tag{6.23}$$

where $\beta(SRA)$ = ratio of drying shrinkage of the concrete containing the shrinkage-reducing admixture to drying shrinkage of the admixture-free concrete, and SRA = dosage of the shrinkage-reducing admixture (%).

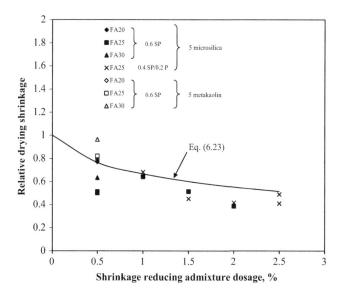

Figure 6.18 Effect of shrinkage-reducing admixture on 100-day relative drying shrinkage of fly ash (FA) concrete containing chemical admixture superplasticizer (SP) and plasticizer (P), and mineral admixtures microsilica and metakaolin [51]; numbers preceding admixture are % content by mass of cementitious material.

As mentioned earlier, the measurement of drying shrinkage of normal-strength concrete exposed to the surrounding environment traditionally includes *autogenous shrinkage* as well as carbonation shrinkage, both of which are regarded as being small compared to drying shrinkage. When exposed to drying at the usual ages, say, from 7 days, most autogenous shrinkage of normal strength has already occurred and therefore does not contribute appreciably to longer-term drying shrinkage. However, in the case of high-strength concrete or high-performance concrete, this is not necessarily true since total measured shrinkage consists of both drying shrinkage and a much more significant proportion of autogenous shrinkage, especially at very early ages of exposure to drying and depending on the type of mineral admixture used with cement to form the total cementitious material. Autogenous shrinkage is considered as a subject in its own right in a later section of this chapter, but at this stage it is appropriate to compare it with drying shrinkage of high-strength concrete determined from tests carried out by Megat Johari [30].

Table 6.1 compares the contributions of drying and autogenous shrinkage of various types of high-strength concrete for two ages of exposure to drying: 1 and 28 days. When exposed at 1 day, the total shrinkage of ordinary Portland cement plain concrete consists of 75% drying shrinkage, whereas for concrete containing mineral admixtures, drying shrinkage is less and may be as low as 42−44% of total shrinkage in the cases of high replacement of cement by metakaolin and ground granulated blast-furnace slag. Except for ordinary Portland cement concrete, the proportion of drying shrinkage is greater for admixture concrete exposed to drying at 28 days but, even so, there is still a significant contribution of autogenous shrinkage to total measured shrinkage ranging from 16% to 46%.

Also listed in Table 6.1 is relative drying shrinkage, i.e., the drying shrinkage ratio of admixture concrete to ordinary Portland cement concrete. In all admixture concretes, the ratio is greater for concrete exposed to drying at the later exposure age of 28 days, thus confirming the greater contribution of drying shrinkage to total measured shrinkage. The shrinkage ratio, together with long-term relative shrinkage obtained from an earlier survey of reported findings for concrete made with different types of mineral admixture [35,52], is also plotted in Figure 6.19 as a function of cement replacement; the survey covered a variety of types and sources of admixture, and shrinkage was determined under a wide range of operating conditions. Where necessary, the shrinkage of the control admixture-free concrete was adjusted according to the relative deformation method presented on p. 160 to allow for any change in mix proportions of the mineral admixture concrete, such as water/cementitious materials ratio. Also, if later-age shrinkage-time data existed, ultimate shrinkage was extrapolated using a hyperbolic expression similar to Eq. (6.11) with the power index equal to 1. It can be seen that Figure 6.19 features large amounts of scatter, which are attributed to variations in fineness, chemical composition, and source of admixture, and probably different times of exposure to drying. Also, in the case of high-strength concrete made with microsilica (Figure 16.19(c)), except for the results of Megat Johari [30], it was not possible to account for any autogenous shrinkage in the analysis and calculated relative shrinkage is based on the total measured shrinkage [35].

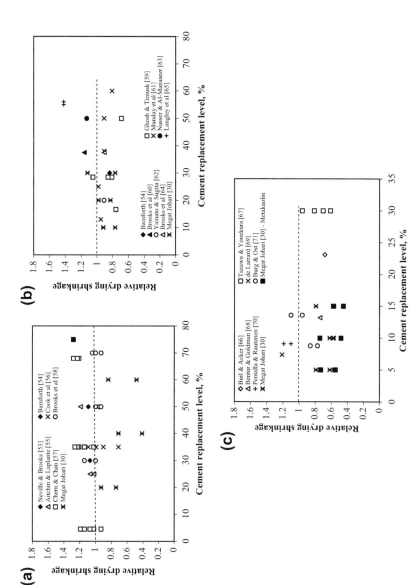

Figure 6.19 Influence of mineral admixtures on drying shrinkage of concrete. (a) Blast-furnace slag. (b) Fly ash. (c) Microsilica and metakaolin.

In the cases of slag and fly ash, the analysis [35] suggested there was no consistent overall influence on drying shrinkage of normal-strength concrete having equal mix proportions except for the mineral admixture, although individual results may have indicated an increase or decrease; in the survey, 28-day cylinder strengths ranged from 20 to 65 MPa. In the case of high strength concrete [30] there is a clear reduction of drying shrinkage for slag (Figure 6.19(a)) but a negligible influence for fly ash (Figure 6.19(b)). For microsilica, the previous analysis covered high-strength concrete ranging from 65 to 118 MPa and the overall trend suggested a decrease of drying shrinkage for higher levels of replacement. Including the results of Megat Johari [30] confirms that there is a reduction of shrinkage, including that for high-strength concrete made with metakaolin (Figure 6.19(c)). As mentioned earlier, the previous analysis of reported data could have included some autogenous shrinkage, which may have affected the admixture and control concretes, and thus relative shrinkage, in a different manner.

Shrinkage of special concretes incorporating mineral and chemical admixtures, *ultra-high-strength concrete* (150−200 MPa) and high-workability *self-consolidating concrete*, is discussed, together with creep behaviour, in Chapter 10.

Shrinkage-Compensating Concrete

To avoid problems associated with drying shrinkage of concrete, it would be advantageous to use cement that does not change its volume due to drying shrinkage or, sometimes, that even expands on hardening. Concrete containing such cement expands in the first few days of its life, and a form of prestress is induced by restraining the expansion with embedded steel reinforcement; steel is in tension and concrete in compression. Restraint by external means is also possible. It is to be noted that expanding cement does not prevent the development of drying shrinkage and cannot produce "shrinkless" concrete, as drying shrinkage occurs after moist curing has ceased, but the magnitude of expansion can be adjusted so that the expansion and subsequent shrinkage are equal and opposite [1]. Expansive cements are used generally to minimize cracking caused by drying shrinkage in concrete slabs and pavements structures and in special circumstances, such as prevention of water leakages.

Expansive cements consist of a mixture of Portland cement, expanding agent, and stabilizer. The expanding agent is obtained by burning a mixture of gypsum, bauxite, and chalk, which form calcium sulfate and calcium aluminate (mainly C_3A). In the presence of water, these compounds react to form calcium sulfoaluminate hydrate (ettringite), with an accompanying expansion of the cement paste. The stabilizer, which is blastfurnace slag, slowly takes up the excess calcium sulfate and brings expansion to an end. Whereas the formation of ettringite in mature concrete is harmful due to its association with sulfate attack and efflorescence [1], a controlled formation of ettringite in the early days after placing of concrete is used to obtain the shrinkage-compensating effect or to obtain an initial prestress arising from restraint by steel reinforcement.

Three main types of expansive cement can be produced—K, M, and S—but only type K is commercially available in the United States. ASTM C 845–04 [72] classifies expansive cements, collectively referred to as type E-1 according to the expansive agent used with Portland cement and calcium sulfate. In each case, the agent is a source of reactive aluminate, which combines with the sulfates of Portland cement to form expansive ettringite. Special expansive cements containing high-alumina cement can be used for situations requiring extremely high expansion [6].

Shrinkage-compensating concrete is the subject of ACI 223-98 [73], where expansion is restrained by steel reinforcement (preferably triaxial) so that compression is induced in the concrete, which offsets the tension in the steel reinforcement induced by restraint of drying shrinkage. It is also possible to use expansive cement to make *self-stressing concrete*, in which there is a residual compressive stress (say, up to 7 MPa) after most of the drying shrinkage has occurred; hence, shrinkage cracking is prevented [1,6].

Autogenous Shrinkage

According to Tazawa [74], autogenous shrinkage is a consequence of *chemical shrinkage*, which arises from the reduction in absolute volumes of the solid and liquid phases of the products of hydration compared with absolute volumes of original unhydrated cement and combined water [75]. Since the volume reduction is restrained by a rigid skeleton of hydrating cement paste, a residual space within the gross volume of the paste is created in the form of voids. When there is no external water supply, the net effect of restrained chemical shrinkage of hydrating cement paste is an apparent volume reduction, which is termed *autogenous volume change* or autogenous shrinkage [3,75].

Figure 6.20 is a diagrammatic representation of the volume contractions of the constituents of cement paste at mixing, at initial set, and after significant hydration. At the time of initial set, the hydrated cement (solid products plus gel water) forms only a small proportion of the total volume, the rest being water-filled capillaries and unhydrated cement. The volume contraction at this stage is due to chemical shrinkage. It is from this stage onward that a rigid skeleton microstructure is formed that restrains contraction due to continuing chemical shrinkage, leading to the formation of voids. As hydration proceeds, the volume of solid products and gel water increases at the expense of unhydrated cement and capillary water. In a sealed system, capillary pores can be empty or full of water depending on the original mix *water/binder ratio* (water/cementitious materials ratio). In totality, therefore, the hydrated cement paste (C-S-H) can be described as a rigid skeleton of solid products of hydration and gel water pores, encompassing voids, water-filled and empty capillaries, and unhydrated cement. In the presence of coarse and fine aggregate, chemical shrinkage is further restrained so that autogenous shrinkage of concrete is less than that of hardened cement paste.

The mechanism responsible for autogenous shrinkage is thought to be the same as that of drying shrinkage and can be explained by capillary tension theory as described in Chapter 15, except that the water is consumed internally due to *self-desiccation*

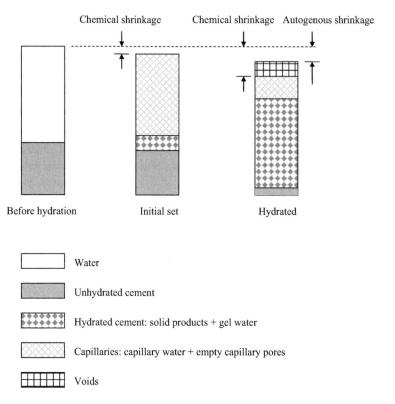

Figure 6.20 Diagrammatic representation of autogenous and chemical shrinkage of hydrating cement paste.

instead of being lost externally to a dry environment, as in the case of drying shrinkage [69,76]. As water is consumed by hydration, the pores empty and the internal relative humidity decreases, causing the curvature of menisci to increase, and the high surface tension imposes compression on the walls of the capillary pores and solid hydrates, thus inducing contraction of the hardened cement paste. The lower the water/cement ratio, the smaller the porosity and diameter of the pores, and the more the curvature of menisci; hence, capillary tension and induced compression on the C-S-H increase.

Measurement

Methods of determining chemical shrinkage and autogenous shrinkage are described by the Japan Concrete Institute [74,75] and by Aitcin [76]. Accurate measurement of autogenous shrinkage from a very early age for the first 24 h requires special procedures. Besides the method developed by the Japanese Concrete Institute [74], which involves casting concrete into a horizontal rectangular square prism, there is a method developed by Megat Johari [30] using a cylindrical mold in which the concrete is cast vertically and then stored horizontally to minimize plastic settlement. The cylindrical mold was chosen to complement standard laboratory molds for specimens used to determine drying shrinkage. With either method, frictional resistance to contraction of the concrete as it sets and hardens is minimized by lining the inside surfaces of the molds with polytetrafluoroethylene (PTFE). In the case of the cylindrical mold, an additional thin sheet of polythene is attached to the PTFE liner to prevent wear and leaks at the joints. A thermocouple located in the middle of the mold monitors changes in temperature during the test in order to correct for any thermal movement. Initial and final setting times of concrete are required to be determined by the penetration resistance test in accordance with ASTM C 403−08 [77]. After casting the concrete cylinder, a top steel plate is fixed to the mold before being turned horizontally.

At the time of initial set, the top and bottom plates are carefully removed and, to facilitate strain measurement, 1-mm-thick metal plates are glued to the ends of the concrete cylinder using a fast-setting chemical metal adhesive. The two halves of the mold are then separated to leave a gap at the joints of about 1.5 mm and any exposed surfaces of concrete are sealed with silicone grease and polythene sheet to prevent moisture loss by evaporation. Horizontal contraction of the concrete cylinder at each end is measured by a linear variable displacement transducer together with the temperature, readings being taken every 15 min and recorded by a data logger. Corrections to converted strains are made for temperature change by assuming the coefficient of thermal expansion of concrete is $10 \times 10^{-6}/°C$. Comparison tests revealed that autogenous shrinkage determined by the cylindrical mold was within 1% of that measured by the prismatic mold test [30]. After demolding at the age of 24 h, cylindrical specimens can be sealed with self-adhesive waterproof tape so that subsequent autogenous shrinkage can be determined by a demountable mechanical strain gauge.

Influencing Factors

Autogenous shrinkage can occur in any type of concrete, regardless of the level of water/binder ratio, and clearly depends on the age at which measurements start. In older publications, tests were probably carried out from the age of demolding of specimens, i.e., 24 h, although autogenous shrinkage is known to start earlier at the time of setting. In normal-strength concrete, this is not particularly important since autogenous shrinkage is very small. For example, Davis [78] observed that the long-term autogenous shrinkage of sealed concrete was of the order $20−130 \times 10^{-6}$ with or without mineral admixtures and having water/binder ratios ranging from 0.61 to 0.94. Due to the relatively small magnitude of autogenous shrinkage, Davis suggested that it was of little practical importance to differentiate between autogenous shrinkage and drying shrinkage of hardened concrete, except in the case of massive structures. For normal-strength concrete, autogenous shrinkage tends to increase at high temperatures, with higher cement content, and possibly with finer cements [6]. Also, at a constant cementitious materials content, a higher content of fly ash leads to a lower autogenous shrinkage [6].

On the other hand, in the case of high-strength or high-performance concrete, autogenous shrinkage becomes more relevant because mineral admixtures partly replace Portland cement and very low water/binder ratios are utilized, adequate work-ability being achieved by the use of superplasticizing chemical admixtures. The outcome is a hardened cementitious materials paste with lower porosity and finer pores compared with normal-strength concrete. If concrete is exposed to the environment, the small amount of available mix water is preferentially required for hydration reactions rather than being lost to the environment. In consequence, self-desiccation and, thus, autogenous shrinkage will be greater for high-strength concrete than in normal-strength concrete. Conversely, drying shrinkage of high-strength concrete will be less than normal-strength concrete. Tazawa and Miyazawa [79] reported that autogenous shrinkage was 700×10^{-6} at the age of 28 days for concrete containing microsilica with a water/binder ratio of 0.17. When restrained, that level of autogenous shrinkage leads to transverse cracking in a reinforced concrete specimen [79] and rapid cracking in sealed concrete specimens [80]. Tazawa and Miyazawa [79] also found that autogenous shrinkage increased as the water/binder ratio decreased from 0.40 to 0.23. Other research [81—83] confirmed the importance of water/binder ratio on the level of autogenous shrinkage.

Besides water/binder ratio, other factors influencing autogenous shrinkage of high-strength concrete are type and fineness of cement/mineral admixture, cementitious materials content, and aggregate content. High-early-strength cement paste exhibits more autogenous shrinkage than ordinary Portland cement paste, whereas moderate-heat cement paste shows less autogenous shrinkage [79,84]. For low-heat Portland cement with high C_2S, very low autogenous shrinkage has been observed, whereas cement with higher amounts of C_3A and C_4AF tend to have higher autogenous shrinkage [79,84].

Autogenous shrinkage is reduced by the use of chemical admixtures, surface tension-reducing agents, water repellent-treated powders and expansive admixtures [84]. Hori et al. [85] reported lower autogenous shrinkage using an additive based on calcium sulfonate than for conventional additives based on calcium sulfoaluminate. In connection with the possibility of reducing shrinkage crack formation in concrete, the effectiveness of a shrinkage-reducing admixture and a *superabsorbent polymer* on early-age shrinkage of ultra-high-performance concrete was investigated by Soliman and Nehdi [86]; strains were measured for 1 week from the age of approximately 6 h. Superabsorbent polymers have the capacity to absorb and retain moisture and act in a similar manner to saturated high-absorptive lightweight aggregates so that there is an internal reservoir available to replenish water lost by self-desiccation during the hydration process and, thus, in sealed or mass concrete, a potential to reduce autogenous shrinkage. Tests revealed that incorporation of the shrinkage-reducing admixture reduced autogenous shrinkage by 25% for a water/binder ratio of 0.25 compared with autogenous shrinkage of a control concrete having the same water/binder ratio, superplasticizer, and 30% of cement replaced by microsilica. The action of the shrinkage-reducing admixture leads to a slower strength development as well as the reduction of autogenous shrinkage of sealed specimens. The reduction of autogenous shrinkage was attributed by the authors to a reduction in surface tension of the pore fluid (water and superplasticizer) leading to lower capillary stress, and less self-desiccation because of a lower drop in pore relative humidity [86]. In the case of concrete containing the superabsorbent polymer, an autogenous shrinkage reduction of 21% was achieved compared with that of the control concrete. Using a combination of both the shrinkage-reducing admixture and superabsorbent polymer yielded an even greater reduction of autogenous shrinkage, namely, 50% of that of the control concrete. In other words, synergy occurred, viz. the combination of chemical admixtures resulted in a lower autogenous shrinkage than the sum of individual autogenous shrinkage reductions of concretes due to each chemical admixture. A synergy effect was also reported for concrete stored under dry conditions of 20 °C and 40% relative humidity [86].

Because of a delayed hardening contribution from the pozzolanic reaction of mineral admixtures, the early-age trends of autogenous shrinkage of cement−pozzolan-based systems are different from those at later ages. This was demonstrated in the tests by Megat Johari [30] with high-strength concrete made with mineral admixtures microsilica, metakaolin, fly ash, and ground granulated blast-furnace slag, which replaced ordinary Portland cement at different levels to make concrete of minimum 28-day cube strength of 80 MPa. There were no changes in mix proportions other than the inclusion of the mineral admixture. The control concrete (admixture-free) had mass proportions of 1:1.5:2.5, with a 0.28 water/cement ratio, a target workability of 100 mm slump being achieved by a high-range water-reducing superplasticizer based on a sulfonated vinyl copolymer. Table 6.2 lists details of initial set, 1-day strength, 28-day strength, and 200-day shrinkage.

Figure 6.21(a) clearly demonstrates that, after 24 h, the effect of all the mineral admixtures was to reduce autogenous shrinkage as determined from initial set, the more so the higher the level of replacement [30]. The reduction of early autogenous shrinkage can be explained by a dilution effect due to replacement of cement by the mineral admixture, namely, a reduction in "actual" cement content and an increase in "effective" water/cement ratio. The dilution effect would be expected to reduce early autogenous shrinkage when there is little or no contribution of the mineral admixture to the hardening process. Moreover, replacement of cement by mineral admixture causes an increase in "effective" superplasticizer dosage, which is likely to retard early-age hydration of cement and thus subdue autogenous shrinkage. Alternatively, at very early ages, the presence of mineral admixture may be considered as "inert aggregate," which restrains autogenous shrinkage of actual hardened cement paste. Figure 6.21(a) indicates that the most effective mineral admixture for reducing early-age autogenous shrinkage is metakaolin and the least effective is slag. With regard to the latter, of all the four admixtures investigated, slag is the only one to have hydraulic properties in its own right and not be dependent on the release of calcium hydroxide to initiate hydration of Portland cement during the process of hardening [6]. Hence, some contribution by slag to early autogenous shrinkage could be possible.

The reduction of early-age autogenous shrinkage of high-strength concrete made with mineral admixtures is confirmed by other test data [87,88], but findings by previous researchers appear to disagree since higher early autogenous shrinkage than that of the control cement pastes occurred [83]. However, other influencing factors have to be considered, such as water/binder ratio, a decrease of which has the effect of increasing early autogenous shrinkage. For example, in the case of Jensen and

Table 6.2 Effect of Replacement of Ordinary Portland Cement by Mineral Admixtures on Initial Setting Time, Strength, Autogenous Shrinkage, and Drying Shrinkage [30]

Concrete Type: Admixture-Replacement, R, %	Initial Set (IS), h	24-h Cube Strength, MPa	28-Day Strength, MPa		24-h Autogenous Shrinkage from IS, 10^{-6}	200-Day Shrinkage, 10^{-6}				Cement Type Factor, γ (Eq. (6.32))
			Cube	Cylinder		Autogenous from IS	Autogenous from 24 h	Drying from 24 h	Total from 24 h	
OPC control	5.0	42.7	86.7	72.6	303	445	142	416	558	1.0
MS-5	6.3	45.5	105.7	92.2	275	431	156	302	458	1.11
MS-10	6.7	45.5	113.9	99.6	224	473	249	237	486	1.14
MS-15	8.8	47.3	117.5	105.8	190	467	277	277	554	1.16
MK-5	6.4	49.3	91.5	78.7	214	485	271	228	499	1.02
MK-10	7.0	44.3	103.7	90.6	163	419	256	199	455	1.08
MK-15	6.5	42.1	103.4	87.8	106	327	221	189	410	1.03
FA-10	6.0	39.8	86.7	73.8	216	372	156	322	478	0.96
FA-20	6.1	28.1	84.3	72.6	132	300	168	345	513	0.90
FA-30	7.8	20.7	82.1	68.6	84	246	162	320	482	0.83
GGBS-20	7.9	31.5	95.3	78.8	246	459	213	307	520	0.95
GGBS-40	11.5	17.5	87.6	68.4	119	345	226	172	398	0.77
GGBS-60	12.4	1.2	86.7	63.2	37	308	271	198	469	0.64

OPC = ordinary Portland cement concrete.
Mineral admixtures: MS = microsilica; MK = metakaolin; FA = fly ash; GGBS = ground granulated blast-furnace slag.

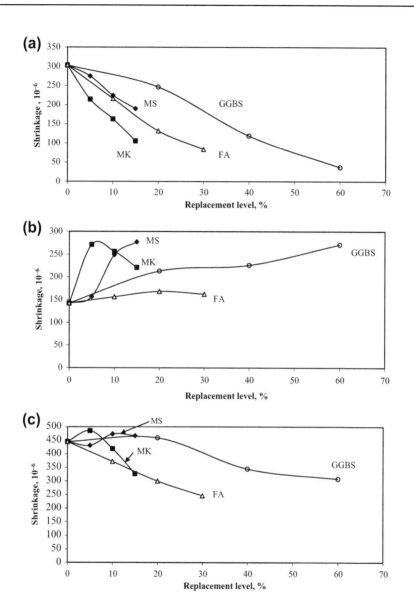

Figure 6.21 Effect of mineral admixtures on autogenous shrinkage [30]. MS = microsilica, MK = metakaolin, FA = fly ash, GGBS = ground granulated blast-furnace slag. (a) 24-h autogenous shrinkage measured from initial set. (b) 200-day autogenous shrinkage measured from 24 h. (c) 200-day total autogenous shrinkage measured from initial set.

Hansen [83], their mixes had the same water/cement ratio so, effectively, the microsilica–cement paste had a lower water/binder ratio, which therefore could explain the higher autogenous shrinkage. Another factor to be considered is fineness of the mineral admixture, an increase of which can lead to more early autogenous shrinkage [84,89].

The trends of 200-day autogenous shrinkage measured from the age of 24 h (later-age autogenous shrinkage) are completely different from those found at early age, as shown in Figure 6.21(b). Autogenous shrinkage increases as the replacement level increases for micosilica, fly ash, and slag but, in the case of metakaolin, there appears to be a reverse trend beyond a level of 5%; Kinuthia et al. [88] also found inconsistent trends using metakaolin at higher replacement levels. An increase of later-age autogenous shrinkage with microsilica was reported previously [84,88], which concurs with the trend in Figure 6.21(b). Other sources report that the effect of fly ash inclusion is different from that in Figure 6.21(b) as it appears to reduce later-age autogenous shrinkage of normal-strength concrete [90,91] and high-performance concrete [92]. These contradictions in trend may be explained by differences in fineness of fly ash, as found by Tangtermsirikul [93]. Other reported data for later-age autogenous shrinkage of concrete containing ground granulated blast-furnace slag [84,87,92] agree with the trends of Figure 6.21(b), the significance of fineness being highlighted.

The general increase of later-age autogenous shrinkage with increase of replacement level of cement by mineral admixture is attributed to higher capillary pore water tension because of reductions in porosity and mean pore size and an increase in the number of finer pores (see Table 10.3). In microstructural tests on mortar samples, the percentage volume of mesopores within the range of $2.5-15$ nm has been found to be greater for cement—microsilica mortar compared with that of ordinary Portland cement mortar [30]. The increase in autogenous shrinkage as the replacement level of microsilica increases is associated with the pozzolanic reaction with the calcium hydroxide released after initial cement hydration has taken place. Sellevold [94] reported that about 24% of microsilica by mass of cement is required to eliminate the calcium hydroxide in a hardened cement—microsilica paste system.

The 200-day total autogenous shrinkage, i.e., the sum of early-age and later-age values, determined in the tests by Megat Johari [30], is shown in Figure 6.21(c) and Table 6.2. In the case of cement—microsilica concretes, there is little effect of changing the level of cement replacement on total autogenous shrinkage, but there are slight increases at low replacements for metakaolin and slag before gradual reductions for higher replacements. It can be seen that fly ash appears to be consistently effective in reducing total autogenous shrinkage at all levels of cement replacement, for example, a 45% reduction at a 30% replacement level. Also of interest and shown in Table 6.2 is the 200-day *total shrinkage*, i.e., the sum of autogenous shrinkage plus drying shrinkage, measured from the age of 24 h. In all cases, the total shrinkage is less than that of the control concrete by $1-30\%$. For the ordinary Portland cement control concrete, from the age of 24 h the contribution of autogenous shrinkage to the total shrinkage is 25%, and the effect of mineral admixture is to reduce drying shrinkage but to increase autogenous shrinkage, the latter's contribution to total shrinkage ranging from 33% (fly ash) to 58% (slag).

The importance of water/binder ratio as an influencing factor on autogenous shrinkage of microsilica—cement concrete has also been investigated [30], especially since the use of microsilica allows a reduction in water/binder ratio, which could result in very high autogenous shrinkage [6,79,88]. The effect is demonstrated in Figure 6.22 for 200-day autogenous shrinkage of high-strength concrete measured from the age of

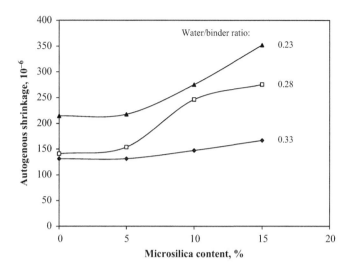

Figure 6.22 Influence of microsilica content and water/binder ratio on 200-day autogenous shrinkage of high-strength concrete measured from the age of 24 h [30].

24 h, where it is apparent that a considerable increase in autogenous shrinkage occurs if the water/binder ratio is reduced from 0.33 to 0.23.

A final comment that can be made regarding the influencing factors is that concerning the *rate of development of autogenous shrinkage*. For ordinary Portland cement concrete, autogenous shrinkage develops at a very fast rate, depending on the water/cement ratio, and quickly reaches an asymptotic level. For example, when measured from initial set, Table 6.2 indicates that 70% of the 200-day value is reached at the age of 24 h. On the other hand, for blended cement concrete, the initial rate is slower because the effective water/cement ratio is higher, and the more so the higher the level of replacement. For example, in the case of concrete with 60% slag, the autogenous shrinkage is only 10% of the 200-day value at the age of 24 h. After the age of 24 h, a reversal of behaviour occurs because of pozzolanicity, so that the blended cement concrete undergoes a higher proportion of 200-day autogenous shrinkage compared with that of Portland cement concrete. For example, in the case of concrete with 60% slag, the proportion of 200-day autogenous shrinkage from the age of 24 h is 88% compared with 33% for ordinary Portland cement concrete.

Prediction

Le Roy et al. [95] refers to the AFREM[a] method for predicting autogenous shrinkage, which can be incorporated in the procedure for the design of prestressed concrete

---

[a] AFREM: Association Française de Recherche et d'Essais sur les Matériaux et les Constructions (French Association for the Research and Testing of Materials and Structures).

structures. It is assumed that autogenous shrinkage before the age of 28 days is related to the degree of hydration and is a function of relative strength, $f_c(t)/f_{c28}$, but autogenous shrinkage does not commence until the age corresponding to $f_c(t)/f_{c28} = 0.1$. After the age of 28 days, autogenous shrinkage is assumed to be a function of time. The relevant expressions are given below.

For $t < 28$ days:

$$\left. \begin{array}{l} \text{when } f_c(t)/f_{c28} < 0.1, \\[6pt] S_{ca} = 0 \\[6pt] \text{when} f_c(t)/f_{c28} = 0.1, \\[6pt] S_{ca} = (f_{c28} - 20)\left[2.2\dfrac{f_c(t)}{f_{c28}} - 0.2\right] \end{array} \right\} \qquad (6.24)$$

For $t \geq 28$ days:

$$S_{ca} = (f_{c28} - 20)[2.8 - 1.1\ \exp(t/96)] \qquad (6.25)$$

If $f_c(t)$ is unknown, it may be estimated from 28-day strength as follows:

$$f_c(t) = f_{c28}\left[\frac{t}{(1.4 + 0.95t)}\right] \qquad (6.26)$$

The method proposed by the Japan Concrete Institute [75] is:

$$S_{ca} = \gamma\left[3070e^{-7.2(w/b)}\right]\left[1 - \exp\left(-a'(t - t_o)^{b'}\right)\right] \qquad (6.27)$$

where $w/b =$ water/binder ratio $= \geq 0.2 \leq 0.5$; $\gamma =$ coefficient for cement type ($= 1$ for ordinary Portland cement); $t_o =$ initial setting time (days); and a' and $b' =$ coefficients that depend on water/binder ratio (Table 6.3).

If $w/b > 0.5$, then

$$S_{ca} = \gamma 80\left[1 - \exp\left(-a(t - t_o)^b\right)\right] \qquad (6.28)$$

Table 6.3 Coefficients a and b for Eq. (6.27)

Water/Binder Ratio, w/b	Coefficient a'	Coefficient b'
0.20	1.20	0.40
0.23	1.50	0.40
0.30	0.60	0.50
0.40	0.10	0.70
More than 0.50	0.03	0.80

Equations (6.27) and (6.28) apply for concrete stored at 20 °C and, for reduced or elevated temperatures, t_o and t are adjusted as follows:

$$t \text{ and } t_o = \sum_{i}^{n} n_i \exp\left[13.65 - \frac{4000}{(273 + T_i)}\right] \tag{6.29}$$

where n_i = number of days when the temperature of the concrete is T_i (°C).

The method of predicting autogenous shrinkage as prescribed by BS EN 1992-1-1: 2004 [96] assumes that autogenous shrinkage is a linear function of concrete strength; the method is based on an updated version of CEB Model Code 90 [34]. Autogenous shrinkage should be considered specifically when new concrete is cast against hardened concrete. The autogenous shrinkage is given by:

$$S_{ca}(t) = 2.5(f_{c28} - 18)\left[1 - \exp\left[-\left(0.2t^{0.5}\right)\right]\right]10^{-6} \tag{6.30}$$

or

$$S_{ca}(t) = 2.5(f_{ck} - 10)\left[1 - \exp\left[-\left(0.2t^{0.5}\right)\right]\right]10^{-6}$$

where t = age in days measured from initial set, f_{c28} = 28-day mean strength = $f_{ck} + 8$, and f_{ck} = characteristic cylinder strength.

It should be remembered that the total shrinkage of concrete exposed to drying comprises drying shrinkage plus autogenous shrinkage, but when reckoned from the start of drying shrinkage the contribution of autogenous shrinkage to total shrinkage is much less, since a significant amount will have occurred before drying shrinkage commences.

According to Tazawa and Miyazawa [97], the AFREM method fails to account for water/binder ratio and type of cement in a satisfactory manner. In addition, both the AFREM and BS EN 1992-1-1: 2004 methods underestimate autogenous shrinkage of ordinary Portland cement concrete in Table 6.2. On the other hand, the Japan Concrete Institute method predicts the autogenous shrinkage of ordinary Portland cement concrete to within 15%, assuming the cement type factor $\gamma = 1$ in Eq. (6.27), but it does not allow for the effects of mineral admixtures as partial replacements of cement. To overcome this deficiency, Megat Johari [30] proposed expressing the cement type factor as a function of replacement level and relative strength ratio as follows:

$$\gamma = 0.5\left[1 - 0.01R + \frac{f_{c28}(R)}{f_{c28}}\right] \tag{6.31}$$

where R = % replacement level of cement by mineral admixture, $f_{c28}(R)$ = 28-day strength of blended cement concrete, and f_{c28} = 28-day strength of ordinary Portland cement concrete (control).

Table 6.2 lists the cement type factors calculated according to Eq. (6.31), and when used to estimate the 200-day autogenous shrinkage from initial setting time, the accuracy of prediction is acceptable, with an average error coefficient of 13% (Eq. (5.4)). It should be noted that Eqs (6.27) and (6.30) are applicable for later ages but are not suitable for estimating early-age autogenous shrinkage of concrete with mineral admixtures, say, for the first 24 h.

Problems

6.1 What are the chemical reactions leading to carbonation shrinkage in concrete?

6.2 List the factors influencing drying shrinkage of concrete.

6.3 Discuss the mechanisms of drying shrinkage and autogenous shrinkage.

6.4 Compare the carbonation shrinkage of concrete: (a) exposed to intermittent rain; (b) protected from rain.

6.5 Explain restrained shrinkage and differential shrinkage.

6.6 How does size and shape of member affect drying shrinkage? Give examples.

6.7 Describe shrinkage-compensating concrete.

6.8 Discuss the influence of aggregate on drying shrinkage of concrete.

6.9 How does water/cement ratio affect drying shrinkage?

6.10 Define hardened cement paste content. How does it influence autogenous shrinkage and drying shrinkage?

6.11 Discuss the influence of chemical admixtures on drying shrinkage of concrete.

6.12 Discuss the influence of mineral admixtures on: (a) autogenous shrinkage; (b) drying shrinkage.

6.13 Is autogenous shrinkage significant? If so, when and for which types of concrete?

6.14 Define what is meant by: (a) chemical shrinkage; (b) autogenous shrinkage.

6.15 Suggest ways of minimizing drying shrinkage of concrete.

6.16 Estimate the autogenous shrinkage of admixture-free ordinary Portland cement concrete at the age of 28 days by the methods of: (a) AFREM; (b) Japan Concrete Institute; and (c) BS EN 1992-1-1. The initial set $= 6$ h, 28-day compressive strength $= 80$ MPa, and the water/cement ratio $= 0.30$.
Answer: (a) 119×10^{-6}; (b) 339×10^{-6}; (c) 101×10^{-6}.

6.17 If the concrete in Question 6.16 contains 10% microsilica as replacement of cement, the water/binder ratio $= 0.3$, the initial set $= 8$ h, and the 28-day compressive strength $= 110$ MPa, estimate the 28-day autogenous shrinkage by the same methods.
Answer: (a) 178×10^{-6}; (b) 352×10^{-6}; (c) 150×10^{-6}.

6.18 Discuss the influence of elevated temperature on drying shrinkage.

References

[1]   Neville AM, Brooks JJ. Concrete technology. 2nd ed. Pearson Prentice Hall; 2010. 422 pp.

[2]   ACI Committee 305R-99. Hot weather concreting, Part 2, manual of concrete practice; 2007.

[3]   L'Hermite R. Volume changes of concrete. In: Proceedings of 4th international symposium on the chemistry of cement, Washington DC; 1960. pp. 659–94.

[4]   Brooks JJ. Dimensional stability and cracking processes in concrete. In: Page CL, Page MM, editors. Chapter 3: durability of concrete and cement composites. Cambridge: Woodhead Publishing Ltd; 2007. pp. 45−85.

[5]   Brooks JJ. 30-year creep and shrinkage of concrete. Mag Concr Res 2005;57(9): 545−56.

[6]   Neville AM. Properties of concrete. 4th ed. Pearson Prentice Hall; 2006. 844 pp.

[7]   Mindess S, Young JF. Concrete, Prentice-Hall, 1981, 671 pp. Japan Concrete Institute, Technical committee on autogenous shrinkage of concrete. In: Tazawa E, editor. Proceedings of International Workshop shrinkage of Concrete, Hiroshima, Part 1. E & F. N. Spon; 1998. pp. 1−63.

[8]   Verbeck GJ. Carbonation of hydrated Portland cement. ASTM Special Publication, No. 205; 1958. pp. 17−36.

[9]   ACI Committee 517.2R-92. Accelerated curing of concrete at atmospheric pressure (withdrawn). American Concrete Institute; 1992. 17 pp.

[10]   Troxell GE, Raphael JM, Davis RE. Long-time creep and shrinkage tests of plain and reinforced concrete. In: Proceedings ASTM, vol. 58; 1958. pp. 1101−20.

[11]   (withdrawn 2012) BS 1881−5. Testing concrete. Methods of testing hardened concrete for other than strength. British Standards Institution; 1970.

[12]   ASTM C 157/C 157/M-08. Test for length change of hardened hydraulic cement mortar and concrete. American Society for Testing and Materials; 2008.

[13]   Roper HR. The influence of cement composition and fineness on concrete shrinkage, tensile creep and cracking tendency. In: Proceedings first Australian conference on engineering materials. University of New South Wales; 1974.

[14]   ACI Committee 209.1R-05. Report on factors affecting shrinkage and creep of hardened concrete. American Concrete Institute Committee 209; 2008. 12 pp.

[15]   Odman STA. Effects of variation in volume, surface area exposed to drying and composition of concrete on shrinkage. In: Proceedings of RILEM/CEMBUREAU international colloquium on the shrinkage of hydraulic concretes, Madrid, vol. 1; 1968. 20 pp.

[16]   Pickett G. Effect of aggregate on shrinkage of concrete and hypothesis concerning shrinkage. J Am Concr Inst 1956;52:581−90.

[17]   Powers TC. Causes and control of volume change. J Portland Cem Assoc Res Dev Lab 1959;1(1):29−39.

[18]   Illston JM, Dinwoodie JM, Smith AA. Concrete, Timber and metals. The Nature and behaviour of structural materials, Von Rostrand Reinhold, International Student Edition, 1979, 663 pp.

[19]   Carlson RW. Drying shrinkage of large members. ACI J. January-February 1937;33: 327−36.

[20]   Bazant Z p, Osman E, Thonguthua W. Practical formulation of shrinkage and creep of concrete. Mater Struct 1976;9(49):395−406.

[21]   Bazant ZP, Najjar LJ. Non-linear water diffusion in non-saturated concrete. Mater Struct 1972;5:3−20.

[22]   Ross AD. Shape, size and shrinkage. Concr Constr Eng; August 1944:193−9.

[23]   Hansen TA, Mattock AH. The influence of size and shape of member on shrinkage and creep of concrete. J Am Concr Inst 1966;63:267−90.

[24]   Keeton JR. Study of creep in concrete. Technical Report No. R333-1, 2. Port Hueneme (Calfornia): US Naval Engineering Laboratory; 1965. Phase 1-5.

[25]   Kesler CE, Wallo EM, Yuan RL. Free shrinkage of concrete mortar. T and AM Report No. 664. Urbana (Illinois): Department of Theoretical and Applied Mechanics, University of Illinois; July 1966.

[26]   Wallo EM, Yuan RL, Lott JL, Kesler CE. Prediction of creep in structural concrete from short time tests. Sixth Progress T. and A. M. Report No. 658. University of Illinois; August 1965. 26 pp.

[27]   Hobbs DW. Influence of specimen geometry on weight change and shrinkage of air-dried concrete specimens. Mag Concr Res 1977;29(99):70−80.

[28]   Bryant AH, Vadhanavikkit C. Creep, shrinkage-size, and age at loading effects. ACI Mater J; March-April 1987:117−23.

[29]   Perenchio WF. The drying shrinkage dilemma − some observations and questions about drying shrinkage and its consequences. Concr Constr 1997;42(4):379−83.

[30]   Megat Johari, MA. Deformation of high strength concrete containing mineral admixtures [Ph.D. thesis]. School of Civil Engineering, University of Leeds; 2000, 296 pp.

[31]   Klieger P. Some aspects of durability and volume change of concrete for prestressing. Research Department Bulletin RX118. Skokie (Illinois): Portland Cement Association; 1960. 15 pp.

[32]   Hanson JA. Prestress loss as affected by type of curing. Prestress Concr Inst J April 1964; 9:69−73.

[33]   CEB-FIP Model Code 1990. CEB Bulletin d'Information No. 213/214. Lausanne: Comite Euro-International du Beton; 1993. pp. 33−41.

[34]   CEB Model Code 90-99. In: Structural concrete-textbook on behaviour, design and performance, updated knowledge on CEB-FIP model code 1990, fip Bulletin 2, vol. 2. Lausanne: Federation Internationale du Beton; 1999. pp. 37−52.

[35]   Brooks JJ. Elasticity, creep and shrinkage of concretes containing admixtures. In: Al-Manaseer Akthem, editor. Proceedings Adam Neville symposium: creep and shrinkage − structural design effects. Atlanta: ACI Special Publication SP − 194; 1997. 2000, pp. 283−360.

[36]   Jessop EL, Ward MA, Neville AM. Influence of water-reducing and set-retarding admixtures on creep of lightweight aggregate concrete. In: Proceedings of RILEM symposium on admixtures for mortar and concrete, Brussels; 1967. pp. 35−46.

[37]   Hope BB, Neville AM, Guruswami A. Influence of admixtures on creep of concrete containing normal weight aggregate. In: Proceedings of RILEM symposium on admixtures for mortar and concrete, Brussels; 1967. pp. 17−32.

[38]   Morgan DR, Welch GB. Influence of admixtures on creep of concrete, third Australian conference on mechanics of structures and materials. New Zealand: University of Auckland; 1971.

[39]   Brooks JJ. Influence of plasticizing admixtures Cormix P7 and Cormix 2000 on time-dependent properties of flowing concretes. Research report. Department of Civil Engineering, University of Leeds; 1984.

[40]   Brooks JJ, Wainwright PJ, Neville AM. Time-dependent properties of concrete containing a superplasticizing admixture. Superplasticizers in concrete. ACI Special Publication SP 62; 1979. pp. 293−314.

[41]   Brooks JJ, Wainwright PJ, Neville AM. Superplasticizer effect on time-dependent properties of air entrained concrete. Concrete 1979;13(6):35−8.

[42]   Brooks JJ, Wainwright PJ, Neville AM. Time-dependent behaviour of high early-strength concrete containing a superplasticizer. Developments in the use of super-plasticizers. ACI Special Publication SP 68; 1981. pp. 81−100.

[43]   Brooks JJ, Wainwright PJ. Properties of ultra high-strength concrete containing a superplasticizer. Mag Concr Res 1983;35(125):205−14.

[44]   Dhir RK, Yap AWF. Superplasticized flowing concrete: strength and deformation properties. Mag Concr Res 1984;36(129):203−15.

[45] Berenjian, J. Superplasticized flowing concrete: microstructure and long-term deformation characteristics [Ph.D. thesis]. Department of Civil Engineering, University of Leeds; 1989, 238 pp.

[46] Alexander KM, Bruere GM, Ivanesc I. The creep and related properties of very high-strength superplasticized concrete. Cem Concr Res 1980;10(2):131−7.

[47] Tokuda H, Shoya M, Kawakami M, Kagaya M. Applications of superplasticizers to reduce shrinkage and thermal cracking in concrete. Developments in the use of superplasticizers. ACI special Publication SP 68; 1981. pp. 101−120.

[48] Ai H, Young JF. Mechanism of shrinkage reduction using a chemical admixture. In: 10th conference on cement chemistry, vol. 6; 1997.

[49] Jiang, X. The effect of creep in tension on cracking resistance of concrete [MSc (Eng.) thesis]. School of Civil Engineering, University of Leeds; 1997, 140 pp.

[50] Kristiawan, SA. Restrained shrinkage cracking of concrete [Ph.D. thesis]. School of Civil Engineering, University of Leeds; 2002, 217 pp.

[51] Al-Manaseer A, Ristanovic S. Predicting drying shrinkage of concrete. Concr Int 2004; 26(8):79−83.

[52] Brooks JJ, Neville AM. Creep and shrinkage of concrete as affected by admixtures and cement replacement materials. Creep and shrinkage of concrete: effect of materials and environment. ACI Special Publication SP-135; 1992. pp. 19−36.

[53] Neville AM, Brooks JJ. Time-dependent behaviour of cemsave concrete. Concrete 1975; 9(3):36−9.

[54] Bamforth PB. An investigation into the influence of partial Portland cement replacement using either fly ash or ground granulated blastfurnace slag on the early age and long-term behaviour of concrete. Research report, 013J/78/2067. Southall (UK): Taywood Engineering; 1978.

[55] Aitchin PC, Laplante R. Volume changes and creep measurements of slag cement concrete. Research report. Universite de Sherbrooke; 1986. 16 pp.

[56] Cook DJ, Hinezak I, Duggan R. Volume changes in Portland-blast furnace slag cement concrete. In: Proceedings second international conference on the use of fly ash, silca fume, slag and natural pozzolans in concrete, supplementary papers volume, Madrid; 1986. 14 pp.

[57] Chern JC, Chan YW. Deformations of concrete made blast-furnace slag cement and ordinary Portland cement. ACI Mater J Proc 1989;86(4):372−82.

[58] Brooks JJ, Wainwright PJ, Boukendakji M. Influences of slag type and replacement level on strength, elasticity, shrinkage and creep of concrete. In: Proceedings fourth international conference on fly ash silica fume, slag and natural pozzolans in concrete, vol. 2. ACI Special Publication SP-132; 1992. pp. 1325−42.

[59] Ghosh RS, Timusk J. Creep of fly ash concrete. ACI J Proc 1981;78. Title No. 78−30.

[60] Brooks JJ, Wainwright PJ, Cripwell JB. Time-dependent properties of concrete containing pulverised fuel ash and a superplasticizer. In: Cabrera JG, Cusens AR, editors. Proceedings international symposium on the use of PFA in concrete, Vol. 1. Dept. of Civil engineering, University of Leeds; 1982. pp. 209−20.

[61] Munday JGL, Ong LT, Wong LB, Dhir RK. Load-independent movements in OPC/PFA concrete. In: Cabrera JG, Cusens AR, editors. Proceedings international symposium on the use of PFA in concrete, vol. 1. Dept. of Civil engineering, University of Leeds; 1982. pp. 243−54.

[62] Yamato T, Sugita H. Shrinkage and creep of mass concrete containing fly ash. In: Proceedings first international conference on the use of fly ash, silica fume, slag and other mineral by-products in concrete. ACI Special Publication SP 79; 1983. pp. 87−102.

[63]   Nasser KW, Al-Manaseer AA. Shrinkage and creep containing 50 percent lignite fly ash at different stress-strength ratios. In: Proceedings second international conference on fly ash, silica fume, slag and natural pozzolans in concrete. ACI Special Publication SP 91; 1986. pp. 443−8.

[64]   Brooks JJ, Gamble AE, Al-Khaja WA. Influence of pulverised fuel ash and a super-plasticizer on time-dependent performance of prestressed concrete beams. In: Proceedings symposium on utilization of high strength concrete, Stavanger, Norway; 1987. pp. 205−14.

[65]   Langley WS, Carette GG, Malhotra VM. Structural concrete incorporating high volumes of ASTM Class F fly ash. ACI Mater J 1989;86(5):507−14.

[66]   Buil M, Acker P. Creep of a silica fume concrete. Cem Concr Res 1985;15:463−6.

[67]   Tazawa E, Yonekura A. Drying shrinkage and creep of concrete with condensed silica fume. In: Proceedings second international conference on fly ash, silica fume, slag and natural pozzolans in concrete. Madrid: ACI Publication SP A1; 1986. pp. 903−21.

[68]   Bentur A, Goldman A. Curing effects, strength and physical properties of high strength silica fume concretes. J Mater Civ Eng 1989;1(1):46−58.

[69]   de Larrard F. Creep and shrinkage of high strength field concretes. In: CANMET/ACI international workshop on the use of silica fume in concrete, Washington (DC); 1991. 22 pp.

[70]   Pentalla V, Rautenan T. Microporosity, creep and shrinkage of high strength concretes. In: CANMET/ACI international workshop on the use of silica fume in concrete, Washington (DC); 1991. 29 pp.

[71]   Burg RG, Ost BW. Engineering properties of commercially available high strength concretes. PCA research and development RD 104T. Skokie (Illinois): Portland Cement Association; 1992. 55 pp.

[72]   ASTM C 845−04. Specification for expansive hydraulic cement. American Society for Testing and Materials; 2004.

[73]   ACI Committee 223-98. Standard practice for the use of shrinkage-compensating concrete, Part 1, ACI manual of concrete practice. American Concrete Institute; 2007.

[74]   Tazawa E. Autogenous shrinkage of concrete. In: Tazawa E, editor. Proceedings of international workshop shrinkage of concrete, Hiroshima. E & F. N. Spon; 1998. 411 pp.

[75]   Japan Concrete Institute. Technical committee on autogenous shrinkage of concrete. In: Tazawa E, editor. Proceedings of international workshop, Hiroshima, Part 1. E & F. N. Spon; 1998. pp. 1−63.

[76]   Aitcin PC. Autogenous shrinkage measurement. In: Tazawa E, editor. Proceedings of international workshop shrinkage of concrete, Hiroshima, Part 3. E & F. N. Spon; 1998. pp. 257−68.

[77]   ASTM C403/C403M − 08. Standard test method for time of setting of concrete mixtures by penetration resistance. American Society for Testing and Materials; 2008.

[78]   Davis HE. Autogenous volume change of concrete. In: Proceedings of 43rd annual meeting, American society of testing and materials; 1940. pp. 1103−13.

[79]   Tazawa E, Miyazawa S. Autogenous shrinkage of concrete and its importance on concrete technology. In: Bazant ZP, Carol I, editors. Creep and shrinkage of concrete. E & F. N. Spon; 1993. pp. 159−68.

[80]   Paillere AM, Buil M, Serrano JJ. Effect of fiber addition on the autogenous shrinkage of silica fume concrete. ACI Mater J 1989;86(2):139−44.

[81]   Le Roy R, de Larrard F. Creep and shrinkage of high-performance concrete: the LCPC experience. In: Bazant ZP, Carol I, editors. Creep and shrinkage of concrete. E & F. N. Spon; 1993. pp. 499−503.

[82] Justnes H, Van Germet A, Verboven E, Sellevold E. Total and chemical shrinkage of low water/cement ratio cement paste. Adv Cem Res 1996;8(31):121−6.

[83] Jensen OM, Hansen FP. Autogenous deformation and change of relative humidity in silica fume-modified cement paste. ACI Mater J 1996;93(6):539−43.

[84] Tazawa E, Miyazawa S. Influence of constituents and composition autogenous shrinkage of cementitious materials. Mag Concr Res 1997;49(178):15−22.

[85] Hori A, Morioka M, Sakai E, Daimon M. Influence of expansive additives on autogenous shrinkage. In: Tazawa E, editor. Proceedings of international workshop on autogenous shrinkage of concrete, Hiroshima, Part 4. E & F. N. Spon; 1998. pp. 187−94.

[86] Soliman A, Nehdi M. Early-age shrinkage of ultra-high-performance concrete under drying/wetting cycles and submerged conditions. ACI Mater J 2012;109(2):131−40.

[87] Hanehara S, Hirao H, Uchikawa H. Relationships between autogenous shrinkage, and the microstructure and humidity changes at inner part of hardened cement paste at early age. In: Tazawa E, editor. Proceedings of international workshop on autogenous shrinkage of concrete, Hiroshima, Part 2. E & F. N. Spon; 1998. pp. 93−104.

[88] Kinuthia JM, Wild S, Sabir BB, Bai J. Influence of metakaolin-PFA blends on the chemical shrinkage properties of cement pastes. In: Proceedings of concrete communication conference 99 − 9th BCA annual conference on higher education and the concrete industry, British cement association. University of Cardiff; 1999. pp. 91−102.

[89] Tazawa E, Miyazawa S. Influence of cement and admixture on autogenous shrinkage of cement paste. Cem Concr Res 1995;25(2):281−7.

[90] Houk IE, Borge OE, Houghton DL. Studies of autogenous volume changes I concrete for Dworshak Dam. ACI J; 1969:560−8. No. 65-65.

[91] Gifford PM, Ward MA. Results of laboratory tests on lean mass concrete using PFA to a high replacement level. In: Cabrera JC, Cusens AR, editors. Proceedings of international symposium on the use of PFA in concrete, vol. 1. Dept. of Civil Engineering, University of Leeds; 1982. pp. 221−30.

[92] Chan YW, Liu CY, Lu YS. Effects of slag and fly ash on the autogenous shrinkage of high performance concrete. In: Tazawa E, editor. Proceedings of international workshop on autogenous shrinkage of concrete, Hiroshima, Part 4. E & F. N. Spon; 1998. pp. 221−8.

[93] Tangtermsirikul S. Effect of chemical composition and particle size of fly ash on autogenous shrinkage of paste. In: Tazawa E, editor. Proceedings of international workshop on autogenous shrinkage of concrete, Hiroshima, Part 4. E & F. N. Spon; 1998. pp. 175−86.

[94] Sellevold EJ. The function of condensed silioca fume in high strength concrete. In: Holand I, Helland S, Jakobsen B, Lenscow R, editors. Proceedings of symposium on utilization of high strength concrete, Stavanger, Norway; 1987. pp. 39−49.

[95] Le Roy R, de Larrard F, Pons G. The after code type model for creep and shrinkage of high-performance concrete. In: Proceedings of the 4th international symposium on utilization of high strength/high performance concrete; 1996. pp. 387−96.

[96] BS EN 1992-1-1. Eurocode 2, design of concrete structures. General rules and rules for buildings (see also UK National Annex to Eurocode 2). British Standards Institution; 2004.

[97] Tazawa E, Miyazawa S. Effect of constituents and curing condition on autogenous shrinkage of concrete. In: Tazawa E, editor. Proceedings of international workshop on autogenous shrinkage of concrete, Hiroshima, Part 5. E & F. N. Spon; 1998. pp. 268−80.

7 Shrinkage of Calcium Silicate and Concrete Masonry

The next three chapters deal with the subject of moisture movement of masonry, which is defined as a time-dependent change of strain caused by moisture migration between the masonry and the environment. As stated in the previous chapter, moisture movement is independent of external load and can take the form of a contraction due to shrinkage or a moisture expansion. This chapter concentrates on shrinkage of calcium silicate masonry and concrete masonry and their component phases: calcium silicate bricks, concrete blocks, and mortar. Chapter 8 deals with moisture movement of clay masonry built from fired clay units that undergo an irreversible moisture expansion on leaving the kiln. Chapter 9 is also relevant to the topic of moisture movement, as it is a special case of irreversible moisture expansion of clay masonry where an enlarged expansion occurs due to cryptoflorescence.

Movements arising from changes in moisture levels in masonry materials are either reversible or irreversible in nature. Reversible moisture movement is caused by changes in seasonal climatic conditions leading to expansion by absorption of moisture (wetting) and contraction by desorption of moisture (drying). Compared with the reversible moisture expansion of clay units, Figure 7.1 shows that the range of reversible shrinkage of calcium silicate and concrete units is much larger and lies

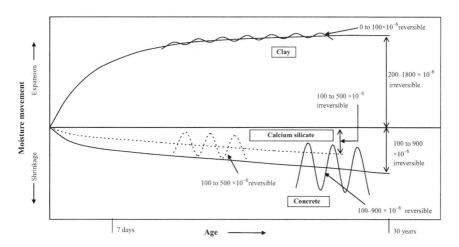

Figure 7.1 Schematic representation of ranges of moisture movement of masonry units: expansion of clay bricks, shrinkage of concrete blocks and calcium silicate bricks.

Concrete and Masonry Movements. http://dx.doi.org/10.1016/B978-0-12-801525-4.00007-8

between 100–500×10^{-6} (0.01–0.05%) and 100–900×10^{-6} (0.01–09%), respectively. For calcium silicate and concrete units, Figure 7.1 indicates that reversible shrinkage is of the same order as the irreversible shrinkage.

Irreversible drying shrinkage is caused mainly by loss of adsorbed water, which is bound within the hardened cement paste or tobermorite structure of the cementitious material component of the unit or mortar (see Chapter 15). However, according to BS 5628-3: 2005 [1], an additional shrinkage of concrete masonry units and mortar can occur as a result of *carbonation* of the cement by atmospheric carbon dioxide (see page 140). The extent of carbonation and the subsequent movement depends on the permeability of the concrete unit or mortar and the ambient relative humidity (RH). In dense masonry units and in autoclaved masonry units, the magnitude of this movement is extremely small and may be neglected. On the other hand, in unprotected open-textured masonry and mortar, shrinkage due to carbonation can be between 20% and 30% of the initial free drying shrinkage.

The content of this chapter consists of a review of publications dealing with shrinkage of calcium silicate and concrete masonry, most of which report test data determined from control walls or piers in tests to determine creep of masonry. Then, influencing factors are discussed, such as mortar type and unit type, with particular attention on how the moisture state of the unit at laying affects subsequent shrinkage of the bonded unit and masonry. The geometry of the cross section of masonry is also an important factor. International Code of Practice methods of estimating shrinkage for design are given and, finally, relationships are developed so as to predict shrinkage of units, mortar, and masonry by composite models.

Type of Mortar

According to BS 5628-3: 2005 [1], typical initial shrinkage of mortars is 400–1000×10^{-6} and subsequent reversible moisture movement is 500–600×10^{-6}. The lower values may be taken as applying to mortars in wetter external walls and the higher values to mortars in drier internal walls. However, the reversible movement of internal walls may be generally neglected since they are unlikely to become wet after drying out initially.

Specific research into the effect of mortar type on shrinkage has been published [2], and the time-dependent curves of part-sealed prisms are shown in Figure 7.2 for mortars having equal consistence, as determined by a dropping ball penetration of 10 mm [3]. Details of designation, mix proportions, water/cement ratio, and strength are listed in Table 7.1. Most shrinkage occurred for the ground granulated blast-furnace slag (GGBS) mortar, followed by the air-entrained mortar (AEA), while the masonry cement (MC) exhibited the least shrinkage. In general terms, for cement–lime mortars, shrinkage was consistently greater for the weaker mortars, but the same shrinkage–strength pattern is not followed by the GGBS, AEA, and MC mortars. Consequently, although convenient, it seems that strength may be only a general indicator of shrinkage for all types of mortar.

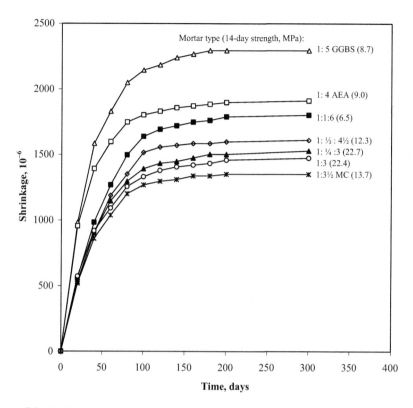

Figure 7.2 Shrinkage of various types of mortar determined using prisms partly sealed to the same volume/surface area ratio as single-leaf wall bed joints (44 mm) [2]; GGBS = ground granulated blast-furnace slag; AEA = air-entraining agent; MC = masonry cement.

Table 7.1 Details of Types of Mortar Used to Determine the Shrinkage-Time Curves of Figure 7.2

Mortar Type			
Cement: Lime: Sand, by Volume	Designation [BS 5628: 1985]	Water/Cement Ratio, by Mass	14-Day Cube Compressive Strength (MPa)
1:0 : 3	(i)	0.56	22.4
1 : ¼ : 3	(i)	0.56	22.7
1 : ½ : 4½	(ii)	0.76	12.3
1:0 : 3½ MC	(ii)	0.54	13.7
1:0 : 4 AEA	(ii)	0.53	9.0
1:1 : 6	(iii)	0.96	6.5
1:0; 5 GGBS	(iii)	0.72	8.7

In the above investigation, shrinkage of mortar was determined on prismatic specimens partly sealed to simulate drying of the mortar bed joints of a single-leaf wall in a storage environment of 65% RH and 21 °C and having a volume/surface area (V/S) ratio = 44 mm (see p. 204). Other investigations [4–8] also reported shrinkage data for mortar using the same experimental procedures, and the results, together with those of Figure 7.2, have been analyzed collectively, as described below.

The development of shrinkage of masonry and its component phases (S_t) with time (t) is readily described by the Ross hyperbolic equation [9]:

$$S_t = \frac{S_\infty t}{aS_\infty + t} \tag{7.1}$$

where S_∞ = ultimate shrinkage and a = reciprocal of initial rate of shrinkage.

In rectified (linear) form, Eq. (7.1) becomes:

$$\frac{t}{S_t} = a + \frac{1}{S_\infty}t \tag{7.2}$$

so that $1/S_\infty$ and a are given, respectively, by the slope and intercept of the graph of t/S_t versus t.

In the case of mortar shrinkage (S_m), analysis [8] revealed that the term $aS_{m\infty}$ was fairly constant, with an average value of 35.7, but $S_{m\infty}$ varied approximately according to the strength of mortar. Now, according to the CEB Model Code for concrete [10], ultimate shrinkage deceases as the 28-day strength increases. In the case of mortar, a similar approximate relationship was found based on 28-day strength of mortar (f_{m28}), but it depended on the curing conditions of the specimens before exposure to drying. For water-cured mortar, the general trend can be expressed as follows:

$$S_{m\infty} = 2600 - 12f_{m28} \tag{7.3}$$

and, for mortar specimens cured under polythene sheet:

$$S_{m\infty} = 2600 - 49f_{m28} \tag{7.4}$$

Alternative correlations based on strength of mortar at the age of exposure to drying or start of shrinkage (f_{mo}) were found. For water-cured specimens:

$$S_{m\infty} = 2600 - 15f_{mo} \tag{7.5}$$

and, for storage under polythene sheet:

$$S_{m\infty} = 2600 - 60f_{mo} \tag{7.6}$$

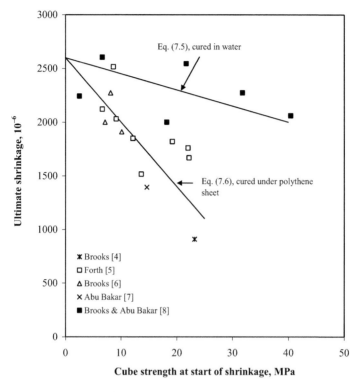

Figure 7.3 Ultimate shrinkage of mortar specimens as a function of strength at age of exposure to drying; specimens partly sealed to represent volume/surface ratio (44 mm) of the bed joint in a single-leaf wall [8].

Equations (7.5) and (7.6) are shown in Figure 7.3, where it is apparent that when cured under polythene sheet prior to exposure to drying, shrinkage is more sensitive to the strength of mortar. In practice, if newly built masonry is covered for 24 h before the start of shrinkage, then Eq. (7.6) would be applicable, with $f_{mo} = 1$-day strength. If the 1-day strength is unknown, it may be estimated from the average development of strength with age of mortar relative to the 28-day strength shown in Table 7.2.

Table 7.2 Strength of Mortar at Different Ages (f_m) Relative to 28-Day Strength (f_{m28}) [8]

Age (days)	1	3	7	14	28	56	200
f_m/f_{m28}	0.4	0.59	0.72	0.83	1	1.04	1.05

Type of Unit

Lenczner [11] reported a vertical shrinkage of 525×10^{-6} after 320 days for block-work piers made with a crushed limestone aggregate; after the same time, shrinkage of the unbonded solid block was 410×10^{-6}. The masonry was built with a 1:1:6 cement–lime mortar and tested at the age of 28 days when the block strength was 5.6 MPa and mortar strength was 8.5 MPa; at the end of testing the block strength had increased to 12.4 MPa.

In later creep and shrinkage tests involving blockwork cavity walls and piers, Lenczner [12] used Aglite lightweight aggregate solid blocks of strength of 3.3 MPa, water absorption of 23.3%, and suction rate of 7.5 kg/min/mm^2. The 28-day strength of the 1:¼:3 cement–lime mortar was 23.4 MPa when strain measurements started on test specimens stored in an environment of 50% RH and 20 °C; after 400 days, vertical shrinkage levels of the wall and pier were 430 and 640×10^{-6}, respectively. As discussed later, this trend was opposite to that expected, since shrinkage of the wall was less than pier, but it could be explained by predrying of blockwork, which was stored at 50% RH before measurements started, so that a faster initial shrinkage of the wall than the pier could have occurred.

Brooks [13] tested a 1–m-high × 2-brick-wide calcium silicate wall and recorded a vertical shrinkage of 210×10^{-6} after 300 days. A 1:½:4½ cement–lime mortar was used having a strength of 12.7 MPa together with solid bricks of strength 27.4 MPa. The walls were covered with polythene sheet for 28 days and then stored at 19 °C and 67% RH. The ultimate vertical shrinkage was estimated from the hyperbolic function (Eqn (7.1)) and found to be 232×10^{-6}.

Information on shrinkage of other types of masonry is scarce. When tested over a 700-day period, CERAM Building Technology [14] found that horizontal shrinkage of carboniferous limestone concrete brick walls built outdoors could reach 600×10^{-6} when shrinkage of the 3-week-old brick itself was 450×10^{-6}. The value for walls exceeded the design values of the now superseded Codes of Practice: BS 5628-3: 1985 [1] and Eurocode 6 [15], but fell within the range of BS EN 1996-1-1: 2005 [16]. When the walls (2.7 m long × 1 m high) were restrained at their ends by returns, shrinkage was reduced by approximately 20%, and reduced by 10% when *bed joint reinforcement* was placed every third course above the damp proof course (dpc). Movements were greater in the lower part of the walls.

CERAM Building Technology [14] also carried out tests on dense and light-weight aggregate concrete masonry units of strengths 20.3 and 17.6 MPa, respectively. After 400 days, the shrinkage of the lightweight blockwork was 470×10^{-6}, and after 850 days, the shrinkage of the dense aggregate blockwork was 75×10^{-6}. Other tests by CERAM Building Technology [14] led to a recommendation that vertical movement of *aircrete* masonry walls should be near the upper end of the range for units given in BS 5628-3: 1985 [1]; *autoclaved aerated concrete (AAC)* masonry units are often referred to as aircrete. The limits of drying shrinkage of various types of *concrete brick*, as determined according to BS 6073-1: 1981 [17], are shown in Table 7.3.

Table 7.3 Shrinkage of Concrete Bricks [14]

Type of Brick	Shrinkage (10^{-6})
Limestone aggregate	250
Semi-perforated	200
Gritstone aggregate	500
Engineering quality	300
100% Oolitic limestone	400
50–90% Oolitic limestone	300
50–90% Carboniferous limestone	350

For precast concrete units and AAC units, the limits of drying shrinkage previously given in BS 6073-1: 1981 [17] were 600 and 900 × 10^{-6}, respectively. That standard has now been superseded by BS EN 771-3: 2003 [18] and BS EN 771-4: 2003 [19], which require the manufacturer to declare values for moisture movement of units.

CERAM Building Technology [14] give the range of shrinkage for AAC units of 400–900 × 10^{-6} as determined from the original dry length, and 200–600 × 10^{-6} for other types of concrete units. The values were obtained from tests carried out according to BS 1881-5: 1970 [20]. CERAM Building Technology emphasized that the figures should not be assumed to be representative of those in a wall as they are moisture strains measured between extreme moisture conditions, namely, on specimens from saturation to oven dry.

Cooper [21] reported that calcium silicate bricks have a potential long-term shrinkage of 400 × 10^{-6}, and the rate of development of shrinkage is slower than for concrete blocks. However, according to CERAM Building Technology [14], it is possible to increase the overall shrinkage under repeated wetting and drying cycles, not all of the shrinkage in each drying cycle being recovered on wetting. By this means it is possible to achieve a total shrinkage that exceeds 400 × 10^{-6}. The same report reiterates that calcium silicate bricks should be kept as dry as practically possible during storage and construction, and the upper limit of 400 × 10^{-6} is consistent with that given in BRE Digest 228 [22]. BS 5628-3: 1985 [1] states that a unit range of 100–400 × 10^{-6} may not be representative of the range of shrinkage that might occur in a wall. CERAM Building Technology [14] reports that the range of shrinkage for calcium silicate units was determined according to the method of BS 1881-5: 1970 [20]. In that test, specimens are dried for a specified period under prescribed conditions of temperature and humidity, which, according to Lea [23] and Neville [24], are equivalent to a long exposure at a RH of 65%.

According to Everett [25], initial irreversible shrinkage of calcium silicate bricks is equal to subsequent reversible moisture movement, namely, 100– 500 × 10^{-6}, and the original BS 187:1978 [26] specified a limit of 400 × 10^{-6} for all classes of brick except that of Class 2 (facing or common bricks). Wet bricks built into long walls tend to crack when they dry, particularly if a strong mortar is used, so again it is recommended that the bricks be kept as dry as possible before and during construction and until construction is complete. If wetting is essential in very hot weather, as little water as possible should be used. Subsequently, BS 187: 1978 was withdrawn and superseded by BS EN 771-2: 2003 [27], but no guidance on moisture movement is offered in the new standard, BS EN 1996-1-1: 2005 [16], which requires that moisture movement should be declared by the manufacturer.

Similarly, there is no guidance given in BS EN 771-5 [28] and BS EN 771-6 [29] on moisture movement of *manufactured stone* and *natural stone* masonry units, respectively, again reliance being placed on the manufacturer to provide data. A method of measurement of moisture movement for aggregate concrete and manufactured stone masonry units is specified by BS EN 772-14 [30]. Here, the total movement is determined as the sum of: (1) measured expansion from the initial condition and after soaking in water and (2) shrinkage from the initial condition and after drying for 21 days in a ventilated oven at 33 °C (see Chapter 16).

Unlike mortar or clay brick, there is little published experimental data on which to ascertain ultimate moisture movement for whole ranges of strength of calcium silicate- and concrete-based units with confidence. However, in the following section, ultimate shrinkage–strength relationships are proposed for calcium silicate, dense aggregate concrete, lightweight aggregate concrete, and AAC units, based on the following assumptions:

- Units are prevented from drying (sealed) until used in construction so that no preshrinkage occurs; otherwise age has to be taken into account.
- Ranges of shrinkage used to develop models are based on those given earlier for the normal class ranges of strength for each type of unit (see below). Also, the ultimate shrinkage values are assumed to represent those obtained for the drying test conditions of BS 1881-5: 1970 [20], which are equivalent to 65% RH and 21 °C [23,24]. A RH of 65% is the average of the United Kingdom (UK) annual average humidity (45%) and UK annual outdoor humidity (85%).
- Ultimate shrinkage applies to masonry units having a V/S ratio = 44 mm, which corresponds to that of a unit embedded within a single-leaf wall.

The proposed ultimate shrinkage– strength relationships are shown in Figure 7.4, the equations being as follows:

Calcium silicate brick (f_{by} = 14–48.5 MPa) and dense aggregate concrete block (f_{by} = 2.8–35.0 MPa):

$$S_{b\infty} = \frac{6000}{f_{by}} \tag{7.7}$$

Lightweight aggregate concrete block (f_{by} = 2.8–10.5 MPa):

$$S_{b\infty} = \frac{1650}{f_{by}} \tag{7.8}$$

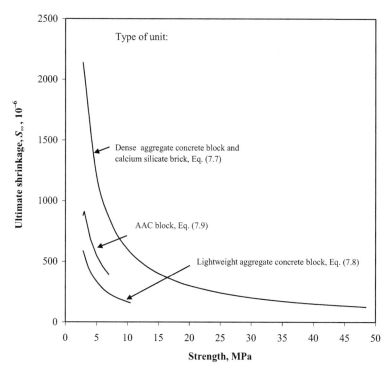

Figure 7.4 Proposed ultimate shrinkage of masonry units as a function of strength; stored at 65% RH; $V/S = 20$ mm (brick) and $V/S = 30$ mm (block).

AAC block ($f_{by} = 2.8$–7.0 MPa):

$$S_{b\infty} = \frac{2730}{f_{by}}$$

(7.9)

It should be emphasized that the validity of the above expressions may be questionable in view of the limited test data on which they are based. In particular, the lack of precise knowledge of the storage conditions prior to use could be a large source of inaccuracy. Nevertheless, when specific data are not available, the expressions are useful inasmuch as they permit approximate estimates of shrinkage of units, say, within 30% of the recorded shrinkage.

Unit Moisture State and Absorption

Tatsa et al. [31] found that presoaking concrete block masonry prisms increased subsequent shrinkage considerably when stored at 20 °C and 60% RH. Compared with dry storage, the 210-day shrinkage of pre-soaked prisms was 63% and 120% greater for hollow block masonry and aerated blocks masonry, respectively.

The importance of moisture state when laying concrete and calcium silicate units was highlighted by CERAM Building Technology [14] since wet units can cause excessive shrinkage of the masonry. Testing of calcium silicate brickwork panels resulted in a range of contraction of $180–270 \times 10^{-6}$, flint–lime brickwork being at the lower limit of the range and sand–lime brickwork at the upper limit. When pre-wetted bricks were used, the shrinkage was greater at 340×10^{-6}. The same report [14] refers to some Dutch research where the occurrence of cracks was influenced by the water content of calcium silicate units at the time of laying. The importance of protection during wet weather to prevent masonry from becoming saturated during construction by covering with polythene sheeting was emphasized.

The effect of prestorage moisture condition on shrinkage of concrete and calcium silicate units has also been mentioned earlier in this chapter and its importance is demonstrated in Figure 7.5, in which three prestorage conditions are considered

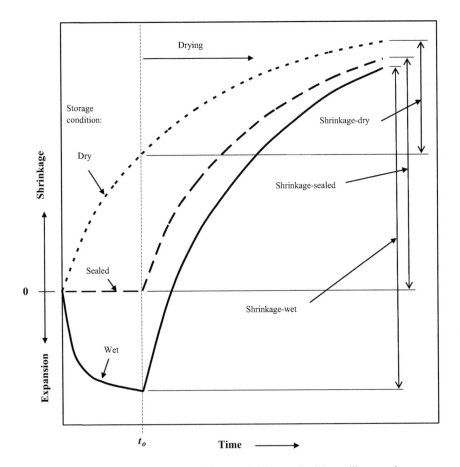

Figure 7.5 Influence of prestorage condition on shrinkage of calcium silicate and concrete units from time t_o

before use in construction at the time of t_o: dry, sealed, and wet. It can be seen that if the units are saturated before use, the subsequent shrinkage is much greater than if the units were stored dry or sealed, the latter representing the usual practical situation where units are stored in shrink-wrapped polythene sheet until required by the brick-layer. Of course, the example applies to shrinkage of unbonded units; however, shrinkage of bonded units within masonry may possibly be affected by absorption of moisture from the fresh mortar.

In order to examine the role of water within the freshly laid mortar on shrinkage of the bonded unit, a *water transfer test* was carried out [32,33] in which the water transport characteristics across the unit/mortar interface of calcium silicate brickwork and concrete blockwork during the setting and hardening processes of mortar were investigated. Tests were performed on brick couplets and block/mortar samples in which the upper brick or block could be removed and weighed periodically after laying the fresh mortar. Figure 7.6 shows the arrangement, which features a polythene mesh grid at the unit/mortar interface to prevent bonding and facilitate easy removal of the unit for weighing; in the case of blocks, a glass plate was used instead of the lower concrete block because of a weight limit on the laboratory balance. After weighing the units, 10 couplets of each type were built so that the upper units could be weighed after periods of up to 70 days. The couplets were cured by sealing under polythene sheet for the first 21 days where the total mass of the system was checked and found to remain constant, indicating there was no loss of water to the outside environment. Subsequently, the couplets were exposed to a drying environment controlled to 65% RH and 21 °C. Figure 7.7 shows that after the initial rapid absorption of water from the freshly laid mortar, there is a slow desorption from the unit, probably due to the demand for water as the cement hydrates. After 21 days, the rate of water loss increases as moisture is lost to the drying environment and, beyond approximately 30 days, additional desorption occurs from the units due to removal of moisture existing in the units prior to bedding with mortar. The pattern of behaviour of the two types of masonry is similar, although the standard water absorption test indicates a higher value for the calcium silicate brick than the concrete block, while the

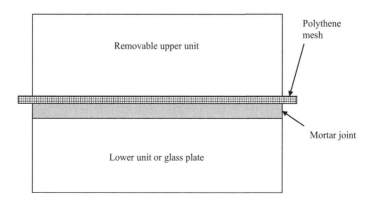

Figure 7.6 Masonry couplet for water transfer test.

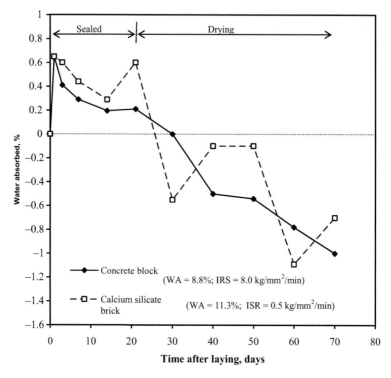

Figure 7.7 Water absorbed by calcium silicate brick and concrete block during the water transfer test; WA = water absorption; ISR = initial suction rate.

standard suction rate test indicates a much greater value for the concrete block than for the calcium silicate brick (Figure 7.7).

The effect of unit moisture content at the time of laying on shrinkage of masonry was also investigated at the same time as the water transfer tests [32,33]. In this case, 3-course-high single-leaf calcium silicate masonry and 2-course-high single-leaf concrete block masonry were instrumented with strain gauges so that bonded units within the masonry wall as well as overall masonry movement could be monitored (see Figure 7.8). Two sets of masonry were constructed: one with dry units and one with units that had been *docked* (wetted under water) for 1 min. Unbonded units and mortar prisms were partly sealed to the same volume/surface ratio as the bonded units and bed face mortar joint in the masonry (see next section). All test samples were stored under polythene for the first 21 days before exposure to drying at 65% RH and 21 °C. Figures 7.9 and 7.10, respectively, show the shrinkage-time characteristics for concrete blockwork and calcium silicate brickwork. During the initial curing (sealed) period, no changes in strain were apparent except for the bonded dry units, which expanded initially by approximately 140×10^{-6} due to the moisture absorbed from the freshly laid mortar. Subsequently, the bonded units underwent shrinkage as moisture transferred back to the mortar, as indicated in Figure 7.7. From the age of

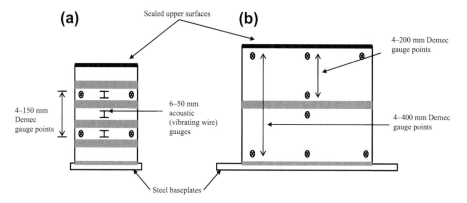

Figure 7.8 Arrangement of test walls for determination of shrinkage in bonded units and mortar (a) 5–stack calcium silicate wall (b) Concrete block couplet.

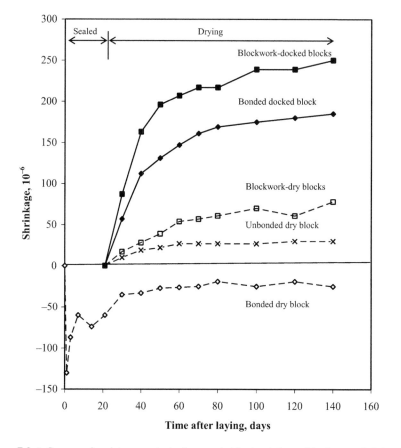

Figure 7.9 Influence of moisture content of concrete blocks at time of laying on shrinkage of bonded blocks and blockwork [32,33].

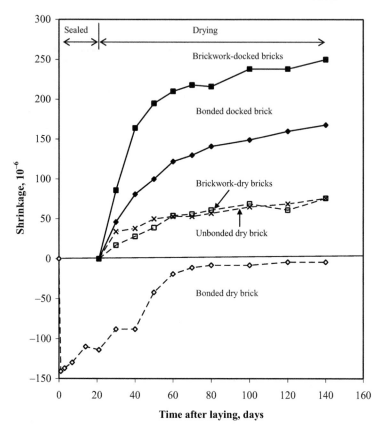

Figure 7.10 Influence of moisture content of calcium silicate bricks at time of laying on shrinkage of bonded bricks and brickwork.

21 days, loss of moisture to the drying environment resulted in the shrinkage-time characteristics shown in Figures 7.9 and 7.10, including further shrinkage of the bonded units. After 140 days, shrinkage of the bonded dry units (measured from 21 days) was greater than shrinkage of the unbonded dry units by 17–40 percent, but the shrinkage of the docked bonded units greatly exceeded that of the bonded dry units by factors of 2.1 and 6.0 for calcium silicate brick and concrete block, respectively. The effect on 140-day shrinkage of concrete blockwork and calcium silicate brickwork built with docked units was an increase by a factor of over three times compared with shrinkage of the same masonry built with dry units, thus demonstrating the importance of ensuring that units are kept in a dry state before being used in construction.

During the same tests [32,33], the shrinkage of the bed mortar joint in the masonry built with dry units was assessed and compared with that of an unbonded mortar prism, partly sealed to the same volume/surface ratio as the mortar bed joint. Because the 10 mm depth of the mortar bed joint was insufficient to install strain measuring devices, its shrinkage (S_{my}) was deduced from the strain measurements of the units and overall strain in the masonry as follows:

$$S_{my} = \frac{gS_{wy} - (g - mn)S_{by}}{mn} \tag{7.10}$$

where g = strain gauge length of masonry, S_{wy} = average shrinkage of the masonry, m = mortar joint thickness, n = number of mortar joints within g, and S_{by} = average shrinkage of the bonded units.

Shrinkage of the 10 mm bed joint of the calcium silicate masonry, shown in Figure 7.8, is given by:

$$S_{my} = \frac{150S_{wy} - 130S_{by}}{20} \tag{7.11}$$

and, for the corresponding concrete block masonry:

$$S_{my} = \frac{400S_{wy} - 390S_{by}}{10} \tag{7.12}$$

Figure 7.11 compares the shrinkage of the mortar bed joint when the masonry is built with dry and docked units, then sealed under polythene sheet for 21 days before being exposed to a drying environment of 65% RH and 21 °C. When laid dry, there is a rapid increase of shrinkage due to the initial rapid absorption of water by the unit from the freshly laid mortar, and then there is an expansion due to transfer of water back from the units as the cement hydrates for the remainder of the curing (sealed) period. The foregoing effect significantly reduces subsequent shrinkage of the mortar bed joint, measured from the age of 21 days. Conversely, the effect of docking the units eliminates preshrinkage during the curing period before allowing full shrinkage to develop on exposure to drying. This situation, of course, also produces maximum shrinkage of unit and masonry (Figures 7.9 and 7.10) which, as stated earlier, is not recommended "good practice" due to the likelihood of cracking.

Figure 7.11 also confirms that docking of units prevents initial water transfer between freshly laid mortar so that the amount of shrinkage of the bed joint, as estimated by Eqs (7.11) and (7.12), is very similar to that determined on separate unbonded prismatic specimens partly sealed to the same volume/surface ratio as the mortar bed joint in the masonry. That observation implies that, in composite modeling of masonry, shrinkage of separate unbonded specimens to represent the unit and mortar phases is not entirely valid for units laid dry, and that water transfer between mortar and unit soon after laying is an important factor in determining shrinkage of calcium silicate and concrete masonry. As discussed earlier, when measured from the start of environmental drying (21 days), shrinkage of bonded dry units is greater than shrinkage of unbonded units. Conversely, in the case of mortar, it can be seen from

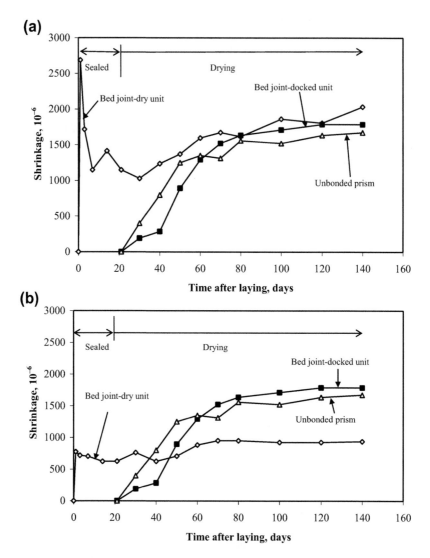

Figure 7.11 Shrinkage of mortar bed joint in masonry built with dry and docked concrete blocks and calcium silicate bricks [33]. (a) Concrete blockwork. (b) Calcium silicate brickwork.

Figure 7.11 that, when measured from a time of 21 days, shrinkage of dry unit mortar bed joint is considerably less than shrinkage of the unbonded prism.

Detailed analysis of the data in Figures 7.9–7.11 revealed that, after approximately 30 days of exposure to drying, the ratio of the dry bonded unit shrinkage to the unbonded unit shrinkage, together with the ratio of bed joint mortar shrinkage to the shrinkage of the unbonded mortar prism, were independent of time. The ratios,

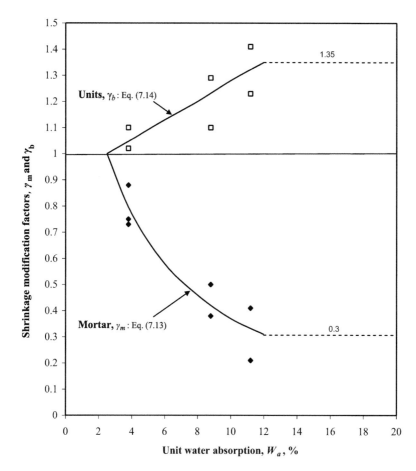

Figure 7.12 Mortar joint and bonded unit shrinkage modification factors due to water absorbed by units when laid dry.

termed the shrinkage modification factors (γ_b, γ_m), are shown in Figure 7.12 and can be conveniently expressed as approximate functions of the unit water absorption (W_a).

The following relationships represent the average trends for the shrinkage modification factors:

Mortar (γ_m):

$$\left. \begin{array}{ll} \text{When } W_a \leq 2.5\%, & \gamma_m = 1 \\[2mm] \text{When } W_a \geq 2.5 \leq 12\%, & \gamma_m = \dfrac{1.81 - 0.036W_a}{1 + 0.29W_a} \\[2mm] \text{When } W_a > 12\%, & \gamma_m = 0.3 \end{array} \right\} \qquad (7.13)$$

Calcium silicate and concrete units (γ_b):

$$\left.\begin{array}{lll} \text{When } W_a \leq 2.5\%, & \gamma_b = 1 \\ \text{When } W_a \geq 2.5 \leq 12\%, & \gamma_b = 1 + 0.037(W_a - 2.5) \\ \text{When } W_a > 12\%, & \gamma_b = 1.35 \end{array}\right\} \qquad (7.14)$$

Figure 7.12 and Eqs (7.13) and (7.14) show that the effect of unit water absorption at the time of bricklaying is to reduce shrinkage of the mortar bed joint, but to increase shrinkage of the bonded unit. The changes are very significant for units of high water absorption, and thus γ_b and γ_m are important factors required for the composite modeling of shrinkage.

Geometry of Cross-Section

Brooks and Bingel [34] showed that geometry of masonry was a factor influencing shrinkage of calcium silicate brickwork and concrete blockwork. Solid calcium silicate bricks of strength 30.0 MPa and lightweight aggregate blocks (Lytag) of strength 8.6 MPa were used with a cement–lime mortar of 7-day strength 6.7 MPa to construct single-leaf walls, cavity wall, hollow piers, and solid piers. The tops of the test specimens were covered, except for the cavity of the cavity walls, to permit drying from the inside surfaces. Two sets of masonry were built: 2-brick-wide × 13-course-high brickwork and 1¼-block-wide × 5-course high blockwork. In addition, 1-brick-wide × 5-course-high model single-leaf brick walls were built and partly sealed, corresponding to the volume to exposed surface area (V/S) ratio of the 13-course-high brickwork so as to simulate the same drying conditions. The results showed that shrinkage decreased as the V/S ratio increased so that, for example, shrinkage of the solid pier was less than that of a single-leaf wall. Also, it was found that shrinkage of the partly sealed model walls satisfactorily represented the shrinkage of the 13-course masonry.

The same geometry effect on shrinkage was confirmed for other types of calcium silicate brickwork and of concrete blockwork [35–37], the sizes and shapes of the masonry being identical to those described above [33]. The strengths of the units were 25.7 MPa for a frogged calcium silicate brick and 13.0 MPa for a dense aggregate solid concrete block, while the strengths of the cement–lime mortar were 7.3 and 6.1 MPa, respectively, for brickwork and blockwork. Figure 7.13 shows a consistent trend of shrinkage with geometry of masonry for calcium silicate brickwork and concrete blockwork throughout the period of testing.

The geometry influence on shrinkage of masonry is similar to that for concrete, which was initially reported in 1966 [38], so that the explanation is the same as that given in the previous chapter dealing with shrinkage of concrete. Essentially, in thicker sections like a solid pier, moisture diffuses over a longer drying path length than in the case of a thin section like a single-leaf wall, and thus shrinkage is slower and less for a solid pier. As mentioned earlier, the geometry factor on shrinkage can be quantified by the V/S ratio, and its general trend for masonry is shown in Figure 7.14

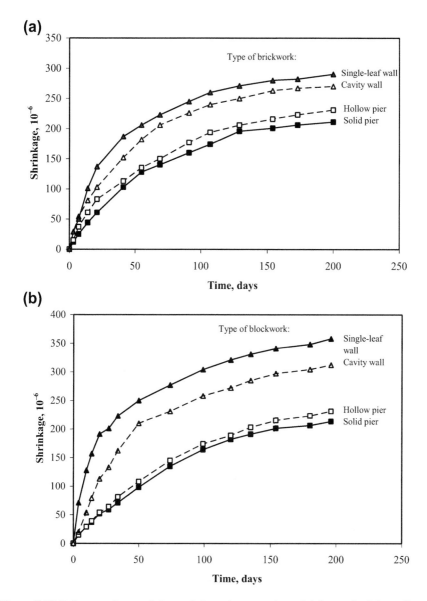

Figure 7.13 Influence of type of size and shape (geometry) on shrinkage of calcium silicate [35–37] (a) Calcium silicate brickwork (b) Concrete blockwork.

for the results of the above-mentioned investigations. Figure 7.14 also demonstrates that shrinkage by simulation of the geometry effect in full-size masonry by the part-sealing of smaller test specimens of 3- or 5-course-high single-leaf walls is in agreement with the general trend, thus suggesting that the testing of smaller sizes of masonry would be more cost-effective than testing large full-scale masonry.

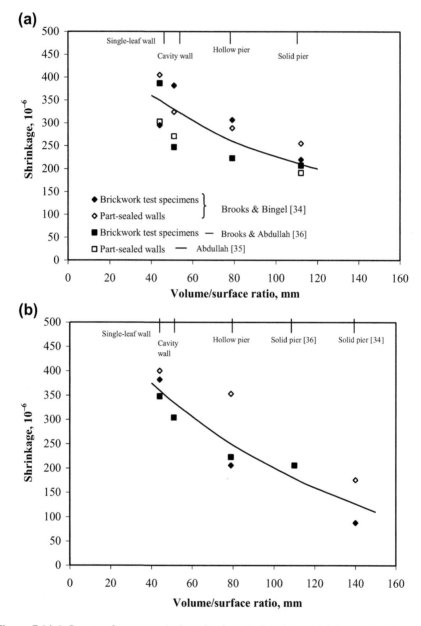

Figure 7.14 Influence of geometry (volume/surface (*V/S*) ratio) on shrinkage of calcium silicate brickwork (a) and concrete blockwork (b) after 180 days.

It is of interest to compare values of V/S for different cross-section geometries of brickwork and blockwork constructed with standard-size units and 10 mm mortar joints. Figure 7.15 shows the comparison in terms of the type of masonry, type of unit, and mortar, the latter two phases being relevant when considering composite modeling of shrinkage. For the examples shown, the V/S ratio ranges from 46 to 138 mm, i.e., by a factor of three and, according to Figure 7.14, the corresponding reduction of shrinkage is similar. In all cases, the V/S ratio of the mortar bed joints is the same as that of the masonry while the V/S ratio of the bonded unit is virtually the same as that of the masonry. For small-section masonry, the V/S of the mortar header joints is similar to that of the bed joints but, for larger sections such as solid piers, the V/S of the mortar header joints can become very large, which implies that, in those instances, the contribution of the shrinkage of the header joints to shrinkage of masonry is very small.

Composite modeling of masonry shrinkage takes into account the influence of geometry on shrinkage of units and mortar by determining the shrinkage of unbonded units and mortar specimens that are partly sealed to simulate the drying conditions of bonded units and mortar joints in masonry. In the case of mortar specimens, typically $200 \times 75 \times 75$ mm unbonded mortar part-sealed prisms are used (see Chapter 16) to represent the V/S ratios of the mortar bed joint in a single-leaf wall, a cavity wall, a hollow pier, and a solid pier; details of the appropriate V/S ratios required are given in Figure 7.15. Some shrinkage-time test results for concrete blocks, calcium silicate bricks, and mortar are shown in Figure 7.16, where it can be seen that simulation of drying in the part-sealed unbonded units results in the same shrinkage trends as for masonry having different geometries (see Figure 7.13). However, as discussed earlier, in the case of undocked units having significant absorption properties, the simulation fails to fully represent shrinkage of the bonded phases because of the water transfer between unit and mortar at the time of bricklaying.

For the family of concrete block, calcium silicate brick. and mortar shrinkage-time curves in Figure 7.16, the coefficients of Eq. (7.1) were obtained by regression analysis of Eq. (7.2). For each material, the ultimate shrinkage, S_∞, and aS_∞ were found to be different functions of V/S [8]; however, when expressed as $R_{S\infty}$ and $R_{aS\infty}$, i.e., ratios relative to those of a single-leaf wall ($V/S = 44$ mm), the functions became reasonably independent of type of unit and mortar. Figure 7.17 shows the relative values of the coefficients of Eq. (7.2), plotted against V/S, which are equal to unity when $V/S = 44$ mm (single-leaf wall); the relationships are as follows:

$$R_{S\infty} = \frac{1.49 + 0.007\left(\frac{V}{S}\right)}{1 + 0.018\left(\frac{V}{S}\right)} \tag{7.15}$$

and

$$R_{aS\infty} = 0.15\left(\frac{V}{S}\right)^{0.5} \tag{7.16}$$

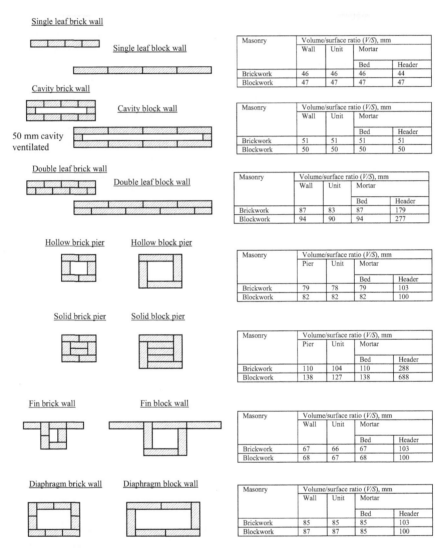

Figure 7.15 Examples of volume/surface (*V/S*) ratio for various types of masonry built with standard-size calcium silicate bricks and standard-size concrete blocks; tops of masonry capped (sealed) except cavity in cavity wall.

The above expressions are applicable to concrete blocks, calcium silicate bricks, and mortar. Values of S_∞ for a single-leaf wall, i.e., $V/S = 44$ mm, are given as a function of strength later and Eq. (7.15) can then be used to estimate the ultimate shrinkage for any other geometry. If it is required to express shrinkage of

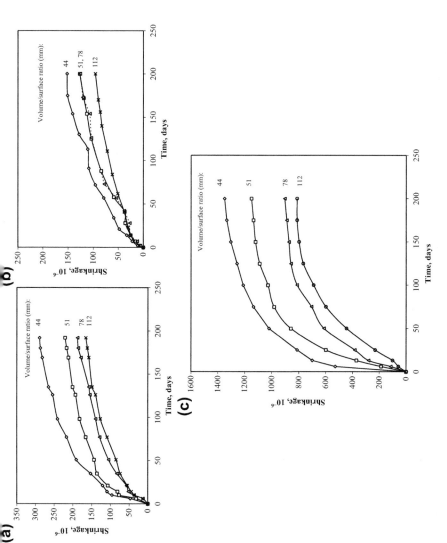

Figure 7.16 Simulation of size effect on shrinkage of units by partly sealing (a) concrete blocks [36], (b) calcium silicate bricks [37], and (c) mortar [36].

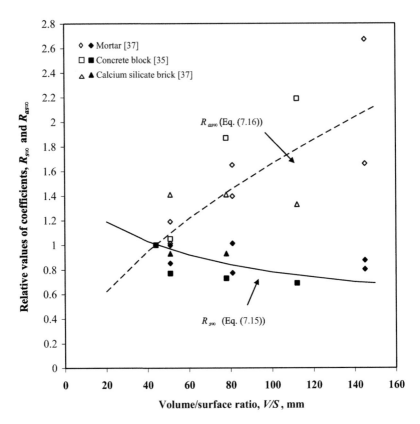

Figure 7.17 Dependency of relative values of coefficients of Eq. (7.1) on volume/surface ratio (V/S) of part-sealed units and mortar specimens; equal to 1 for a $V/S = 44$ mm.

mortar or unit as a function of time, the term aS_∞ is required, and the average values for part-sealed specimens in Table 7.4 may be assumed. For other geometries, aS_∞ can be adjusted according to Eq. (7.16), for example, the right-hand column of Table 7.4 gives aS_∞ for unsealed standard-size units and an unsealed

Table 7.4 Average Values of Coefficient (aS_∞) for Eq. (7.1)

	Coefficient (aS_∞)	
Component of Masonry	**Part-Sealed ($\equiv$ Bonded in Single-Leaf Wall)[a]**	**Unsealed Unbonded Unit or Specimen**
Concrete block	24 ($V/S = 44$ mm)	19 ($V/S = 30$ mm)
Calcium silicate brick	72 ($V/S = 44$ mm)	40 ($V/S = 17$ mm)
Mortar	36 ($V/S = 44$ mm)	19 ($V/S = 16$)

[a]Ends of specimens fully sealed.

mortar prism ($75 \times 75 \times 200$ mm), which would be used for laboratory tests to determine shrinkage.

Anisotropy

In tests on calcium silicate brickwork [4,13], the unbonded brick shrinkage revealed some degree of anisotropy, with shrinkage between bed faces being slightly less ($\approx 85\%$) than shrinkage between header faces. Nevertheless, the corresponding lateral or horizontal shrinkage of the single-leaf wall was almost the same as vertical shrinkage (Figure 7.18). However, more detailed tests described below do suggest some small degree of anisotropy of calcium silicate brickwork. According to the composite model theory presented in Chapter 3, for isotropy of unit shrinkage, shrinkage should be slightly less in the horizontal direction than in the vertical direction, depending on the level of mortar shrinkage (see Figure 3.14). Test results generally confirm that statement to be true in most cases for the calcium silicate brickwork and also for concrete blockwork [35,37], the average value of horizontal shrinkage being 93% of the vertical shrinkage.

Prediction of Shrinkage

Code of Practice Recommendations

According to BS 5628-2:2005 [1], to calculate the loss of prestress in the tendons of posttensioned masonry, the minimum shrinkage of concrete and calcium silicate masonry should be assumed to be 500×10^{-6}.

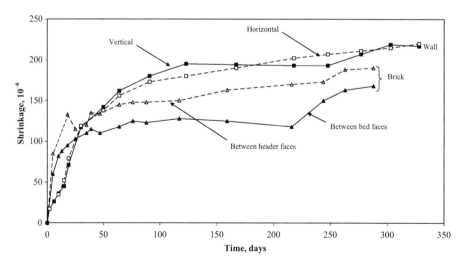

Figure 7.18 Comparison of shrinkage of calcium silicate single-leaf wall and unbonded bricks in the vertical and horizontal directions [4,13].

The current standard, BS EN 1996-1-1:2005 [16], specifies that the shrinkage of masonry should be determined by test and the final value by evaluation of test data. The ranges for masonry (UK National Annex value) are:

- Calcium silicate: 400–1000 $\times$ 10^{-6} (200 $\times$ 10^{-6}).
- Dense aggregate concrete and manufactured stone: 600–1000 $\times$ 10^{-6} (200 $\times$ 10^{-6}).
- Lightweight aggregate concrete: 1000–2000 $\times$ 10^{-6} (400 $\times$ 10^{-6}).
- Autoclaved aerated concrete: $-$200–400 $\times$ 10^{-6} (200 $\times$ 10^{-6}).
- Natural stone: $-$700–400 $\times$ 10^{-6} (100 $\times$ 10^{-6}).

ACI 530.1R-05 [39] relates the shrinkage of concrete masonry (S_w) to shrinkage of unit (S_b) as follows:

- $S_w = 0.15 S_b$, but not more than 650 $\times$ 10^{-6}, for masonry made of moisture-controlled concrete units.
- $S_w = 0.5 S_b$ for masonry made of nonmoisture-controlled concrete units.

Composite Model

The vertical shrinkage of calcium silicate brickwork and solid concrete blockwork is given by Eqs (3.80) and (3.82), respectively (see also Tables 3.1 and 3.2), viz.:

$$S_{wy} = 0.86 S_{by} + 0.14 S_m \qquad (7.17)$$

$$S_{wy} = 0.952 S_{by} + 0.048 S_m \qquad (7.18)$$

In the above expressions, S_{by} and S_m are the shrinkage of unbonded units and mortar, respectively, which, in the theoretical derivation of composite models, are assumed equal to the corresponding shrinkage of bonded units and mortar joints. However, it was demonstrated in the water transfer test on p. 197 that, after laying, dry units absorb water from the fresh mortar so that the subsequent shrinkage of the bonded units is increased, but there is a corresponding reduction of shrinkage of the mortar joints. To allow for these features, shrinkage modification factors, γ_b and γ_m, are incorporated in Eqs (7.17) and (7.18) as follows:

$$S_{wy} = 0.86 \gamma_b S_{by} + 0.14 \gamma_m S_m \qquad (7.19)$$

$$S_{wy} = 0.952 \gamma_b S_{by} + 0.048 \gamma_m S_m \qquad (7.20)$$

The shrinkage modification factors are related to the water absorption of the units, as given by Eqs (7.13) and (7.14), respectively, or given by Table 7.5.

Now consider first the equations for single-leaf masonry, in which both mortar and units have V/S ratios $= 44$ mm and the masonry is stored in a drying environment of 65% RH and 21 °C. Shrinkage of an unbonded mortar specimen at any time t (S_m)

Table 7.5 Shrinkage Modification Factors for Mortar Joints and Bonded Units

Water Absorption of Unit (%)	Shrinkage Modification Factor	
	Mortar, γ_m (Eq. (7.13))	Units, γ_b (Eq. (7.14))
≤2.5	1	1
3	0.91	1.02
4	0.77	1.06
6	0.58	1.13
8	0.46	1.20
10	0.37	1.28
12	0.31	1.35
>12	0.30	1.35

after curing under polythene sheet is given by Eq. (7.1), with $aS_{m\infty} = 35.7$ (Table 7.4) and $S_{m\infty}$ from Eq. (7.6):

$$S_m = \frac{(2600 - 60f_m)t}{35.7 + t} \tag{7.21}$$

Assuming the shrinkage of units is isotropic, viz. $S_{by} = S_{bx} = S_b$, the shrinkage-time expression of <u>unbonded</u> concrete blocks or calcium silicate bricks is also given by Eq. (7.1), but with $aS_{b\infty} = 23.5$ or 71.9 (Table 7.4) and $S_{b\infty}$ from Eqs (7.7)–(7.9). In the case of calcium silicate units, shrinkage at any time is given by:

$$S_b = \frac{\left(\frac{6000}{f_{by}}\right)t}{71.9 + t} \tag{7.22}$$

and, for dense aggregate concrete units:

$$S_b = \frac{\left(\frac{6000}{f_{by}}\right)t}{23.5 + t} \tag{7.23}$$

For lightweight aggregate concrete units, shrinkage at any time t is given by:

$$S_b = \frac{\left(\frac{1650}{f_{by}}\right)t}{23.5 + t} \tag{7.24}$$

and, lastly, shrinkage of AAC units:

$$S_b = \frac{\left(\frac{2730}{f_{by}}\right)t}{23.5 + t} \tag{7.25}$$

To allow for the geometry effect on shrinkage of masonry other than for single-leaf walls ($V/S = 45$ mm), then S_∞ and aS_∞ in the foregoing equations have to be multiplied by $R_{S\infty}$ (Eq. (7.16)) and $R_{aS\infty}$ (Eq. (7.17)), respectively. Furthermore, for other storage conditions having a RH different from 65%, but similar temperature, the following factor may be used, as recommended by the CEB 1990 Model Code expression for shrinkage of concrete [10]:

$$R_{RH} = 1.379\left[1 - (0.01RH)^3\right] \qquad (7.26)$$

where R_{RH} = relative humidity factor (= 1 when RH = 65%).

The shrinkage-time expressions for shrinkage of mortar and units at any time can now be obtained by combining Eqs (7.15), (7.16), and (7.26) with Eqs (7.21)–(7.25). The resulting equations for any time and for ultimate shrinkage are listed in Table 7.6 (Eqs (7.27–7.31)). After determining the shrinkage modification factors γ_m and γ_b from Eqs (7.13) and (7.14), respectively, the shrinkage of masonry can now be obtained from either Eqs (7.19) or (7.20) by substituting the value for S_m from Eq. (7.27) and the appropriate value of S_b from one of Eqs (7.28)–(7.31).

The accuracy of predicting shrinkage of unbonded mortar specimens by Eq. (7.27a) has been assessed for a variety of curing conditions prior to the start of shrinkage measurements: (1) cured moist, under polythene sheet; (2) cured dry, pre-shrinkage allowed; and (3) cured in water. Figure 7.19 shows the comparison between predicted and measured shrinkage, the overall accuracy as determined by the error coefficient (Eq. (5.4)) being 16% for an average shrinkage of 1575×10^{-6}. Corresponding assessment of accuracy of predicting shrinkage of units has not been possible because, as stated earlier, very limited experimental results are available from which to derive unit shrinkage–strength relationships. A similar situation exists for assessing the accuracy of predicting the overall shrinkage of masonry by the composite models. Even when data do exist, publications often lack important experimental details of unit water absorption and prestorage conditions of the masonry before the start of shrinkage measurements. However, where reasonable assumptions can be made with shrinkage-time data [4,11,34,35], accuracy of shrinkage as determined by the error coefficient is within 25%. No test data are available for assessing the accuracy of predicting shrinkage of AAC masonry.

To solve the composite model equations for the estimation of shrinkage of masonry, the following input data are required:

- f_m = strength of mortar.
- f_{by} = strength of unit.
- V/S = volume/surface ratio for type of masonry.
- RH = relative humidity of drying environment.
- t_o = age at first exposure to drying.
- t_s = age at start of shrinkage measurement.
- t = age at which shrinkage is required.

The following examples illustrate the application of the composite model to estimate shrinkage of calcium silicate and concrete masonry.

Table 7.6 Equations for Predicting Shrinkage of Mortar, Calcium Silicate, and Concrete Units

Material	Shrinkage at Any Time (10^{-6})		Ultimate Shrinkage (10^{-6})	
	Equation	No.	Equation	No.
Mortar	$S_m = \left\{ \dfrac{27.58(130 - 3f_{mo})t}{3.57\left(\frac{V}{S}\right)^{0.61} + t} \right\}\left[1.2 - 0.02\left(\frac{V}{S}\right)^{0.61}\right]\left[1 - (0.01RH)^3\right]$	(7.27a)	$S_{mo} = 27.58(130 - 3f_{mo})\left[1.2 - 0.02\left(\frac{V}{S}\right)^{0.61}\right]\left[1 - (0.01RH)^3\right]$	(7.27b)
Calcium silicate unit	$S_b = \left\{ \dfrac{\left(\frac{8274}{f_{by}}\right)t}{7.19\left(\frac{V}{S}\right)^{0.61} + t} \right\}\left[1.2 - 0.02\left[\frac{V}{S}\right]^{0.61}\right]\left[1 - (0.01RH)^3\right]$	(7.28a)	$S_{bo} = \left(\frac{8274}{f_{by}}\right)\left[1.2 - 0.02\left(\frac{V}{S}\right)^{0.61}\right]\left[1 - (0.01RH^3)\right]$	(7.28b)
Dense aggregate concrete unit	$S_b = \left\{ \dfrac{\left(\frac{8274}{f_{by}}\right)t}{2.35\left(\frac{V}{S}\right)^{0.61} + t} \right\}\left[1.2 - 0.02\left[\frac{V}{S}\right]^{0.61}\right]\left[1 - (0.01RH)^3\right]$	(7.29a)	$S_{bo} = \left[\frac{8274}{f_{by}}\right]\left[1.2 - 0.02\left(\frac{V}{S}\right)^{0.61}\right]\left[1 - (0.01RH)^3\right]$	(7.29b)
Lightweight aggregate unit	$S_b = \left\{ \dfrac{\left(\frac{2275}{f_{by}}\right)t}{2.35\left(\frac{V}{S}\right)^{0.61} + t} \right\}\left[1.2 - 0.02\left(\frac{V}{S}\right)^{0.61}\right]\left[1 - (0.01RH)^3\right]$	(7.30a)	$S_{bo} = \left[\frac{2275}{f_{by}}\right]\left[1.2 - 0.02\left(\frac{V}{S}\right)^{0.61}\right]\left[1 - (0.01RH^3)\right]$	(7.30b)
AAC unit	$S_b = \left\{ \dfrac{\left(\frac{3765}{f_{by}}\right)t}{2.35\left(\frac{V}{S}\right)^{0.61} + t} \right\}\left[1.2 - 0.02\left(\frac{V}{S}\right)^{0.61}\right]\left[1 - (0.01RH)^3\right]$	(7.31a)	$S_{bo} = \left[\frac{3765}{f_{by}}\right]\left[1.2 - 0.02\left(\frac{V}{S}\right)^{0.61}\right]\left[1 - (0.01RH^3)\right]$	(7.31b)

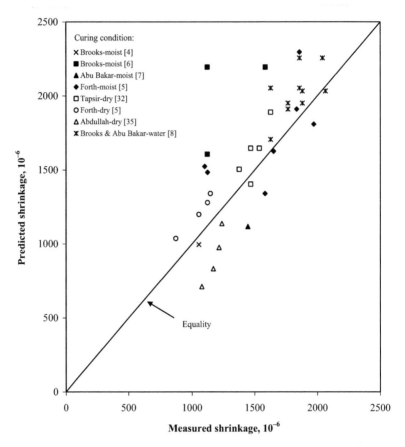

Figure 7.19 Comparison of predicted and measured shrinkage of mortar specimens cured under different conditions; prediction based on strength at the age of first exposure to drying [8].

Examples

1. It is required to estimate the ultimate shrinkage movement of concrete blockwork that forms the inside leaf of a cavity wall of a recreational building, 8 m long and 3 m high. The units are of a dense aggregate concrete type of strength 15 MPa and water absorption of 12%, bonded with a Class M6 designation (ii) mortar as specified by BS 5628: 2005. Calculate both the vertical and horizontal movement after 50 years.

Solution

Assuming the age at first exposure to drying, $t_o = 1$ day, $f_m = 1$-day strength of mortar is given by Table 7.2 as 0.4×28-day strength $= 0.4 \times 6 = 2.4$ MPa. Also, Figure 7.15 gives the V/S ratio for the mortar bed joint and unit of a single-leaf block wall as 47 mm. Assuming the average RH $= 45\%$ for the

Examples—cont'd

interior of an air-conditioned building and ultimate shrinkage occurs after 50 years, substitution in Eq. (7.27b) from Table 7.6 yields $S_{m\infty}$ as follows:

$$S_{m\infty} = [7.58(130 - 3 \times 2.4)]\left[1.2 - 0.02(45^{0.61})\right]$$
$$\left[1 - (0.01 \times 45)^3\right]$$
$$= 842.7 \times 10^{-6}$$

From Eq. (7.29b), the ultimate shrinkage of a dense aggregate concrete unit is:

$$S_{b\infty} = \left[\frac{8274}{15}\right][1.2 - 0.02(45^{0.61})]\left[1 - (0.01 \times 45)^3\right]$$
$$= 499.4 \times 10^{-6}$$

Now, for a water absorption $= 10\%$, the mortar and unit modification factors are given by Eqs (7.13) and (7.14) as $\gamma_m = 0.37$ and $\gamma_b = 1.28$, respectively. Substitution of $S_{m\infty}$, $S_{b\infty}$, γ_m, and γ_b in Eq. (7.20) gives:

$$S_{wy} = 0.952 \times 1.28 \times 499.4 + 0.048 \times 0.37 \times 842.7$$
$$= 624 \times 10^{-6}$$

Remembering that shrinkage is measured in units of microstrain (10^{-6}), for a height of 3 m, the 50-year contraction due to shrinkage is given by $624 \times 10^{-6} \times 3 \times 1000 = \underline{1.9 \text{ mm}}$. Assuming that horizontal shrinkage is the same as vertical shrinkage, the 50-year horizontal contraction of the concrete block wall is $624 \times 10^{-6} \times 5 \times 1000 = \underline{3.1 \text{ mm}}$.

2. A diaphragm wall is constructed from calcium silicate brick of strength 25 MPa using a cement–lime mortar of 28-day strength of 12 MPa. Post-tensioning is to be carried out after 6 weeks after construction and it is required to calculate the shrinkage of the calcium silicate brickwork at the age of 10 years to allow for the loss of prestress in the posttension bars. Assume the average RH = 65% and the water absorption = 8%.

Solution

The shrinkage resulting in the loss of prestress is the 10-year shrinkage minus the shrinkage at the age of 6 weeks. Assuming shrinkage starts 24 h after construction, shrinkage of mortar at the age of 6 weeks is given by Eq. (7.27a), with $t = 41$ days and $f_m = 0.4 \times 12 = 4.8$ MPa:

Continued

Examples—cont'd

$$S_m = \left[\frac{27.58(130 - 3 \times 4.8)41}{3.57(85^{0.61}) + 41}\right]\left[1.2 - 0.02(85)^{0.61}\right]\left[1 - (0.01 \times 65)^3\right]$$

$$= 901.0 \times 10^{-6}$$

Shrinkage of mortar after 10 years is given by Eq. (7.27a), with $t = 10 \times (365 - 1) = 3649$ days, namely, 2047.8×10^{-6}. Hence, deducting preshrinkage prior to posttensioning, shrinkage of mortar affecting loss of prestress $S'_m = 1146.8 \times 10^{-6}$.

Shrinkage of the calcium silicate units at the age of 6 weeks is given by Eq. (7.28a):

$$S_b = \left[\frac{\left(\frac{8742}{25}\right)41}{7.19(85)^{0.61} + 41}\right]\left[1.2 - 0.02(85)^{0.61}\right]\left[1 - (0.01 \times 65)^3\right]$$

$$= 82.0 \times 10^{-6}$$

and at the age of 10 years is 339.6×10^{-6}. Deduction of shrinkage at the age of 6 weeks leaves $S'_b = 257.6 \times 10^{-6}$, which affects the loss of prestress.

Now, the shrinkage modification factors arising from the transfer of water from fresh mortar to newly laid unit for a unit water absorption $= 8\%$ are $\gamma_m = 0.46$ and $\gamma_b = 1.2$ (Table 7.5). Substitution of the shrinkage modification factors, together with S'_m and S'_b, in Eq. (7.19) yields the shrinkage of the diaphragm wall contributing to loss of prestress, $S'_{wy} = \underline{340 \times 10^{-6}}$

Concluding Remarks

The main factors affecting shrinkage of calcium silicate brickwork and concrete blockwork have been identified and quantified. In addition to the individual shrinkage-time characteristics of unbonded unit and mortar, an important influence arises from the initial transfer of water between mortar and unit due to suction/absorption of the freshly laid unit, which has the effect of increasing shrinkage of the bonded unit and reducing shrinkage of the mortar joints.

Ultimate values of shrinkage of unbonded mortar and units depend on the average RH of the environment, strength, and geometry or volume/surface area ratio. The water transfer effect is quantified by shrinkage modification factors that can be related to the water absorption of the unit. The rate development of shrinkage with time of units and mortar also depends on the geometry of the masonry. The importance of keeping the units dry prior to construction and protecting masonry during construction is emphasized because of the significant influence of moisture state on potential shrinkage of units. For example, appreciable enhancement in shrinkage of calcium silicate and concrete units can occur as a result of prewetting or docking.

The ultimate shrinkage–strength characteristics of some concrete units given in this chapter should be regarded as tentative in view of the lack of experimental data. Consequently, confirmation of prediction of corresponding shrinkage of masonry by composite models remains outstanding. On the other hand, for calcium silicate and dense aggregate concrete units for which more shrinkage–strength data exist, estimates of shrinkage of corresponding masonry by composite models are acceptable.

Problems

1. Does shrinkage due to carbonation contribute to shrinkage of masonry?
2. How does type of curing affect the shrinkage of mortar?
3. Compare the potential shrinkage of concrete units for the following prestorage conditions: (a) wet units; (b) sealed units; and (c) dry units.
4. What is a potential problem with wetting calcium silicate and concrete units prior to laying? Does the same problem occur with clay units?
5. Describe the moisture movement of a calcium silicate brick after laying in fresh mortar for sealed storage conditions followed by drying.
6. For the same conditions, describe the corresponding moisture movement for mortar.
7. Define the mortar and unit shrinkage modification factors.
8. Describe a test in which you could determine shrinkage modification factor.
9. Explain the so-called geometry effect on shrinkage of masonry.
10. How can the geometry effect be quantified? Give examples for a single-leaf wall and a solid pier.
11. Is the rate of development of shrinkage of a solid pier different from that of a single-leaf wall? Explain your answer.
12. Is shrinkage of calcium silicate or concrete masonry isotropic?
13. Calculate the ultimate shrinkage of calcium silicate double-leaf wall for a 28-day mortar strength $= 15$ MPa, brick strength $= 23$ MPa, and water absorption $= 10\%$. The wall is protected with polythene sheet for 3 days before exposure to an environment with an RH $= 65\%$.
    Answer: 356×10^{-6}

References

[1] BS 5628. Code of practice for the use of masonry, Part 3: materials and components, design and workmanship. British Standards Institution; 1985 (withdrawn) and 2005 (withdrawn).

[2] Forth JP, Brooks JJ. Influence of mortar type on the long-term deformation of single leaf clay brick masonry. In: Proceedings of the fourth international masonry conference. British Masonry Society; 1995. pp. 157–61.

[3]   BS 4551. Methods of testing mortars, screeds and plasters. British Standards Institution; 1980.

[4]   Brooks JJ. Composite models for predicting elastic and long-term movements in brickwork walls. In: Proceedings of the British Masonry Society, Stoke-on-Trent, No. 1; 1986. pp. 20–3 (Presented at eighth international conference on loadbearing brickwork (building materials section), British Ceramic Society, 1983).

[5]   Forth JP. Influence of mortar and brick on long-term movements of clay brick masonry [Ph.D. thesis]. School of Civil Engineering, University of Leeds; 1995.

[6]   Brooks, JJ. Time-dependent behaviour of masonry and its component phases, Final report, EC science and technology programme, Contract No. C11-0926, Brussels; 1996.

[7]   Abu Bakar BH. Influence of anisotropy and curing on deformation of masonry. [Ph.D. thesis]. School of Civil Engineering, University of Leeds;1998.

[8]   Brooks JJ, Abu Bakar BH. Shrinkage and creep of masonry mortar. Mater Struct 2004;37:177–83.

[9]   Ross AD. Concrete creep data. Struct Eng 1937;15(8):314–26.

[10] CEB-FIP Model Code for Concrete Structures 1990. Evaluation of the time-dependent behaviour of concrete, Bulletin d'Information No. 199. Lausanne: Comite Europeen du Beton/Federation Internationale de la Precontrainte; 1991.

[11] Lenczner D. Creep in concrete blockwork piers. Struct Eng 1974;52(3):97–101.

[12] Lenczner D. Creep in brickwork and blockwork cavity walls and piers. Proc Brick Ceram Soc 1979;27:53–66.

[13] Brooks JJ. Time-dependent behaviour of calcium silicate and Fletton clay brickwork walls. In: West WHW, editor. Proceedings of the British Masonry Society, Stoke-on Trent, No.1; 1986. pp. 17–9 (Presented at eighth international conference on load-bearing brickwork (building materials section), British Ceramic Society, 1983).

[14] CERAM Building Technology. Movement in masonry. Partners Technol Rep; 1996:119.

[15] DD ENV 1996-1-1. Design of masonry structures, Part1.1: General rules for reinforced and unreinforced masonry. British Standards Institution; 1996, Eurocode 6 (withdrawn).

[16] BS EN 1996-1-1. Design of masonry structures, general rules for reinforced and unreinforced masonry structures (see also NA BS EN 1996-1-1: 2005). British Standards Institution; 2005, Eurocode 6.

[17] BS 6073–6081. Precast concrete masonry units, Part1: Specification for precast concrete units. British Standards Institution; 1981 (withdrawn).

[18] BS EN 771–773. Specification for masonry units, aggregate concrete masonry units (dense and lightweight aggregates). Includes UK NA, British Standards Institution; 2003.

[19] BS EN 771–774. Specification for masonry units, autoclaved aerated concrete. British Standards Institution; 2003.

[20] BS 1881–1885. Testing concrete, Methods of testing hardened concrete for other than strength. British Standards Institution; 1970.

[21] Cooper P. Movement and cracking in long masonry walls. CIRIA Practice Note, Special Publication; 1986. 44.

[22] Building Research Establishment. Estimation of thermal and movement stresses: Part 2, Digest No. 228; 1979.

[23] Lea FM. The chemistry of cement and concrete. London: Arnold; 1970.

[24] Neville AM. Properties of concrete. 4th ed. Pearson Prentice Hall; 2006. 844 pp.

[25] Everett A. Mitchell's materials. 5th ed. Longman Scientific & Technical; 1994. 257 pp.

[26] BS 187. Specification for calcium silicate (sandlime and flintlime) bricks. British Standards Institution; 1978 (withdrawn).

[27] BS EN 771–772. Specification for masonry units. Calcium silicate masonry units. British Standards Institution; 2003.

[28] BS EN 771–775. Specification for masonry units. Manufactured stone masonry units. British Standards Institution; 2003.

[29] BS EN 771–776. Specification for masonry units. Natural stone masonry units. British Standards Institution; 2003.

[30] BS EN 772–814. Methods of test for masonry units. Determination moisture movement aggregate concrete manufactured stone masonry units. British Standards Institution; 2002.

[31] Tatsa E, Yishai O, Levy M. Loss of steel prestress in prestressed concrete blockwork walls. Struct Eng 1973;51(5):177–82.

[32] Tapsir SH. Time-dependent loss of prestress in post-tensioned diaphragm and fin masonry walls [Ph.D. thesis]. Department of Civil Engineering, University of Leeds; 1994: 272 pp.

[33] Forth JP, Brooks JJ, Tapsir SH. The effect of unit water absorption on long-term movements of masonry. Cem Concr Compos 2000;22:273–80.

[34] Brooks JJ, Bingel PR. Influence of size on moisture movements in unrestrained masonry. Mason Int 1985;(4):36–44.

[35] Abdullah CS. Influence of geometry on creep and moisture movement of clay, calcium silicate and concrete masonry [Ph.D. thesis]. Department of Civil Engineering, University of Leeds; 1989: 290 pp.

[36] Brooks JJ, Abdullah CS. Creep and drying shrinkage of concrete blockwork. Mag Concr Res 1990;42(150):15–22.

[37] Brooks JJ, Abdullah CS. Geometry effect on creep and moisture movement of brickwork. Mason Int 1990;3(3):111–4.

[38] Hanson TC, Mattock AH. Influence of size and shape of member on shrinkage and creep of concrete. ACI J 1966;63:267–90.

[39] ACI Committee 530.1R-05. Commentary on specifications for masonry structures, ACI manual of concrete practice. American Concrete Institute; 2007. Part 6.

8 Moisture Movement of Clay Brick Masonry

As stated in the previous chapter, movements arising from changes in moisture levels in masonry materials are either reversible or irreversible in nature but, although calcium silicate and concrete units undergo shrinkage, clay units exhibit moisture expansion. The general reversible moisture movement of clay bricks is less than 100×10^{-6} (0.01%) and is regarded as negligible in design compared with the irreversible moisture expansion of clay units, which ranges from 200 to 1800×10^{-6} (0.02–0.18%) depending on the type of clay (see Figure 7.1). The irreversible moisture expansion of fired clay units is a unique property that begins as soon as the bricks have cooled after leaving the kiln, expanding rapidly at first, then slowly over along period of time.

Moisture movement of masonry is the net result of the combined moisture movements of units and mortar. Like concrete, mortar exhibits both reversible and irreversible shrinkage, and its effect in clay masonry is to restrain the expansion of the clay bricks; the restrained effect often results in irreversible expansion of clay masonry, but sometimes shrinkage of clay masonry can occur. The moisture movement of clay masonry depends on the combination of factors that affect the shrinkage of mortar, such as strength, as well as the factors that affect irreversible expansion of brick, such as the type and age of the fired clay. Moreover, the transfer of moisture across the mortar/brick interface can play an influential role.

Compared with calcium silicate and concrete units, moisture movement of fired-clay bricks has received far more attention due to the particular property of irreversible moisture expansion in the design of brickwork with the need to avoid cracking and spalling. In this chapter, the topic is reviewed historically, then influencing factors are discussed before presenting a new analytical method for estimating moisture movement of clay units. A corresponding review of irreversible moisture movement of masonry is then presented, influencing factors are identified, and design code of practice guidelines are given. Finally, a prediction method is presented and validated with independent, published test data.

Irreversible Moisture Expansion of Clay Units

According to Hosking et al. [1], irreversible moisture expansion is caused by chemical reactions between water and certain constituents of ceramic bodies, rather than due to the physical absorption of water. Specifically, the reactions are the hydration of amorphous silicates, silica, glasses, and alumina constituents of the fired clay; in addition, free lime and some salts are capable of hydration and, therefore, of contributing to

Concrete and Masonry Movements. http://dx.doi.org/10.1016/B978-0-12-801525-4.00008-X

expansion. Jessop [2] states that expansion cannot be produced by capillary-absorbed water because this water would be in hydrostatic tension, thus tending to produce a contraction. On the other hand, physical adsorption of water could reduce intermolecular-induced compressive stress within the solid clay, resulting in elastic expansion. However, that physical phenomenon cannot be the main cause of expansion since adsorbed water can be removed by heating to 100 °C and, to reverse moisture expansion, temperatures of 600 °C or more are required. Jessop [2] reiterates the opinions of Hosking et al. that the most significant reactions causing "permanent" expansion are between amorphous silica, γ- alumina, and glass, since it has been shown that these reactions are not reversed at ordinary temperatures. Also, some reactions are far more active in steam than in water vapor.

In a review of early work on irreversible moisture expansion, West [3] states that kiln-fresh briquettes expand at a rate that is rapid at first and then decreases with time. The glass phase in a ceramic body is responsible for moisture expansion, the availability of liquid water being not essential; as, for example, bricks in air at 30% relative humidity expand more slowly, but in a similar manner to bricks soaked in water. High-pressure steam accelerates moisture expansion by a different mechanism to that of natural expansion so that such treatment does not provide a reliable basis for a moisture expansion test, unlike the use of hot water. There is a firing temperature at which expansion is a maximum, which for most clays is between 900 and 1000 °C. West [3] also states that the moisture expansion of a whole wire-cut brick is not the same as that of a smaller briquette cut from it.

About the same time, comprehensive test data were reported by Freeman and Smith [4] for 10 different clays fired to a range of temperatures; nine of the clays were used in brick manufacture in the United Kingdom. The main findings were: changing the relative humidity of storage from 65 to 90% increased the moisture expansion by 20%; high pressure steam treatment caused a large expansion, which did not correlate well with natural expansion, a finding that was in agreement with the finding of West [3]. On the other hand, steam treatment at atmospheric pressure did correlate well with expansion occurring under normal conditions. Expansion occurring in the first day after firing was large and not closely related to subsequent 127-day expansion suggesting different mechanisms; for example, surface energy release caused by physical or chemical adsorption (rapid) followed by a slower hydration of constituent of the fired body. There was no significant difference in expansion between solid and perforated bricks.

In a paper dealing with the influence of composition on the same test data, Freeman [5] found that porosity had a dominating effect on expansion. Crystallization of the Ca- and Mg-bearing phases tended to be associated with moderate or very low expansions that were little affected by changes in firing temperature, whereas the absence of such phases was associated with high expansions at low firing temperatures that could be greatly reduced by harder firing. Differences in clay mineralogy appear to have an insignificant effect on expansion.

For bricks stored outside, Laird and Wickens [6] found that expansions were greater for exposed conditions than for storage under cover, and expansions of bricks standing for 1 month prior to testing were significantly less than expansions of kiln-fresh bricks.

Jessop [2] states that increasing the relative humidity of the atmosphere tends to increase the rate of moisture expansion. Also, cyclic wetting and drying of bricks at 21 °C and drying at 100 °C results in far greater expansions than if the bricks are continuously stored at 21 °C or subjected to cyclic wetting and drying at 18 °C.

Smith [7] reported the long-term expansion results of the original test program started by Freeman and Smith [4]; the average expansion of bricks stored at 65% and 90% relative humidity (RH) was reported and the time of zero expansion was taken as day 1, thus eliminating expansion occurring in the first 24 h. The 7.5-year expansion was approximately 1.75 times that at 127 days. Of the clays used mostly in the manufacture of bricks, which were all made by the extrusion process, the largest expansions occurred with Weald clay fired to 950 °C, Carboniferous shale fired to 950 and 1025 °C, and Devonian shale fired to 1025 °C. The London stock clay bricks had low expansions at all firing temperatures. It can be seen in Figure 8.1 that irreversible moisture expansion of some clays is very sensitive to firing temperature. An implication is that expansions may be four times greater for bricks drawn from different parts of the kiln, even though the actual temperature difference is not greater than 60 °C [2].

Freeman and Smith [4], and McDowell and Birtwistle [8] found linear relationships between natural expansion after 4−6 months and short-term expansion in steam (4−6 h). Subsequently, Lomax and Ford [9] correlated accelerated expansion using steam at atmospheric pressure with the natural expansion occurring after 5 years, which was about 1.6 times the accelerated expansion after 4 h. At the start of testing, all bricks were less than 2 weeks old and were reconditioned to the "kiln-fresh" state

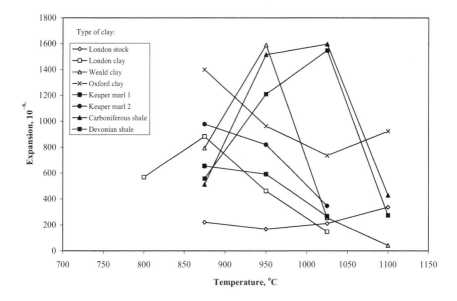

Figure 8.1 Influence of clay type on long-term irreversible expansion of bricks fired to different temperatures [4].

by a desorption procedure. The natural expansion storage conditions were outdoor and undercover, apart from taking measurements when bricks were stored at constant temperature for 24 h beforehand. Assuming that natural expansion increased linearly with logarithm of time, the 50-year expansion was found to be 2.5 times the 4-h expansion in steam. Figure 8.2 shows the relationship between accelerated expansion in steam (M_s) and measured natural 5-year expansion. Also shown is the extrapolated expansion (M_{50}) as a function of expansion in steam with a 99% upper confidence limit, the relationship in microstrain being:

$$M_{50} = 2.5M_s + 0.25 \tag{8.1}$$

Using the upper confidence limit, Lomax and Ford [9] categorized expansion of bricks into low, L (<0.4 mm/m or $<400 \times 10^{-6}$); medium, M ($0.4{-}0.8$ mm/m or 400 to 800×10^{-6}); and high, H (>0.8 mm/m or $>800 \times 10^{-6}$). The categorization of irreversible moisture expansion by L, M, and H is used by U.K. brick manufacturers. Using Smith's results [7], Foster and Tovey [10] proposed a similar procedure to classify irreversible moisture expansion of clay units into three groups: low ($0{-}500 \times 10^{-6}$), medium ($500{-}1000 \times 10^{-6}$), and high ($1000{-}1600 \times 10^{-6}$).

Brooks and Forth [11] reported test data for irreversible moisture expansion of 20 different types of clay brick and, together with previous published results, developed a model to predict the expansion of any type of clay brick fired at any temperature and age. The types of brick were chosen to have a range of irreversible moisture expansion and strength according to the manufacturer's data. Table 8.1 gives a description of the

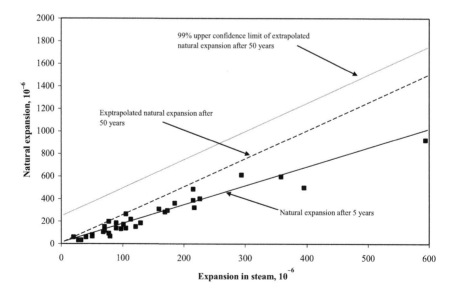

Figure 8.2 Relationships between natural expansion and expansion in steam after 4 h [9].

Table 8.1 Typical Properties of Clay Bricks as Supplied by Manufacturers [11]

Name	Description	Clay Type	Firing Temp. °C	Soluble Salts			
				Mg	Na	K	SO$_4$
Capel multi red	Frogged	Weald	980	0.006	0.008	0.064	0.450
Chatsworth gray	Perforated – 3 holes	Keuper marl (lower)	1045	0.003	0.11	0.023	0.060
Chesterton smooth red	Perforated – 9 holes	Etruria marl	1130	0.04	0.001	0.004	0.013
Fletton commons	Frogged	Lower oxford	1000	0.002	0.007	<0.01	0.650
Funton 2nd mild stock	Frogged	Brickearth/chalk breeze	1015	0.001	0.004	0.008	0.059
Gloucester golden multi	Perforated – 3 holes	Ball clays/clay shale	1100	0.050	0.001	0.008	0.037
Heather countryside straw	Perforated – 3 holes	Keuper marl (lower)	1075	0.008	0.003	0.160	0.020
Highbury buff	Perforated – 3 holes	Fireclay/shale	1040	0.001	0.003	0.002	<0.01
West hoathley dark blue	Frogged	Wadhurst	1100	0.007	0.010	0.023	0.135
Jacobean blue/brown	Frogged	Fireclay mixture	1130	0.001	0.002	0.003	<0.01
Leyburn buff	Perforated – 3 holes	Fireclay mixture	1120	<0.01	0.010	0.010	0.110
MH 1st hard stock	Frogged	Brickearth	1040	0.001	0.001	0.004	0.150
MH kentish multi	Frogged	Gault/shale	1040	0.003	0.001	0.037	0.450
Ravenhead red rustic	Perforated – 3 holes	Mudstone/clay shale	1050	0.030	0.003	0.005	0.024
Ridings red multi	Perforated – 3 holes	Coal measure shale	1035	<0.01	0.01	<0.01	0.050
Roughdales golden rustic	Perforated – 3 holes	Fireclay/shale	1055	0.090	0.002	0.009	0.014
Surrey red multi	Perforated – 3 holes	Weald	1050	0.060	0.003	0.005	0.020
Throckley class 'B' engineering	Perforated – 10 holes	Coal measure/fireclay	1070	0.070	0.004	0.004	0.026
Waingroves red smooth	Perforated – 10 holes	Coal measure shale	1080	0.001	0.003	0.002	0.010
Wickes facing	Perforated – 3 holes	Coal measure shale	1000	<0.01	<0.01	<0.01	0.030

bricks, type of clay, firing temperature, and soluble salt analysis. Bricks, shrink-wrapped in polythene sheet, were collected from the factories and then stored indoors for 3 weeks before being transferred to the test laboratory controlled to a temperature of 21 ± 2 °C and RH of $65 \pm 5\%$. At the time of testing, the bricks were approximately 1 month old. Irreversible moisture expansion was measured over a period of 1 year. Although in previous research, clay brick expansion was measured just between header faces to correspond with horizontal movements in masonry, this investigation [11] obtained data for bed face expansion as well as header face expansion of bricks to correspond with vertical movements of masonry. The mean expansions of four bricks were determined between bed faces and between header faces. The 1-year values of expansion are given in Table 8.2 together with strength, water absorption, and initial suction rate. Although the standard deviation of expansion of some bricks was high ($\pm 30\%$), the overall average standard deviation for 1-year expansion was reasonable at approximately 14% for both bed and header faces. That standard deviation was less than that of initial suction rate (29%), but the variability of expansion was greater than variability of strength or water absorption, which had average levels of 9 and 6%, respectively.

According to Jessop [2], extruded bricks have the same expansions regardless of whether they are perforated or solid, and bricks made by a dry press process show only half the expansion of those made by an extrusion process. Furthermore, even though the extrusion process tends to orientate the plate-like particles of clay in the direction of extrusion (between bed faces), there is no difference between bed and header face expansions [2]. Freeman & Smith [4] also reported no significant difference between header face and bed face expansions of perforated or extruded bricks after 128 days. On the other hand, Brooks and Forth [11] found that it wasn't possible to detect any influence of the method of manufacture on expansion, although most bricks exhibited anisotropic behaviour as described in the next paragraph.

Figure 8.3 shows that expansion between bed faces is greater than that between header faces. The bed face/header expansion ratio varies with type of brick, the average and range of values quickly decreasing with time up to 200 days before becoming fairly constant. Extrapolation of the curve of Figure 8.3 suggests that the average long-term bed face/header face ratio ≈ 2. However, since the range of values is so large, it is difficult to justify a generalized value for all types of brick. Although the results of the above investigation [11] revealed moisture expansion to be aniso-tropic, there was no consistent difference in bed face/header face expansion ratio due to the manufacturing process, namely, between pressed and extruded bricks.

Irreversible Expansion Model

Analysis of the results shown in Table 8.2 reveals that there are no simple correlations of expansion with strength, water absorption, initial suction rate, or soluble salt content for either the bed face or header face expansion. Consequently, none of those param-eters can be considered suitable for developing a model for estimating irreversible moisture expansion of clay bricks.

Table 8.2 Measured Average Properties of Clay Bricks; Figures in Parenthesis are Standard Deviations [11]

Brick	Strength, MPa	Water Absorption, %	Initial Suction Rate, kg/m²/min	1-year Expansion, 10^{-6}	
				Bed Face	Header Face
Capel multi red	18.9 (0.8)	13.6 (0.9)	1.4 (0.48)	305.5 (5.1)	176.2 (13.9)
Chatsworth gray	21.3 (2.7)	34.1 (0.6)	3.8 (0.37)	275.2 (65.2)	104.7 (8.1)
Chesterton smooth red	71.4 (4.5)	5.1 (0.3)	0.3 (0.04)	181.8 (51.4)	61.0 (15.3)
Fletton commons	25.6 (3.3)	20.3 (1.0)	1.8 (0.48)	386.3 (69.2)	223.9 (52.4)
Funton 2nd mild stock	28.5 (10.6)	23.3 (2.0)	3.8 (1.90)	287.9 (29.1)	94.1 (29.3)
Gloucester golden multi	54.1 (3.1)	7.7 (0.7)	0.7 (0.09)	262.6 (68.0)	125.9 (11.7)
Heather countryside straw	34.1 (3.8)	24.0 (1.1)	2.0 (0.30)	217.2 (53.2)	49.0 (17.0)
Highbury buff	63.7 (3.8)	12.7 (0.8)	1.0 (0.12)	376.2 (50.4)	300.8 (23.8)
West hoathley dark blue	18.2 (2.3)	14.6 (1.1)	0.7 (2.30)	234.8 (32.4)	80.8 (23.8)
Jacobean Blue/Brown	75.3 (4.6)	7.5 (0.2)	1.3 (0.21)	219.7 (54.3)	95.4 (15.6)
Leyburn buff	75.7 (10.2)	6.6 (0.5)	0.3 (0.08)	444.4 (21.8)	123.2 (20.0)
MH 1st hard stock	18.2 (2.3)	25.1 (1.1)	3.4 (0.16)	237.4 (37.3)	98.1 (16.5)
MH kentish multi	22.1 (2.2)	19.0 (1.3)	1.5 (0.58)	219.7 (23.9)	79.5 (15.6)
Ravenhead red rustic	69.5 (6.4)	6.7 (0.4)	0.6 (0.10)	250.0 (27.7)	119.3 (20.0)
Ridings red multi	72.9 (6.8)	6.1 (0.3)	0.6 (0.04)	345.9 (100.0)	123.2 (26.5)
Roughdales golden rustic	46.6 (2.2)	9.6 (0.2)	0.8 (0.13)	429.3 (47.2)	275.6 (30.3)
Surrey red multi	77.9 (4.7)	8.0 (0.7)	0.6 (0.07)	275.2 (39.1)	84.8 (18.4)
Throckley class 'B' engineering	64.0 (2.5)	5.1 (0.4)	0.7 (0.04)	287.9 (17.6)	229.2 (8.9)
Waingroves red smooth	120.7 (11.5)	4.2 (0.5)	0.2 (0.02)	207.1 (10.1)	112.6 (8.7)
Wickes facing	70.5 (5.3)	6.8 (0.6)	0.3 (0.06)	270.0 (12.7)	125.9 (14.6)

The model that was eventually adopted began development by revisiting the results of Smith [7], which consisted of comprehensive header-face expansion-time data for 1-day-old bricks made from a variety of clays and fired at different temperatures, ranging from 875 to 1100 °C, according to the type of clay. The results presented in

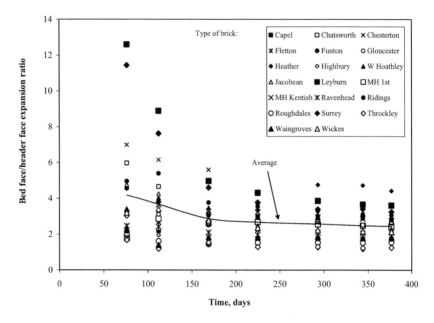

Figure 8.3 Ratio of bed face expansion to header face expansion as a function of time of unbonded clay bricks [11].

Figure 8.1 are the average expansion of bricks stored at 65% RH and 90% RH, for which there was an actual difference in expansion of 20% as stated earlier. For the range of temperature and clays, there was no consistent trend of expansion with temperature as, in some cases, expansion was less for a higher firing temperature, whereas in others, the opposite trend occurred. However, when a different range of firing temperature was considered, and one that was more typical of manufactured bricks, viz 980−1130 °C, a suggested pattern of behaviour emerged as can be seen in Figure 8.4. That figure expresses relative expansion as a function of firing temperature, the relative expansion being defined as the ratio of expansion at any temperature to that at a temperature of 1050 °C. The temperature of 1050 °C was chosen as a midrange value and corresponding expansions were obtained by interpolation of Smith's expansion vs temperature curves [7].

Figure 8.4 indicates that, although the Oxford clay and London stock have the smallest changes in expansion over the temperature range, the Keuper Marls and London clay are very sensitive to changes in temperature, showing a rapid decrease in expansion as the temperature increases from 980 to 1050 °C. A similar trend occurs for the Weald clay, but for the whole temperature range of 980−1100 °C. On the other hand, the Carboniferous and Devonian shales exhibit bilinear trends: a slower decrease in expansion from 980 to 1050 °C and then, like the Weald clay, a rapid decrease from 1050 to 1100 °C. The foregoing features were used to derive a simplified relative

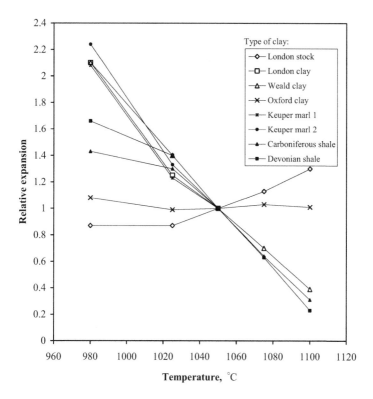

Figure 8.4 Relative expansion of bricks as a function of temperature between 980 and 1100 for the data of Smith [7] (= 1 at 1050 °C).

expansion chart, as shown in Figure 8.5, with different types of clay allocated to different categories, as explained in the following paragraphs.

Using the relative expansion factors from Figure 8.5, Smith's [7] expansion-time results at different temperatures were adjusted to the chosen standard temperature of 1050 °C to form a series of expansion-time curves for different types of clay. The same procedure was carried out for other test results obtained by CERAM Building Technology [12]. Like Smith's tests [7], their tests used 1-day-old bricks made from Keuper Marl and Weald clays but, in addition, they included bricks made from clays: Etruria Marl, Coal Measure Shale, and Fireclay. For the Keuper Marl and Weald clay bricks, the header face expansion-time results were found to be similar to those of Smith after they were adjusted to the standard temperature of 1050 °C by using relative expansion factors from Figure 8.5. In addition, after temperature adjustment, the expansion of the Etruria Marl bricks was found to be similar to the expansion of the Keuper Marl bricks. The expansion of the Coal Measure Shale bricks appeared to be unaffected by firing temperature when CERAM Building Technology's results [12] were compared with those of Brooks and Forth [11] over the same period of time. As the Fireclay bricks were fired 1057 °C, there was no need to adjust the associated expansion-time curve.

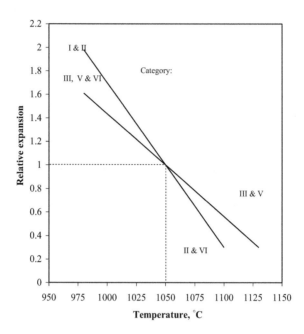

Figure 8.5 Simplified chart to allow for the effect of firing temperature on relative expansion of bricks (= 1 at 1050 °C); see Table 8.3 for categories of clay.

By grouping those clays having similar expansion-time trends, six expansion-time categories for a standard firing temperature were established, as shown in Figure 8.6. When the actual firing temperature differs from 1050 °C, the expansion may be adjusted according to Figure 8.5. The types of clay allocated to each category are listed in Table 8.3, together with ultimate values of expansion and coefficients (a and b) of the following expansion-time expression:

$$\varepsilon_t = \frac{t^{0.5}}{a + bt^{0.5}} \tag{8.2}$$

where ε_t = expansion (10^{-6}) at a firing temperature = 1050 °C after time t (days, measured from 1 day) .

The form of Eq. (8.2) is commonly used to express the development of drying shrinkage of concrete with time [13]. It was also found to be the most suitable for representing the expansion-age curves of Figure 8.6, and conveniently has the advantage of having an ultimate expansion = $1/b$. In general, for all bricks made from any type of clay, Eq. (8.2) indicates that the rate of development of ultimate expansion is 25% in the first month and 50% in the first year.

Table 8.3 shows that not all brick expansions are assumed to be influenced by firing temperature; namely, all those made from clays in Category IV and Oxford clay in Category V. All the other types are influenced by temperature, and Table 8.3 gives

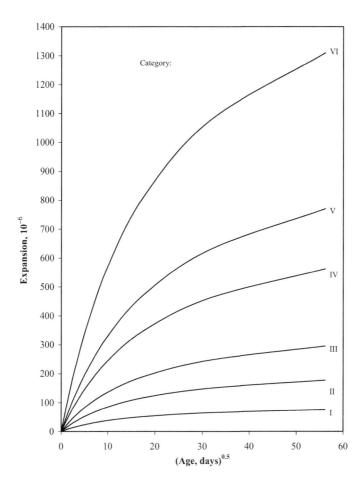

Figure 8.6 Categories of irreversible moisture expansion of clay bricks fired to a temperature of 1050 °C as a function of age; measured from 1 day.

the equations for relative expansion vs firing temperature for the appropriate category. The equations correspond to the curves of Figure 8.6.

Table 8.3 also includes the types of clay used by Brooks and Forth [11] that were not covered by the tests of Smith [7] and the tests of CERAM Building Technology [12]. Those clays were categorized on the basis of trial and error by matching the experimental results (header expansions) with Figure 8.6 after allowing for age and firing temperature.

Application of the model demonstrates that there is reasonable agreement between estimates and measured irreversible moisture expansions. The comparisons are shown in Figures 8.7–8.12. In the case of Smith [7], estimated expansion-time curves cover a 7-year period (Figures 8.7 and 8.8). Similarly, as shown in Figure 8.9, the model gives

Table 8.3 Influence of Firing Temperature on Moisture Expansion for Different Categories of Clay Bricks [11]

Category (see Figure 8.6)	Clay Type	Ultimate Expansion for $T = 1050\,°C$, 10^{-6}	Coefficients of Eq. (8.2) for $T = 1050\,°C$		Equations for Relative Expansion, R_e (Figure 8.5) ($R_e = 1$ for $T = 1050\,°C$)
			a	b	
I	London clay (T)	70	0.23	14.0×10^{-3}	$R_e = 16 - \dfrac{T}{70}$ $(T \leq 1050)$
II	Weald (T), brickearth (T), gault/ shale (T), brickearth/Chalk breeze (T)	250	0.075	4.0×10^{-3}	$R_e = 16 - \dfrac{T}{70}$
III	Keuper marl (T), etruria marl (T), mudstone/clay shale (T), wadhurst (T)	300	0.043	3.35×10^{-3}	$R_e = 10 - \dfrac{0.6T}{70}$
IV	Coal measure shale, ball clays/ clay shale	440	0.023	2.28×10^{-3}	$R_e = 1$ (no temp. influence)
V	Oxford clay, fireclay mixture (T), fireclay shale (T), coal measure/fireclay (T).	1000	0.019	1.0×10^{-3}	$R_e = 1$ (Oxford clay); $R_e = 10 - \dfrac{0.6T}{70}$
VI	Fireclay (T), carboniferous shale (T), devonian shale (T)	1820	0.012	0.55×10^{-3}	$R_e = 16 - \dfrac{T}{70}$ $(T \geq 1050)$ $R_e = 10 - \dfrac{0.6T}{70}$ $(T \leq 1050)$

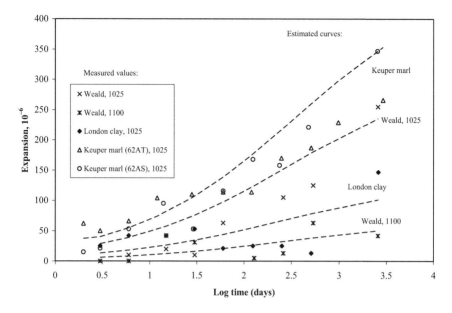

Figure 8.7 Comparison of estimated moisture expansion by Eq. (8.2) and measured moisture expansion by Smith [7] for clay bricks in Categories II and III.

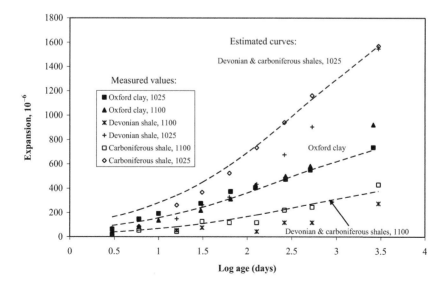

Figure 8.8 Comparison of estimated moisture expansion by Eq. (8.2) and measured moisture expansion by Smith [7] for Category V clay bricks.

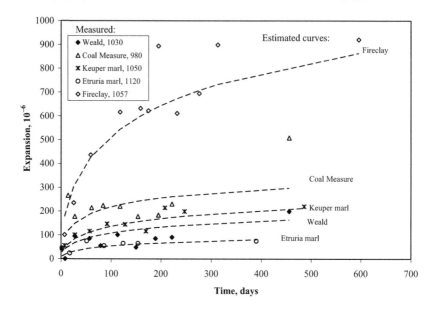

Figure 8.9 Comparison of estimated expansion of clay bricks using Eq. (8.2) and measured moisture expansion by CERAM Buliding Technology [12].

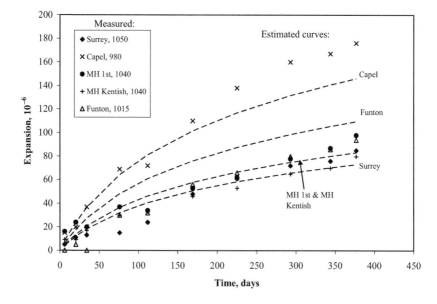

Figure 8.10 Comparison of estimated and measured moisture expansions of clay bricks from Category II by Brooks and Forth [11].

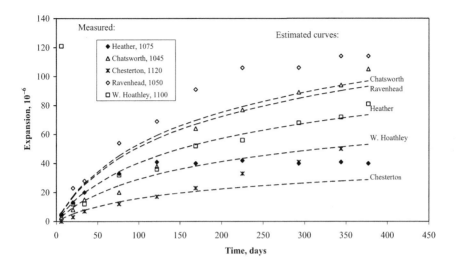

Figure 8.11 Comparison of estimated and measured moisture expansions of clay bricks from Category II by Brooks and Forth [11].

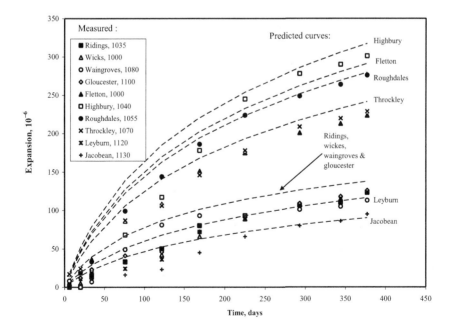

Figure 8.12 Comparison of estimated and measured moisture expansions of clay bricks in Categories IV and V by Brooks and Forth [11].

a good representation of the results of CERAM Building Technology [12] measured up to 1 or 2 years, although longer-term experimental points are few and have a large scatter.

Whereas the foregoing estimations were for 1-day-old bricks, the tests of Brooks and Forth [11] used 30-day-old bricks. In their investigation, the prediction model was applied by estimating the expansion from the age of 1 day and then deducting the estimated 30-day expansions. The estimated curves are compared with experimental points in Figures 8.10−8.12. As Figure 8.10 shows, the estimations of the long-term expansions for the Capel and Funton bricks are slightly underestimated and overestimated, respectively, but the estimates for the other Category II bricks are satisfactory. For Category III bricks (Figure 8.11), longer-term expansions of the Ravenhead and W. Hoathley bricks are underestimated by the model, whereas the Heather brick is overestimated. However, the differences are considered acceptable because of the overall low levels of expansion. As shown in Figure 8.12, satisfactory estimates are achieved for most of the expansions for bricks in Categories IV and V, the exception being the Fletton brick expansion, which is overestimated. On the other

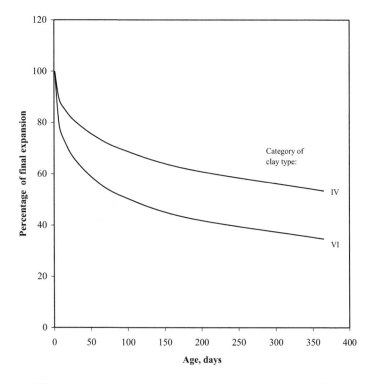

Figure 8.13 Effect of age before laying on reduction of long-term irreversible moisture expansion of clay bricks; the two curves are constructed using Eq. (8.2) and represent the range of all types of clay given in Table 8.3.

hand, the model gives an accurate estimate of moisture expansion for Fletton bricks as reported by Smith [14], who measured an average 4-year expansion of 690×10^{-6} for kiln-fresh bricks for a range of storage conditions—from Table 8.3 using a Category V clay with $R_e = 1$ and the coefficients for Eq. (8.2), the irreversible expansion model yields a prediction of 670×10^{-6}.

The influence of aging of the clay brick on subsequent expansion is apparent in the tests of Brooks and Forth [11], where bricks were about 1 month old; expansions are less with than in the case of Smith [7] and CERAM Building Technology [12], both of which utilized kiln-fresh bricks. Figure 8.13 demonstrates the general effect of reducing the long-term irreversible moisture expansion of clay bricks by storing them before being used in construction. For example, on average, storing bricks for 1 month would reduce its long-term expansion by approximately 30%. For moisture expansion of aged Fletton bricks (at least 1 year old), Smith [14] recorded a 61% reduction in expansion after 4 years compared with the expansion of kiln-fresh bricks after 4 years. As mentioned earlier, Laird and Wickens [6] also demonstrated significantly lower expansions of different types of 1-month-old clay bricks, compared with kiln-fresh bricks.

Taking into account the inherent variability of irreversible moisture expansion of clay bricks and test data due to different storage conditions, viz. relative humidity influence, the overall accuracy of the prediction model is considered to be satisfactory. It allows the expansion-time characteristics to be estimated for different types of clay brick fired to temperatures between 980 and 1130 °C. Provided that the age of the brick is known, expansion can be calculated by Eq. (8.2) with the appropriate coefficients taken from Table 8.3 and, if required, an adjustment made for firing temperature according to Table 8.3. Alternatively, expansion may be estimated from Figures 8.5 and 8.6. The following example illustrates the two methods:

Example

Suppose it is required to estimate the 50-year irreversible expansion of a clay brick made from Wadhurst clay fired to a temperature of 1100 °C. At the time of construction, the bricks were 2 months old.

According to Table 8.3, Wadhurst clay is in Category III and is temperature sensitive. The required expansion can be estimated from the equations given in Table 8.3 or from Figures 8.5 and 8.6:

Using equations

For a firing temperature $= 1050$ °C, the appropriate expansion-time expression from Eq. (8.2) and Table 8.3 is:

$$\varepsilon_t = \frac{t^{0.5}}{0.043 + 0.00335 t^{0.5}}$$

when $t = 50$ years $= 18{,}250$ days; and expansion (ε_t) from the age of 1 day - $= 273 \times 10^{-6}$. The expansion occurring in the first 2 months or 60 days is

---

Continued

Example—cont'd

112×10^{-6}. Therefore, the expansion occurring from 2 months to 50 years is $(273-112) \times 10^{-6} = 161 \times 10^{-6}$. However, that value is for a temperature $= 1050\,°C$, and for other temperatures the relative expansion (Table 8.3) is given by:

$$R_e = 10 - \frac{0.6T}{70}$$

when $T = 1100\,°C$, $R_e = 0.57$, so that the required expansion $= 0.57 \times 161 \times 10^{-6}$ $= \underline{92} \times \underline{10^{-6}}$

Graphical method

Table 8.3 gives the ultimate expansion for Category III clay as 300×10^{-6} and, from Figure 8.6, the estimated expansion after 60 days ($t^{0.5} = 7.75$) is 120×10^{-6}. Therefore, for a standard firing temperature of $1050\,°C$, the expansion occurring from 2 months is $(300-120) \times 10^{-6} = 180 \times 10^{-6}$. The relative expansion for a firing temperature of $1100\,°C$ is 0.56 (Figure 8.5) so that the corrected expansion is $180 \times 10^{-6} \times 0.56 = \underline{101} \times \underline{10^{-6}}$.

In the example above, estimates for 50-year irreversible moisture expansions apply to the header face direction of clay bricks, and are appropriate for estimating horizontal moisture movement of masonry. In the case of estimating vertical moisture movement of masonry, in the absence of specific test data for the type of clay in question, irreversible moisture expansion between bed faces of clay bricks may be assumed to be equal to the header face expansion.

Clay Brickwork

Before dealing with relevant factors that influence the moisture movement of clay brickwork, the brickwork/brick expansion ratio is discussed because a value is often recommended in design documents as a convenient way of estimating expansion of clay brickwork. Consequently, it is pertinent to review relevant background information.

Brickwork/Brick Expansion Ratio

As mentioned in the previous section, Smith [14] measured moisture expansion of Fletton bricks for 4 years, the bricks being used to build 12-course-high, closed-in cavity walls, 2.29 m long on a damp-proof course so that they were unrestrained. Expansions were measured horizontally on walls located: outdoors-exposed, outdoors-sheltered, and indoors controlled to 75% RH. In general, the brickwork/brick expansion ratio was found to be approximately 0.6.

According to Foster and Tovey [10], other results by Smith [14] for half brick walls less than 1 m high and built off bituminous damp-proof courses with no restraint other than their own weight, were also about 60% of the unbonded brick expansions. In view of all the evidence therefore, for design purposes, Foster and Johnston [15] suggested the adoption of that value, namely, a brickwork/brick expansion ratio of approximately 0.6 for all types of brickwork. However, test results for different types of clay brick by CERAM Building Technology [12] indicated that a range of 0.5−1.0 was more appropriate.

In contradiction to the findings of CERAM Building Technology [12], the review by West [3] had reported a greater expansion of brickwork panels in the vertical direction rather than the horizontal direction, which suggested an interaction between bricks and mortar and that the mortar itself had expanded. *Sulfate attack* was attributed as the cause of unusually large expansions observed by Smith [14] in other tests on walls exposed to the weather. In some cases, brickwork/brick expansion ratios greater than unity were also observed by CERAM Building Technology [12]. Except for a solid pier, very high vertical expansions were measured by Brooks and Bingel [16] in 1-m-high × 2-brick-wide walls and hollow piers built in Fletton brick. Similarly, large vertical expansions and brickwork/brick expansion ratios greater than unity were measured by Forth and Brooks [17] in 1-m-high × 2-brick-wide single-leaf walls built from a variety of types of clay brick. In those situations, the *enlarged expansion* was attributed to the phenomenon of *cryptoflorescence*, a topic that is dealt with in Chapter 9. In this chapter it is emphasized that we are concerned with moisture movement of brickwork in which that phenomenon does <u>not</u> occur. There are other factors that affect the brickwork/brick expansion ratio, which will become apparent after considering the composite modeling of moisture movement.

Composite Model Expressions

Appropriate expressions are given in Table 3.1 (Eq. (3.80) and (3.77)). In the vertical direction the moisture movement of brickwork is:

$$S_{wy} = 0.86S_{by} + 0.14\gamma_m S_m \qquad (8.3)$$

and, in the horizontal direction, the moisture movement is:

$$S_{wx} = \frac{0.955S_{bx} + \left[0.046 + 0.152\frac{E_m}{E_{bx}}\right]\gamma_m S_m}{\left[1 + 0.152\frac{E_m}{E_{bx}}\right]} \qquad (8.4)$$

In fact, the above expressions are slightly different from those given in Table 3.1 because they include the water absorption factor, γ_m, to allow for the reduction in shrinkage of the mortar bed joint due to absorption of water by units when the latter are laid dry (see Chapter 7). It is therefore relevant to consider a corresponding water absorption factor to allow for possible changes in moisture expansion characteristics of clay units due to water absorption at the time of laying. As could be expected, both expressions indicate that shrinkage of mortar is an influencing factor in moisture movement of brickwork, but the difference in form of the equations suggests anisotropy as another possible factor and, in the case of the horizontal moisture movement expression, there is also a dependency on the mortar/brick modulus ratio, E_m/E_{bx}. The importance of those factors will now be considered together with age of the clay unit at the time of construction.

Clay Unit Moisture State and Absorption

In the previous chapter, the effects of water absorption by dry units from freshly-laid mortar on the shrinkage of the mortar bed joint, calcium silicate, and concrete units of the unit were shown to be significant. The assessment of those effects involved measuring shrinkage of masonry and shrinkage of the bonded unit so that the mortar bed joint shrinkage could be deduced. The shrinkage of the bonded unit and mortar bed joint were then compared with corresponding shrinkage of unbonded units and shrinkage of independent mortar specimens. The ratios of bonded shrinkage to unbonded shrinkage, termed in the water absorption factors, were then used to adjust the shrinkage contributions of mortar and unit in the composite model equations for estimating the shrinkage of masonry.

The same procedure could not be adopted for deducing the shrinkage of mortar bed joints in clay brickwork undergoing moisture movement because of the likelihood of enlarged moisture expansions due to cryptoflorescence occurring in small walls built with dry clay units of high water absorption. Consequently, in the following analysis, it was assumed that water absorption factors for mortar bed joints in clay brickwork would be the same as those deduced for masonry built from calcium silicate and concrete units.

Water absorption factors for bed face and header face directions of clay units were obtained by analysis of other test data [18] consisting of strain measurements of bricks embedded in 1-m-high × 2-brick-wide single-leaf walls together with corresponding measurements of unbonded bricks. Figure 8.14 shows the clay unit water absorption factor plotted as a function of unit water absorption. The large amount of scatter renders the establishment of definitive trends difficult, although, interestingly, the overall mean of the bonded/unbonded bed face ratio of 1.19 is slightly greater than the mean of the bonded/unbonded header face ratio of 0.92, which suggests some anisotropic effect. Anisotropy may arise from differential restraint of unit moisture expansion by shrinkage of mortar joints, there being more horizontal restraint to header face expansion than vertical restraint to bed face expansion. Nevertheless, in view of the scatter and for the simplicity of composite modeling of moisture movement of clay masonry, when laid dry, the moisture expansion clay units are assumed to be unaffected by water absorbed from the freshly-laid mortar, i.e., $\gamma_b = 1$. Moreover, since good site practice requires that units have high initial suction rates (and hence probably high water

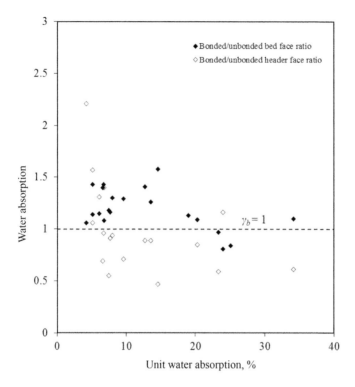

Figure 8.14 Effect of water absorbed from freshly-laid mortar on moisture of different types of bonded brick in clay brickwork; expressed as water absorption factor = ratio of bonded brick expansion to unbonded brick expansion.

absorptions) and be wetted or docked before laying in mortar, the assumption of $\gamma_b = 1$ is also appropriate for that requirement. For the same reason, it can be assumed that the mortar absorption factor $\gamma_m = 1$.

Mortar/Brick Modulus Ratio

The mortar/brick modulus ratio, (E_m/E_{bx}), of Eq. (8.4) is an effective modulus or reduced modulus ratio that allows for creep of mortar and brick due to stresses induced by internal restraint to moisture movement, and it has a significant effect on horizontal moisture movement through the numerator term of Eq. (8.4). In Chapter 2, the effective modulus of elasticity is simply defined as the stress divided by the elastic strain-plus-creep on first loading. The effective modulus approach works well for predicting strains of concrete resulting from small variations in stress provided the concrete is fairly mature, such as that subjected to external load from the age of 28 days. However, in the case of moisture movement of brickwork starting very soon after construction, the normal effective modulus approach is not really applicable because both the strength of the mortar and level of induced stresses change significantly at early ages [19]. For example, the effective modulus of mortar as determined, say, from the age of 1 day, would be extremely small due to the large elastic strain and very high creep but, as the mortar rapidly gains strength, this would be offset by a decrease in elastic strain (increase in elastic modulus) and reduced creep. By contrast, the effective modulus of brick remains largely unchanged because of its relatively small creep.

As stated in Chapter 3, to simplify matters and to assign a realistic value to the modulus ratio for horizontal moisture expansion, it is convenient to assume that the modulus ratio remains constant and equal to the elastic modulus ratio of mortar and brick based on standard strength measurements.

Anisotropy

The mortar/brick modulus ratio influences the degree of anisotropy of moisture movement in clay masonry because it quantifies the horizontal restraint of the expanding brick by the shrinking mortar bed joint. In Figure 3.14, the theoretical case of masonry built with an expanding clay brick showed that horizontal movement was less than vertical movement, depending on the level of mortar shrinkage. When the modulus ratio is assumed to equal 0.5 and mortar shrinkage is high, brickwork contracts due to shrinkage with a high degree of anisotropy. In contrast, for low-shrinkage mortar, the brickwork expands with a lower degree of anisotropy. A lower modulus ratio results in a more anisotropy and, in fact, depending on the shrinkage of mortar, this can result in vertical shrinkage and horizontal expansion as verified in tests by Abdullah [20]. His findings, using 1-m-high × 2-brick-wide brickwork built with a 3-month-old Class B Engineering brick and a 1:½:4½ cement—lime mortar, having a 28-day strength of 6.5 MPa, are shown in Figure 8.15.

Geometry

In common with the influence on shrinkage of mortar, calcium silicate, and concrete masonry, Figure 8.15 also demonstrates that the geometry or size and shape of the cross section is a factor when clay brickwork undergoes shrinkage in the vertical direction. Thus, solid piers of larger cross-sections undergo shrinkage at a slower rate and have less overall vertical shrinkage than single-leaf walls of thinner cross sections. In other words, vertical shrinkage depends on the average drying path length, which can be quantified by the volume/surface area (V/S) ratio. Figure 8.15 shows a consistent trend of vertical shrinkage with V/S ratio but, for the smaller horizontal moisture movement, the trend of moisture expansion with V/S is somewhat inconsistent.

Age of Brick

Earlier in this chapter, the effect of the length of storage or age of bricks before use on reducing their subsequent moisture expansion was discussed (see Figure 8.13)

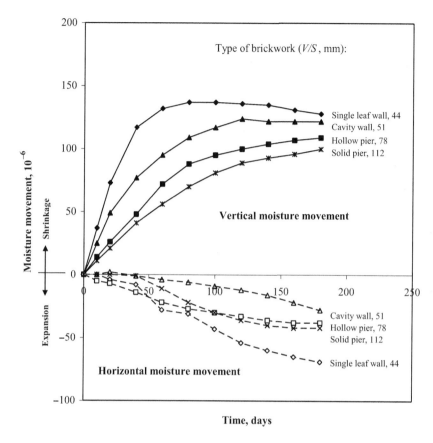

Figure 8.15 Influence of geometry of clay brickwork on vertical and horizontal moisture movement [20]; *V/S* = volume/surface area ratio.

and, consequently, age of brick can be expected to be a factor in the moisture movement of clay brickwork. In this respect, compared with walls built with kiln-fresh bricks, Thomas [21] reported a 36% reduction in expansion after 300 days for 215-mm-thick walls built with 15-day-old carboniferous shale bricks and boulder clay bricks. To illustrate the effect in more detail, consider the example of the Fletton brick cavity wall as used by Smith [7], full details of which are given later on page 248. For different periods of storage, the effect age of brick on the horizontal moisture movement of brickwork after 4 years, S_{wx}, is shown in Table 8.4. As the age of the brick increases, it can be seen that there is a corresponding reduction on moisture movement of the wall and, in fact, after a 1-year period of storage there is a small negative moisture expansion, i.e., shrinkage. Table 8.4 also indicates that the brickwork/brick expansion ratio decreases as the age of the brick increases.

Table 8.4 Estimated Effect of Age of Brick on Horizontal Moisture Movement of Fletton Clay Brickwork After 4 years

Period of Storage or Age of Brick	Expansion of Brick, S_{bx}, 10^{-6}	Horizontal Moisture Expansion of Wall, S_{wx}, 10^{-6}	Brickwork/Brick Expansion Ratio, S_{wx}/S_{bx}
Kiln fresh (1 day)	668	407	0.61
14 days	504	262	0.52
28 days	452	215	0.48
3 months	392	162	0.41
1 year	191	−16	−0.08

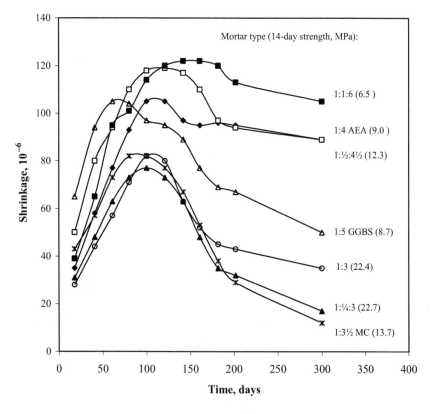

Figure 8.16 Influence of type of mortar on vertical moisture movement of clay brickwork [23]; MC = masonry cement, GGBS = ground granulated blast-furnace slag, AEA = air entraining agent.

Mortar

When considering the influence of mortar on the moisture movement of brickwork, Beard et al. [22] found expansion to be greater for brickwork constructed with a higher-strength 1:3 cement mortar than with a lower-strength 1:1:6 cement-lime mortar, but similar moisture expansions were observed by Smith [14] for walls built with 1: 1:6 and 1:$\frac{1}{4}$:3 cement-lime mortars. Forth and Brooks [23] investigated the effect of different mortar types, having 28-day strengths varying from 9.6 to 27.3 MPa, on moisture movement of 1-m-high × 2-brick-wide single-leaf walls built with 3-month-old Class B Engineering bricks. The walls were covered with polythene sheet for 14 days prior to taking readings of vertical moisture movement. In all cases, the pattern behaviour of moisture movement was an initial shrinkage, followed by expansion from approximately 100 days, suggesting that shrinkage of the mortar joint dominated at early stages, but later, expansion of the brick becomes more influential. Figure 8.16 shows the test results, the most moisture movement occurring for the low strength cement-lime mortar and the least for masonry cement mortar. For the cement-lime mortars, moisture movement decreases as the strength increases, whereas for the other types of mortar, the pattern of moisture movement is as follows:

Masonry cement (MC) < Ground granulated blast-furnace slag (GGBS) < Plasticised or air entrained (AEA).

Composite Model Prediction

Expressions derived by composite modeling of moisture movements in the vertical and horizontal directions have already been given by Eq. (8.3) and (8.4), respectively. In the following example, Eq. (8.4) will be used to predict the expansion-time curve of the cavity wall used by Smith [14], which was referred to earlier in this chapter.

Example

For a 1:1:6 mortar, the assumed strength $(f_m) = 5$ MPa, and from Eq. (5.5), this yields an elastic modulus of elasticity $(E_m) = 4$ GPa.

The strength of the Fletton brick (f_{by}) is assumed to be 20 MPa, and from Eq. (5.7), this yields a modulus of elasticity between header faces $(E_{bx}) = 10$ GPa. Hence, $E_m/E_{bx} = 0.4$, so that substitution in Eq (8.4) gives:

$$S_{wx} = 0.888S_{bx} + 0.112S_m \qquad (8.5)$$

Calculation of moisture movements of mortar and brick are as follows:

Mortar, S_m

Relative humidity of storage (RH) = 75%; Fletton bricks docked and therefore mortar water absorption factor, $\gamma_m = 1$.

Continued

Example—cont'd

The calculated volume/surface ratio (V/S) of closed cavity wall (no drying from inside cavity) $= 69$ mm and so the shrinkage of mortar (S_m) at any time (t) is given by substituting f_m, V/S, and RH in Eq. (7.26a). Hence:

$$S_m = 1714 \left[\frac{t}{47.25 + t} \right] \times 10^{-6}$$

Brick, S_{bx}

For moisture expansion of a kiln-fresh Fletton brick (Lower Oxford clay), Table 8.3 gives the details for Category V as $S_{b\infty} = -1000 \times 10^{-6}$, $a = 0.019$, and $b = 0.1 \times 10^{-3}$. Hence, from Eq. (8.2):

$$S_{bx} = -\frac{t^{0.5}}{0.019 + 0.1 \times 10^{-3} t^{0.5}} \times 10^{-6}$$

Substitution of S_m and S_{bx} in the above Eq. (8.5) now yields values of S_{wx} for various times.

Figure 8.17 shows that the prediction of the indoors—75% RH curve is satisfactory. The prediction yields a brickwork/brick expansion ratio after 4 years of 0.61 (see Table 8.4), which agrees with Smith's [14] general observed value of about 0.6. However, Smith reported that, for walls exposed to the weather, the rates of expansion increased markedly from about the age of 150 days, so that the brickwork/brick expansion ratio exceeded 1.0; chemical analysis attributed the cause of increased expansion to sulfate attack of the mortar.

It is of interest to note that, in the previous the example of Smith [14], the water absorption of the Fletton brick was 20% and, if the bricks had been laid dry, shrinkage of the mortar joints would have been less by a factor of 0.3 (Eq. (8.12)). In consequence, according to Eq. (8.4), the 4-year horizontal expansion of brickwork, S_{wx}, would have increased significantly from 407 to 574×10^{-6}, i.e., by 41%. That example demonstrates the importance of prewetting high-absorptive clay units with regard to irreversible moisture expansion of brickwork.

Some other estimates of horizontal moisture movement for unrestrained cavity walls tested by CERAM Building Technology [12] are shown in Figure 8.18, together with the measured values over a period of 15 months. Five walls were built from different types of clay brick with a 1:1:6 cement-lime mortar. Since the bricks were not wetted or docked prior to being laid, predictions by Eq. (8.4) included allowances for the water absorption factor, $\gamma_{w\alpha}$, according to Eq. (7.12). With the exception of the Keuper Marl brick wall, which showed an initial shrinkage, measured horizontal moisture movements were expansions throughout the period of testing. It can be seen from Figure 8.16 that the predictions for the Fireclay and Keuper Marl clay brick walls are satisfactory, but 400-day estimates for the Weald, Coal Measure, and Etruria Marl

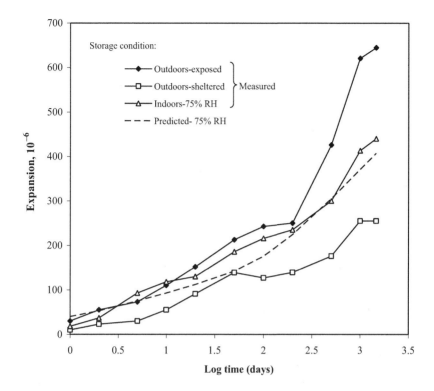

Figure 8.17 Measured horizontal moisture expansions of cavity Fletton brick walls by Smith [14] compared with predicted expansion for 75% RH using Eq. (8.4).

brick walls are underestimates. However, Ceram Building Technology [12] reported a higher measured brickwork/brick expansion ratio than expected for the Etruria Marl brick wall (3.0), and the ratio for the Weald wall was also slightly high (1.0). With the exception of the Etruria Marl wall, which is negative, the other predicted brickwork/brick expansion ratios after 500 days were as expected, namely, less than 1.0 ranging from 0.41 (Weald) to 0.80 (Fireclay). The high measured expansion ratios suggest the occurrence of enlarged expansions rather than just irreversible moisture expansion.

In the same tests, CERAM Building Technology [12] monitored vertical moisture movement, which apparently showed no significant variation in magnitude to the measured horizontal moisture movement. The composite model prediction by Eq. (8.3) also confirmed that observation and by assuming the moisture expansion of bricks was isotropic, i.e., $S_{by} = S_{bx}$. In fact, the 500-day calculated moisture movement of brickwork in the vertical direction was only slightly less than that predicted in the horizontal direction. That statement might seem at odds with the earlier discussion of anisotropy (p. 244), which, theoretically, is significant, especially at high levels of mortar shrinkage. However, in the case of tests of CERAM Building Technology [12],

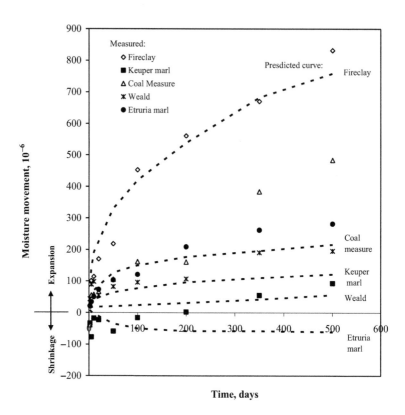

Figure 8.18 Comparison of predicted and measured horizontal moisture movement of cavity walls built from different types of clay brick; average measured values from CERAM Building Technology [12] and predicted curves by Eq. (8.4).

bricks were laid dry and, therefore, allowing for water absorption factors in Eq. (8.3) had the effect of reducing the shrinkage of mortar bed joint.

Design Code Guidelines

In 1973, design recommendations for moisture movement in masonry, Code of Practice, CP 121: Part 1 [24] provided information on the moisture expansion of brickwork with respect to the type of clay from which the corresponding unbonded brick was made. Based on Freeman and Smith's results [4], the same Code also provided details of the moisture expansion of unbonded bricks for the type of clay from which they were manufactured. Subsequently, CP 121 was superseded by BS 5628: Part 3: 1985 [25], which prescribed a single maximum design unit expansion of 1000×10^{-6}, with a qualification that it was dependent on the type of clay used and firing temperature, and could be modified at the designer's discretion. That standard was then superseded by BS 5628-3: 2005 [26], which recommended movements to be considered as a whole and stressed the importance of incorporating movement joints in the design. For clay masonry units, there were no specific values for expansion given, since "the amount of expansion depends on the type of clay and the degree of firing", and the actual movement in a wall depends on the degree of restraint and the mortar properties. It is of interest to note that before being withdrawn, BS 5628-2: 2005 [27] disregarded the effect of moisture expansion on the force in the prestressing tendons in structural prestressed masonry. BS 5628-3: 2005 has also now been withdrawn and replaced by the current standard BS EN 1996-1-1: 2005 [28] in which a range of moisture movement of clay masonry is specified as 200×10^{-6} (shrinkage) to 1000×10^{-6} (expansion), with the U.K. National Annex [28] recommending a moisture expansion of 500×10^{-6}. In the United States, unless otherwise determined by test, ACI 530-05 [29] expresses the irreversible moisture expansion as the *coefficient of moisture expansion* and recommends a value of 300×10^{-6}.

Concluding Remarks

The considerations of previous research findings and the development of prediction models have revealed the main influencing factors involved in irreversible moisture expansion of clay bricks and moisture movement of brickwork. In the case of irreversible moisture expansion of clay bricks, the main factors are:

- Type of clay.
- Firing temperature.
- Age.
- Degree of anisotropy.

In the case of moisture movement of clay brickwork, the additional factors are:

- Water absorption of the brick.
- Degree of wetting or docking of brick prior to laying.
- Type of mortar.
- Anisotropy, i.e., whether vertical or horizontal movement.
- Brick/mortar modulus ratio in the case of horizontal movement.

The use of kiln-fresh clay bricks and bricks of a young age will lead to horizontal irreversible moisture expansion in brickwork, but the use of older bricks can lead to shrinkage of clay brickwork, or lead to shrinkage followed by expansion especially in the vertical direction. These latter situations can arise in brickwork used to determine creep in the laboratory, which requires a control moisture movement wall identical to the wall being subjected to load for creep determination.

As far as design is concerned, the contents of this chapter can be used to provide more precise estimates of moisture movement of clay masonry, say, when considering the type of clay brick, age of brick at the time of construction, and the effect of mortar type. The brickwork/brick expansion ratio is not constant and depends on several factors, particularly the age of the brick. However, a maximum value of 0.6 may be assumed for brickwork built with kiln-fresh bricks of high absorption combined with a low-shrinkage mortar. In addition, the maximum brickwork/brick expansion ratio is a useful indicator to confirm that enlarged expansion due to cryptoflorescence is absent. This latter topic is discussed fully in Chapter 9.

Problems

1. Estimate the irreversible expansion of a clay brick after 50 years. The brick is made from Coal Measure Shale, fired to a temperature of 1080 °C, and is to be used in construction when it is 1 month old.
   Answer: 254×10^{-6}.
2. What is thought to be the cause of irreversible moisture expansion of clay bricks?
3. Are clay bricks anisotropic with regard to irreversible moisture expansion? If so, in what way?
4. Describe an accelerated expansion test for clay bricks.
5. List three main factors influencing irreversible moisture expansion of clay bricks.
6. Does the type of mortar affect moisture movement of clay brickwork?
7. Why is the water absorption of the brick important with regard to long-term moisture movement of brickwork?
8. Does docking or wetting of clay bricks before construction affect moisture movement of brickwork?
9. Does clay brickwork always expand in the long term?
10. Besides the type and age of clay brick, list other factors affecting moisture movement of brickwork.
11. The brickwork/brick expansion ratio can be assumed to be equal to 0.6. Is this assumption valid?
12. How is irreversible moisture movement of clay masonry dealt with in Codes of Practice?

References

[1]   Hosking RG, Hueber HV, Waters EH, Lewis RE. The permanent moisture expansion of clay products I. Bricks, CSIRO Aust Div Build Res Tech; 1959. Paper No. 6.

[2]   Jessop KL. Moisture, thermal, elastic and creep properties of masonry: a state-of-the-art report. In: Suter GT, Keller HU, editors. Proceedings Second Canadian Symposium, Ottawa; 1980. pp. 505—20.

[3]   West HWH. Moisture movement of bricks ands brickwork. Transactions Br Ceram Soc April 1967;66:137—60.

[4]   Freeman IL, Smith RG. Moisture expansion of structural ceramics, I. Unrestrained expansion. Transactions Br Ceram Soc 1967;66:13—36.

[5]   Freeman IL. Moisture expansion of structural ceramics, II. In: The influence of composition. Xth International Ceramic Congress in Stockholm, Sweden; 1966. pp. 141—53.

[6]   Laird RT, Wickens AA. The moisture expansion of full-size bricks. Trans Br Mason Soc 1968;67:629—38.

[7]   Smith RG. Moisture expansion of structural ceramics III. Long-term unrestrained expansion of test bricks. Trans J Br Ceram Soc 1973;72:1—6.

[8]   McDowell IC, Birtwistle R. In: West HWH, Speed KH, editors. Sibmac Proceedings. Stoke-on-Trent: British Ceramic Research Association; 1971. p. 75.

[9] Lomax J, Ford RW. Investigations into a method of assessing the long-term expansion of clay bricks. BCRL Tech Note 1983;348. 11 pp.

[10] Foster D, Tovey AK. Brickwork and blockwork, Specification 93. 87th ed. MBC Architectural Press and Building Publications; 1989.

[11] Brooks JJ, Forth JP. Categorisation of irreversible expansion of Clay bricks. Mason Int 2007;20(3):129—40.

[12] CERAM Building Technology. Design for movement, Project Ref. BDA/MOVE; 1993. 30 pp.

[13] CEB-FIP. Model code for concrete structures, evaluation of time-dependent properties of concrete, Bulletin d'information 199. Lausanne: Comite European du Beton/Federation Internationale de la Precontrainte; 1990.

[14] Smith RG. Moisture expansion of structural ceramics IV. Expansion of unrestrained Fletton brickwork. Trans J Br Ceram Soc 1974;73:191—8.

[15] Foster D, Johnston CD. Design for movement in clay brickwork in the UK. In: Proceedings of the British ceramic Society, No. 30; 1982. pp. 1—12.

[16] Brooks JJ, Bingel PR. Moisture expansion of Fletton brickwork. In: Proceedings of the first International masonry Conference. Stoke-on-Trent: British Masonry Society, No. 2; 1988. pp. 12—4.

[17] Forth JP, Brooks JJ. Influence of clay unit on the moisture expansion of masonry. In: Proceedings Fourth International masonry Conference, Vol. 1. London: British Masonry Society; 1995. pp. 80—4.

[18] Bremner, A., The origins and effects of cryptoflorescence in fired-clay masonry, PhD Thesis, School of Civil Engineering, University of Leeds, 2002

[19] Neville AM, Dilger WH, Brooks JJ. Creep of plain and structural concrete. London and New York: Construction Press; 1983. 361 pp.

[20] Abdullah, C. S., Influence of geometry on creep and moisture movement of clay, calcium silicate and concrete masonry, PhD Thesis, Department of Civil Engineering, University of Leeds, 1989.

[21] Thomas A. Moisture expansion in brickwork. Trans J Br Ceram Soc 1971;70:35—8.

[22] Beard R, Dinnie A, Richards R. Movement of brickwork. Trans Br Ceram Soc 1969; 68(73):73—90.

[23] Forth JP, Brooks JJ. Influence of mortar type on the long-term deformation of single leaf clay brick masonry. In: Proceedings Fourth International masonry Conference, vol. 1. London: British Masonry Society; 1995. pp. 157—61.

[24] CP 121: Part 1: 1973, Code of Practice for Walling, Part1: Brick and Block Masonry, British Standards Institution (Withdrawn).

[25] BS 5628: Part 3: 1985, Code of Practice for Use of Masonry. Materials and components, design and workmanship, British Standards Institution (Withdrawn).

[26] BS 5628—3: 2005, Code of Practice for Use of Masonry. Materials and components, design and workmanship, British Standards Institution (Withdrawn).

[27] BS 5628—2: 2005 (withdrawn), Code of Practice for Use of Masonry: Part 2: Structural use of reinforced and prestressed masonry, British Standards Institution.

[28] BS EN 1996-1-1, 2005, Eurocode No. 6: Design of Masonry Structures. General rules for reinforced and unreinforced masonry structures, British Standards Institution. See also: UK National Annex to BS EN 1996-1-1: 2005.

[29] ACI Committee 530-05. Building code Requirements for masonry Structures, ACI Manual of Concrete practice, Part 6. American Concrete Institute; 2007.

9 Enlarged Moisture Expansion due to Cryptoflorescence

In Chapter 8, reference was made to instances of an *enlarged moisture expansion* in clay brickwork; namely, an expansion in excess of the irreversible expansion of the clay brick used to construct the brickwork. Typically, but not always, it occurs in small, unrestrained walls; some types of clay masonry built with low strength, high water absorption, and high initial suction rate. Enlarged expansion may be regarded as a transitional zone effect at the mortar/brick interface, the product of which is crystallization or, more specifically, *cryptoflorescence*. The phenomenon is similar in nature to the more commonly-known surface crystallization known as *efflorescence*. Those two modes of crystallization are really different manifestations of the same phenomena, collectively known as *florescence*, and are usually associated with problems arising from volume change that can cause disruption and decay of building materials.

This chapter reviews the nature of florescence and discusses the potential sources of soluble salts leading to crystallization and mechanisms involved that could be responsible for enlarged moisture expansion. Influencing factors are discussed, particularly that of in-plane restraint, together with the implications of cryptoflorescence occurring in a load-free control specimen when assessing creep in the laboratory.

Nature of Florescence

In 1932, Schaffer [1] described the crystallization of salts within the pores of building materials as cryptoflorescence, and crystallization of salts on the surface of materials, usually white, as efflorescence. The latter is well known due to the first drying of newly-laid clay brickwork that has been exposed to rain. However, whereas with some materials efflorescence always appears on the surface and after a time is washed way by rain leaving the material none the worse, with other materials cryptoflorescence is predominant and considerable damage may result. Therefore, although efflorescence is of no major detriment to the brickwork, cryptoflorescence can lead to volume expansion that can cause disintegration of the surface of clay bricks. The pattern of failure is similar to that of frost attack.

Soluble salts crystallize out of solution when water is removed by drying. If the external surface is sealed, internal drying can take place from pores, and crystallization occurs from beneath the sealed surface. Cryptoflorescence occurs when the rate of movement of the salt solution through the material pores is less than the rate of evaporation of water from the external surface, so that a drying zone may form beneath the surface of the material [2].

Concrete and Masonry Movements. http://dx.doi.org/10.1016/B978-0-12-801525-4.00009-1

In the manufacturing of cement, gypsum is added to the clinker in order to prevent flash set by the hydration of tricalcium aluminate (C_3A). Gypsum quickly reacts with C_3A to produce ettringite, which is harmless at this stage because concrete, for example, is in a semiplastic state so that resulting expansion can be accommodated. A similar reaction takes place when hardened concrete is exposed to sulfate from external sources; for example, groundwater containing sodium, calcium, or magnesium sulfates. The sulfates react with both $Ca(OH)_2$, and the hydrated C_3A to form gypsum and ettringite, respectively, resulting in an increase of volume. In the case of concrete, disruption forces can result, and this is generally known as sulfate attack [3]. In the case of masonry, the external source of sulfates could be the clay brick, and sulfate attack of the mortar can occur.

Bonnell and Nottage [4] studied the crystallization of salts within a porous medium of fine sand confined in a mold by a compressive pressure of up to 8 MPa. The change in volume of the medium was observed during heating and cooling cycles over a range of temperature, which was about the transition temperature of various crystallized salts; namely, the temperature at which dissociation into salt and water occurs. The hydrates tested alone registered an expansion on heating, as witnessed by the dissociation of crystals of magnesium sulfate heptahydrate ($MgSO_4.7H_2O$) into magnesium sulfate hexahydrate ($MgSO_4.6H_2O$) and water. On cooling, there was a gradual decrease in volume as recrystallization occurred. On the other hand, when the same crystals of magnesium sulfate were tested with sand, the opposite behaviour occurred, viz. on heating, there was a decrease in bulk volume as dissociation occurred. The hydrates with sand then registered an expansion on cooling, i.e., crystallization of salts was accompanied by an expansion of the mixture. Similar trends occurred with a mixture of sand and crystals of sodium sulfate, although it is known that crystallization of the hydrated salt in an aqueous solution is accompanied by a decrease in volume.

The conclusions drawn by Bonnell and Nottage [4] were that the anhydrous salt, or lower hydrate, may be further hydrated even against moderately high stresses, which are well above the tensile strength of normal porous building materials. The implication is that the salt crystals could exert a sufficient force to bring about disintegration of the material. The suggested mechanism was that crystals can form when they are restrained on more than one side and the expansion is not restricted to the unrestrained directions and, regardless of the confining conditions, crystals will still form with the same size and shape.

Butterworth [5] stated that there are three main sources of soluble salts in bricks. Occasionally, the clay from which bricks are made contains salt, usually gypsum, but it can be easily separated from the unfired clay. Another possibility is pyrites (iron sulfide), which on heating can decompose to form sulfates after reacting with clay bases. A third source is the sulfur in the coal used for firing the bricks; here, oxides are formed that can react with the clay, and salts generated in this way usually decompose at higher firing temperatures. Many bricks contain calcium sulfate, or calcium carbonate, which can be converted into calcium sulfate during firing, and calcium sulfate is less decomposed by heat than other sulfates found in bricks. Many bricks contain upwards of 3.0% of calcium sulfate. Provided that bricks contain no other soluble matter, the presence of this quantity of calcium sulfate does not affect the appearance or durability of brickwork that is normally protected by a damp-proof course and a roof. Calcium sulfate has a low solubility (being soluble in water to the extent of only one part in 500) and when it occurs alone it hardly ever gives rise to efflorescence. It cannot, however, be entirely ignored because if another salt such as potassium sulfate is present, it may form a double salt, a form it is more soluble so that appreciable amounts may be found in efflorescences.

According to Butterworth [5], magnesium sulfate, which causes the most serous kind of efflorescence failures, is rarely present in amounts exceeding 0.5%, and this is the maximum that can be tolerated. Magnesium sulfate is more easily decomposed in the firing of bricks than is calcium sulfate, and a firing range of 1000–1050 °C is sufficient for its elimination. Sodium sulfate and potassium sulfate can also lead to heavy efflorescence if they are present in quantity, but they are more easily decomposed in firing than are calcium and magnesium sulfates. Ferrous sulfate, like magnesium sulfate, is not found in the majority of bricks, and when it is found it is in very small amounts (less than 0.05%).

Butterworth [5] lists the factors affecting the ability to efflorescence as:

- Solubility. The weight of any given soluble salt that can be dissolved in a fixed weight of water at any given temperature is limited. And when that weight has been dissolved, the solution is said to be saturated. Efflorescence is more likely to appear on drying when the solution is near to saturation; temperature is also a factor.
- Pore structure of the brick. It affects the rate of evaporation of water upon drying.
- Shape of crystals. They may affect movement of salt solution and crystal growth.
- Distribution of salts within a brick.
- Water movement in newly-built brickwork, suction rate, degree of wetting, or docking affects absorption of alkali sulfates from mortar; degree of protection from rain.

According to Bowler [6], sulfates that are significantly water soluble may enter mortar joints and react with calcium hydroxide, which is a soluble salt of low solubility derived from the hydrated cement or lime to form calcium sulfate (gypsum), viz.

$$Ca(OH)_2 + Na_2SO_4 = CaSO_4 + 2NaOH$$

In the above reaction, a theoretical expansion of more than two is involved. The most widely recognized form of sulfate attack involves the chemical reaction of soluble sulfates with hardened cement pastes, especially the C_3A component. Regardless of the specific sulfate concerned, the end products of this reaction are calcium sulfo-aluminates:

$$3CaO.Al_2.6H_2O + CaSO_4 + water \rightarrow 3CaO.Al_2O_3.CaSO_4.12H_2O$$
$$(\text{monosulfate form})$$

or

$$3CaO.Al_2.6H_2O + 3CaSO_4 + water \rightarrow 3CaO.Al_2O_3.3CaSO_4.31H_2O$$
$$(\text{ettringite})$$

In this case, there is a greater theoretical expansion of 4.8 compared with the formation of gypsum.

A review by Gaskill [7] lists the principal compounds causing florescence in brickwork in order of frequency of occurrence as:

- Sodium sulfate (Na_2SO_4).
- Potassium sodium sulfate ($K_3Na(SO_4))_2$.
- Calcium sulfate dihydrate ($CaSO_4.2H_2O$).
- Magnesium sulfate hydrates ($MgSO_4.6H_2O$; $MgSO_4.7H_2O$; $MgSO_4.4H_2O$).

In building materials, salt-saturated solutions undergo an increase in volume on crystallization into the above compounds. Although water experiences a 10% increase in volume when it converts from a liquid to a crystalline state upon freezing, anhydrous sodium sulfate (Na_2SO_4) becomes decahydrate sodium sulfate ($Na_2SO_4.10H_2O$) when the brick becomes wet and the accompanying change in volume is about 300% [7]. The most aggressive salt on crystallization is generally assumed to be magnesium sulfate ($Mg SO_4.7H_2O$), and has long been accepted as a primary cause of the surface failure of bricks. However, Binda and Baronio [8] are of the opinion that sodium sulfate (Na_2SO_4) is the most aggressive salt. In fact, crystallization of sodium sulfate is used for a durability test in South Africa [9]. The test indicates that the higher the concentration of the salt solution, the quicker the brick suffers decay, and the higher the firing temperature, the greater the resistance to decay from crystallization of salts.

Gaskill's review [7] refers to a paper by Laurie and Milne [10], who had found that crystallization of calcium sulfate was responsible for the decay of a brick as a result of being drawn from the mortar. Similarly, from investigations into causes of decay on certain buildings in the United States [11], chemical analysis revealed the presence of calcium sulfate in fairly large quantities, and crystalline particles of calcium sulfate were found in the spalled sections of the brick. Gaskill [7] points out that calcium sulfate may be produced from acid rain in which sulfuric acid reacts with lime ($Ca(OH_2)$) in the masonry; sulfuric acid is formed from the sulfur in burning coal. Analyses of samples of efflorescence on the surface of a wall by X-ray diffraction revealed the presence of mirabilite (sodium sulfate decahydrate) and thernardite (anhydrous sodium sulfate). The conversion of the latter into the former involves a threefold expansion, and the implication is that it can cause disruptive pressures [12], although the processes involved are complex and by no means universally agreed [13].

Hydration of Cement and Moisture Transfer Across the Brick/Mortar Interface

Cement hydration begins as soon as water is added to the cement [14]. Then, depending on the type of cement and temperature, from approximately 1 h stiffening occurs as measured by the initial set. After about 2 h, the water has become saturated with lime and there is evidence of hydration with ettringite and calcium silicate hydrate on the surface on cement grains [15]. Final set occurs approximately from about 3 h and marks the beginning of hardening and strength development. In the hydration of calcium silicates, it is the tricalcium silicate (C_3S) that sets first to form C–S–H together with calcium hydroxide ($Ca(OH)_2$). The other calcium silicate, C_2S, reacts in a similar manner with water, but more slowly, and it produces less $Ca(OH)_2$. Thus, it is apparent that $Ca(OH)_2$ is formed in quantity quite early in the setting and hardening process of cement mortar.

The water transport characteristics across the clay brick/mortar interface during the setting and hardening processes of mortar have been investigated [16,17]. Tests were performed on sealed brick couplets, as shown in Figure. 7.6, in which the upper brick could be removed and weighed periodically during the first 24 h after laying the fresh mortar. Figure 9.1 compares the water absorption histories of a high water absorption Fletton brick, and a low-water absorption Class B Engineering brick over the first 24-h period after lying. In the case of the Fletton brick, there is a rapid initial absorption within 30 min, followed by a small desorption before absorption reoccurs between 1 and 7 h; the latter period corresponds approximately to the period of setting. From about 7 h onwards, the Fletton brick shows desorption due to the take-up of water by the hydrating mortar. As expected, the Class B brick has a lower overall level of absorption and, after an initial rapid increase, has a gradual increase before reaching a constant level of absorption from a time of 3 h.

As the cement paste continues to hydrate, large crystals of calcium hydroxide $Ca(OH)_2$ appear after 24 h [15] together with other crystalline products (ettringite, C_3A and C_4AF) formed by hydration of both calcium silicate hydrates, C_2S and C_3S. The crystals reinforce a continuous gel of C–S–H. In the case of the sealed brick/mortar couplet (Figure 9.2), transfer of moisture from brick to mortar occurs from day 1 to day 3, probably to assist in further hydration. Then, the situation reverses and

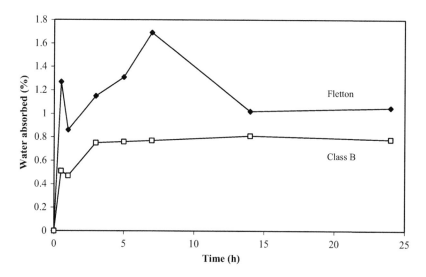

Figure 9.1 Water absorbed by freshly-laid brick in brickwork couplet over 24 h [17].

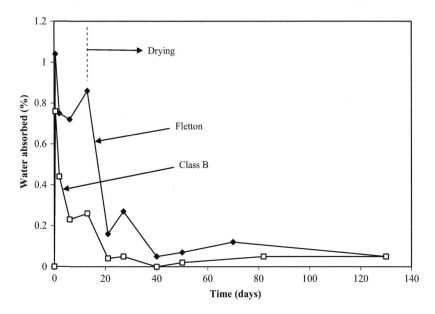

Figure 9.2 Water absorption history of freshly-laid brick in brickwork couplet cured under polythene for 14 days, then exposed to air at 65% RH and 21 °C [17].

moisture is transported back to the brick, although that feature is less apparent in the case of the Class B brick.

From the brick couplet water absorption histories in Figures 9.1 and 9.2, it is apparent that in both the first 24 h and subsequent 13 days there is ample opportunity for water or salts in solution to be transferred across the interface from mortar to brick, and vice-versa. The pattern of water movement depends on the size and distribution of pores in both brick and mortar at the interface, since those pores determine the opposing suction forces of mortar and brick. The situation will be continuously changing as the products of hydration are deposited in the surface pores of the brick to create bonding and the demand for water for cement hydration of the mortar.

After the age of 14 days, when the couplet is exposed to drying, there are quite rapid losses of moisture due to evaporation from all surfaces of the brick couplet (Figure 9.2). This situation is conducive to internal crystallization of any salts in solution in the interfacial pores, and any microcracks, since moisture cannot be easily transported across the interface to the outer drying surfaces in order to effloresce.

Factors Influencing Enlarged Expansion

In discussing the nature of florescence, it is apparent that the type of brick or type of clay, firing temperature, and possibly type of mortar are factors affecting the ability of brickwork to have the potential for enlarged expansion. This section identifies further factors by reviewing previous research in which unusually large expansions were measured, and describes attempts to identify salts in the brick and mortar that could be responsible for cryptoflorescence.

In a review by West [18], a reference was made to research by Clews [19] in which vertical expansions in excess of those of the bricks were found in some types of clay brickwork walls. West suggested that there was an interaction between bricks and mortar and that the mortar itself was expanding. An unusually large, horizontal moisture expansion was reported by Smith [20] in tests with wet Fletton brickwork, which was attributed to sulfate attack of the mortar. However, compared with Fletton brickwork built with Portland cement mortar, even larger vertical and horizontal expansions were found by Beard [21], using high alumina cement mortar with a low C_3A content, which inferred that sulfate attack was not responsible.

Enlarged moisture movements of clay masonry were observed during experiments to study the influence of geometry on moisture movement of 13-course-high × 2-brick-wide brickwork, and 6-course-high × 1-block-wide blockwork [22]; smaller 5- or 6-stack-high walls were also tested. The geometry of the walls and piers was quantified in terms of the volume/surface ratio, V/S. In the cases of shrinkage of calcium silicate brickwork and concrete blockwork, the trend was as expected in that vertical shrinkage decreased linearly as the V/S increased or, in other words, as the cross section of masonry increased. In the case of Fletton clay brickwork, small expansions occurred initially but, surprisingly, after some 50 days or so, the moisture expansion in the vertical direction of the walls and hollow pier increased rapidly; horizontal movements were consistently small throughout the period of testing. As Figure 9.3 shows, expansions were much smaller in the case of 13-course-high solid

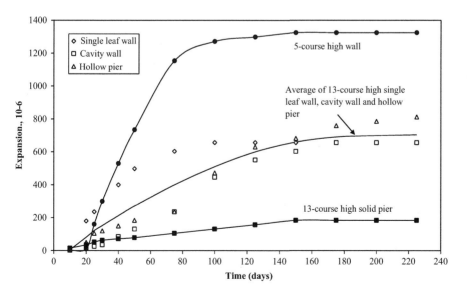

Figure 9.3 Enlarged moisture expansion of Fletton brickwork [22].

pier but, in the smaller 5-course-high test walls, an even greater enlarged moisture expansion occurred. It should be mentioned that the results shown in Figure 9.3 are average values and, in fact, very large ranges of enlarged moisture expansion actually occurred on the different sides of the masonry [22].

At the time, no explanation was forthcoming as to the cause of the very large expansion, except to say that there appeared to be an influence of size that was different from the normal drying effect through the V/S influence. Also, it was noted that the bricks were about 2 years old, and they could have possessed significant anisotropy of irreversible moisture expansion, namely, the bed-face expansion could have been much greater than the header-face expansion. Subsequently, the experiments were repeated with 13-course-high brickwork built with young bricks (2 weeks old) [23]. The behaviour was similar to that of the previous investigation [22], although enlarged expansions occurred slightly later, i.e., approximately 70 days, and expansions of the cavity wall and hollow pier were less in the repeat tests. Although significant anisotropy of irreversible moisture expansion existed in the Fletton bricks, it was insufficient to account for the extent of enlarged vertical expansion in the brickwork. Moreover, chemical analysis of the bed joint mortar revealed the sulfate content to be within the limits of BS 4551: 1980, which suggested that sulfate attack of the mortar was not responsible for the observed expansion.

A further investigation by Forth and Brooks [24] listed specific previous publications where enlarged expansion had been reported, as quantified when the brickwork/brick expansion ratio was approximately ≥ 1. Table 9.1 gives the details of those publications, including those of a later investigation by Bremner et al. [28]. Forth [17] had found significant enlarged expansion of Fletton brickwork to confirm that reported in the earlier work discussed earlier [22,23], especially for smaller five-stack-high walls. Moreover, Forth et al. [25–27] also found enlarged expansion for other types of clay brick that had strengths of less than 50 MPa and water absorptions of between 15 and 20%.

The same investigation [24] was mainly concerned with the attempt to determine the source of enlarged moisture expansion in Fletton brickwork. Experiments involved five-stack-high test walls constructed with a cement-lime mortar and Fletton bricks, and the same mortar with Class B Engineering bricks that were known not to undergo enlarged moisture expansion. The walls were stored under polythene for 14 days, at which time strain measurements began and then continued for 170 days. The wall strain and individual bonded bricks were measured at the positions shown in Figure 9.4, and the strain between header faces of unbonded bricks was also monitored. At the end of testing, samples of brick and mortar were taken at the position where the maximum expansion occurred in the Fletton wall (see Figure 9.4). In fact, wall expansions varied enormously from 200 to 2000×10^{-6} across the four strain gauge positions, with an average of 1300×10^{-6}. In contrast, the moisture movement of the wall built with the Class B brick was a shrinkage of 200×10^{-6}.

Figure 9.5 compares the average strain measured in the bricks that were bonded in the walls and the average strain of the unbonded bricks. It can be seen that the difference in moisture expansion between the bonded and unbonded Class B bricks is small. On the other hand, although the unbonded Fletton brick underwent a gradual

Table 9.1 Enlarged Vertical Moisture Expansion of Previous Investigators when the Brickwork/Brick Expansion Ratios are Greater than 1

Source	Brick/Clay Type	Age of Brick	Curing/Storage Conditions	Type of Brickwork	Test Duration	Brickwork/Brick Expansion Ratio
West [18]	Brick earth Boulder Oxford Keuper	–	None/outdoors	Two-leaf solid wall (0.9 × 0.9 m)	1 year	1.8 2.5 1.8 1.8
Beard et al. [21]	Fletton	Kiln-fresh 2 weeks	None/indoors	Single-leaf wall, (3.2 × 1 m)	6 years	2.7 1.8
Brooks and Bingel [22]	Fletton	Aged	14 days under polythene/indoors (uncontrolled)	Single-leaf wall Cavity wall Hollow pier Solid pier (13-course high × 2-brick wide)	240 days	9.7 9.0 11.5 2.5
Brooks and Bingel [23]	Fletton	2 weeks	14 days under polythene/indoors (uncontrolled)	Single leaf wall Cavity wall Hollow pier Solid pier (13-course high × 2-brick wide)	400 days	7.0 2.3 2.0 0.9
Forth [17]	Fletton Dorket Honeygold Fletton	Aged	14 days under polythene, 14 days at 65% RH and 21 °C/ 65% RH and 21 °C	Single-leaf wall Cavity wall Hollow pier Single-leaf wall (13-course high × 2-brick wide) 5-course-high, single-leaf wall	160 days	3.9 2.3 2.0 2.4 12.3

Continued

Table 9.1 Enlarged Vertical Moisture Expansion of Previous Investigators when the Brickwork/Brick Expansion Ratios are Greater than 1 —cont'd

Source	Brick/Clay Type	Age of Brick	Curing/Storage Conditions	Type of Brickwork	Test Duration	Brickwork/Brick Expansion Ratio
Forth et al. [25]	Oxon Gold	2 months	3 days under polythene/indoors[a] 7 days under polythene/indoors[a] 14 days under polythene/indoors[a]	Single-leaf wall (13-course high × 2-brick wide)	140 days	6.6 2.7 2.2
Forth et al. [26]	Dorket Honeygold	Aged	2 days under polythene/indoors[a] 7 days under polythene/indoors[a]	Single-leaf wall (13-course high × 2-brick wide)	1 year	4.9 3.5
Forth et al. [27]	Kempston	4 days	2 days under polythene/indoors[a]	Single-leaf wall (13-course high × 2-brick wide)	240 days	7.6
Bremner et al. [28]	Coal measure Fireclay (Throckley Class B).	1 month	7 days Under polythene/indoors[a]	Single-leaf wall (13-Course high × 2-brick wide)	1 year	0.9
	Fireclay shale. (Roughdales Golden Rustic).					1.5

Material	
Etruria Marl (Chesterton Smooth red).	1.1
Brick-earth/Chalk Breeze (Funton 2nd Mild Stock).	3.0
Lower Oxford (Fletton). Fireclay Shale (Highbury Buff).	2.3
Weald (Capel Multi red).	1.8
Fireclay mixture (Jacobean Blue/Brown).	1.8
Brick Earth (MH 1st hard Stock).	2.5
Keuper Marl lower (Chatsworth Gray).	3.6
	2.0
Keuper Marl lower (Heather Countryside Straw).	4.5
Gault Shale (MH Kentish Multi).	2.4

[a]Storage indoors at 65% RH and 21 °C.

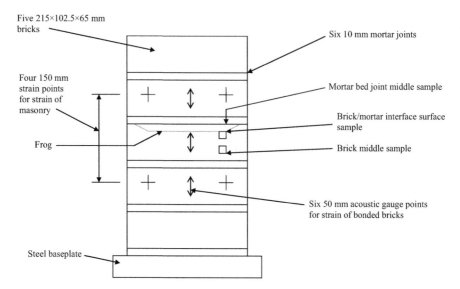

Figure 9.4 Strain measurement and sample points of five-stack masonry test wall [24].

expansion, the bonded brick showed an initial shrinkage before starting to expand after about 60 days. The initial shrinkage is thought to be due to a loss of moisture from the bonded brick, which had initially absorbed water from the plastic mortar. Overall, compared with the average wall expansion, the strains of the bonded bricks were small, which implied that the location of enlarged expansion was near or at the brick/mortar interface.

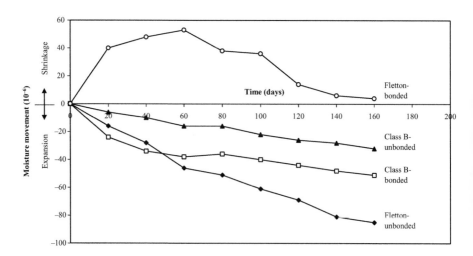

Figure 9.5 Moisture movement of Fletton and Engineering Class B bonded and unbonded bricks [24].

Chemical analysis was carried out on samples of brick and mortar taken at the brick/mortar interface, and in the middle of the mortar joint and middle of the brick. X-ray diffraction revealed that the levels of calcium sulfate and ettringite in the middle of the mortar joint were low and similar for both types of masonry, thus confirming the earlier finding [21] that enlarged expansion was not associated with sulfate attack of the main body of the mortar bed joint. Comparison of the interface and middle samples of the Fletton mortar bed joint showed calcium hydroxide (portlandite) to be present in the middle, but not at the interface. The interface sample had a greater concentration of calcium carbonate (calcite), possibly caused by carbonation, i.e., the reaction of calcium hydroxide with carbon dioxide, possibly from the outside air percolating through large capillary pores and plastic shrinkage cracks at the interface. Levels of ettringite were almost nonexistent at the interface, but ettringite could have been present in the surface pore of the brick [17].

A comparison of the diffraction patterns of the brick samples showed that the interface surface sample contained calcium sulfate (anhydrite) and calcium carbonate. Those products were not present in the middle sample of the bonded brick, nor in samples of the unbonded brick. However, gypsum (hydrated or crystallized calcium sulfate) was found in the surface sample of the bonded Fletton brick. Microprobe analysis of a representative pore confirmed that calcium sulfate was concentrated around the inside of the pore material, suggesting that crystallization of the calcium sulfate salt may be the reason for the enlarged moisture expansion observed in the Fletton masonry.

The later investigation by Bremner et al. [28] compared the vertical moisture movement of 13-course-high brickwork walls constructed from 20 different types of clay brick, many of which had a brickwork/brick expansion ratio exceeding unity after 1 year (see Table 9.1), In general, the relatively weaker bricks that had high water absorption and initial suction rates produced greater brickwork/brick expansion ratios, but there were exceptions to that trend. As postulated by West [18], Bremner et al. [25] confirmed restraint to be factor in enlarged expansion, as in 70% of cases, the strain in the lower section of the walls was less due to the dead weight of upper section of brickwork. Enlarged expansions also occurred in the horizontal direction, but in fewer cases than in the vertical direction, and not necessarily for the same type of brick.

Bremner [29] found that the type of mortar did not have a significant influence of enlarged moisture expansion, which is demonstrated in Figure 9.6, where the 300-day brickwork/brick expansion ratio is plotted against the initial suction rate of the brick. Compared with compressive strength and water absorption, the correlation with the initial suction rate was found to be the best. Nevertheless, for any given initial suction rate, there exists considerable scatter of the expansion ratio in Figure 9.6. However, the average trend suggests that the potential for enlarged moisture expansion (brickwork/ brick ratio ≥ 1) is when bricks have an initial suction rate greater than 0.7 $kg/m^2/min$. The finding of a critical suction rate may suggest that docking or wetting of the brick prior to laying may offer a possibility of reducing the effects of cryptoflorescence and enlarged expansion. However, it is normally recommended that docking should be limited to bricks having initial suction rates exceeding 1.5 $kg/m^2/min$, otherwise the

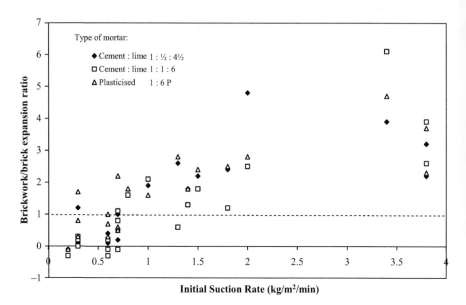

Figure 9.6 Brickwork/brick expansion ratio after 300 days as a function of initial suction rate of brick for different types of mortar.

development of bond strength may be impaired. It should be emphasized that the expansion ratios of Figure 9.6 are for 13-course-high × 2-brick-wide laboratory test walls, and that the unbonded brick expansion involved is that between bed faces. Therefore, the results are not necessarily applicable for all types of brickwork and, in fact, it will be shown later that enlarged expansion depends very much upon height of brickwork or, more specifically, the in-plane restraint.

Bremner [29] also investigated the chemical and physical nature of the brick/mortar interface using mercury intrusion porosity (MIP), scanning electro microscopy (SEM), and X-ray diffraction (XRD). MIP tests on unbonded units revealed that those units involved in enlarged moisture expansion of brickwork had larger pores and a larger pore volume when compared with bricks that did not yield an enlarged expansion. That finding confirmed the general tendency of brickwork to undergo enlarged moisture expansion when built with bricks of lower strength, higher water absorption, and higher initial suction rate. Watson et al. [30] found that a coarse, pore-size distribution in the brick produced the least efflorescence in masonry, because the consequential smaller capillary stresses would cause the evaporation point of saturated salts to be transferred to a point below the surface, whereas higher capillary stress in finer pores would lead to efflorescence at the surface. However, the resistance to flow is a factor, which is markedly less in coarser pores than in finer pores, and it is likely that resistance to flow dominates the initial transfer of water and cementitious material from mortar to brick in freshly-laid brickwork [30].

In Bremner's analysis [29], crystalline phases were identified by XRD from samples taken at the brick/mortar interface after various times up to 140 days. The brickwork walls, which were chosen for investigation, had previously exhibited the greatest enlarged expansion. They were built with a Keuper Marl clay brick and a Fletton brick. No ettringite was detected after 7 days, but thereafter it was found after 21 days, the magnitude remaining the same up to 70 days before decreasing at 140 days. SEM revealed the presence of calcium sulfate ($Ca(SO)_4$) within unbonded brick pores across a wide range of pore diameters and in large quantities, possibly arising from the reaction of pyrites (FeS_2) and calcium carbonate ($CaCO)_3$ contained in the clay [30].

Enlarged expansion was, therefore, attributed by Bremner [29] to the crystallization of ettringite at the brick/mortar interface, rather than crystallization of anhydrite or calcium sulfate as proposed by Forth and Brooks [24]. The reaction between calcium sulfate in the brick surface pores and the C_3A in the mortar at the interface probably caused the precipitation of ettringite at the interface, and the low solubility of calcium sulfate allowing the reaction products to be formed over a longer period of time. In a later paper [31], it was suggested that the presence of calcium sulfate at later ages could be caused by oxidation of ettringite.

Summary of Cryptoflorescence Mechanism

From the previous review and chemical analysis of samples from the brick/mortar interface, the mechanism by which cryptoflorescence can cause an enlarged moisture expansion of brickwork is summarized as follows:

- The initial absorption of water from the fresh mortar by the brick transfers unhydrated cementitious material in the surface pores of the brick. The loss of water can result in plastic shrinkage microcracks in the mortar, whereas water gain in the brick allows soluble sulfates present to pass into solution.
- Since the masonry is cured under a polythene membrane for 14 days, most of the absorbed water is held within the brick and, as a first stage, some salt solution can react with the hydrating cementitious material in the surface pores of the brick. In a second stage, some salt solution can return to the hydrating mortar joint via the interface after only a few hours. Both stages can allow sulfates to react with $Ca(OH)_2$ to form calcium sulfate and then ettringite.
- Over the next few days, some of those products in solution return to the brick/mortar interface, and, together with products from the first stage, probably locate in the brick surface pores and microcracks in the mortar. After removal of the polythene curing membrane at 14 days, drying of the outer brick surfaces occurs, and the evaporation of water promotes crystallization at the interface. The resulting volume change manifests itself as an enlarged expansion.

In-Plane Restraint

On several occasions in this chapter, reference has been made to the enlarged moisture expansion exhibited by Fletton clay brickwork. Yet, in some early test results involving the assessment of creep of Fletton brickwork, no unusual vertical expansion of the control wall was observed [32]. With the benefit of hindsight, it was realized that the important differences between that control wall and those used later in creep work were that: (1) the bricks were wetted before laying, and (2) the wall was capped with a 50-mm-deep steel header plate to be made identical to the creep wall, except for the applied load. In fact, the control wall initially underwent shrinkage before expanding by 40×10^{-6} after 300 days. When the result was compared with later findings of enlarged expansion, it became clear that wetting the bricks reduced the suction properties, and the steel capping header plate acted as a restraint, thus possibly restraining or even preventing cryptoflorescence and enlarged expansion.

Test results using other types of clay brickwork, known to exhibit enlarged moisture expansion, have been compared with and without steel capping plates bedded to the top surface of walls [31]. The walls were cured under polythene for 1 day and then exposed to drying in air controlled at 65% RH; measurements started at the age of 7 days. Although the dead weight of the header plates only imposed a small compressive stress of 0.02 MPa on the wall, it dramatically reduced the moisture movements, most of which were enlarged expansions due to cryptoflorescence, as can be seen in Figure 9.7 In fact, two types of brickwork exhibited a small shrinkage when the walls were capped

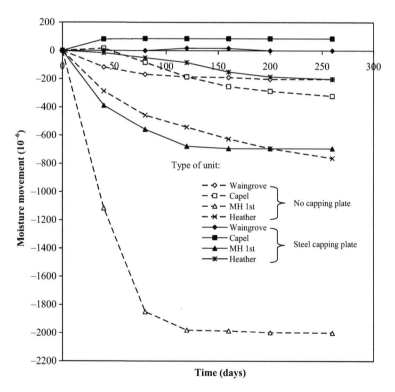

Figure 9.7 Influence of steel capping plate on enlarged moisture expansion due to cryptoflorescence occuring in 13-course-high, single-leaf walls built with different types of clay unit [31].

with headers plates, thus demonstrating the effectiveness of in-plane restraint in suppressing cryptoflorescence. A full list of properties of the units and brickwork used in the tests is given in Table 12.1 of Chapter 12.

Another series of tests involved the measurement of stress induced by fully restraining clay brick walls from expanding [26], when the accompanying load-free control walls exhibited enlarged expansion. A Dorket Honeygold brick was used to build 13-course-high × 2-brick-wide brickwork, and measurements started at the age of 2 days. Moisture expansion of an uncapped control wall after 1 year was 700×10^{-6} compared with the unbonded brick expansion of 150×10^{-6}, thus yielding a brickwork/brick expansion ratio of 4.7. According to the level of enlarged moisture expansion, the stress induced would be expected to be much greater than 0.2 MPa, which was actually measured. An even smaller induced stress of less than 0.1 MPa was measured with restrained Melford Yellow clay brickwork in other tests [27]. Those low stresses occurred despite the uncapped control wall having a brickwork/brick expansion ratio of at least five. Again, those findings suggested that the source of enlarged expansion was absent in the restrained walls.

The influence of partial restraint to vertical enlarged expansion of clay brickwork was investigated in more detail by comparing the vertical strain over the full height of 25-course, 13-course, and 5-course single-leaf walls as shown in Figure 9.8 [33]. In addition, two walls of 13- and 5-course brickwork were capped with steel header plates, and moisture movement of unbonded bricks was measured between bed faces. When built with Fletton bricks, Figure 9.9 clearly demonstrates that all wall strain measurements indicate expansions in excess of that of the unbonded brick, and therefore the existence of enlarged expansion. However, the extent of enlarged expansion depended upon the mass of brickwork and steel capping plate above the strain gauge measuring position. The greatest expansions occurred for the uncapped walls and the top 750-mm gauge position of the 25-course-high wall, and the lowest expansion occurred for the bottom 750-mm strain gauge position. These results demonstrate that enlarged expansion depends on in-plane restraint or vertical dead-weight stress. When those expansions are compared with that of the unbonded brick and estimated wall expansion, it can be seen that the level of stress generated in these walls was insufficient to completely suppress or prevent cryptoflorescence. The estimated or "true" expansion of the wall shown in Figure 9.9 was calculated using Eq. (7.6).

Figure 9.10 shows moisture movement as a function of stress due to the dead load for the above-mentioned Fletton walls, as well as for complementary walls built with a Class A Engineering brick and walls built with a concrete brick. In the latter cases, it appears that there is no apparent influence of in-plane restraint on moisture movement. The overall levels agreed with the estimated moisture movements given by Eq. (7.6), which are, in fact, shrinkage strains, notwithstanding that the unbonded Class A brick underwent a small, irreversible moisture expansion of 120×10^{-6} (between bed faces) after 300 days; the corresponding movement of the unbonded concrete brick was a shrinkage of 280×10^{-6}.

On the other hand, Figure 9.10 indicates a clear the influence of in-plane restraint or dead-weight load on total moisture expansion of the Fletton walls, with the average

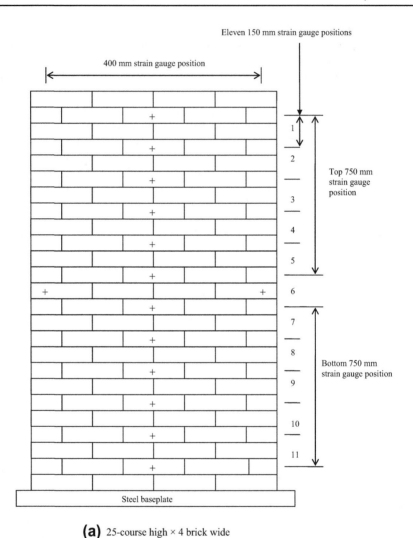

(a) 25-course high × 4 brick wide
(uncapped)

Figure 9.8 Arrangement of strain gauge measuring positions for restraint tests on single-leaf walls [33], (a) 25-course high × 4-brick wide (uncapped), (b) 13-course high × 2-brick wide (uncapped), (c) 5-stack high (uncapped), (d) 13-course high × 2 brick-wide (capped), (e) 5-stack high (capped).

relationship between total expansion (ε_e) and stress due to dead-weight load (σ) being represented as follows:

$$\varepsilon_e = \frac{[1.47 - 9.76\sigma]10^4}{7 + 430\sigma} \qquad (9.1)$$

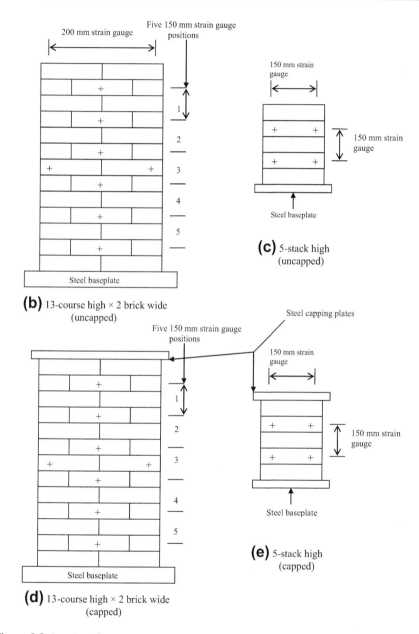

Figure 9.8 (*continued*).

The total moisture expansion comprises the true moisture expansion and the enlarged expansion. According to Eq. (9.1), the critical compressive stress at which the Fletton expansion–stress curve intersects with the estimated "true" moisture expansion of 270×10^{-6} is 0.06 MPa. In other words, the stress required to prevent enlarged

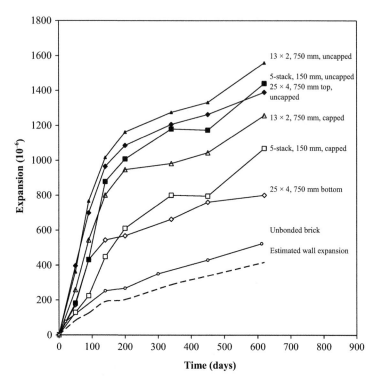

Figure 9.9 Vertical expansion of the Fletton walls shown in Figure 9.8 compared with unbonded brick (between bed faces) and estimated wall expansion.

moisture expansion of Fletton brickwork due to cryptoflorescence is 0.06 MPa. Interestingly, further extrapolation of the expansion–stress curve for the case when $\varepsilon_e = 0$ represents the case of full restraint to total moisture movement brickwork and, according to Eq. (9.1), that occurs when the stress = 0.15 MPa.

The stress required to prevent enlarged expansion due to cryptoflorescence of clay brickwork is of general interest in practical situations for the design of walls of low height, and is of particular interest in laboratory tests to determine creep of brickwork for reasons demonstrated in the next section. With regard to full restraint of Fletton brickwork, the estimated stress of 0.15 MPa may be compared with the test results mentioned earlier involving full restraint to moisture movement of 13-course-high × 2-brick-wide walls [26,27]. A similar range (0.05–0.15 MPa) of induced stress was measured by Bremner [29] for fully restrained walls built with other types of brick, which, when used to build un-restrained walls, exhibited enlarged expansions. Hence, the test results demon-strate that enlarged expansion does not occur in fully restrained walls where induced stresses are greater than the critical compressive stress required to suppress cryptoflorescence.

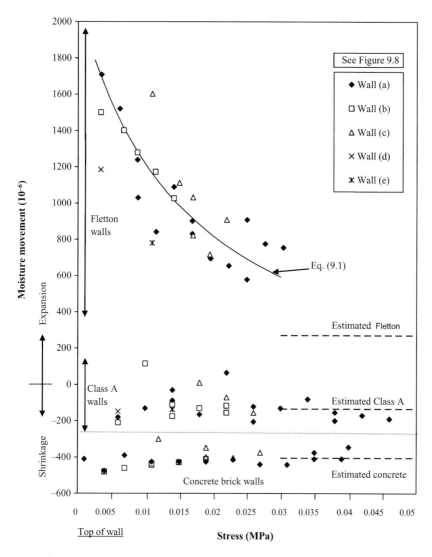

Figure 9.10 Vertical moisture movement of fletton clay, Class A clay, and concrete brick walls measured at different heights, as a function of stress due to dead load after 300 days.

Quantification of Creep

The topic of creep of masonry is fully dealt with in Chapter 12, but at this stage it is relevant to stress the influence of enlarged expansion when determining creep in the laboratory. For concrete or masonry, the definition of creep given in Chapter 2 requires that a load-free control specimen, which is identical to the specimen subject to load, is used to allow for any time-dependent strains incurred that are not associated with the compressive load. To isolate the time-dependent strain purely due to the load, i.e., creep, those strains not associated with load are added or deducted form the strain measured on the specimen under load according to whether they are expansions or contractions. Typically, load-free strains are those due to environmental changes in temperature and humidity. Clearly, if chemical or physical phenomena occur to a different extent in the control specimen than in the specimen under load, resulting in additional strain in the control specimen, then creep could be erroneously quantified.

The above situation arises in the case of clay brickwork when the control specimen undergoes enlarged expansion due to cryptoflorescence [31]. Figure 9.11 gives a schematic example of time-dependent strains measured in (a) a wall under constant load, and (b) a control wall. In wall (a), the total measured strain consists of the elastic strain plus creep plus moisture movement strain. In wall (b), two curves for moisture movement strain are represented: (1) the "true" expansion, and (2) the total expansion including the enlarged expansion. Figure 9.11(c) shows the deduced creep after taking into account the two moisture expansion cases, and it can be seen that using the moisture expansion with enlarged expansion results in a greater level of creep (curve 2).

Consequently, when carrying out tests to assess creep of brickwork in which the control specimen is prone to cryptoflorescence, the measured moisture movement of the control brickwork (corresponding to curve 2 of Figure 9.11) has to be carefully considered. Complementary tests to determine the moisture movement of the unbonded brick would confirm whether the brickwork/brick expansion ratio exceeds unity and, therefore, confirm the presence of enlarged expansion in the control wall. It is not possible to separate "true" moisture movement from enlarged expansion by testing. For example, the control wall could be restrained to a stress that will suppress cryptoflorescence, so that only the true moisture expansion is developed, but the precise level of stress required would not be known without prior testing. Too much restraint will, of course, prevent some "true" moisture movement as well as enlarged expansion. Also, the stress required would have to be applied externally, as in a creep test, because stresses due to self-weight would require very large control walls with, for example, a heavy steel header plate. In view of those circumstances, it would not be possible to quantify creep in a satisfactory manner.

On the other hand, estimates of the "true" vertical moisture movement can be made by the composite model as discussed in the previous chapter (Eq. 7.6), which requires knowledge of brick irreversible moisture expansion and mortar shrinkage. The estimates may be made without the need for testing, but additional tests carried out at the same time as the brickwork tests using the actual unbonded bricks and mortar prisms cast during the brick-laying process would yield more accurate estimates of moisture movement of the brickwork.

Besides being required for quantification of creep, knowledge of "true" moisture movement is desirable for design purposes in its own right since, compared with laboratory specimens, practical brickwork would be much more robust with dead-weight forces sufficiently high enough to subdue cryptoflorescence. In some situations, however, such as unrestrained low walls and at the top of high walls, enlarged

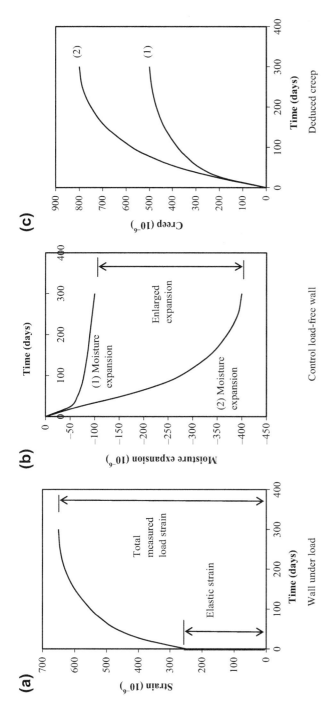

Figure 9.11 Effect of enlarged expansion due to cryptoflorescence in the control wall on quantifying creep of brickwork. The deduced creep in (c) is given by the total measured strain in (a) plus the moisture expansion in (b).

expansion is possible, and suitable substantial capping or coping would be required. Horizontal movement joints may have to accommodate the additional vertical movement in outer leafs of cavity walls supported by shelf-angles.

Concluding Remarks

In addition to moisture movement, an enlarged moisture expansion can occur in some types of clay brickwork due to cryptoflorescence. The phenomenon is more likely to occur in small, unrestrained brickwork built from low strength bricks, such as that used in laboratory creep tests. The effect is more apparent in the vertical direction than in the horizontal direction and mainly depends upon:

- Type of clay brick.
- Initial suction rate of the brick when in excess of $0.7 \ kg/m^2/min$.
- Degree of docking or wetting of the brick prior to laying.
- Length of curing under polythene membrane.
- In-plane restraint or height of brickwork.
- Geometry.

Enlarged expansion can be reduced by wetting bricks with a suction rate greater than $1.5 \ kg/m^2/min$ and, if practical, prolonging the curing period by covering newly-laid brickwork with polythene sheet. Moreover, enlarged expansion is less likely in robust clay brickwork having a high volume/drying surface area ratio. A compressive stress of approximately 0.06 MPa, as provided by dead load, capping, or coping, should be sufficient to prevent the occurrence of cryptoflorescence in most types of clay brickwork.

Since the enlarged expansion cannot be separated from the total measured moisture movement, the "true" moisture movement of small control brickwork specimens, as required to quantify creep in laboratory tests, should be estimated by the composite models presented in Chapter 7.

Problems

9.1 Define enlarged moisture expansion.
9.2 When is it likely to occur?
9.3 What is the cause of enlarged expansion?
9.4 Explain the difference between efflorescence and cryptoflorescence.
9.5 What are typical soluble salts in clay bricks?
9.6 Give the chemical reactions that yield gypsum and ettringite.
9.7 State the main influencing factors affecting enlarged moisture expansion.
9.8 How would you minimize enlarged moisture expansion?
9.9 What happens when brickwork, which is prone to cryptoflorescence, is fully restrained?
9.10 How would you allow for moisture movement strain in order to quantify creep of brickwork that exhibits enlarged expansion when unrestrained?

References

[1] Schaffer RJ. The weathering of natural building stones. Department of Scientific and Industrial Research, Building Research, special Report No. 18; 1932. 56–72.

[2] Building Research Board. Report of the Director of Building Research. Department of Scientific and Industrial Research; 1930.

[3] Neville AM, Brooks JJ. Concrete technology. Second Pearson Prentice Hall; 2010. 438 pp.

[4] Bonnell DGR, Nottage ME. Studies in porous materials with special reference to building materials. I. The crystallisation of salts in porous materials. Trans J Soc Chem Ind 1939;LVIII:16–21.

[5] Butterworth B. Efflorescence and staining of brickwork. The Brick Bulletin; December 1962. 8 pp.

[6] Bowler GK. Deterioration of mortar by chemical and physical action, the response of masonry to the environment. British Masonry Society, Autumn Meeting; December 1991. 10 pp.

[7] Gaskill RL. Decay of masonry materials through water and salt crystallisation [B.Sc. dissertation]. University of Salford, 1995, 84 pp.

[8] Binda L, Baronio G. Mechanisms of masonry decay due to salt crystallisation. Durability Build Mater, 4. Elsevier Science Publishers; 1987.

[9] Boucher PS, Loubster PJ, Coetzee JC. Progress on the development of a salt durability test for fired clay bricks. In: South African Ceramic Society, SACS Symposium: Durability of Ceramic Products, Pretoria; 1987. pp. 19–24.

[10] Laurie AP, Milne J. The evaporation of water and salt solutions from surface of brick, stone and mortar. Proc R Soc Edinburgh 1926;XLVII-Part 1(4):52–68.

[11] Hardesty JM. Disintegration of face bricks by dissolved salts. Bell Lab Rec 1944;22(5):222–4.

[12] Hime WD, Martinek RA, Backus LA, Marusin SL. Salt hydration distress. Concr Int October 2001;23(10):43–50.

[13] Neville AM. Efflorescence-surface blemish or internal problem? Part 1: the knowledge. Concr Int; August 2002:86–90.

[14] Mindess S, Young FJ. Concrete. New Jersey: Prentice-Hall; 1981. 671 pp.

[15] Illston JM, Dinwoodie JM, Smith AA. Concrete, timber and metals. Van Nostrand Reinhold; 1979. 663 pp.

[16] Forth JP, Brooks JJ, Tapsir SH. The effect of unit absorption on long-term movements of masonry. Cem Concr Compos 2000;22:273–80.

[17] Forth JP. Influence of mortar and brick on long-term movements of clay brick masonry [Ph.D. thesis]. School of Civil Engineering, University of Leeds, 1995, 300 pp.

[18] West HWH. Moisture movement of bricks and brickwork. Trans Br Ceram Soc 1967;66(4):137–60.

[19] Clews FH. Experiments to assess the durability of bricks and brickwork. Proc Br Ceram Soc 1965;4:93–108.

[20] Smith RG. Moisture expansion of structural ceramics, IV Expansion of unrestrained Fletton brickwork. Trans Br Ceram Soc 1974;73(6):191–8.

[21] Beard R, Dinnie A, Richards R. Movement of brickwork. Trans Br Ceram Soc 1969;68(3):73.

[22] Brooks JJ, Bingel PR. Influence of size on moisture movements in unrestrained masonry. Masonry Int 1985;4:36–44.

[23] Brooks JJ, Bingel PR. Moisture expansion of Fletton brickwork. In: West HWH, editor. Proceedings of the First International Masonry Conference, 2. Stoke-on-Trent: British Masonry Society; 1988. pp. 12–4.

[24] Forth JP, Brooks JJ. Cryptoflorescence and its role in the moisture expansion of clay brick masonry. Masonry Int 2000;14(2):55–60.

[25] Forth JP, Bingel PR, Brooks JJ. Influence of age at loading on long-term movements of clay brick and concrete block masonry. In: Proceedings 7th North American Masonry Conference, South Bend; 1996. pp. 811–21.

[26] Forth JP, Bingel PR, Brooks JJ. Response of clay masonry panels subject to restraint of irreversible vertical moisture expansion. Weimar: Ibausil, Bauhaus-Universitat; 1997. 1-0825–1-0835.

[27] Forth JP, Bingel PR, Brooks JJ. Stresses in restrained masonry panels. In: 11th International Brick/Block Masonry Conference, Shangai; 1997. pp. 685–92.

[28] Bremner A, Brooks JJ, Forth JP, Bingel PR. Irreversible moisture expansion of unbonded clay brick units and brickwork panels. 9th Canadian Masonry Symposium. University of New Brunswick; 2001. 11pp.

[29] Bremner A. The origins and effects of cryptoflorescence in fired-clay masonry [Ph.D. thesis]. School of Civil Engineering, University of Leeds, 2002. 232 pp.

[30] Watson A, May JO, Butterworth B. Studies of pore size distribution: I. Apparatus and preliminary results. Trans Br Ceram Soc 1957;56(2):37–52.

[31] Forth JP, Brooks JJ. Creep of masonry exhibiting cryptoflorescence. Mater Struct 2008;41:909–20.

[32] Brooks JJ. Time-dependent behaviour of calcium silicate and Fletton clay brickwork walls. In: Proceedings of the 8th International Conference on Loadbearing Brickwork, Building Materials Section, Stoke-on-Trent. British Ceramic Society; 1983. 9 pp.

[33] Bingel PR, Brooks JJ, Forth JP, Bremner A. Moisture expansion of Fletton clay brickwork panels. In: Proceedings of the 12th International Brick/Block Masonry Conference, Madrid; 2000. pp. 259–66.

10 Creep of Concrete

This chapter deals exclusively with creep of plain concrete, which, together with elastic deformation, shrinkage, and thermal movement, designers have to take into account to analyze reinforced and prestressed concrete structures to ensure long-term serviceability. Creep of concrete is a manifestation of the fact that the relation between stress and strain is a function of time and, since moisture movement readily occurs in concrete under normal ambient storage conditions, there are different types or categories of creep. In the first instance, those categories are explained before proceeding to discuss in detail the many influencing factors affecting creep in compression: type and content of aggregate, water/cement ratio, stress/strength ratio, type of cement, age at loading, size and shape of member, storage environment, type of load, time under load, chemical and mineral admixtures, and temperature. After dealing with reversibility of creep or creep recovery and Poisson's ratio, creep under different types of loading is discussed: tensile, cyclic, and other types of load. Standard methods of predicting creep under static loading and drying shrinkage are given in Chapter 11.

Categories of Creep

The complete spectrum where creep is possible for concrete subjected to any level of sustained, constant loading expressed as a fraction of the ultimate short-term strength is illustrated in Figures 10.1 and 10.2. The stress–strain curve for short-term compressive strength test in Figure 10.1 yields the elastic strain for any level of stress. At stress levels greater than approximately 0.5, there is a likelihood of nonlinear elastic behaviour as explained in Chapter 4. Selecting a level of stress and sustaining it yields the characteristics shown with creep continuing at a decreasing rate for up to 50 years. For sustained stresses up to approximately 0.5 of the short-term strength, creep is assumed to be proportional to stress (linear creep), but at higher stress, nonlinearity gradually increases until at a stress/strength ratio beyond approximately 0.8, time-dependent failure is likely, i.e., creep rupture occurs. Experimental results corresponding to Figure 10.1 for concrete loaded in compression were obtained by Rusch [1]. Concrete subjected to tensile loading behaves essentially in the same manner but with a lower failure threshold of approximately 0.6 [2].

Figure 10.2 shows the strain-time curves corresponding to the stress–strain curves of Figure 10.1. The stress–strain curve leading to failure by creep rupture is characterized by three stages: initial rate of creep (primary), steady rate of creep (secondary), and finally, creep with unstable crack growth (tertiary). For stresses below the failure threshold, primary and part-secondary stages occur, but the latter does not lead to creep

Concrete and Masonry Movements. http://dx.doi.org/10.1016/B978-0-12-801525-4.00010-8

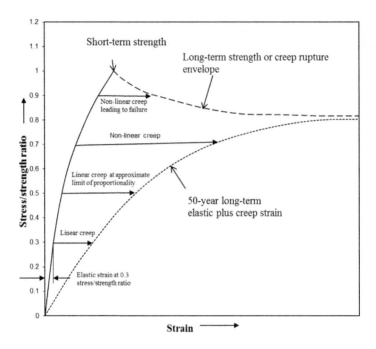

Figure 10.1 Schematic representation of stress−elastic strain plus creep behaviour of concrete under different levels of sustained loading up to failure.

rupture and, in fact, like shrinkage, creep proceeds for many years [3]. This chapter is concerned with factors influencing linear creep of concrete, i.e., for sustained stresses below the limit of proportionality, so that creep can be quantified as creep per unit stress, namely, *specific creep* or *creep compliance*, with units of 10^{-6} per MPa. A review of experimental results confirming linearity of creep for stresses up to approximately 50% of the strength is given by Neville et al. [4].

In consequence, creep is defined as the increase in strain with time under a sustained constant stress and is reckoned from the initial elastic strain given by the secant modulus of elasticity (see p. 72) at the age at loading, as illustrated in Figure 10.2. Strictly speaking, however, creep should be reckoned from the elastic strain at the time when creep is determined, since the elastic strain decreases with age due to an increases in modulus of elasticity (see Chapter 4). However, for simplicity and convenience, this effect is assumed to be small and ignored. It may be recalled in Chapter 4 that since the curvature of the short-term stress−strain curve for concrete is rate-dependent, the demarcation between the start of creep and elastic strain is not clearly defined. For this reason, total strain per unit of stress or *creep function* or *compliance* (elastic strain plus creep) is used to quantify creep by some methods of prediction (Chapter 11). Other methods use the term *creep coefficient* or *creep factor*, which expresses creep as a fraction of the elastic strain.

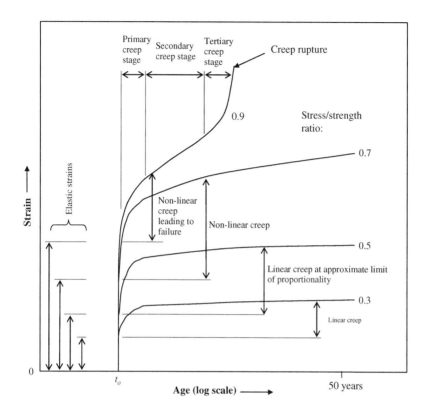

Figure 10.2 Schematic representation of creep behaviour of concrete for different levels of sustained loading from age t_o.

If there are other time-dependent deformations occurring at the same time that are not associated with the applied stress, then they have to be taken into account when determining creep, viz. shrinkage, swelling, and thermal movement due to temperature changes. This is illustrated by considering the following situations in which concrete is loaded to a compressive stress σ at the age t_o and sustained until some later age t. In all cases, concrete is cured in water until age t_o and subsequently creep tests carried out in different storage environments, which induce load-independent deformations other than creep. Suppose the secant modulus of elasticity is E at the age t_o, then the elastic strain at loading in all cases is σ/E.

1. Concrete sealed from the age t_o

At age t, the measured strain (ε_a) is comprised of elastic strain (σ/E), creep (c_a), and autogenous shrinkage (S_a). Hence, creep is:

$$c_a = \varepsilon_a - \frac{\sigma}{E} - S_a \tag{10.1}$$

In this case, sealed concrete simulates mass or large volume concrete in which moisture loss to the environment is minimal. Here, creep is often categorized as *basic creep* since autogenous shrinkage is small particularly for normal-strength concrete. However, for high-strength concrete, autogenous shrinkage is far more significant, especially for concrete loaded at early ages (see Chapter 6).

2. Concrete Allowed to Dry from Age t_o

At age t, the measured strain (ε_b) is comprised of elastic strain (σ/E), creep (c_b), and drying shrinkage (S_h). Hence:

$$c_b = \varepsilon_b - \frac{\sigma}{E} - S_h \qquad (10.2)$$

This situation is common for structural concrete members, such as beams and columns stored in drying indoor and outdoor environments Here, creep is usually much greater than basic creep in case (1) and it is known as *total creep* since it consists of *drying creep* as a consequence of moisture loss, as well as basic creep.

3. Concrete stored in water from age t_o

At age t, the measured strain, ε_c, is comprised of the same elastic strain as in the previous cases, creep c_c and swelling S_w. Since swelling is an expansion:

$$c_c = \varepsilon_c - \frac{\sigma}{E} + S_w \qquad (10.3)$$

Compared with drying shrinkage, swelling of normal weight aggregate concrete is much smaller, so that this case is often regarded as approximating to basic creep. However, this may not be the case for lightweight concrete. In practice, case (3) corresponds to submerged concrete, such as dams and bridge piers partly submerged in water. In the laboratory, it is often convenient to determine basic creep of concrete by using specimens immersed in water (see Chapter 16).

4. Concrete sealed and subjected to a rise in temperature from age t_o.

Assuming autogenous shrinkage to be negligible, the measured strain, ε_d, is comprised of the same initial elastic strain, creep c_d, and thermal expansion S_T. Hence:

$$c_d = \varepsilon_d - \frac{\sigma}{E} + S_T \qquad (10.4)$$

In practice, case (4) represents normal-strength mass concrete undergoing a temperature rise due to heat of hydration, concrete used in nuclear shields, and concrete exposed to fire. In normal-strength concrete, the assumption of negligible autogenous shrinkage is valid, but that is not the case in high-strength concrete, so that S_T would probably include some autogenous shrinkage.

From the above expressions, it can be seen that in order to determine creep, separate measurements of S_a, S_h, S_w, and S_T are required on load-free specimens. An interesting fact is that creep is defined on an additive basis, i.e., it is assumed that those deformations of load-free specimens also occur in the specimens under load and are not affected by stress. If this assumption were correct, then creep in all four cases would be the same; however, as it will be demonstrated later, this is not the case. For example, in case (2) it has already been mentioned that creep is greater than in case (1) even though drying shrinkage has been taken into account. In general, the order of creep is: $c_b > c_T > c_w > c_a$.

Factors Influencing Creep in Compression

Aggregate

In normal-weight concrete, the source of creep is the hardened cement paste because the aggregate is not liable to creep at the level of stress to which concrete is subjected. Because the aggregate is stiffer than the cement paste, the main role of aggregate is to resist the creep of cement paste, the effect depending on the elastic modulus of aggregate and its volumetric proportion. In fact, the role of aggregate in creep of concrete is the same as that in shrinkage of concrete discussed in Chapter 6. Hence, the stiffer the aggregate, the lower the creep, and the greater the aggregate content, the lower the creep. Figure 10.3 shows that creep of concrete is sensitive to low values of elastic modulus of aggregate, but beyond approximately 70 GPa, the effect of aggregate modulus is constant.

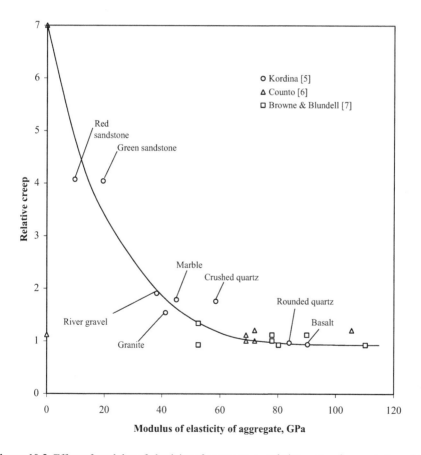

Figure 10.3 Effect of modulus of elasticity of aggregate on relative creep of concrete (equal to one for an aggregate with modulus = 69 GPa) [8].

Several authors have used *composite models* to quantify influencing factors on modulus of elasticity of concrete; they are presented in Chapter 3 and their prediction performance is compared in Chapter 4. For the modeling of specific creep, Counto [6] used Eq. (3.7), which was derived from the model for elasticity with the use of an effective modulus of elasticity (Eq. (3.6)) to allow for time-dependent strain. To verify the creep model, Counto performed tests on concrete made with various types and concentrations of aggregate, the results of which are listed in Table 10.1, together with those predicted by other models given in Chapter 3, viz. parallel model or composite hard (Eq. 3.10), series model or composite soft (Eq. (3.2)), and Hirst/Dougall model (Eq. (3.4)) and England's model [9]. Series 1 tests refer to a 1:2.06 mortar with a water/cement ratio of 0.33 to which the aggregates were added. Series 2 consisted of a cement paste mix with a water/cement ratio of 0.5 to which the aggregates were added. Compared to the other models, Counto's model shows the best overall agreement for predicting creep of concrete and, furthermore, the same model also has been shown to represent creep recovery satisfactorily [6]; creep recovery is discussed later in this chapter.

Table 10.1 demonstrates how the modulus of elasticity of concrete is affected by aggregate type and fractional volume of aggregate, g, or fully hydrated cement paste content, which is equal to $(1-g)$. By analogy with the dependency of drying shrinkage on hardened cement paste content, Neville [10] demonstrated that creep at a constant stress/strength ratio was similarly dependent. A theoretical relationship for drying shrinkage was derived by Pickett [11], which is presented in Chapter 6 (Eq. (6.3)), and then adapted for creep by Neville, viz.

$$c = c_p(1 - g)^{\alpha} \tag{10.5}$$

where $c =$ creep of concrete at a constant stress/strength ratio, $c_p =$ creep of cement paste at a constant stress/strength ratio, $g =$ total aggregate content plus unhydrated cement by volume, and $\alpha =$ parameter reflecting the properties of aggregate as given by Eq. (6.2).

Typically, α varies from one to two depending on type of aggregate, time under load, and storage conditions, but in the long-term, α becomes independent of storage conditions. Equation (10.5) applies to lightweight aggregate concrete [10] as well as normal-weight aggregate concrete, the latter dependency being illustrated in Figure 10.4.

According to Neville [13], porosity of the aggregate has also been found to affect creep of concrete but, because aggregates with a higher porosity generally have a lower modulus of elasticity, it is possible that porosity of aggregate is not an independent factor in creep. However, it is likely that absorption of the aggregate plays a role in the transfer of moisture within the concrete, and thus influences creep. Such a process may explain the high initial creep occurring with some lightweight aggregates batched in a dry condition.

Water/Cement Ratio, Stress/Strength Ratio, Type of Cement, and Age at Loading

The effect of a change in water/cement ratio on creep of concrete is twofold. First, the volumetric cement paste content changes and, second, the strength or maturity changes. The effect of the first has been dealt with in the previous section and is

Table 10.1 Comparison of Creep Predicted by Composite Models with Experimental Values After 195 days Under Load (Series 1) and 367 days under Load (Series 1): Stored at 17 °C and 93% RH [6]

Coarse Aggregate				Specific Creep, C_c (10^{-6} per MPa) as Predicted by Model					Measured Specific Creep, 10^{-6} per MPa
Type	Modulus of Elasticity, E_a, GPa	Fractional Volume, g	Modulus of Elasticity of Concrete, E_c, GPa	Parallel (Eq. (3.1)/(3.6))	Series (Eq. (3.2)/(3.6))	Hirsch/Dougill (Eq. (3.4)/(3.6))	Counto (Eq. (3.7))	England (ref. [9])	
Series 1 (mortar: $E_m = 40.5$ Gpa, $C_m = 38.6 \times 10^{-6}$ per MPa)									
Cast iron	104.8	0.50	71.7	2.8	19.3	11.0	12.1	12.2	14.3
		0.25	54.3	8.6	28.9	18.8	20.7	22.6	19.0
Flint gravel	74.5	0.50	55.4	4.8	19.3	12.0	12.7	13.2	12.2
		0.25	47.2	12.4	28.9	20.7	21.7	23.9	19.1
Glass	72.4	0.50	54.2	5.0	19.3	12.1	12.8	13.3	15.1
		0.25	46.9	12.8	28.9	20.8	21.8	24.1	23.1
Polythene[a]	0.293	0.50	14.5	75.2	19.3	47.2	98.9	105.9	108.3
		0.25	22.9	51.0	28.9	40.0	56.9	74.2	49.4
Series 2 (cement paste: $E_m = 10.6$ Gpa, $C_m = 315 \times 10^{-6}$ per MPa)									
Steel	206.8	0.55	33.4	0.3	141.8	71.0	81.4	—	76.9
Flint gravel	72.4	0.55	27.4	1.9	141.8	71.7	81.8	—	65.9
Glass	72.4	0/55	26.5	2.0	141.8	71.9	81.8	—	122.1

[a]An allowance was made for creep of polythene.

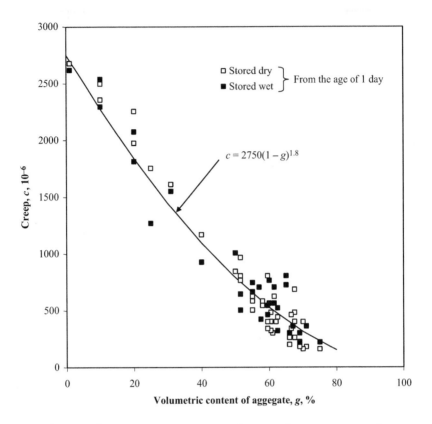

Figure 10.4 Effect of aggregate content on creep of concrete for a constant water/cement ratio [10].

quantified by Eq. (10.5). With regard to the second influence, in 1940, Lorman [14] suggested that creep is approximately proportional to the square of the water/cement ratio, other factors being constant. For a constant volume of aggregate or cement paste content, creep increases as the water/cement ratio increases, as demonstrated in Figure 10.5. Since an increase in the water/cement ratio causes the porosity to increase and the strength to decrease, it can be expected that creep is related to both those parameters. Indeed, in the case of strength, it has been found for a wide range of mixes that creep is approximately inversely proportional to strength at the time of application of load [16]. Since, creep is also proportional to the applied stress (provided it is less than 0.5 of the strength), Neville [16] proposed the *stress/strength ratio rule*, which states that for constant mix proportions and the same type of aggregate, creep is approximately proportional to the applied stress and inversely proportional to the strength at the time of application of load. The rule is satisfactory for mature concrete loaded at later ages but, at early ages of loading, creep is also affected by the change in strength while under load [4]. As a proportion of the initial strength, a

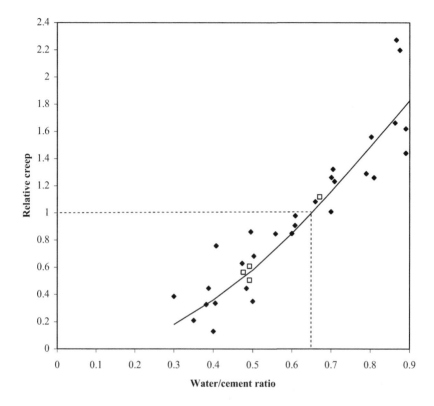

Figure 10.5 Ultimate specific creep of concrete as a function of water/cement ratio expressed relative to that for a water/cement ratio = 0.65; data of several investigations adjusted to a constant cement paste content = 0.2 (by mass) [15].

low water/cement ratio concrete has a smaller development of strength than a high water/cement ratio concrete. Hence, applying the stress/strength ratio rule, creep at a constant initial stress/strength ratio would be greater for a low water cement ratio concrete than for a high water/cement ratio concrete.

Some long-term creep results are shown in Figure 10.6 for concrete cured in water for 14 days, then subjected to load and subsequently stored in air at 65% relative humidity as well as in water [3]. The concrete specimens were made with different water/cement ratios and the applied compressive stress was 0.3 of the 14-day strength. Hence, the creep-time curves are for equality of stress/strength ratio to allow for the strength influence arising from a change in water/cement ratio. However, it appears that there is still an influence of water/cement ratio for both total creep and basic creep, but this is actually attributed to the different volumetric cement paste contents that, assuming full hydration and no air voids, varied from 31% to 38% for the respective water/cement ratios of 0.5−0.8. When creep per unit stress is considered, i.e., specific creep, the total effect of changing the water/cement ratio on either relative total creep or relative basic creep is similar to the overall trend of Figure 10.5.

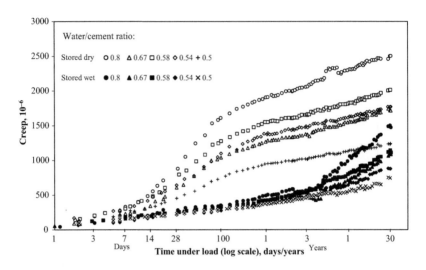

Figure 10.6 Long-term creep at a constant stress/strength ratio of normal weight aggregate concrete stored in air and in water from the age of 14 days [3]; concrete made with rapid-hardening Portland cement, quartzitic sand, and gravel in the proportions 1:1.71:3.04.

Creep of concrete made with different cements was investigated by Washa and Fluck [17], and generally creep is affected by the *type of cement* insofar as it influences the strength of concrete at the time of application of load. On the basis of equality of stress/strength ratio, most Portland cements sensibly lead to the same creep. On the other hand, on the basis of equality of stress, the specific creep increases in the order of type of cement: high alumina cement, rapid-hardening cement (Type III), and ordinary Portland (Type I). The order of magnitude of creep of Portland blast-furnace (Type IS), low-hear Portland (Type IV), and Portland pozzolan (Type IP and P) cements is less clear. Fineness of cement affects strength development at early ages, and thus affects creep but not creep at a constant stress/strength ratio; contradictory results may be due to the indirect influence of gypsum [13]. The finer the cement, the higher its gypsum requirement, so that regrinding a cement in the laboratory without the addition of gypsum produces a improperly retarded cement that exhibits high shrinkage and high creep [13].

In the BS EN 1992-1-1: 2004 method of predicting creep of concrete, the different rates of hardening of various types of cement are taken into account by adjusting the age at loading factor relative to that calculated for normal hardening cement (see Chapter 11, Eq. (11.15)).

Creep of concrete is higher with expansive cement than when made with Portland cement, whether creep is expressed in terms of specific creep or creep at a constant stress/strength ratio [4]. Expansive cements are used to make *shrinkage-compensating concrete*, which is described in Chapter 6. Russell [18] obtained creep data on plain and reinforced lightweight aggregate concrete slabs (1220 × 610 × 152 mm) with

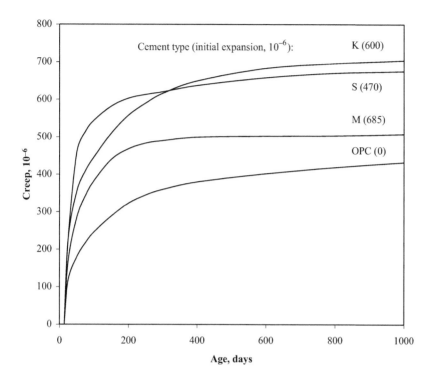

Figure 10.7 Total creep of shrinkage-compensating concrete made with expansive cement types K, S, and M compared with creep of ordinary Portland cement concrete loaded at the age of 14 days [18]; stored at 21 °C and 55%. RH.

and without steel reinforcement made using expansive cement. Maximum expansion was reached after 3 days of curing under polythene, after which slabs were stored at 21 °C and 55% RH so that shrinkage occurred. The slabs were subjected to a uniaxial stress of 6.9 MPa applied to the ends of the slabs at the age of 14 days. Figure 10.7 shows that creep of concrete made with Types K and S expansive cements is considerably greater than for ordinary Portland cement concrete, there being no obvious correlation with initial expansion or with subsequent shrinkage, which was approximately $500 \times 10\text{-}6$ at the age of 3 years. Similar tends in creep behaviour occurred with reinforced concrete slabs [18].

Creep of *high alumina cement* is affected by structural changes that take place in the hydrated high alumina cement with time [4]. The changes are due to *conversion* of metastable calcium aluminate hydrates from hexagonal to cubic form, which results in a lowering of strength due to increased porosity. The conversion, encouraged by a temperature higher than normal and by the presence of moisture, results in a higher creep, particularly for basic creep [4,19].

Polymer and resin concretes exhibit much higher creep than Portland cement concrete but, for epoxy concretes, only a moderate increase in creep related to the amount of resin in the mix is generally found [4]. Creep of *polyester resin concrete* with sand as the fine aggregate is affected adversely by elevated temperature, and so are strength and modulus of elasticity, both of which decrease as the temperature increases [4]. When cement-filled or aggregate-filled, creep of polyester resin concrete is similar to Portland cement concrete but sensitive to small temperature changes. *Polymer impregnated concrete (PIC)* exhibits little creep depending on the level of polymer content and the requirement that the process of polymerization involves prior dehydration of concrete; removal of evaporable water causes a reduction in creep and renders the concrete impermeable so that, under drying conditions, there is no moisture movement to the surrounding medium. At higher levels of polymer content, the polymer or resin becomes the more creep-sensitive phase [4].

The influence on creep of mineral admixtures blended with Portland cement is discussed later in this chapter, along with chemical admixtures such as water-reducers or plasticizers and high range water-reducers or superplasticizers.

Since strength increases as the concrete ages or matures due to hydration of cement, application of the stress/strength ratio rule implies that creep will decrease as *the age at loading* increases or as the *period of moist-curing* increases. Analysis of reported experimental data by L'Hermite [20] confirmed that statement when based on relative creep (see Figure 10.8), i.e., the ratio of creep of concrete subjected to load at any age relative to creep loaded at the age of 7 days; creep consistently reduces linearly with the logarithm of age of loading for ages at loading of 7−300 days. However, the behaviour at ages of less than 7 days does not always follow the same trend, as implied by the results of Brooks and Farrugia [26] using fly ash and ordinary Portland cement concretes; very early-age creep behaviour may be affected by the concurrent high rate of hardening or gain of strength [4]. Data of Figure 10.8 apply for concrete stored at ambient temperature, and early-age creep of mass concrete undergoing a temperature rise due to heat of hydration, which is discussed later in this Chapter. In connection with a temperature effect on creep, it is relevant to note that in high-pressure steam-cured (autoclaved) concrete, the effect of age at loading on creep is virtually absent since the structure of hydrated cement paste is not further modified by the passage of time. In fact, the structure of autoclaved hydrated paste is microcrystalline, which accounts for its different creep behaviour [27].

Size and Shape of Member

The influence of the size and shape of concrete member on creep is similar to the influence on drying shrinkage as described in Chapter 6. The relation between creep coefficient and volume/surface ratio derived by Hanson and Mattock is shown in Figure 10.9. The decrease of creep with an increase of size is smaller than in the case of drying shrinkage; also, the actual shape of the specimen is of lesser importance than in the case of drying shrinkage (see Figure 6.11). Strictly speaking, creep in this case is total creep according to the definition of creep given earlier for concrete stored in drying conditions (Eq. (10.2)). For large concrete members having a large volume/surface area ratio, total creep becomes less and tends to approach the basic creep of sealed concrete or mass concrete, as indicated in Figure 10.9. Since drying creep is

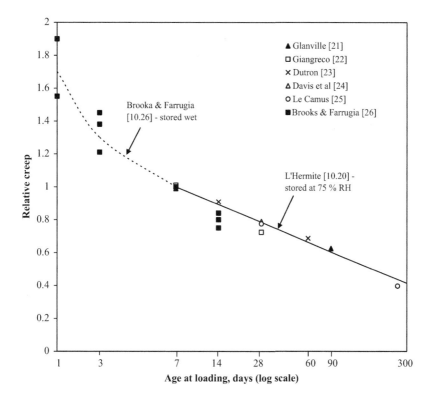

Figure 10.8 Influence of age at application of load on creep of concrete relative to creep of concrete stored at ambient temperature and loaded at 7 days.

the difference between total creep and basic creep, the size and shape of the member only influence the drying creep component of total creep.

As in the case of drying shrinkage, Bryant and Vadhanavikkit [29] proposed the use of an equivalent thickness term to quantify size and shape of member on creep, the equivalent thickness being based on the average drying path length of moisture diffusion from the inside to the outer surface of the concrete member. In Chapter 11, the methods of predicting creep allow for size of the concrete member in terms of volume/surface ratio, average thickness, or effective (theoretical) thickness. All those terms, together with the alternative term of equivalent thickness, are defined in Eqs. (6.6)–(6.10) and quantified for various sizes and shapes in Table 6.13.

Storage Environment

Like the influence on drying shrinkage, an important external influencing factor on creep is the *relative humidity* of the air surrounding the concrete. Generally, creep is higher the lower the relative humidity, as illustrated in Figure 10.10. Thus, even

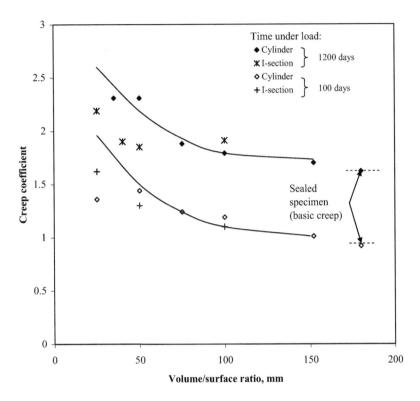

Figure 10.9 Effect of size and shape of specimen on creep coefficient of drying concrete stored in air at a relative humidity of 50% [28].

though drying shrinkage has been taken into account in determining creep (Eq. (10.2)), there is still an influence on creep of moisture loss due to drying. In fact, the difference between total creep at, say, 50% relative humidity and basic creep at 100% relative humidity is, as previously mentioned, termed drying creep. Figure 10.10 applies to concrete cured in fog, but if concrete specimens are allowed to dry out prior to application of load, creep is less. If predrying is such that hygral equilibrium exists under load, then creep is much reduced. However, such practice is not normally recommended as a means or reducing creep, especially for young concrete, because inadequate curing will lead to low tensile strength and possibly the formation of shrinkage-induced cracks (see Chapter 14).

Figure 10.10 suggests that long-term creep is inversely proportional to relative humidity of storage. However, for a lower relative humidity of less than 50%, creep has been found to be less than would be expected from linearity [31], because the removal of evaporable water becomes more difficult as it is held more firmly within the hydrated cement paste. Hence, the rate of diffusion is slower so that over a wider range of relative humidity, creep actually varies inversely in a nonlinear manner.

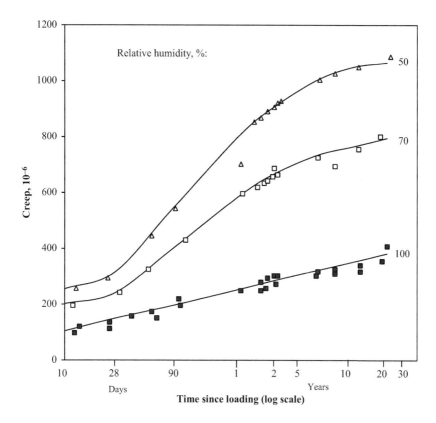

Figure 10.10 Creep of concrete stored at different relative humidities after curing in fog for 28 days [30].

The above discussion deals with the influence of storage environment after application of load on creep concrete that has been previously cured in water or at 100% relative humidity. In fact, besides depending on loss of evaporable water under load, creep depends also on the amount actually present when the load is applied. Thus, the *relative humidity of storage during the curing period* is of interest, especially from a practical point of view when it is not always possible to avoid some degree of drying prior to concrete members being subject to load. Dutron [23] and then Hanson [32] investigated relative humidity on creep before and after loading; the results of the latter are shown in Table 10.2.

The table shows the influence on creep of both the amount of water present in the concrete at the time of application of load and moisture loss during creep after 100 days, the results being expressed as creep relative to creep of concrete cured and stored in water. Generally, a smaller amount of water present when the concrete is subject to load means less creep potential. For example, concrete predried at a relative humidity of 10% before sealing had significantly less creep than water-cured concrete, even though the former was less hydrated and had a lower strength. Tests 1−3 all

Table 10.2 Effect of Relative Humidity of Storage before and During Loading on Creep of Concrete (Reproduced from Hanson's paper [32], Table 5.2.2 entitled creep of drying concrete, p. 80).

Series	Initial Curing 8–28 Days	Treatment at 28 Days of Age	Storage 28–128 Resp. 28–228 Days	Loaded at Days	Creep Relative to Creep in Continuous Water Storage
1	Air 70% RH 20°C		Air 50% RH 20°C	28	1.70
2	Air 70% RH 20°C		Air 60% RH 20°C	28	1.39
3	Air 70% RH 20°C		Air 70% RH 20°C	28	1.29
4	Air 70% RH 20°C		Water 20°C	28	1.31*
5	Water 20°C		Water 20°C	28	1.0
6	Air 70% RH 20°C	Sealed	Liquid paraffin 20°C	28	0.70
7	Air 10% RH 20°C	Sealed	Liquid paraffin 20°C	28	0.58
1a	Air at 70% RH		Air 50% RH 20°C	128	0.68**
2a	Air at 70% RH		Air 60% RH 20°C	128	0.59**
3a	Air at 70% RH		Air 70% RH 20°C	128	0.57**
4a	Air at 70% RH		Water 20°C	128	-
5a	Water 20°C		Water 20°C	128	1.0**
6a	Air 70% RH 20°C	Sealed	Liquid paraffin 20°C	128	0.48**
7a	Air 10% RH 20°C	Sealed	Liquid paraffin 20°C	128	0.44**

* Estimated.
**Based on creep values obtained after 28 days of sustained loading.

showed higher relative creep that, in fact, is total creep induced by drying under load, i.e., basic creep plus drying creep (see Eq. (10.2)), whereas tests 5–6 are basic creep according to Eq. (10.1). Test 5 is a special case of *wetting creep*, which is discussed shortly. In the case of specimens loaded at the later age of 128 days, the effect of prolonged predrying means there is a lower evaporable water content at the time of loading and less moisture loss while under load; hence, relative creep is much reduced.

Confirmation of reduced creep with a lower water content at the time of application of load is apparent by several researchers [33–35]; both basic and total creep are reduced and it seems that no creep takes place in concrete containing no evaporable water [36,37]. Ross [38] even suggested that predrying followed by wetting at loading, in order to produce shrinkless and creepless concrete, be used as small elements for prestressed members.

Rewetting

Although dry concrete creeps little or not at all, its creep capacity can be restored by rewetting, and creep recovery can also be restarted by rewetting [39]. According to Ali and Kesler [40], the direction of moisture movement is immaterial as far as its effect on creep is concerned. Although drying shrinkage is mostly reversible on rewetting, total creep shows an apparent increase when concrete is rewetted. This increase is termed *wetting creep* in accordance with the adopted additive definition of creep (Eq. (10.2)). The phenomenon is explained schematically in Figure 10.11(a), which shows

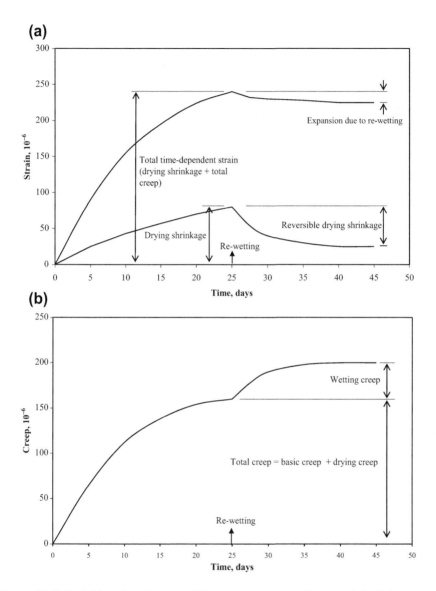

Figure 10.11 Definition of wetting creep [4]; concrete cured wet then stored dry before rewetting. (a) Strains of loaded specimen and control specimen re-wetted at the age of 25 days. (b) Creep of loaded specimen re-wetted from 25 days (total time-dependent strain minus drying shrinkage).
Source: Creep of Plain and Structural Concrete, A. M. Neville, W. H. Dilger and J. J. Brooks, Pearson Education Ltd. © A. M. Neville 1983.

measured strain—time-dependent curves for a concrete specimen under load and a control specimen stored in a drying environment before rewetting at the age of 25 days. On rewetting, the reversible shrinkage of the control specimen is greater than expansion due to wetting of the total time-dependent deformation of the specimen under load, so that there is an apparent increase in total creep Figure 10.11(b). In actuality, it is the drying creep component of total creep that undergoes expansion, basic creep being unaffected [4]. Renewed creep on rewetting has also been encountered in torsion tests, where not only renewed creep but cracking also occurred [40]. A factor in renewal of creep on rewetting is that new gel is formed and, since it is in a virgin state, its rate of creep is high [4].

Carbonation

It is known that the process of carbonation results in an increase of drying shrinkage of concrete, as explained in Chapter 6. Carbonation is due to the formation of carbonates in the hardened cement paste caused by the reaction with carbon dioxide in the atmosphere in the presence of moisture. Although occurring in normal outdoor conditions, it is significantly greater in an unventilated laboratory where the concentration of carbon dioxide may be much greater. Moreover, since only the surface layer of concrete becomes carbonated, it is likely that small laboratory specimens will be affected more than full-size concrete members used in structures. These points are of importance in considering the application of laboratory test data to structural behaviour.

Carbonation of concrete before application of load reduces subsequent creep [42], which is probably due to carbonation products being deposited in small pores and pores left by conversion of calcium hydroxide to calcium carbonate; the process reduces porosity and increases strength before loading, thus causing creep to be reduced [43]. On the other hand, a significant increase in creep due to carbonation of hardened cement paste specimens 3 days after application of load is also reported by Parrott [43]. His tests were carried out for using specimens predried for 18 weeks at 65% RH before loading and subsequently stored at 65% RH after loading, and the increase in creep was attributed to a decrease in load-bearing capacity under load during the chemical conversion of calcium hydroxide to calcium carbonate.

Alternating Humidity

Actual concrete structures are exposed to outdoor climatic conditions and, therefore, are subjected to alternating ambient humidity so that the relation between creep under those conditions and creep under constant humidity is required. Experimental evidence suggests that creep is increased under alternating humidity between two limits, and is greater than creep at constant humidity within the same limits, but only when the load is applied prior to first drying out [44—47]. If the load is applied after first drying out, which is usually the case in practice, creep is lower than when concrete is loaded during the first drying out, and is approximately equal to creep at the constant upper level of humidity [45]. However, in general, it is recommended that under alternating humidity, a small allowance for additional creep is desirable, corresponding to a relative humidity somewhat lower than the actual lower limit of site exposure [4].

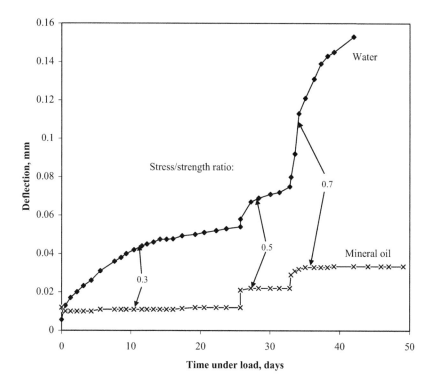

Figure 10.12 Creep of unreinforced mortar beams in mineral oil and in water [34].

Other Environments

Cilosani [34] reported creep results for storage in mineral oil, which are shown in Figure 10.12 in terms of deflection (elastic plus creep) of mortar beams subjected to different stress/strength ratios at various stages of loading. Compared with storage in water, mortar stored in mineral oil does not creep even at a high stress/strength ratio of 0.7 when a high creep rate could be expected as in the case of water-stored mortar. On the other hand, Hansen's tests [48] indicated that for mortar beams stored in paraffin oil, creep continues at a higher rate after 5 years than would be expected for storage in water or in air, because adsorbed water molecules are partially replaced by hydrocarbon molecules.

Hannant [49] found no influence on creep when specimens were stored in benzene or carbon tetrachloride. Therefore, although liquids with a large molecular size have no effect on creep, methyl alcohol, which has a similar size to that of water (0.35 nm dia.), does has some effect. This effect suggests that molecular size is a significant factor; that influence is sometimes known as the *molecular sieve effect* by Mills, who investigated the influence of size of molecule on strength of concrete [50].

Any influence of hydrocarbons on creep of concrete is of interest in connection with concrete structures used in offshore oil and gas production. Crude oils may contain a

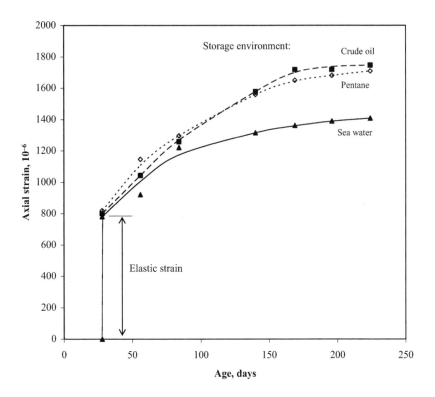

Figure 10.13 Elastic strain and creep of concrete under a uniaxial compressive stress of 21 MPa stored in various environments with a hydrostatic stress of 21 MPa [51].

significant proportion of pentane and low-molecular weight hydrocarbons, which may affect concrete because of their ability to penetrate the cement paste structure, especially during the curing stage [4]. Creep-time test data have been obtained for sea water, pentane, and crude oil [51]. After curing in fog for 28 days, cylindrical specimens were subjected to axial compressive stress of 21 MPa and immersed in each of the environments under a hydrostatic stress of 21 MPa. Thus, the specimens were subjected to triaxial loading representing environmental conditions in deep seawater. Figure 10.13 shows that elastic strains were similar on application of load for all storage conditions, but subsequent creep in hydrocarbon environments was appreciably greater than in seawater. Nevertheless, the authors concluded that hydrocarbons are not detrimental to the time-dependent properties of concrete, provided that it is of low permeability.

Development with Time

Figures 10.10 and 10.14 show examples of creep-time curves over a period of 30 years and, like drying shrinkage, creep is a gradual process, developing rapidly at first and

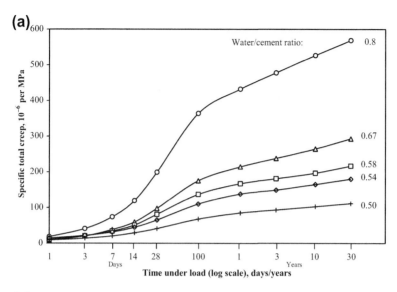

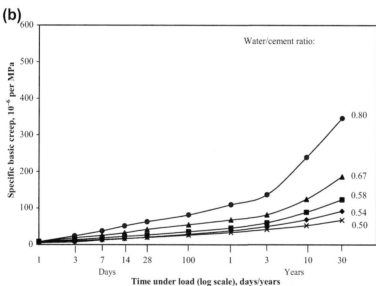

Figure 10.14 Long-term specific creep of normal weight aggregate concrete calculated from experimental data of Figure 10.6 [3] (a) stored dry (b) stored wet.

then more slowly toward an asymptotic value. In general, for a whole range of concrete types and operating conditions:

- 20–30% of 30-year creep occurs in the first 2 weeks after applying the load.
- 35–55% of 30-year creep occurs in 3 months.
- 45–75% of 30-year creep occurs in 1 year.

The ranges of creep reflect the influence of factors in creep of concrete that have been described. In Chapter 11, methods of prediction generally quantify the development of creep in terms of the creep coefficient (creep/elastic strain ratio) $\phi_c(t, t_o)$, with time under load by some form of the following hyperbolic expression:

$$\phi_c(t, t_o) = \phi_{c\infty} \left[\frac{(t - t_o)}{a_c \phi_{c\infty} + (t - t_o)} \right]^n \tag{10.6}$$

where t = age of concrete, t_o = age at application of load after moist curing, $(t - t_o)$ = time under load, $\phi_{c\infty}$ = ultimate creep coefficient, a_c = a coefficient approximately equal to the reciprocal of 1-day creep coefficient, and $n = 0.3$ or 0.5 (see page 327). Depending on the particular method of prediction, many of the influencing factors discussed earlier are taken into account by the term $\phi_{c\infty}$, such as type of concrete, age at loading, size and shape of the concrete member (expressed in terms of volume/surface ratio (V/S)), and ambient relative humidity of storage. The coefficient a_c is also a function of the size and shape of member.

Admixtures

The use of admixtures in concrete is often accompanied by a change in mix proportions in order to achieve the required property. Consequently, compared with concrete without the admixture, if there is a change in creep, it is not known whether the change is due to the incorporation of the admixture or to the change in mix proportions. This situation often occurs in research publications in which the control or plain concrete has a different water/cementitious materials ratio (water/binder ratio) than the concrete containing the admixture. An example where this occurs is *air-entrained concrete*. Like drying shrinkage, *air-entraining agents* might be expected to increase creep if bubbles of air are regarded as aggregate with zero modulus of elasticity, and therefore reduce resistance to deformation of concrete. Indeed tests appear to confirm that fact, but only when compared with creep of a control concrete having the same mix proportions (except for the air entraining agent [4]). However, in practice, it may appear that air entrainment is not a factor in creep because workability is improved so that a lower water/cement or leaner mix can be used, which has the effect of reducing creep, thus offsetting any increase in creep due to entrained air.

In general, to isolate any influence of any admixture on creep of concrete, the *relative deformation method* can be used to adjust the creep of the control concrete; this method is fully explained in Chapter 6 and its application to creep is now described.

For the same types of coarse and fine aggregates, a change in mix proportions involves a change in water/cement ratio and the volume of cement paste. For a constant water/cement (w/c) ratio, the relation between creep and volume of cement paste $(1-g)$ is given by Eq. (10.5) so that expressing creep relative to creep of concrete having standard mix proportions of $w/c = 0.5$ and $(1-g) = 0.3$, Eq. (6.3) can be rewritten as:

$$R_{cc} = R_{pc} \left(\frac{1-g}{0.3} \right)^n = R_{pc} R_g \tag{10.7}$$

where R_{cc} = relative creep of concrete, R_{pc} = creep of cement paste of w/c relative to creep of cement paste with a $w/c = 0.5$, and R_g = relative volume fraction of cement paste. It should be remembered that creep is that at a constant stress/strength ratio [10].

As for drying shrinkage, analysis of published data [52] revealed average values of $n = 1.8$ for normal weight aggregates and 1.0 for lightweight aggregates. Moreover, since Eq. (10.5) applies for creep at a constant stress/strength ratio, the effect of a change in w/c ratio (and therefore strength) is already been taken into account so that $R_{pc} = 1$. Therefore, for normal weight aggregate concrete:

$$R_{cc} = R_{gc} = 8.73(1 - g)^{1.8} \tag{10.8}$$

and, for lightweight aggregate concrete:

$$R_{cc} = 3.33(1 - g) \tag{10.9}$$

Equations (10.8) and (10.9) are the same as the cement paste coefficients for drying shrinkage, and are shown graphically in Figure 6.15(a). The accuracy of predicting creep of plain concrete by the relative deformation method has been found to be satisfactory with an error coefficient (Eq. (5.4)) of 16% [52]. Thus, provided that creep of a particular type of concrete is known together with its cement paste content or mix proportions, the relative deformation method permits an estimate of creep at the same stress/strength ratio for another type of concrete made from the same aggregates but having different mix proportions. Cement paste volume fraction content may be estimated from mix proportions by mass using Eq. (6.22), using appropriate values for density, water/cementitious materials ratio, and specific gravity of cementitious material when the mix contains mineral admixtures (see p. 162). Example I is now presented to illustrate the relative deformation method to estimate creep of concrete, having known mix proportions from a known creep of concrete of different mix proportions.

Example I

Using the same concrete mixes stipulated in the example on drying shrinkage on p. 162 in Chapter 6, the ultimate creep at a 0.2 stress/strength ratio is 800×10^{-6} for concrete mix 1, which has a total aggregate/cement ratio = 5.5 and a water/cement ratio = 0.55. The ultimate creep at the same stress/strength ratio is required for concrete mix 2, which has an aggregate/cement ratio = 4.0 and a water/cement ratio = 0.45. Assume that the air content is zero and the density of concrete for both mixes is 2400 kg/m^3.

Solution

As detailed previously on p. 162, the cement paste contents of the two concrete mixes are estimated from Eq. (6.22), viz. $(1-g_1) = 0.295$ and $(1-g_2) = 0.319$. Hence, from Eq. (10.8), the coefficients are $R_{cc1} = 0.970$ and $R_{cc2} = 1.116$,

Continued

Example I—cont'd

respectively, for the two mixes. Recalling that R_{cc} is defined as the ratio of creep with any $(1-g)$ to creep of a reference mix with $(1-g) = 0.3$, then creep of the reference mix is $800 \times 10^{-6} \div R_{cc1} = 800 \times 10^{-6} \div 0.970 = 825 \times 10^{-6}$. Hence, the creep of concrete mix 2 is $825 \times 10^{-6} \times R_{cc2} = 825 \times 10^{-6} \times 1.116 = \underline{921} \times \underline{10^{-6}}$.

The relative deformation method has been used to assess the influence of chemical admixtures on creep of concrete in reported cases where mix proportions of the admixture concrete differed from that of the control plain concrete [52]. The admixtures investigated were: *plasticizers* (*water-reducers*) and *superplasticizers* (*high-range water-reducers*). The procedure was to adjust the creep of the plain control concrete in proportion to R_{cc} calculated for the admixture concrete, so that the difference between the observed creep of the admixture concrete and the adjusted creep of the control concrete could be attributed to only the presence of the admixture. In cases where the admixture was used to make *high workability* or *flowing concrete* without any change in mix composition from that of the control concrete, no adjustment to creep was necessary [52].

The outcome of the above assessment is shown in Figure 10.15. In the case of plasticizers, 40 data sets of basic creep and 15 data sets of total creep were analyzed, whereas corresponding figures for superplasticizers were 16 and 21, respectively, for basic and total creep. The types of plasticizer were: lignosulfonate [53–55] and carboxylic acid [53–57], and the types of superplasticizers were: sulfonated naphthalene formaldehyde condensate [58–61], sulfonated melamine formaldehyde condensate [62,65], and copolymer [57,63]. Figure 10.15 indicates that no individual type of admixture behaves differently, and there is no obvious difference in basic and total creep. However, there is an overall trend that indicates an average increase of 20% of creep at a constant stress/strength ratio when plasticizers and superplasticizers are use to make high workability or flowing concrete. In other words, the presence of the admixture causes a general 20% increase of creep, and that admixture effect is similar to that on drying shrinkage (Figure 6.16) with the standard deviation indicating a variation of approximately 25%. As in the case of drying shrinkage, the overall increase in creep may be associated with the ability of the admixture to entrain air (see p. 162).

Example II illustrates application of the relative deformation method for estimating creep of concrete containing a superplasticizing admixture.

Example II

In Example I, imagine that concrete mix 2 contains a superplasticizer and the respective strengths at the age of loading are 40 and 50 MPa for concrete mixes 1 and 2. If the applied stress is 0.2 of the strength, calculate the specific creep of the two concretes.

Example II—cont'd

Solution

The effect of the superplasticizer on creep at a constant stress/strength ratio of mix 2 is an increase of 20%, i.e., $(921 + 184) \times 10^{-6} = 1105 \times 10^{-6}$. Since the strength of mix 2 at the age of loading is 50 MPa, the applied stress $= 0.2 \times 50 = 10$ MPa, so the specific creep $= 1105 \times 10^{-6} \div 10 = \underline{111 \times 10^{-6}}$ per MPa. The stress applied to mix 1 concrete is $0.2 \times 40 = 8$ MPa, so that the specific creep is $800 \times 10^{-6} \div 8 = \underline{100 \times 10^{-6}}$ per MPa.

A *shrinkage-reducing admixture* appears to reduce specific creep as well as drying shrinkage (see p. 162) when compared with an admixture-free control concrete having identical mix proportions [60]. Figure 10.16 shows some short-term data acquired during an investigation into *restrained shrinkage cracking* of concrete, and it seems that

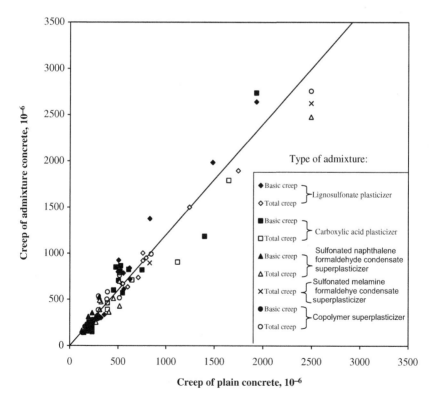

Figure 10.15 Comparison of creep of concrete at constant stress/strength ratio with and without plasticizing and superplasticizing admixtures; mean percentage increase, and standard deviation of creep of admixture concrete $= 120 \pm 25$ [52].

total creep or creep in tension under drying conditions (Figure 10.16(b)) is reduced by the admixture in a similar manner to that of drying shrinkage (Figure 10.16(a)). Similar trends, but to a lesser extent, were found by Kristiawan [67] and also for short-term total creep in compression. However, confirmation of the reduced-creep effect is required from longer-term experimental data.

An analysis of the influence of mineral admixtures: blast-furnace slag, fly ash, and microsilica on creep of concrete was reported in 2000 [68]. Creep was quantified in terms of relative creep, i.e., creep of the admixture concrete as a fraction of creep of the plain (admixture-free) concrete loaded to the same stress/strength ratio. If the mix

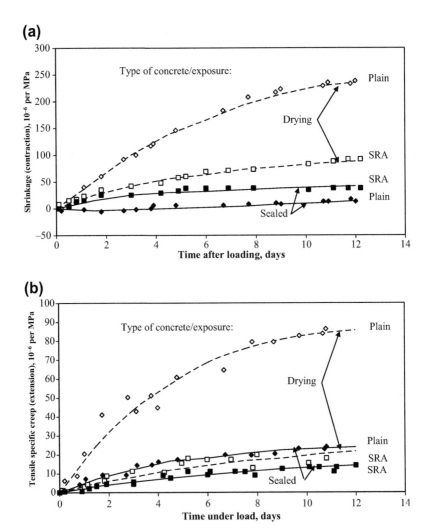

Figure 10.16 Influence of shrinkage reducing admixture (SRA) on shrinkage and tensile creep of concrete [66] (a) Shrinkage with and without SRA (b) Tensile creep with and without SRA.

proportions between the admixture and plain concretes differed, creep of the plain concrete was adjusted according to the relative deformation method explained earlier. Where later-age creep-time data were available, ultimate creep was estimated by extrapolation using a hyperbolic equation similar to Eq. (10.6) with $n = 1$. Figure 10.17 shows the outcome of the analysis together with creep data obtained for high strength concrete [69]. Although there is a high degree of scatter, the overall trends suggest that the effect

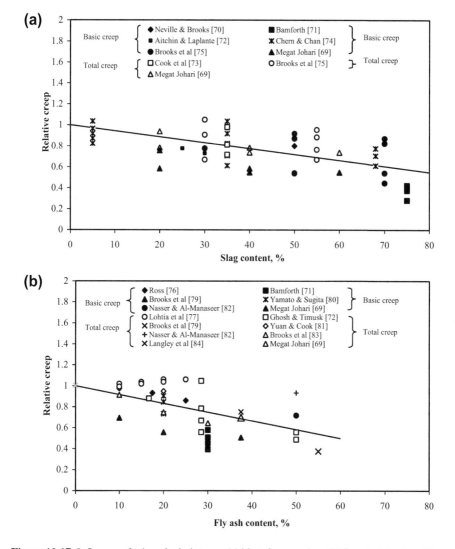

Figure 10.17 Influence of mineral admixtures: (a) blast-furnace slag, (b) fly ash, (c) microsilica, and (d) metakaolin on creep of concrete; relative creep is ratio of creep at a constant stress/ strength ratio of admixture concrete to that of admixture-free concrete [68].

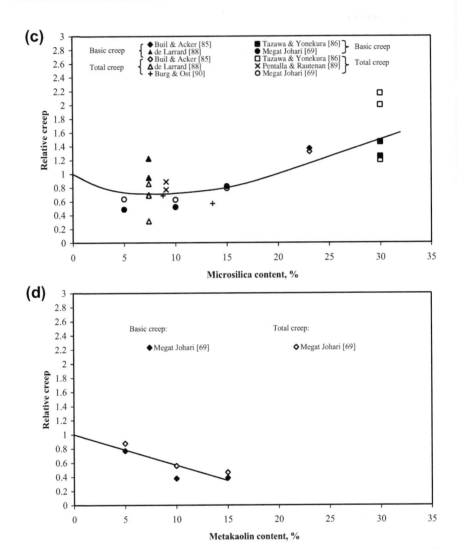

Figure 10.17 cont'd.

of increasing the replacement level of cement by either blast-furnace slag or fly ash is to reduce creep, there being no consistent difference between basic and total creep (Figure 10.17 (a) and (b)). As the replacement level of fly ash increases, creep reduces at a faster rate than for slag. In the case of finer mineral admixtures, microsilica and metakaolin, which are used in smaller quantities than slag and fly ash, there are larger reductions in creep for an increasing replacement level up to 15%, but then for micro-silica creep increases for higher replacement levels (Figure 10.17(c)); Bilodeau et al. reported a similar trend [91]. The use of metakaolin appears to reduce creep even more than for microsilica, but limited experimental data are available (Figure 10.17(d)).

Example III below illustrates the use of the relative deformation method to estimate creep of air entrained lightweight aggregate concrete containing blast-furnace slag.

Example III

Specific creep of air-entrained concrete made with ordinary Portland cement and lightweight coarse and fine aggregate is known to be 60×10^{-6} per MPa after 20 years under load. The mix constituent details are: air content $= 5.5\%$, the total aggregate/cement ratio $= 2.78$, a water/cement ratio $= 0.55$, specific gravity of aggregate $= 1.45$, specific gravity of cement $= 3.15$, age at loading $= 28$ days when compressive strength $= 25$ MPa, and density of concrete $= 1600$ kg/m^3.

It is required to compare the 20-year specific creep of concrete when 40% of mass of cement is replaced by ground granulated blast-furnace slag of specific gravity $= 2.9$ and the water/cementitious materials ratio $= 0.45$; the load is applied at the same age and is equal 0.3 of the 28-day strength $= 20$ MPa and the density of concrete $= 1550$ kg/m^3. Assume that the air content and total aggregate/cementitious materials ratio are unchanged.

Solution
From the mix details, cement past content can be calculated using Eq. (6.20). For the Portland cement lightweight aggregate concrete (control):

$$(1 - g_1) = \frac{5.5}{100} + \frac{1600 \times 10^{-3}}{(1 + 2.78 + 0.55)} \left[\frac{1}{3.15} + 0.55 \right] = 0.376$$

and, for the cement/slag concrete:

$$(1 - g_2) = \frac{5.5}{100} + \frac{1550}{(0.6 + 0.4 + 2.78 + 0.45)} \left[\left(\frac{0.6}{3.15} + \frac{0.4}{2.9} \right) + 0.45 \right]$$
$$= 0.340$$

Now creep of lightweight aggregate concrete for any cement paste content relative to that at a cement paste content $= 0.3$ is given by Eq. (10.9), so that for the two concretes:

$$R_{cc1} = 3.33 \times 0.376 = 1.252$$

$$R_{cc2} + 3.33 \times 0.340 = 1.132$$

Also, the specific creep of the control concrete is given so that creep at 0.3 stress/strength ratio $= 60 \times 10^{-6} \times (0.3 \times 25) = 450 \times 10^{-6}$, which occurs with $(1-g_1) = 0.376$. Hence, creep at 0.3 stress/strength ratio for concrete with $(1-g_2) = 0.340$ is:

$$450 \times 10^{-6} \times \frac{1.132}{1.252} = 406.9 \times 10^{-6}$$

Continued

Example III—cont'd

Since some of ordinary Portland cement has been replaced by blast-furnace slag, this has the effect of reducing creep at a constant stress/strength ratio, which is estimated from Figure 10.16(a) as relative creep $= 0.80$ for 40% replacement level. Hence, 20-year creep becomes $0.8 \times 406.9 \times 10^{-6} = 326 \times 10^{-6}$ and, for an applied stress $= (0.3 \times 20) = 6$ MPa, and specific creep $= 326 \times 10^{-6} \div 6 = \underline{54.3 \times 10^{-6}}$ per MPa.

It is emphasized that the above discussion on the influence of mineral admixtures on creep is based on creep determined at a constant stress/strength ratio and, as illustrated in Example III, corresponding trends in specific creep may not be the same since specific creep depends on strength, and hence, applied stress at the age of load application. In another situation, consider creep of high-strength concrete with and without mineral admixtures as shown in Table 10.3 [69]; details are given for 200-day basic creep and total creep as measured at 0.2 stress/strength ratio, and corresponding specific creep calculated from the applied stress. It can be seen that both relative creep at a constant stress/strength ratio and relative specific creep reduce as the mineral admixture content increases, but more so for relative specific creep when strength is appreciably greater than that of the control concrete because of the higher applied stress. This is particularly so for concrete containing admixtures, microsilica and metakaolin, whereas for slag and fly ash, relative creep at a constant stress/strength ratio and relative specific creep are similar. Creep of the foregoing high strength concrete was determined at the loading age of 28 days and creep behaviour of mineral admixture concrete at earlier ages (when strength is significantly less) appears to be different.

Basic creep results of normal-strength concrete containing the mineral admixtures loaded at early ages at loading were reported by Botassi et al. [92]. The cement replacements of admixtures used were: metakaolin (9%), calcined clay (26%), and blast-furnace slag (49%), sealed specimens being subjected to load at ages ranging from 1 to 7 days. At ages of loading of 3 and 7 days, the admixture concretes showed lower specific creep than that of the reference or control plain concrete having identical mix proportions (except for the admixture). However, when loaded at the age of 1 day, the opposite trend occurred, i.e., specific creep of admixture concrete was greater than that of the reference mix. At that early age, the strengths of the admixture concretes were appreciably less than the strength of the reference concrete, whereas at the later ages, strengths of the admixture concretes and reference concrete were similar.

Many theories of creep and shrinkage theories of concrete relate to microstructure of the hardened cement paste, so that the effects of mineral admixtures on porosity and the size of pores of high-strength concrete are of particular interest. Using mortar samples made with different admixtures, the pore structure has been investigated by the method of mercury intrusion porosimetry (MIP) [69]. The pore size classification used was that recommended by the International Union of Applied Chemistry, namely,

Table 10.3 Creep of High Strength Concrete after 200 days under Load; Moist-Cured and Loaded at the Age of 28 days [69]

Concrete type		28-day Cylinder Strength, MPa	Applied Stress[a], MPa	Basic Creep				Total Creep			
				0.2 Stress/Strength Ratio		Specific Creep		0.2 Stress/Strength Ratio		Specific Creep	
Admixture	R, %			10^{-6}	Relative Creep	10^{-6} Per MPa	Relative Creep	10^{-6}	Relative Creep	10^{-6} Per MPa	Relative Creep
Control (admixture-free)		72.6	13.9	285	1	20.5	1	358	1	25.8	1
GGBS	20	87.9	15.6	244	0.86	15.6	0.76	329	0.92	21.1	0.82
	40	82.0	13.9	212	0.74	15.3	0.75	290	0.81	20.9	0.81
	60	75.0	15.6	204	0.72	15.0	0.73	284	0.79	20.9	0.81
Fly ash	10	86.0	14.3	246	0.86	17.2	0.84	351	0.98	24.6	0.95
	20	85.8	13.5	192	0.67	14.2	0.69	298	0.83	22.1	0.86
	30	84.4	12.6	163	0.57	12.9	0.63	253	0.76	20.1	0.78
Microsilica	5	92.2	18.3	177	0.62	9.7	0.47	245	0.68	13.4	0.52
	10	99.6	19.8	179	0.63	9.0	0.44	234	0.65	11.8	0.46
	15	105.8	20.8	269	0.94	12.9	0.63	302	0.84	14.5	0.56
Metakaolin	5	108.7	16.8	235	0.82	14.0	0.68	312	0.87	18.6	0.72
	10	90.6	17.7	128	0.45	7.2	0.35	201	0.56	11.4	0.44
	15	87.8	17.9	115	0.40	6.4	0.31	171	0.48	9.6	0.37

R = mass replacement of cement by admixture.
GGBS = ground granulated blast-furnace slag.
[a]0.2 of creep cylinder strength (76 dia. × 200 mm).

Table 10.4 Influence of Mineral Admixtures on Porosity and Pore Sizes of Mortar at the Age of 28 days [69]

Mortar type			Volume of Pores in the Size Range (%):				
			Mesopores				Macropores
Admixture	R, %	Porosity, %	25−150 A	150−300 A	300−500 A	Total A	>500 A
Control		13.0	6.1	7.4	43.3	56.7	43.3
GGBS	20	11.1	11.8	44.9	24.8	81.7	18.5
	40	9.0	11.6	46.1	30.0	87.2	12.8
	60	9.2	36.3	27.1	9.7	73.1	26.9
Fly ash	10	11.5	9.1	29.8	43.3	82.2	17.8
	20	13.4	25.4	26.7	34.9	86.9	13.1
	30	11.8	27.1	36.4	20.2	83.8	16.2
Microsilica	5	9.7	43.5	27.0	18.1	88.6	11.4
	10	8.9	60.3	11.9	7.7	79.7	20.3
	15	8.2	52.7	14.6	9.0	76.3	23.7
Metakaolin	5	10.0	29.2	48.2	14.5	92.0	8.0
	10	9.7	57.3	32.6	4.2	94.1	5.9
	15	9.4	70.5	16.7	3.3	90.5	9.5

A = Anstron unit = 0.1 nm.
R = mass replacement of cement by admixture.

micropores (<25 A (2.5 nm)), mesopores (25−500 A (2.5−50 nm)), and macropores (>500 A (>50 nm)). Except for the fly ash mortar, there was a consistent trend of lower porosity as the admixture content increased, but the trend was different for each type of admixture. The MIP tests revealed no micropores for any of the mixes, and the highest volume of macropores occurred for the ordinary Portland cement control mortar. It is clear from Table 10.4 that the effect of all mineral admixtures is to reduce the volume of macropores and increase the volume of mesopores. When mesopores are further subdivided, microsilica and metakaolin concretes are seen to have greater volume of smaller mesopores in the ranges of 25−150 A and 150−300 A, whereas the majority of mesopores in slag and fly ash concrete are in the ranges of 150−300 A and 300−500 A. This observation may be explained by fineness, which is greater for microsilica and metakaolin than for slag and fly ash. When those observations are compared to specific creep given in Table 10.4, it is apparent that, generally, the lowest basic creep and lowest total creep occur for the microsilica and metakaolin concretes, i.e., for concrete having the smallest pores, although there is no consistent trend with level of cement replacement (R).

Developments in the combined use of mineral and chemical admixtures to make concrete have improved strength and durability performance so much so that accompanying deformation characteristics cannot be simply assessed by comparison with a conventional "control" concrete, i.e., a mix having the same constituents except for the admixtures. Examples of such special types of concrete are *ultra high-strength*

Table 10.5 Properties of Ultra High-Strength Cement/Microsilica Mortar Matrix with and without Steel Fiber Reinforcement [93]

Property	Sealed		Unsealed	
	Plain Matrix	Fiber Reinforced Matrix	Plain Matrix	Fiber Reinforced Matrix
Strength, MPa	158.8	195.4	158.8	195.4
Static modulus of elasticity, GPa	58.3	62.3	58.3	62.3
Secant modulus of elasticity[a], GPa	50.0	59.4	50.7	55.6
250-day shrinkage, 10^{-6}	210	180	325	163
250-day specific creep, 10^{-6} Per MPa	6.5	6.7	6.7	9.0
250-day creep coefficient	0.33	0.40	0.34	0.50

[a]At loading in creep tests.

concrete and *self-consolidating concrete (SCC)*. In the case of ultra high-strength concrete, Brooks and Hynes [93] investigated the creep of the matrix of a structural material known as "Compresit", which has a compressive strength of 150−200 MPa. The mortar matrix consists of a cementitious binder of 19 of % microsilica, 81% of ordinary Portland cement with a sand/binder ratio of 1.27, and a water/binder ratio of 0.18. A highly viscous workable mix is achieved by incorporating a superplasticizer dispersing agent with prolonged mixing times under high frequency vibration. The hardened mortar matrix itself is very brittle and, for practical application, the corresponding concrete requires sufficient ductility, which is achieved by incorporating 6% and 15% volume of steel fibers and deformed steel bars, respectively. Table 10.5 gives deformation properties obtained on the plain and fiber reinforced mortar matrix $50 \times 50 \times 100$ mm specimens subjected to uniaxial short-term and long-term compressive loading. The age of the specimens used in the tests ranged from 0.75 to 2 years.

Table 10.5 indicates that the effect of fibers is to significantly increase strength, modulus of elasticity, and reduce shrinkage. The total load strain (elastic + creep + shrinkage) is also reduced, but not to the same extent as shrinkage [93]. This has the effect of indicating a higher specific creep and creep coefficient for the fiber reinforced matrix compared with the plain matrix. The relation between static modulus of elasticity and compressive strength agrees with the ACI 318-05 expression given by Eq. 4.10. The majority of shrinkage is due to autogenous shrinkage with drying shrinkage as well as drying creep being small. There was no detectable increase in strength and modulus with age of the matrix, and yet the presence of autogenous shrinkage due to pozzolanic reaction of microsilica implied that hydration was taking place over the period of testing [93]. The noninfluence on strength may be explained by Sellevold's observation that microsilica does not decrease total porosity (and thus does not increase strength) but subdivides pores into smaller ones while keeping the total pore volume the same [94]. As stated on p. 176, the same author also reported that about 24% by mass of microsilica is required to use all calcium hydroxide from cement paste, so that with a replacement level of 19%, a delayed pozzolanic reaction seems to have been possible. When exposed to the environment, drying shrinkage and drying creep are small because water is mainly retained and available for hydration in the small pores rather than being lost to the environment.

Self-consolidating concrete or self-compacting concrete (SCC) is a highly flowable (slump flow of around 650 mm), nonsegregating concrete that does not require any mechanical consolidation after placement, and it differs from conventional vibrated concrete of normal consistency (slump ≈ 150 mm). Typically, self-consolidating concrete has a lower water/cementious materials ratio, a higher cementitious materials paste content, a greater quantity of smaller fines, less coarse aggregate, and a smaller nominal size of coarse aggregate. Those differences affect creep and shrinkage properties of concrete and, therefore, may result in greater loss of prestress and increased deflection of structural concrete elements.

A review of previous research by Long and Khayat [95] revealed that, as for conventional concrete, creep of SCC is greater in tension than in compression (see p. 330). However, there was inconsistency in the findings of previous investigators as to whether creep of SCC concrete was different from conventional concrete compacted by vibration, but it seems that variations in creep could be explained by differences in mix proportions since concretes were not compared on the basis of equality of mix constituents and proportions. An earlier literature review by Khayat and Long [96] had revealed that a similar situation existed for shrinkage of self-consolidated concrete. In their own tests [95,96], high workability of several self-consolidating mixes were achieved using a polycarboxylate-based superplasticizing admixture (high range water reducing agent (HRWRA)), and an organic thickening type viscosity-modifying admixture (VMA). Creep specimens were loaded to 40% of the 18-h compressive strength after steam curing, and then stored at 50% RH and 23 °C. Autogenous shrinkage was determined on sealed specimens stored at 23 °C, and total shrinkage (autogenous shrinkage + drying shrinkage) was determined on specimens stored with the creep specimens.

Long and Khayat [95] stated that 250-day SCC total creep was greater by 10—20% than creep of conventional high performance concrete of normal consistency made with a similar water/cementitious materials ratio, using a normal dosage of superplasticizer with no viscosity-modifying admixture. Autogenous shrinkage was found to be similar, but self-consolidating concrete exhibited 5—30% more drying shrinkage than conventional, high-performance vibrated concrete [96]. Satisfactory predictions of creep and drying shrinkage were achieved using the CEB—FIP MC 90 model [97], but it was necessary to include a cement type factor modification factor to yield satisfactory estimates by the AASHTO 2007 model [98].

Temperature

The influence of temperature on creep of concrete is of interest in mass concrete undergoing a temperature cycle due to heat of hydration, in nuclear pressure vessels, in the performance of bridges, and in fire conditions. Quantifying temperature effects on creep can be rather complex such that in some cases, they are insufficiently under-stood to be included in prediction methods. Creep can be determined in laboratory tests under changing temperature after allowing for thermal expansion or contraction deter-mined on companion load-free specimen as expressed by Eq. (10.4). It may also be noted that the elastic strain term of Eq. (10.4) will be also affected by temperature so that creep is reckoned from the temperature-affected elastic strain when concrete is loaded after being heated or cooled. However, if concrete is loaded before being heated or cooled, creep is reckoned from ambient temperature elastic strain so that creep will include some temperature-affected elastic strain. The influence of tempera-ture on elastic stain and modulus of elasticity of concrete is discussed in Chapter 4.

The time at which the temperature of concrete rises relative to the time at which the load is applied affects the creep-temperature relation. For concrete stored at elevated temperature throughout its life, Figure 10.18 indicates that the increase in creep rela-tive to creep at 21 °C appears to be the greatest at a temperature of approximately 70 °C [99]; however, this is not always the case since other authors have reported maximum creep at 90 °C [104] and even as high as 150 °C [100]. Mindess and Young [101] suggest that conflicting data may be a result of different preloading treatments

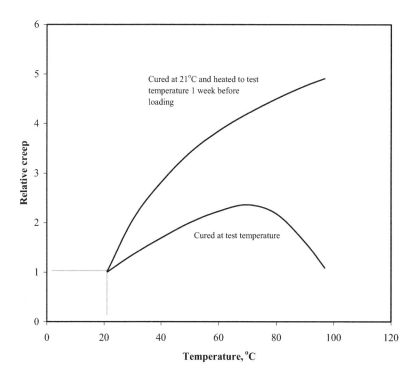

Figure 10.18 Influence of temperature on basic creep of saturated concrete relative to basic creep at 21 °C after 225 days under load; specimens cured at the stated temperature from 1 day until loading at 1 year [99].

(time and temperature), which affect the degree of hydration and strength at the age of loading. Rate of development of strength under load may also play a role in determining the pattern of creep behaviour [4]. The influence of elevated temperature on strength is usually one of accelerating early strength, but long-term strength tends to be less than strength at normal temperature, because rapid initial hydration causes a nonuniform distribution of cement gel with a more porous physical structure compared with structure at normal temperature. With high initial temperature, there is insufficient time available for hydration products to diffuse away from cement grains and for uniform precipitation in interstitial space. As a result, a concentration of hydration products is built up in the vicinity of the hydrating cement grains, a process that retards subsequent hydration and thus development of longer-term strength [102].

Figure 10.18 also shows that, for any elevated temperature, basic creep is significantly less for concrete stored continuously at the elevated temperature than when the temperature is raised only a week before application of load. This observation demonstrates that *maturity* is a factor in creep; for example, prolonged application of heat before loading increases maturity and reduces creep. Maturity, M, can be expressed as:

$$M = \sum_{0}^{t_o} (T + 10)\Delta t_o \tag{10.10}$$

where T = average temperature (°C) in the age interval Δt_o and t_o = age at loading.

Alternatively, maturity may be expressed in terms of an *equivalent hydration period* (t_{oT}) i.e., equivalent to curing at a constant temperature of 20 °C:

$$t_{oT} = \sum_{0}^{t_o} \left[\exp\left(\frac{4000}{\theta_0} - \frac{4000}{\theta} \right) \right] \Delta t_o \tag{10.11}$$

where $\theta_o = 293$ K and $\theta = (T + 273)$°K.

In practice, examples of the influence of maturity on creep occur when concrete is *heat-cured* or *steam-cured*. ACI Committee 209 [103] states that basic creep and drying creep are reduced by as much as 30% of concrete because of increases of strength at early ages due to heat or steam curing. Table 10.6 shows results obtained by Hanson [27] for lightweight and normal weight concretes, and there were significant reductions in creep when either specific creep or creep for equality of strength at the time of loading were used as the basis for comparison. Low or high-pressure steam curing reduces creep more when Type III Portland cement is used than with Type I Portland cement, and reductions are greater for autoclaving than for steam curing.

The influence of elevated temperature on total creep of unsealed concrete, i.e., concrete drying at the same time as undergoing creep, is shown in Figure 10.19. When subjected to temperature at the same time or just prior to application of load, there is a rapid increase in total creep as the temperature is raised to about 50 °C, which is similar to the behaviour of basic creep of sealed concrete; then, a decrease in total creep as the temperature is reduced to about 120 °C followed by another increase in

Table 10.6 Percentage Reductions in 3-year Creep for Steam-Cured and Autoclaved Concrete Compared with Moist-Cured Concrete [27]

	Steam Curing		Autoclaving	
Aggregate Type	Type I Cement	Type III Cement	Type I Cement	Type III Cement
Equal Compressive Strength at Loading				
Expanded shale	20	30	62	76
Sand and gravel	18	34	76	82
Specific Creep at 28 days Age of Loading				
Expanded shale	14	31	59	75
Sand and gravel	16	28	75	83

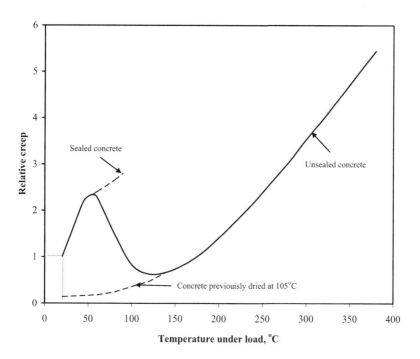

Figure 10.19 Influence of temperature on total creep of unsealed concrete relative to total creep at 20 °C; specimens moist-cured for 1 year and then heated to test temperature 15 days before loading [104].

total creep as the temperature is raised to at least 400 °C. The initial increase in total creep is due to a rapid expulsion of evaporable water, and when all of that water has been removed, total creep is greatly reduced and becomes equal to that of predried (desiccated) concrete.

So far, the effects of temperature on creep may be conveniently summarized by the following example based on the CEB MC90 prediction model [97], which expresses the creep coefficient as follows (see also Eq. (11.13)):

$$\phi_{28}(t, t_o) = \phi_{RH}\beta(f_{cm28})\beta(t_o)\beta_c(t - t_o) \tag{10.12}$$

where $\phi_{28}(t, t_o)$ = creep coefficient of concrete at any age t after loading at age t_o based on the temperature-affected modulus of elasticity at the age of 28 days, $\beta(f_{cm28})$ = coefficient based on 28-day mean strength (Eq. (11.15), $\beta(t_o)$ = age at loading coefficient, $\phi_{RH}(h)$ = coefficient depending on relative humidity of storage and member size (Eq. (11.19), and $\beta_c(t-t_o)$ = coefficient describing development of creep with time under load $(t-t_o)$.

Considering first the situation where concrete is cured at any constant temperature over the range 5−80 °C before loading and then subsequent storage at 20 °C, the only coefficient of Eq. (10.12) affected by temperature is $\beta(t_o)$, which is given by CEB MC 90 as :

$$\beta(t_o) = \frac{1}{0.1 + t_{oT}^{0.2}} \tag{10.13}$$

where t_{oT} = equivalent age at loading as given by Eq. (10.11)

Now, to compare the effect of temperature on creep coefficient, the relative creep coefficient may be used, i.e., ratio of creep coefficient at any temperature (T) to creep coefficient when $T = 20$ °C, viz. from Eq. (10.12):

$$R_{\phi28}(t, t_o, T) = \frac{\beta(t_{oT})}{\beta(t_{o20})} \tag{10.14}$$

Consider next the effect of elevated storage temperature on creep coefficient over the range of 5−80 °C from the start of creep after moist curing at 20 °C. According to the CEB Model Code 90, the affected coefficients of Eq (10.12) are as follows:

$$\phi_{28}(t, t_o, T) = \varphi_{RH,T}\beta(f_{cm28})\beta(t_o)\beta_{cT}(t - t_o) \tag{10.15}$$

where $\phi_{RH,T} = \phi_T + [\phi_{RH}(h) - 1]\phi_T^{1.2}$

$\phi_T = \exp[0.015(T - 20)]$

$$\beta_{cT}(t - t_o) = \left[\frac{(t - t_o)}{\beta_H\beta_T + (t - t_o)}\right]^{0.3}$$

$$\beta_T = \exp\left[\frac{1500}{(273 + T)} - 5.12\right]$$

In the above expressions, $\phi_{28}(t,t_o,T)$ = creep coefficient of concrete stored at temperature T at any age t after loading at age t_o based modulus of elasticity at the age of 28 days at 20 °C, $\beta(f_{cm28})$ = coefficient based on 28-day mean strength (Eq. (11.16)), $\beta(t_o)$ = age at loading coefficient at 20 °C, $\phi_{RH}(h)$ = coefficient depending on relative humidity of storage and member size (Eq. (11.15)), ϕ_T = temperature correction factor, $\beta_{cT}(t-t_o)$ = coefficient describing development of creep with time under load $(t-t_o)$, and β_T = temperature correction factor.

Proceeding as before, the effect of temperature may be assessed by the relative creep coefficient that is obtained from Eq. (10.15):

$$R_{\phi28} = \frac{\phi_{RH,T}\beta_{cT}(t - t_o)}{\phi_{RH,20}\beta_{C20}(t - t_o)} \tag{10.16}$$

The relative creep coefficient given by Eq. (10.14) is shown in Figure 10.20(a), which confirms that the creep coefficient decreases as the temperature of curing and, hence, the maturity increases; for example, compared with curing at 20 °C, there is a 36% reduction in creep coefficient at 80 °C. The corresponding predicted influence of temperature of storage given by the relative creep coefficient of Eq. (10.16) of a concrete member after moist curing at 20 °C is shown in Figure 10.20(b). It can be seen that relative creep coefficient increases and is slightly more sensitive to temperature in the early stages of creep; in the long term, the 50-year creep coefficient is some 2.6 times greater at 80 °C than when stored at 20 °C. The last case considered is when concrete is stored at elevated temperature after being cured at the same elevated temperature, which is given by the product of the two relative creep coefficients, given by Eq. (10.14) and Eq. (10.16). The result, shown in Figure 10.20(c), is that the relative creep coefficient of Figure 10.20(b) is reduced due to the increase in maturity at loading; for example, the relative creep coefficient after 50 years is 1.6 times greater at 80 °C than when stored at 20 °C.

It should be remembered that in the CEB MC90 model [97], creep coefficient, ϕ_{28}, is the ratio of creep at any age at loading divided by the elastic strain at the age of 28 days. It is related to creep coefficient, ϕ_{to}, based on elastic strain at the same age at loading as the start of creep as follows:

$$\phi_{to} = \phi_{28}\frac{E_{to}}{E_{28}} \tag{10.17}$$

where E_{to} and E_{28} = modulus of elasticity at the ages of t_o and 28 days, respectively.

If unknown, the modulus of elasticity may be estimated from strength by Eq. (11.26) as recommended by the CEB MC90. The choice of relative creep coefficient is unaffected by the effect of temperature, i.e., whether based on the age at loading of 28 days or on any age at loading t_o.

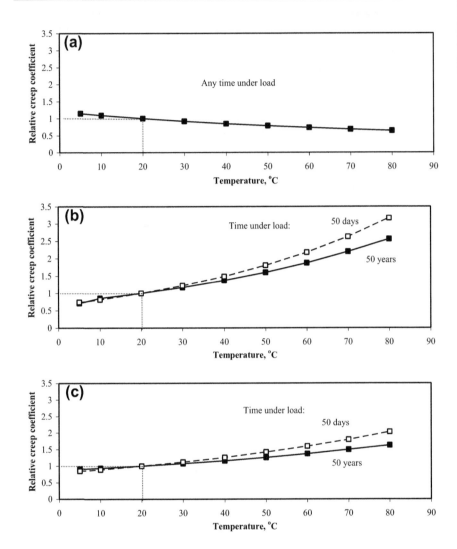

Figure 10.20 Effect of temperature on 50-year creep coefficient relative to that at 20 °C according to CEB MC90 [97]; concrete member $V/S = 50$ mm, moist-cured for 14 days before loading, and then drying in air at 65% RH (a) Cured at different temperature and stored at 20 °C. (b) Cured at 20 °C and stored at different temperature. (c) Cured and stored at different temperature.

Creep is affected in a different manner by the time when concrete is heated relative to application of load, which is demonstrated as follows: Consider basic creep of sealed or saturated concrete cured and heated to test temperature just before the load is applied, and let it be compared with basic creep of concrete cured at normal temperature and then heated just after application of load. The resulting creep-temperature characteristics are shown in Figure 10.21 for ordinary Portland cement concrete and the same concrete, but with 50% of cement partially replaced with blast-furnace

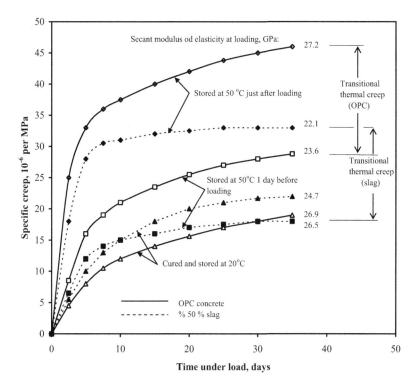

Figure 10.21 Influence of temperature change before and after application of load at the age of 14 days on basic creep of ordinary Portland cement (OPC) concrete and blast-furnace slag—OPC concrete [105].

slag. For both types of concrete, there are considerable increases in creep when heat is applied shortly after loading at the age of 14 days, compared with creep when heat is applied 1 day before the load was applied. The additional creep was labeled *transitional thermal creep* by Illston and Sanders [106]. Previously, Taylor and Williams [107] had reported an increase in the rate of creep when concrete already under load was heated, which was probably due to transitional thermal creep.

Although transitional thermal creep was defined as creep in excess of constant-temperature creep by Illston and Sanders, Hansen and Eriksson [108], and Arthanari and Yu [109] were probably the first to observe and report the phenomenon. Illston and Sanders carried out tests on saturated mortar specimens subjected to torsional loading while undergoing cycles of temperature. They concluded that transitional thermal creep occurs rapidly, is approximately independent of maturity, is zero when the temperature decreases or is raised to the same level a second or subsequent time, and is irrecoverable when the load is removed. Similar observations had been made earlier [108], and were later confirmed by Parrott [110] from measurements on hardened cement paste under uniaxial compression.

Fahmi et al. [111] recognized that some of the transitional thermal creep is due to an increase in temperature-induced elastic strain. They also reported appreciable increases in basic creep with additional cycles of temperature but not for total creep. Some of the findings by Illston and Sanders were questioned by Bamforth [114], who heated 3-year-old concrete to 65 °C that had already been subjected to an early-age temperature cycle in order to simulate conditions in newly-cast mass concrete. It was found that transitional thermal creep occurred even though the concrete had been subjected to a similar temperature in the early-age temperature cycle. It was also found that transitional thermal creep did not develop rapidly, but was time-dependent, taking 2 weeks to fully develop. Hence it seems that the occurrence and rate of development of transitional thermal creep may depend on the age of concrete when heat is applied, as suggested in a review by Khoury et al. [113].

Illston and Sanders [114] found specific transitional thermal creep of mortar under torsion to be related to temperature as follows:

$$C_{ttc} = 2.46(T - 20) - 0.0082(T - 20)^2 \qquad (10.18)$$

where C_{ttc} = specific transitional creep (10^{-6} per MPa) and T = temperature (°C).

For example, when $T = 60$ °C, $C_{ttc} = 85.3 \times 10^{-6}$ per MPa, which is similar to the value obtained by Parrott [92] for mortar subjected to compression. However, when $T = 50$ °C, Eq. (10.18) yields $C_{ttc} = 66.4 \times 10^{-6}$ per MPa, which is somewhat greater than values for concrete indicated by Figure 10.21 where C_{ttc} is approximately 17×10^{-6} per MPa for both ordinary Portland cement concrete and 50% slag concrete. Thus, since transitional thermal creep is lower for concrete than for mortar, it appears that aggregate content may be an influencing factor.

Khoury et al. [115,116] found that very high strains were developed in drying concrete heated to a high temperature of 600 °C after an application of load to simulate temperatures reached in fire conditions; they labeled the additional strain as *transient thermal strain*. Presently, the universal term for transitional thermal creep and transient thermal creep seems to be *transient thermal creep*, and CEB Model Code 90 method of prediction [97] quantifies this additional creep occurring due to a temperature increase in terms of a *transient thermal creep coefficient* as follows:

$$\Delta\phi_{T,trans} = 0.0004(T - 20)^2 \qquad (10.19)$$

where T = temperature (°C).

When $T = 50$ °C, Eq. (10.19) yields a transient creep coefficient of 0.36, which is slightly less than those indicated by the experimental data of Figure 10.21, multiplying the transitional thermal creep (17×10^{-6} per MPa) by the secant modulus of elasticity yields an average transient creep coefficient of approximately 0.43 for ordinary Portland cement concrete and slag-cement concrete.

Although, the effects of very high temperature on creep such as in fire conditions are not the main topic of this chapter, some experimental data are of interest, such as those presented in Table 10.7.

Table 10.7 Effect of High Temperature on Elasticity and Creep of Concrete [104,117]

Temperature, °C	Relative Elastic Modulus	Relative Elastic Strain	Relative Creep
23	1.0	1.0	1.0
150	0.81	1.23	3.3
425	0.56	1.79	6.4
480	0.46	2.17	14.9
650	0.36	2.78	32.6

It can be seen that there are considerable increases in elastic deformation and creep, especially the latter at very high temperature. Further information on concrete exposed to fire can be found in ACI 216R [118] and BS EN 1992: 2004 [119].

Cyclic variation in temperature down from a steady value has no effect on creep if sealed specimens have been heated to the steady temperature before the application of load [120]. However, a heating cycle during the sustained load causes the deformation to increase due to some elastic strain and transient thermal creep [32] but, following a drop in temperature, there is no sudden change the rate of creep or evidence of creep recovery [106].

Service conditions of concrete pressure vessels in nuclear reactors operate at elevated temperature, steep temperature gradients, and various combinations of temperature and pressure. As the vessel walls are in excess of 4.5-m thick, they are virtually mass concrete so that temperature-affected basic creep is involved. In this connection, creep of concrete is thought to be unaffected by *irradiation*, and any practical significance of damage to concrete properties by radiation that may influence creep would be limited to the inner surface of vessel walls [4].

Creep at Freezing Temperatures

Short-term basic creep concrete stored at temperatures down to −20 °C was studied by Johanson and Best [121]. Sealed concrete was cured for 42 days at 20 °C and then preconditioned at the test temperature for 3 days before loading. Figure 10.20 shows 24-day basic creep at low temperature relative to basic creep at 20 °C. Between −20 and −10 °C creep was small and independent of temperature, but as the temperature increased, the presence of ice led to a higher creep. However, this dropped to zero before ice-free concrete exhibited the expected trend of increasing basic creep with an increase of temperature. At the ice point there was instability, probably because of large changes in energy of the system as a result of freezing and thawing [4]. Marzouk [122] reported higher basic creep of high-strength silica fume/fly ash concrete stored at low temperatures in ocean water. The results suggest a different creep-temperature trend as indicated in Figure 10.22. At −20 °C, creep was less than basic creep of concrete stored at 20 °C, which was attributed to the slow rate of strength development and maturity of concrete as a result of the reduction of the secondary hydration process

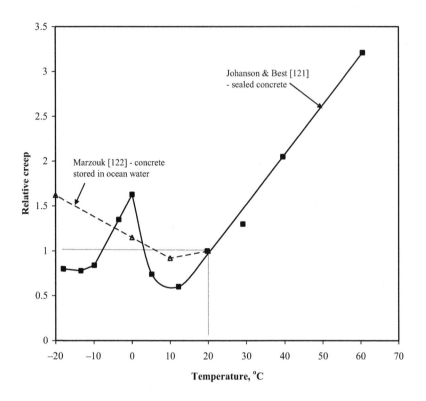

Figure 10.22 Basic creep of concrete at low temperatures relative to basic creep at 20 °C.

between calcium hydroxide and microsilica at low temperature. A comparison of data in Figure 10.22 indicates that the type of cement or cementitious material could be a factor in creep at freezing temperatures.

Extensive flexural tests on creep of frozen concrete by Podvalnyi [123] revealed much higher basic creep than on concrete stored in saturated air at 20 °C, and also found saturated concrete subjected to cycles of freezing and thawing to exhibit more basic creep than the same concrete stored at 20 °C.

Early-Age Creep of Heat-Cured Mass Concrete

This topic is of particular interest because of a possibility of thermal cracking (see Chapter 14) and has been investigated by Bamforth [71] and by Brooks et al. [124]. The latter investigators included tensile creep tests as well as compressive creep tests on concrete stored in water tanks, whose temperature was controlled to match that of an insulated, 300-mm freshly-cast concrete cube to simulate the adiabatic rise in temperature due to heat of hydration of cement. On reaching peak temperature, the control system was switched off and the water-stored specimens were allowed to cool until reaching the ambient temperature of the laboratory.

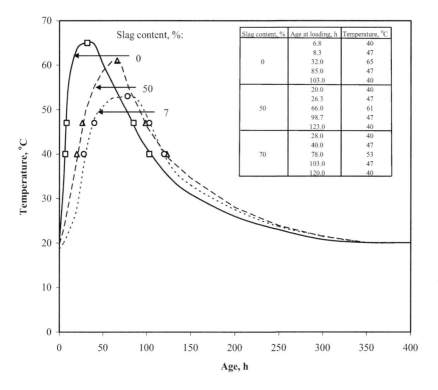

Slag content, %	Age at loading, h	Temperature, °C
	6.8	40
	8.3	47
0	32.0	65
	85.0	47
	103.0	40
	20.0	40
	26.3	47
50	66.0	61
	98.7	47
	123.0	40
	28.0	40
	40.0	47
70	78.0	53
	103.0	47
	120.0	40

Figure 10.23 Temperature-matched curing cycles of ordinary Portland cement concrete with and without slag; symbols and table show ages at loading storage temperature of creep tests [124].

Temperature profiles of concrete with and without ground granulated blast-furnace slag are shown in Figure 10.23, where it can be seen that the effect of replacement of cement by slag is to reduce the rate of rate of temperature rise and peak temperature. For each type of concrete, two series of creep tests were carried out. First, after curing in the temperature-matched tanks until the ages specified on the ascending and descending temperature-time curves, specimens were loaded and transferred to three water tanks preheated and controlled to temperatures of 40 °C, 47 °C, and peak temperature, respectively. In the second series of creep tests, loads were applied (1): at the time to reach 40 °C, and then creep specimens subjected to the continuing temperature rise up to the peak temperature and beyond; and (2) at the peak temperature, and then creep specimens subjected to the descending history of Figure 10.23. The second series was intended to quantify the effect of varying temperature on creep, and in particular transient thermal creep.

Over the short testing period of about 20 days, Figure 10.24 compares creep function (elastic strain plus creep per unit of stress) of ordinary Portland cement concrete in compression and tension after being loaded and stored at the ages and temperature

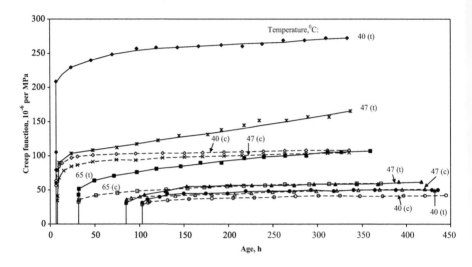

Figure 10.24 Early-age compressive (c) and tensile (t) creep functions of ordinary Portland cement concrete stored at constant elevated temperature after loading at ages specified in Figure 10.23 of the temperature-matched curing cycle [124].

specified in Figure 10.23. It can be seen that creep function in tension is greater than creep function in compression, as was also the situation for most slag-cement concretes [124]. It should be noted that the creep functions of Figure 10.24 are based on assumed linearity of stress and strain, i.e., calculated from quotients of measured strain and actual applied stress, the latter being equal to 0.3 of strength at the specified ages of loading listed in Figure 10.23; cube compressive strength ranged from 10 to 55 MPa, whereas tensile strength varied from 0.6 to 4.3 MPa so that applied stress on the creep specimens in compression was much greater than in tension. The validity of assumed linearity in the relation between creep in tension and compression is discussed in the next section.

Results of the second series of tests are given in Figure 10.25, where creep of concrete was loaded in compression at the age when the temperature first reached 40 °C and then remained constant. These results are compared with creep of concrete loaded at the same age, but subsequently the temperature was allowed to increase to the peak value (Figure 10.25(a)). It is apparent that the differences are small and inconsistent, which implies that transient thermal creep (as defined earlier) is insignificant; that implication contradicts earlier observations for more mature concrete. Therefore, it seems that transient thermal creep is associated with the degree of maturity and, possibly, the rate of increase in temperature. Figure 10.25(b) also demonstrates that allowing temperature to decrease after the application of load at peak temperature results in creep that is similar to creep at constant peak temperature. The latter observations concur with those of Illston and Sanders [106] and Hansen [32] for more mature mortar and concrete.

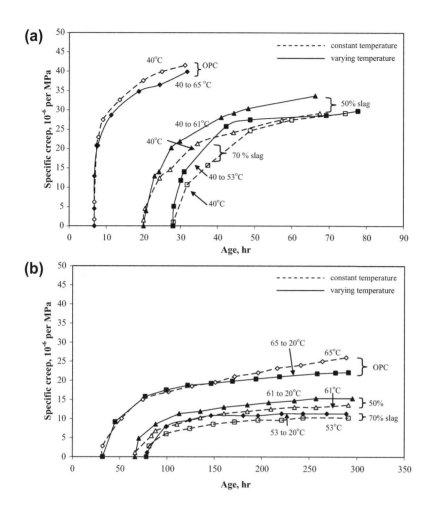

Figure 10.25 Comparison of compressive creep of concrete subjected to constant temperature and varying temperature of the temperature-matched curing cycle of Figure 10.21: (a) ascending temperature and (b) descending temperature [124].

Since early-age creep of temperature-matched cured concrete is accompanied by large changes in elastic strain, creep is best quantified by creep coefficient, and the creep coefficient−time characteristic has been found to be described by an exponential-power expression of the type [125]:

$$\phi(t, t_o) = \phi_m \exp\left[-A(3 - \log_{10}(t - t_o))^B\right] \tag{10.20}$$

where $\phi_m = $ 40-day creep coefficient, A and B are coefficients, all of which depend on the equivalent hydration period given by Eq. (10.11).

Average values of the coefficients of Eq. (10.20), which are applicable to ordinary Portland cement and slag-cement concrete, are obtained as follows:

$$
\left.
\begin{aligned}
\phi_m &= 1.0 - 0.8\left(\exp\left[-0.15(4 - \log_{10}t_{oe})^3\right]\right) \\
A &= 0.13\log_{10}t_{oe} \\
B &= 0.87A^{-0.45}
\end{aligned}
\right\}
\qquad (10.21)
$$

Equations (10.20) and (10.21) apply to the compressive creep coefficient, which may be converted into specific creep by multiplying by the elastic strain per unit of stress. Since elastic strain is also affected by temperature in the heat-curing process, it can be estimated from the modulus of elasticity-strength relationship given in Figure 4.16(b).

Corresponding early-age tensile creep coefficients (ϕ_T) for heat-cured ordinary Portland cement and slag-cement concretes may be estimated from compressive creep coefficients (ϕ_C) as follows:

$$
\phi_T = [4.1 - 0.55 \log_{10}(t - t_o)]\phi_C^{[1.3+0.15(\log_{10}(t-t_o))]}
\qquad (10.22)
$$

where $(t-t_o)$ = time under load (h)

It will be observed that Eq. (10.22) is time-dependent and yields greater creep coefficients in tension than in compression, especially at early times under load. The relationship between tensile and compressive creep for concrete cured at a normal temperature is discussed subsequently.

Creep Recovery

Referring to the section on categories of creep presented at the beginning of this chapter, a comparison of cases (2) and (3) is shown in Figure 10.26, where the total load strain (elastic strain plus creep) is plotted for concrete stored in water and in air, and then unloaded at some later age. On load removal, there is an immediate elastic strain response, or *instantaneous recovery*, which is generally smaller than the initial elastic strain at loading because the modulus of elasticity increases with age. The instantaneous recovery is followed by time-dependent *creep recovery* curve, which is similar in shape to the preceding creep curve, but is smaller and approaches a maximum value rapidly. The creep recovery is always smaller than creep so that there is a *residual deformation* even after a very short period under load of 1 day only. It is thus apparent that creep is not a completely reversible phenomenon, and the residual deformation may be viewed as *irreversible creep*. After 30 years, only 1–30% of creep was found to be reversible [3]. Figure 10.26 indicates that creep recovery for the two storage conditions is similar, which implies that basic creep is partly reversible and that drying creep is irreversible.

Creep recovery, or reversible creep, is of interest in predicting the behaviour of concrete under decreasing stress and for elucidating mechanisms of creep. Creep recovery

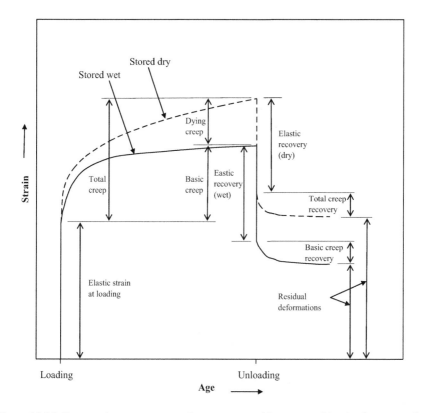

Figure 10.26 Creep and creep recovery of concrete stored in water and in air after removal of load.

is affected by several factors, details of which may be found in a review by Neville et al. [4]. In general, the consensus of opinion is:

- Creep recovery is greater for stronger concrete, for a lower water/cement ratio, and for a greater age at loading.
- Creep recovery is proportional to the stress removed.
- Creep recovery increases with an increase in the modulus of elasticity of aggregate.
- Hardened cement paste exhibits creep recovery, but is considerably slower and smaller than that of concrete.

Creep Poisson's Ratio

Under uniaxial compression, creep occurs not only in the axial direction, but also in the lateral or normal directions. This is referred to as *lateral creep*. By analogy to elastic strains, the ratio of lateral creep to axial creep is termed creep Poisson's ratio. Under axial compression, lateral creep is an extension, but, like elastic Poisson's ratio, the change of sign is ignored when quantifying creep Poisson's ratio.

In general, published research suggests that basic creep Poisson's ratio is equal to or very similar to the elastic Poisson's ratio in the case of sealed or mass concrete [4]. Under drying conditions, total creep Poisson's ratio is lower than when basic creep takes place because the effect of drying lateral creep is very small or absent in the case of drying concrete. For example, Gopalakrishnan et al. [125, 126] reported average values of 0.17–0.20 for wet-stored concrete, which were very close to the range of elastic Poisson's ratio (0.16–0.19). Under drying conditions, the same authors found total creep Poisson's ratio for concrete to be much less (0.07) than the elastic Poisson's ratio (0.19), and corresponding values for hardened cement paste were 0.08 and 0.22, respectively. Thus, the fact that the presence of normal weight aggregate has the effect of reducing Poisson's ratio while the reducing effect of lateral drying creep on total creep Poisson's ratio, is confirmed. There appears to be no consistency of the effect of time under load on creep Poisson's ratio, with some research observing an increase and some a decrease, depending on the type of aggregate [4]. Comprehensive test data for creep Poisson's ratio of concrete made with a wide range of aggregates have been reported by Kordina [5].

Tensile Creep

Creep in tension is of interest in estimating the possibility of cracking due to shrinkage or thermal stress (see Chapter 14), in the calculation of tensile stress in prestress concrete beams, and in the design of water retaining structures. Application of truly uniaxial tension, even in short-term strength tests, presents some degree of difficulty (see Chapter 16), and accurate measurement of tensile strain poses a problem because tensile stresses are small compared with compressive strain resulting from the normal range of stress. Moreover, when there is concurrent drying under load, the magnitude of drying shrinkage of the companion test specimen is similar to the strain of the specimen under load, so that there is a possibility of greater error in the computed tensile creep than in equivalent compressive creep tests.

Traditionally, for design purposes, it has been assumed that tensile creep is equal to compressive creep, but experimental evidence appears to suggest tensile creep is the greater of the two. Tensile creep is known to be influenced by the same factors as compressive creep, as demonstrated by Bissonnette et al. [127]; there can be a difference with regard to the extent of the influence of some of those factors. A simple approach to quantifying tensile creep, therefore, is to compare and relate it to compressive creep of concrete tested under identical conditions. Clearly, any comparison should be made on the basis of equal stress, usually in terms of specific creep, preferably with compressive creep determined under low stress similar to the tensile stress used to measure tensile creep. However, if specific compressive creep is calculated from compressive creep measured under normal stress, say, at 0.3 stress/strength ratio, then it is assumed that the compressive creep–stress relationship is linear, and this may not always be the case.

Glanville and Thomas [128] found total specific creep in compression and in tension to be the same under an equal applied stress, i.e., low compressive stress equal to the tensile stress. The equality of creep in the case of basic creep was partially confirmed by the U.S. Bureau of Reclamation [129] in tests on mass-cured concrete subjected to load at 28 and at 90 days. On the other hand, Davis et al. [130] found tensile creep to be appreciably greater than compressive creep for both mass-cured concrete and to concrete drying under load; different types of cement and ages at loading were investigated. Mamillan's tests [131] indicated specific total creep of neat cement paste to be about five times higher in tension than in compression. For moist-cured concrete loaded at the age of 7 days, Illston [132] confirmed a higher total creep in tension than in compression under equal stress, and in the same paper, a higher creep in tension of saturated concrete was reported. Brooks and Neville [133] confirmed a higher creep in tension for moist-cured concrete loaded at 28 and 56 days, but when predried for 28 days before loading at 56 days, compressive creep was higher; these comparisons were based on compressive specific creep determined at 0.3 stress/strength ratio. In all cases of replacement of cement by blast-furnace slag, tensile specific creep was higher than compressive specific creep for concrete loaded to the same stress/strength ratio at the age of 14 days [105]. Similarly, Li et al. [134] also reported higher tensile specific creep for concrete made with different types of cement, mineral, and chemical admixtures subjected to load at the age of 3 days; after 120 days under load, the tensile specific creep exceeded compressive specific creep by a factor ranging from two to almost five.

The review of the above findings revealed that in the majority of cases, on the basis of unit stress, the ratio of tensile creep to compressive creep was reasonably independent of time under load, and was higher for earlier ages at loading as shown in Figure 10.27. Although there is considerable scatter, the tensile/compressive specific creep ratio (R_c) is nearly always greater than unity, and the average trend with age at loading (t_o) can be represented by:

$$R_c = \frac{3.51 + 0.53t_o}{0.15 + 0.45t_o} \tag{10.23}$$

Thus, tensile-specific creep (C_{st}) may be estimated from a known compressive specific creep (C_{sc}) by the use of Eq. (10.23). The corresponding tensile/compressive creep coefficient ratio (R_ϕ) is given by:

$$R_\phi = \frac{E_t}{E_c} \frac{C_{st}}{C_{sc}} = \frac{E_t}{E_c} R_c \tag{10.24}$$

where E_t and E_c are the tensile and compressive moduli of elasticity, respectively.

A general relationship between tensile and compressive moduli of elasticity is shown in Figure 10.28, which has been deduced from modulus-strength equations presented in Chapter 4. Although there is likely to be significant scatter, the general trend suggests a higher modulus in tension than in compression, but the difference between

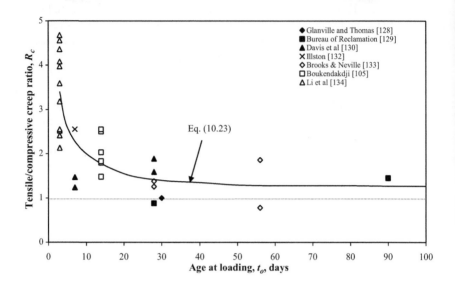

Figure 10.27 Ratio of tensile creep to compressive creep as a function of age at loading for concrete cured at a normal temperature.

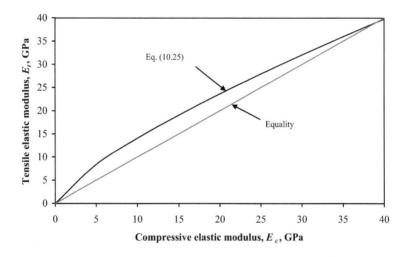

Figure 10.28 General trend of tensile modulus of elasticity as a function of compressive modulus of elasticity for concrete cured at a normal and elevated temperature.

the two moduli reduces as the strength of concrete increases. The general trend of Figure 10.28 is:

$$E_t = 2.5E_c^{0.75} \qquad (10.25)$$

The effect of the modulus influence is that the tensile/compressive creep coefficient ratio is slightly greater the than the tensile/compressive creep ratio, which applies to concrete cured at normal temperature. For heat-cured concrete loaded at very-early age, tensile modulus is also greater than the compressive modulus of elasticity so that Eq. (10.18) is applicable. Moreover, the tensile creep coefficient is also greater than compressive creep coefficient, but, in this case, the relationship depends on (equivalent) age at loading and time under load (see Eq. (10.11)).

Cyclic Creep

Creep under alternating or cyclic loading is of interest in structures such as bridges, pavements, and in shoreline and offshore structures due to wave action. Generally, cyclic loads cause an increase in creep compared with creep under a sustained load equal to the mean stress of the cyclic load. *Cyclic creep* is defined as the increase in creep under cyclic loading and is illustrated in Figure 10.29. It is important to realize that cyclic creep is measured relative to creep under a static load equal to the <u>mean</u> cyclic stress, and not the creep under a static stress equal to the upper cyclic stress. The definition is based on cyclic stresses ranging between σ_1 and σ_2 and is of the form:

$$\sigma = \sigma_m + \frac{(\sigma_2 - \sigma_1)}{2} \sin[2\pi\omega t] \qquad (10.26)$$

where σ_m = mean stress = $0.5(\sigma_2 + \sigma_1)$, $(\sigma_2 - \sigma_1)$ = range of stress, ω = frequency of loading (Hz), and t = time (sec).

Thus, creep should be determined at the value of mean stress, σ_m, i.e., at the midpoint of the cycle, since measurements of either will yield different values of creep if any change of modulus of elasticity occurs within the range of stress. Figure 10.30 compares creep of concrete subjected to different frequencies when measured at the minimum, maximum, and mean stresses. A summary of the main factors influencing creep under cyclic loading is as follows:

- Fatigue failure can occur when the upper maximum stress exceeds 0.55 of the static strength at loading [135,136]. For example, see Figure 10.29.
- Deformation under a cyclic load is greater than that under a static load equal to the mean cyclic load and increases as the range of stress increases, as shown in Figure 10.29.
- There is no increase in deformation when cyclic stress follows a static stress equal to the upper cyclic stress, but when the opposite occurs, i.e., a cyclic stress followed by a static load equal to the upper cyclic stress, the deformation increases [138,139].
- Uniform cycling of load causes less creep than an irregular pattern within the same range of stress [137].

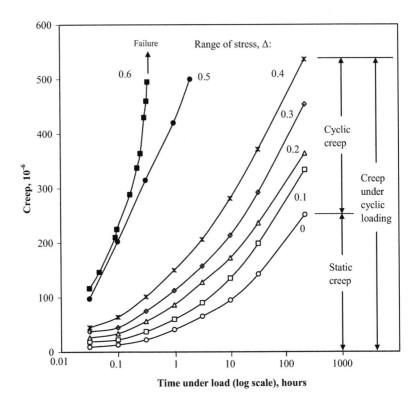

Figure 10.29 Influence of range of stress on creep of concrete under a cyclic load at a mean stress $\sigma_m = 0.35$; Δ and σ_m expressed as a fraction of 14-day strength [135].

- Cyclic creep is irrecoverable [135].
- Cyclic creep is approximately proportional to mean stress and to the range of stress [135, 139].
- Cycling of load results in a higher rate of creep at early ages and also leads to a greater long-term creep [139].
- Cyclic creep increases with an increase of frequency of loading, but only at frequencies greater than approximately 30 cycles per hour, depending on the time since loading. Below that frequency, cyclic creep is approximately constant (see Figure 10.30).
- The mechanism of cyclic creep is thought to be associated with microcracking since there is no additional corresponding relaxation of stress under cyclic straining [141].
- When measured at the mean stress, cyclic creep is unaffected by humidity of storage, and therefore total cyclic creep = basic cyclic creep, or there is no drying creep component of cyclic creep (see Figure 10.31). However, drying creep under a cyclic stress is approximately the same as drying creep measured under a static stress equal to the mean cyclic stress, which confirms the observation by Bazant and Panula [142].
- The ratio of creep under a cyclic stress to creep under a static stress equal to the mean cyclic stress is greater after short periods under load, and is greater for sealed concrete than for drying concrete, as shown in Figure 10.32.

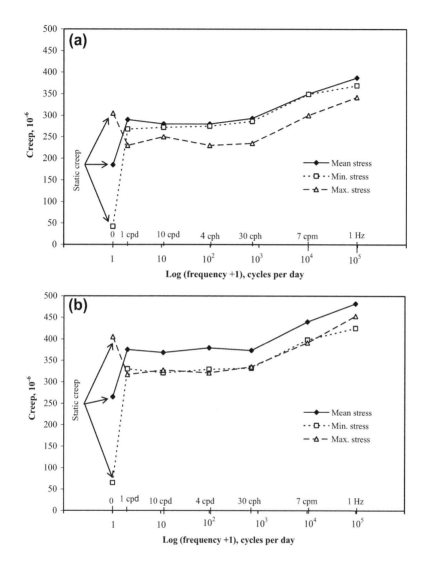

Figure 10.30 Influence of frequency of cyclic loading on 5-day creep of concrete measured at mean stress, minimum stress, and maximum stress; mean stress and range of stress = 0.3 and 0.4 of 14-day strength, respectively (a) Sealed concrete (basic creep) (b) Drying concrete (total creep) [140].

- *Cyclic modulus of elasticity*, i.e., the relation between range of stress and range of strain, is not affected by frequency of loading between one cycle per day (cpd) and 1 Hz. Cyclic modulus of elasticity of sealed concrete is approximately 5% greater than that of drying concrete, but the difference may be ignored for practical purposes. It may be assumed to equal the secant modulus of elasticity applicable to the initial loading of a static stress equal to the mean cyclic stress [140].

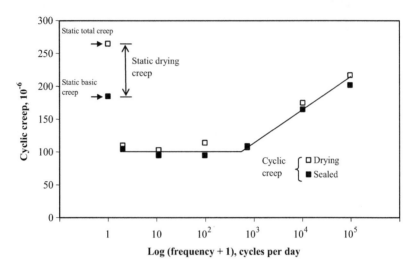

Figure 10.31 Comparison of cyclic creep and static creep of drying and sealed concrete after 5 days under load; mean stress, and range of stress = 0.3 and 0.4 of 14-day strength, respectively [140].

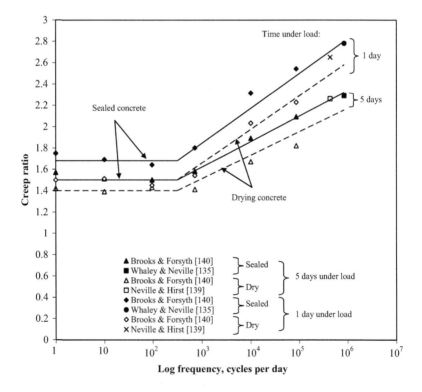

Figure 10.32 The influence of frequency of loading on the ration of creep under a cyclic stress measured at the mean stress to creep under a static stress equal to the mean cyclic stress; mean stress and range of stress = 0.3 and 0.4 of 14-day strength, respectively [140].

Prediction of Cyclic Creep

The standard methods of predicting creep presented in Chapter 11 do not include creep under cyclic loading. Bazant and Panula [142] have proposed a method, but it is restricted to a range of frequency between 5 and 25 Hz. Based on the foregoing observations, a method has been derived that is based on the relative creep ratio (R_{cy}) i.e., the ratio of creep under a mean cyclic stress to the static creep under a stress equal to the mean cyclic stress. Thus, in order to estimate creep under a cyclic stress, the static creep is required to be known, which may be estimated by standard methods or obtained by short-term measurement and extrapolation, as described in Chapter 11.

For a given cyclic mean stress and range of stress, it has been shown [140] that the dependency of the relative basic creep ratio of sealed concrete, $(R_{cyb}(t, t_0))$ on time under load $(t-t_o)$ and frequency of loading (see Figure 10.33) can be expressed as:

$$R_{cyb}(t, t_0) = A(\omega)\ln(t - t_0)^{B(\omega)} \qquad (10.27)$$

where $A(\omega)$ and $B(\omega)$ are dependent on frequency of loading.

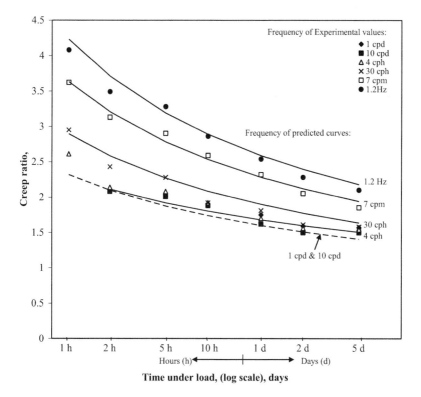

Figure 10.33 Ratio of creep under a cyclic stress to creep under a static stress equal to the mean cyclic stress (R_{cyc}) as a function of time under load for sealed concrete subjected to mean cyclic stress and range of stress of 0.3 and 0.4 of the 14-day strength, respectively; comparison of experimental results with curves predicted by Eq. (10.31).

The relative basic cyclic creep ratio may also be written as:

$$R_{cyb}(t, t_o) = \frac{c_{cy}(t, t_o) + c_{sb}(t, t_o)}{c_{sb}(t, t_o)} = 1 + \frac{c_{cy}(t, t_o)}{c_{sb}(t \cdot t_o)} \tag{10.28}$$

where $c_{cy}(t, t_o) =$ cyclic creep at a constant stress/strength ratio and $c_s(t, t_o) =$ static creep at a constant stress/strength ratio.

Now for a mean stress $= 0.30$ and range of stress $= 0.4$ of the strength at loading, regression analysis on Eq. (10.27) yields that when $\omega \leq 30$ cycles per day:

$$A(\omega) = 5.5 \text{ and } B(\omega) = -0.55 \tag{10.29}$$

and when $\omega \geq 30$ cycles per day:

$$\left. \begin{aligned} A(\omega) &= 0.255 + 2.791(\lg_{10}\omega) \\ B(\omega) &= -\left[0.55 + 0.177(\lg_{10}\omega - 1.7)^{0.47}\right] \end{aligned} \right\} \tag{10.30}$$

In the original investigation [140], coefficients A and B of Eq. (10.27) were determined for $(t - t_o)$ in minutes so that, after converting into days and changing natural logarithm to $\log_{10}$, Eq. (10.27) becomes:

$$R_{cyb}(t, t_o) = A(\omega)[7.272 + 2.303 \lg_{10}(t - t_o)]^{B(\omega)} \tag{10.31}$$

Figure 10.33 indicates good agreement of experimental and predicted creep ratios by using Eq. (10.31) for sealed concrete subjected to a mean stress/strength ratio of 0.3 and range of stress/strength ratio of 0.4. In general terms, for different values of mean stress and range of stress within the range of proportionality, cyclic creep will be different. However, assuming mean stress affects static creep and cyclic creep in a similar manner, only the effect of range of stress, Δ, on cyclic creep needs be considered. From Eq. (10.28) and Eq. (10.31), cyclic creep is now given by:

$$c_{cy}(t, t_o) = c_{sb}(t, t)\frac{\Delta}{0.4}\left\{A(\omega)[7.272 + 2.303 \lg_{10}(t - t_o)]^{B(\omega)} - 1\right\} \tag{10.32}$$

and it follows that basic creep under a cyclic load is:

$$c_{cy}(t, t_o) + c_{sb}(t, t_o) = c_{sb}(t, t_0)$$
$$\left[1 + \frac{\Delta}{0.4}\left\{A(\omega)[7.272 + 2.303 \lg_{10}(t - t_o)]^{B(\omega)} - 1\right\}\right] \tag{10.33}$$

In the case of drying concrete, the corresponding, relative total creep ratio under cyclic loading, $R_{cyt}(t, t_o)$, is less than that for sealed concrete since there is no drying creep component of cyclic creep. Thus, cyclic creep is the same for any storage condition, and when drying of concrete occurs the total creep ratio is given by:

$$R_{cyt}(t, t_o) = \frac{c_{cy}(t, t_o) + c_{st}(t, t_o)}{c_{st}(t, t_o)}$$

(10.34)

where $C_{st}(t, t_o)$ = total static creep and $c_{cy}(t, t_o)$ is given by Eq. (10.32).
Consequently, total creep under a cyclic load is given by:

$$c_{cy}(t, t_o) + c_{st}(t, t_o) = c_{st} + \frac{\Delta}{0.4} c_{sb} \left\{ A(\omega)[7.272 + 2.303 \lg_{10}(t - t_o)]^{B(\omega)} - 1 \right\}$$

(10.35)

Thus, provided static basic creep for sealed or mass concrete and total static creep of drying concrete are known, total creep under cyclic loading may be estimated by Eq. (10.35).

In the long term, compared with creep of sealed concrete under static loading, according to Eq. (10.33), the long-term basic creep ratio suggests increases in creep under cyclic loading of approximately 23% assuming $\Delta = 0.4$ and $\omega = 1$ Hz. However, long-term test data are really desirable for verification of that projection.

Also of interest from a practical viewpoint is the maximum strain resulting from a cyclic load. This can be estimated from knowing the cyclic modulus of elasticity, which is assumed to be equal to the secant modulus resulting from application of a static strain. For instance, in the reported tests [140], the 14-day strength and secant modulus of elasticity, E_c, at the age of loading were 39.4 MPa and 34.0 GPa, respectively. For a mean stress of 0.25 and range of stress 0.3 of the strength, the actual applied stresses were $\sigma_m = 9.9$ MPa and $\Delta = 11.8$ MPa. After 5 days under a cyclic load at a frequency of 50 cycles per day, the basic specific creep of sealed concrete, $C_{bs} = 22.4 \times 10^{-6}$, so that the maximum load strain (or creep function or *compliance*), $J(t, t_o)$, comprises the initial elastic strain due to applying the mean stress plus 5-day creep at the mean cyclic stress (calculated from Eq. (10.31)) plus the amplitude elastic strain due to a stress of $0.5 (\sigma_2 - \sigma_1)$, viz.

$$J(t, t_o) = \frac{\sigma_m}{E_c} + \left[R_{cyb} \sigma_m C_{bs} \right] + 0.5 \times (\sigma_2 - \sigma_1) \times \frac{1}{E_c}$$

$$= \frac{9.9}{34.0 \times 10^3} + \left[1.504 \times 9.9 \times 22.4 \times 10^{-6} \right] + \frac{0.5 \times 11.8}{34 \times 10^{-6}}$$

$$= \underline{798 \times 10^{-6}}$$

For long-term estimates, the cyclic modulus of elasticity may be assumed to increase with age in the same manner as secant or static modulus of elasticity (see Chapter 4).

Other Types of Load

Little information is available on *creep in torsion*, but it is believed to be qualitatively similar to creep in compression in that it is affected by the same factors [4]. Likewise, *creep in bond* between concrete and reinforcement has been investigated only very little, and even its existence is in doubt; creep in bond is thought to be advantageous because it would reduce induced stress in steel due to shrinkage of concrete [4].

The influence of stress and strain gradients on creep of concrete has not been studied in detail, such as in the case of *flexural creep*. Generally, in a modulus of rupture test, a strain gradient retards mortar cracking and increases the maximum strain that can be reached before failure, so it follows that the use of stress–strain data obtained from uniaxial tests would lead to conservative results when applied to strength in flexure [4]. Consequently, a similar situation could be expected when applying uniaxial creep data to predict flexural creep deflection. The advantage of using the flexural test for creep investigations is that drying shrinkage does not cause any bending and deflection so there is no need for a control specimen. On the other hand, flexural creep is complicated by the fact that the lower side of a vertical loaded horizontal beam specimen is in tension and the upper side is in compression. It has already been demonstrated that unixial tensile creep is greater than compressive creep depending on the time under load, but this may not apply to flexural creep with a strain gradient under drying conditions, as indicated by the tests of Davis et al. [130]. Nevertheless, generally, flexural creep deflection of concrete is influenced qualitatively in the same manner by factors involved in creep under unixial compressive stress [4].

Multiaxial Stress

Knowledge of Poisson's ratio is of particular interest in the case of concrete subjected to multiaxial stress since, in any direction, creep occurs due to the stress applied in that direction, and creep also occurs due to the Poisson's ratio effect of creep strains in the two normal directions. The question is whether all these strains occur independently, in which case they can be deduced from uniaxial creep data, or whether the behaviour is more complex. The answer to the question is the latter, as shown by the tests carried out by Gopalakrishnan et al. [4,125]. They found that creep under multiaxial compression is less than under uniaxial compression of the same magnitude and in the same direction. Even under hydrostatic compression there is considerable creep. Furthermore, Poisson creep in any direction was affected by the applied stress in that direction so that the *effective creep Poisson's ratio* under multiaxial compression was less than the uniaxial creep Poisson's ratio. In fact, the effective Poisson's ratio was a function of the overall state of stress applied to the concrete specimen. Consequently, creep under multiaxial stress cannot be simply predicted from uniaxial creep so that the principle of superposition does not apply, and creep can be predicted accurately only by empirically determined, effective creep Poisson's ratios [4].

Problems

10.1. Define creep of concrete.

10.2. Discuss the different categories of creep.

10.3. Describe the role of aggregate in creep of concrete.

10.4. List the main factors affecting creep of concrete.

10.5. Compare creep of mass concrete with creep of a structural member exposed to drying.

10.6. What is the approximate relation between creep, stress, and strength of concrete?

10.7. What terms are used to quantify the size and shape of a concrete member?

10.8. What are the effects of replacing cement with fly ash and blast-furnace slag on creep of concrete: (a) for constant mix proportions, and (b) for constant workability?

10.9. What are the effects of (a) shrinkage reducing admixture, and (b) dosage level of shrinkage-reducing admixture on creep of concrete?

10.10. What is transient thermal creep?

10.11. How does elevated temperature before loading affect creep?

10.12. What does the relation between tensile and compressive creep depend on?

10.13. Describe how you would measure creep under a cyclic load.

10.14. What is the difference between creep under a cyclic load and cyclic creep?

10.15. Describe the influence of frequency of loading on cyclic creep of sealed and drying concrete.

10.16. Comment on creep of shrinkage-compensating concrete.

10.17. Compare with creep of concrete cured and stored in water: (a) creep of concrete dried before loading and sealed after loading, and (b) creep of concrete sealed before loading and dried after loading.

10.18. Describe wetting creep.

10.19. What is the general effect on creep of using a high range water-reducing admixture to produce (a) flowing concrete, and (b) high strength concrete?

10.20. Using a standard method of prediction, the 50-year creep of concrete is estimated to be 800×10^{-6}. The concrete is made with ordinary Portland cement, a total aggregate/cement ratio of 5.8, water/cement ratio of 0.54, air content of 5%, and density $= 2350 \text{ kg/m}^3$. It is subjected to a sustained compressive stress of 8 MPa at the age of 28 days when the mean strength is 32 MPa. Using the relative deformation method, calculate the 50-year specific creep for concrete subjected to the same stress/strength ratio, but the 28-day strength at loading $= 28$ MPa when 30% of cement is replaced by fly ash and the water/binder ratio $= 0.48$. Assume the specific gravity of fly ash $= 2.35$ and concrete density and air content are unchanged.
Answer: 82.5×10^{-6} per MPa.

10.21. Under uniaxial compression, the secant modulus of elasticity of concrete is 25 GPa when loaded at the age of 7 days, and 5-year specific creep is 50×10^{-6} per MPa. Estimate the corresponding tensile specific creep and tensile creep coefficient.
Answer: (a) 109.4×10^{-6} per MPa, (b) 3.0.

10.22. Creep of concrete of large cross-section after 2 years under a static uniaxial compressive stress $= 0.25$ of the strength is 200×10^{-6}. Estimate the corresponding (a) cyclic creep; and (b) creep under a cyclic load, when the frequency is 20 cycles per day at the same mean static stress and range of stress $= 0.3$ of the strength.
Answer: (a) 44×10^{-6}; (b) 244×10^{-6}.

References

[1]   Rusch H. Researches toward a general flexural theory for structural concrete. ACI J 1960;75(1):1−28.

[2]   Domone PL. Uniaxial tensile creep and failure of concrete. Magazine of Concrete Research; vol. 26, No. 88: 144−152.

[3]   Brooks JJ. 30-year creep and shrinkage of concrete. Mag Concr Res 2005;57(9): 545−56.

[4]   Neville AM, Dilger WH, Brooks JJ. Creep of plain and structural concrete. London and New York: Construction Press; 1983. 361 pp.

[5]   Kordina K. Experiments on the influence of the mineralogical character of aggregates on the creep of concrete. RILEM Bull; March 1960:7−22. Paris, No. 6.

[6]   Counto UJ. The effect of the elastic modulus of the aggregate on the elastic modulus, creep and creep recovery of concrete. Mag Concr Res 1964;16(48):129−38.

[7]   Browne RD, Blundell R. The behaviour of concrete in prestressed concrete pressure vessels. In: Proceedings first international conference on structural mechanics in reactor technology, Berlin, Sept, 1971. Nuclear engineering and design, vol. 20, no. 2; 1972. pp. 429−75.

[8]   Concrete Society. The creep of structural concrete. Technical paper No. 101; 1973. 47 pp.

[9]   England GL. Method of estimating creep and shrinkage strains in concrete from properties of constituent materials. ACI J 1965;62:1411−20.

[10] Neville AM. Creep of concrete as a function of its cement paste content. Mag Concr Res 1964;16(46):21−30.

[11] Pickett G. Effect of aggregate on shrinkage of concrete and hypothesis concerning shrinkage. J Am Concr Inst 1956;52:581−90.

[12] Ward MA, Jessop EL, Neville AM. Some factors in creep of lightweight aggregate concrete. In: Proceedings of RILEM symposium on lightweight aggregate concrete, Budapest; 1967. pp. 745−59.

[13] Neville AM. Properties of concrete. 4th ed. Pearson Prentice Hall; 2006. 844 pp.

[14] Lorman W,R. The theory of concrete creep. Proc ASTM 1940;40:1082−102.

[15] Wagner O. Das kriechen unbewehrten betons, Deutcher Ausschuss fur Stahbeton. No. 131, Berlin; 1958. 74 pp.

[16] Neville AM. Role of cement in creep of mortar. ACI J 1959;55:963−84.

[17] Washa GW, Fluck PG. Effect of sustained loading on compressive strength and modulus of elasticity of concrete. ACI J 1950;45:93−100.

[18] Russell HG. Performance of shrinkage-compensating concrete in slabs. Research and development bulletin RD057.01D. Skokie, Illinois: Portland Cement Association; 1978. 12 pp.

[19] Neville AM, Wainwright PJ. High-aluminma cement concrete. Lancaster/New York: Construction Press; 1975. 201 pp.

[20] L'Hermite R. What do we know about plastic deformation and creep of concrete? Paris, No. 1. RILEM Bulletin; March 1959. pp. 1−25.

[21] Glanville WH. Creep of concrete under load. Structural Eng 1933;II(2):54−73.

[22] Giangreco E. Recherches experimentales sur la fluage des ciments. Paris: Annals Institut Technique du Batiment et des Travaux Publics; 1954. No. 79-80, pp. 665−676.

[23] Dutron R. Creep in concretes, RILEM Bulletin, Paris, No. 34: pp. 11−33.

[24] Davis RE, Davis HE, Hamilton HS. Plastic flow of concrete under sustained stress. Proc ASTM 1934;34(Part 2):354−86.

[25] Le Camus. Recherches experimentales sur la deformation du beton et du beton arme: deformations lentes. Paris: Institut Technique du Batiment et des Travaux Publics; Jan. 1947. 10 pp.

[26] Brooks JJ, Farrugia R. Early-age load deformations of PFA concrete, third international conference on the Use of fly ash, silica fume, slag and natural pozzolans in concrete. Supplementary papers volume. Trondheim, Norway: ACI Special Publication, SP-114; 1989. pp. 237—51.

[27] Hanson JA. Prestress loss as affected by type of curing. Prestress Concr Inst J April 1964;9:69—93.

[28] Hanson TC, Mattock AH. Influence of size and shape of member on the shrinkage and creep of concrete. ACI J 1966;63:267—90.

[29] Bryant AH, Vadhanavikkit C. Creep, shrinkage-size, and age at loading effects. ACI Mater J; March-April, 1987:117—23.

[30] Troxell GE, Raphael JM, Davis RE. Long-time creep and shrinkage tests of plain and reinforced concrete. Proc ASTM 1958;58:1101—20.

[31] L'Hermite RG, Mamillan M. Further results of creep and shrinkage tests. In: Proceedings of an international conference on the structure of concrete. London: Cement and Concrete Association; 1968. pp. 423—33.

[32] Hansen TC. Creep and stress relaxation of concrete. Proceedings No. 31. Stockholm: Swedish Cement and Concrete Research Institute; 1960. 112 pp.

[33] Weirig HJ. Einflussauf dasa biegekriechen von zementmortal. Schweiz Bauztg 1964; 82(29):512—5.

[34] Cilosani ZN. On the probable mechanism of creep of concrete. Beton I Zhelezobet Mosc; 1964:75—8. No. 2.

[35] Ruetz W. A hypothesis for creep of hardened cement paste and the influence of simultaneous shrinkage. In: Proceedings of an international conference on the structure of concrete. London: Cement and Concrete Association; 1968. pp. 365—87.

[36] Glucklich J, Ishai O. Creep mechanism in cement mortar. ACI J 1962;59:923—48.

[37] Mullen WG, Dolch WL. Creep of Portland cement paste. Proc ASTM 1964;64: 1146—70.

[38] Ross AD. Shrinkless and creepless concrete. Civ Eng Public Works Rev 1951;46(545): 853—4.

[39] L'Hermite RG. Volume changes of concrete. In: Proceedings fourth international symposium on the chemistry of cement, vol. 2, Washington DC; 1960. pp. 659—94.

[40] Ali I, Kesler CE. Mechanisms of creep in concrete, symposium on creep in concrete. ACI Special Publication No. 9; 1964. pp. 35—57.

[41] Ishai O, Glucklich J. The effect of extreme hygrometric changes on the isotropy and deformability of mortar and concrete specimens. In: Proceedings RILEM-CIB Symposium on moisture problems in Buildings; Otaniemi, Finland; 1965. p. 26.

[42] Alexander J. Influence de la carbonation sur la fluage en compression du beton. Rev Mater Constr; Nov. 1973:22—9. Paris, No. 684.

[43] Parrott LJ. Increase in creep of hardened cement paste due to carbonation under load. Mag Concr Res 1975;27(92).

[44] Hanson TC. Creep of concrete. Bulletin No. 33. Stockholm: Swedish Cement and Concrete Research Institute; 1968. 48 pp.

[45] Hanson TC. Creep of concrete, the effect of variations in the humidity of the ambient atmosphere, sixth congress of the international association for bridge and structural engineering. Stockholm: Preliminary Publication; 1960. pp. 57—65.

[46] Al-Alusi HR, Bertero VV, Polivka M. Einfluss der feuche auf schwinden und kreichen von beton. Bet Stahlbetonbau 1978;73(1):18−23.

[47] Muller HS, Pristl M. Creep and shrinkage of concrete, at variable ambient conditions, creep and shrinkage of concrete. In: Bazant ZP, Carol I,E, Spon FN, editors. Proceedings fifth international RILEM symposium, London; 1993. pp. 15−26.

[48] Hansen TC. Creep of oil-saturated concrete. Mater Struct; 1969:145−8. Paris, No. 2.

[49] Hannant DJ. The mechanism of creep in concrete. Mater Struct; 1968:403−10. Paris, No. 5.

[50] Mills RH. Molecular sieve effect in concrete. In: Proceedings fifth international symposium on chemistry of cement, Tokyo, III; 1968. pp. 74−85.

[51] Maxson OG, Achenbach GD. properties of concrete in contact with pressurized hydrocarbons and sea water; 1976. OTC 2662, offshore technology conference, Houston, Texas.

[52] Brooks JJ. Influence of mix proportions, plasticizers and superplasticizers on creep and drying shrinkage of concrete. Mag Concr Res 1989;41(148):145−53.

[53] Jessop EL, Ward MA, Neville AM. Influence of water-reducing and set-retarding admixtures on creep of lightweight aggregate concrete. In: Proceedings RILEM Symposium on admixtures for mortar and Concrete, Btussels; 1967. pp. 35−46.

[54] Hope BB, Neville AM, Guruswami A. Influence of admixtures on creep of concrete containing normal weight aggregate. In: Proceedings RILEM Symposium on admixtures for mortar and Concrete, Btussels; 1967. pp. 17−32.

[55] Morgan DR, Welch GB. Influence of admixtures on creep of concrete. In: Proceedings Third Australian Conference on Mechanics of structures and materials. New Zealand: University of Auckland; 1971.

[56] Neville AM, Brooks JJ. Time-dependent behaviour of concrete containing a plasticizer. Concrete 1975;9(10):33−5.

[57] Brooks JJ. Influence of plasticizing admixtures: Cormix P7 and Cormix 2000 on time-dependent properties of flowing concretes. Research report. Department of Civil Engineering, University of Leeds; 1984.

[58] Brooks JJ, Wainwright PJ, Neville AM. Time-dependent properties of concrete containing a superplasticizing admixture. In: Superplasticizers in concrete. ACI Special Publication SP 62; 1979. pp. 293−314.

[59] Brooks JJ, Wainwright PJ, Neville AM. Superplasticizer effect on time-dependent properties of air entrained concrete. Concrete 1979;13(6):35−8.

[60] Brooks JJ, Wainwright PJ, Neville AM. Time-dependent behaviour of high early-strength concrete containing a superplasticizer. In: Developments in the use of superplasticizers. ACI Special Publication SP 68; 1981. pp. 81−100.

[61] Brooks JJ, Wainwright PJ. Properties of ultra-high strength concrete containing a superplasticizer. Mag Concr Res 1983;35(125):205−14.

[62] Dhir RK, Yap AWF. Superplasticized flowing concrete: strength and deformation properties. Mag Concr Res 1984;36(129):202−15.

[63] Berenjian J. Superplasticized flowing concrete: microstructure and long-term deformation properties [Ph.D. thesis]. Department of Civil Engineering, University of Leeds; 1989, 238 pp.

[64] Tokuda H, Shoya M, Kawakami M, Kagaya M. Applications of superplasticizers to reduce shrinkage and thermal cracking in concrete. In: Developments in the use of superpasticizers. ACI Special Publication SP 68; 1981. pp. 101−20.

[65] Alexander KM, Bruere GM, Ivanesc I. The creep and related properties of very high strength superplasticized concrete. Cem Concr Res 1980;10(2):131−7.

[66] Jiang X. The effect of creep in tension on cracking resistance of concrete, MSc (Eng) thesis. School of Engineering, University of Leeds; 1997, 140 pp.

[67] Kristiawan SA. Restrained shrinkage cracking of concrete [Ph.D. thesis]. School of Engineering, University of Leeds; 2002, 217 pp.

[68] Brooks JJ. Elasticity, shrinkage and creep of concretes containing admixtures. In: Al-Manseer A, editor. Proceedings Adam neville Symposium: creep and shrinkage-structural design effects. Atlanta: ACI Special Publication SP 194; 2000. pp. 283−360.

[69] Megat Johari MA. Deformation of high strength concrete containing mineral admixtures [Ph.D. thesis]. School of Engineering, University of Leeds; 2000, 296 pp.

[70] Neville AM, Brooks JJ. Time-dependent behaviour Cemsave concrete. Concrete 1979; 9(3):36−9.

[71] Bamforth PB. An investigation into the influence of partial Portland cement replacement using either fly ash or ground granulated blast-furnace slag on the early age and long-term behaviour of concrete. Southall, UK: Taywood Engineering; 1978. Research report 013J/78/2067.

[72] Aitchin PC, Laplante R. Volume changes and creep measurements of slag cement concrete. Research report. Universite de Sherbrooke; 1986. 16 pp.

[73] Cook DJ, Hinezak I, Duggan R. Volume changes in Portland blast-furnace slag cement concrete. In: Proceedings second international conference on the use of fly ash, silica fume, slag, and natural pozzolans in concrete, supplementary papers volume, Madrid; 1986. p. 14.

[74] Chern JC, Chan YW. Deformations of concrete made from blast-furnace slag cement and ordinary Portland cement. ACI Mater J Proc 1989;86(4):372−82.

[75] Brooks JJ, Wainwright PJ, Boukendakji M. Influences of slag type and replacement level on strength, elasticity, shrinkage and creep of concrete. In: Proceedings fourth international conference on fly Ah silica fume, slag and natural pozzolans in concrete, 2. ACI Special Publication SP-132; 1992. pp. 1335−42.

[76] Ross AD. Some problems in concrete construction. Mag Concr Res 1960;12(34): 27−34.

[77] Lohtia RP, Nautiyal BD, Jain OP. Creep of fly ash concrete. ACI digest paper No. 73-79 ACI J; Aug. 1976:469−72.

[78] Ghosh RS, Timusk J. Creep of fly ash concrete. ACI J Proc 1981;78. Title No. 78−30.

[79] Brooks JJ, Wainwright PJ, Cripwell JB. Time-dependent properties of concrete containing pulverised fuel ash and a superplasticizer. In: Cabrera JC, Cusens AR, editors. Proceedings international symposium on the use of PFA in concrete, vol. 1. Department of Civil Engineering, University of Leeds; 1982. pp. 209−20.

[80] Yamato T, Sugita H. Shrinkage and creep of mass concrete containing fly ash. In: Proceedings first international conference on the use of fly ash, silica fume, slag and other mineral by-products. ACI Special Publication SP 79; 1983. pp. 87−102.

[81] Yuan RL, Cook JE. Time-dependent deformations of high strength fly ash concrete. In: Cabrera JC, Cusens AR, editors. Proceedings international Symposium on the Use of PFA in Concrete, vol. 1. Department of Civil Engineering, University of Leeds; 1982. pp. 255−60.

[82] Nasser KW, Al-Manaseer A. Shrinkage and creep containing 50 per cent lignite fly ash at different stress/strength ratios. In: Proceedings second international conference on fly Ah silica fume, slag and natural pozzolans in concrete. ACI Special Publication SP 91; 1986. pp. 443−8.

[83] Brooks JJ, Gamble AE, Al-Khaja WA. Influence of pulverised fuel ash and a superplasticizer on time-dependent performance of prestressed concrete beams. In:

        Proceedings symposium on utilization of high strength concrete, Stavanger, Norway;
        1987. pp. 205−14.

[84]   Langley WS, Carette GG, Malhotra VM. Structural concrete incorporating high vol-
        umes of ASTM Class fly ash. ACI Mater J 1989;86(5):505−14.

[85]   Buil M, Acker P. Creep of silica fume concrete. Cem Concr Res 1985;15:463−6.

[86]   Tazawa E, Yonekura A. Drying shrinkage and creep of concrete with condensed silica
        fume. In: Proceedings second international conference on fly ash, silica fume, slag and
        natural pozzolans in concrete. Madrid: ACI Publication SP 91; 1986. pp. 903−21.

[87]   Bentur A, Goldman A. Curing effects, strength and physical properties of high strength
        silica fume concretes. J Mater Civ Eng 1989;1(1):46−58.

[88]   de Larrard F. Creep and shrinkage of high strength field concretes. International
        workshop on the use of silica fume in concrete, Washington DC. CANMET/ACI; 1991.
        22 pp.

[89]   Pentalla V, Rautenan T. Microporosity, creep and shrinkage of high strength concretes.
        International workshop on the use of silica fume in concrete, Washington DC.
        CANMET/ACI; 1991. 29 pp.

[90]   Burg RG, Ost BW. Engineering properties of commercially available high strength
        concretes. PCA Research and Development RD 104T, Portland Cement Association;
        1992. 55 pp.

[91]   Bilodeau A, Carette GG, Malhotra VM. Mechanical properties of non-air entrained,
        high strength concrete incorporating supplementary cementing materials. Mineral sci-
        ences laboratories division report MSL 89−129. Ottawa: CANMET; 1989. 30 pp.

[92]   Botassi dos Santos S, Silva Filho LCP, Calmon J l. Early-age creep of mass concrete:
        effects of chemical and mineral admixtures. ACI Mater J 2012;109(5):537−44.

[93]   Brooks JJ, Hynes JP. Creep and shrinkage of ultra high-strength silica fume concrete.
        In: Bazant ZP, Carol I,E, Spon FN, editors. Proceedings Fifth international RILEM
        symposium on creep and shrinkage of concrete, Barcelona; 1993. pp. 493−8.

[94]   Sellevold EJ. The function of condensed silioca fume in high strength concrete. In:
        Holand I, Helland S, Jakobsen B, Lenscow R, editors. Proceedings of symposium on
        utilization of high strength concrete; Stavanger, Norway; 1987. pp. 39−49.

[95]   Long Wu-Jian, Khayat KH. Creep of prestressed self-consolidating concrete. ACI
        Mater J 2011;108(5):476−84.

[96]   Khayat KH, Long Wu-Jian. Shrinkage of pre-cast, prestressed self-consolidating con-
        crete. ACI Mater J 2010;107(3):231−8.

[97]   CEB-FIP, CEB Model Code 1990, CEB Bulletin d'information No. 213/214, Comite-
        Euro International du Beton, Lausanne; 1993, 437 pp.

[98]   AASHTO. AASHTO LRFD bridge design Specifications. 4th ed. Washington, DC:
        American Association of State and Highway Officials; 2007. 1518 pp.

[99]   Nasser KW, Neville AM. Creep of old concrete at normal and elevated temperature.
        ACI J 1967;64:97−103.

[100]   Nasser KW, Lohtia RP. Creep of mass concrete at high temperature. ACI J 1971;68:
        276−81.

[101]   Mindess S, Young FY. Concrete. New Jersey: Prentice-Hall; 1981. 671 pp.

[102]   Neville AM, Brooks JJ. Concrete Technology. 2nd ed. Pearson Prentice Hall; 2010. 442
        pp.

[103]   ACI Committee 209.1R-05. Report on factors affecting shrinkage and creep of hard-
        ened concrete. ACI Manual of Concrete Practice, Part 1, American Concrete Institute;
        2012.

[104]   Marechal JC. Le fluage du beton en function de la temperature. Mater Struct 1969;2(8): 111—5.

[105]   Boukendakdji M. Mechanical properties and long-term deformation of slag cement concrete [Ph.D. thesis]. Department of Civil Engineering, University of Leeds; 1989, 290 pp.

[106]   Illston JM, Sanders PD. The effect of temperature change on the creep of mortar under torsional loading. Mag Concr Res 1973;25(84):135—44.

[107]   Taylor RS, Williams AJ. The design of prestressed concrete pressure vessels with particular reference to Wylfa. In: Proceedings of third international conference on the peaceful uses of atomic energy, Geneva; 1964. pp. 445—654.

[108]   Hansen TC, Erikson L. Temperature change effect on behaviour of cement paste mortar and concrete under load. ACI J Proc 1966;63(4):489—502.

[109]   Arthanari S, Yu CW. Creep of concrete under uniaxial and biaxial stresses at elevated temperatures. Mag Concr Res 1967;19(60):149—56.

[110]   Parrott LJ. A study of transitional thermal creep in hardened cement paste. Mag Concr Res 1979;31(107):90—103.

[111]   Fahmi HM, Polivka M, Bresler B. Effect of sustained and cyclic temperature on creep of concrete. Cem Concr Res 1972;2(25):591—606.

[112]   Bamforth P. The effect of temperature variation on the creep of concrete. Technical note UTN19. Construction Industry Research and Information Association (CIRIA); July 1980. 44 pp.

[113]   Khoury GA, Grainger BN, Sullivan PJE. Transient thermal strain of concrete: literature review, conditions within specimen and behaviour of individual constituents. Mag Concr Res 1985;37(132):131—44.

[114]   Illston JM, Sanders PD. Characteristics and prediction of creep of saturated mortar under variable temperature. Mag Concr Res 1974;26(88):169—79.

[115]   Khoury GA, Grainger BN, Sullivan PJE. Strain of concrete during first heating to 600°C under load. Mag Concr Res 1985;37(132):195—215.

[116]   Khoury GA, Grainger BN, Sullivan PJE. Strain of concrete during first cooling from 600oC under load. Mag Concr Res 1986;38(144):3—12.

[117]   Cruz CR. Apparatus for measuring creep of concrete at high temperatures. PCA Res Dev Bull 225, Portland Cem Assoc 1968;10(3):36—42.

[118]   ACI Committee 216-07, Code requirements for determining fire resistance of concrete and masonry assemblages, ACI manual of Concrete Practice, Part 1, American Concrete Institute, 2012.

[119]   BS EN 1992-1-1: 2004, Eurocode 2. Design of Concrete structures, general rules, British Standards Institution.

[120]   Hannant DJ. The strain behaviour of concrete under compressive stress at elevated temperatures. Report No. RD/L/N/67/66. Research and Development Department, Central Electricity Generating Board; June 1966. 30 pp.

[121]   Johansen R, Best CH. Creep of concrete with and without ice in the system. No. 16. RILEM Bull; 1962. pp. 44—57.

[122]   Marzouk HM. Effect of low temperature on the creep behaviour of high strength concrete. In: Daye MA, editor. Creep and shrinkage of Concrete: effect of materials and environment. ACI Special Publication, SP135-4; 1992. pp. 51—64.

[123]   Podvalnyi AM. Creep of freezing concrete. Dokl Akad Nauk SSSR 1963;148(5): 1148—51.

[124]   Brooks JJ, Wainwright PJ, Al-Kaisi AF. Compressive and tensile creep of heat-cured ordinary Portland and slag cement concretes. Mag Concr Res 1991;42(154):1—12.

[125] Gopalakrishnan KS, Neville AM, Ghali A. Creep Poisson's ratio of concrete under multiaxial; compression. ACI J 1969;66:1008—20.

[126] Gopalakrishnan KS, Neville AM, Ghali A. A hypothesis on the mechanism of creep of concrete with reference to multiaxial compression. ACI J 1970;67:29—35.

[127] Bisonnette B, Pigeon M, Vaysburd AM. Tensile creep of concrete: study of its sensitivity to basic parameters. ACI Mater J 2007;104(4):360—8.

[128] Glanville WH, Thomas FG. Studies in reinforced concrete-IV. Further investigations on creep or flow of concrete under load. Building research technical paper No. 21, London; 1939. 44 pp.

[129] US Bureau of Reclamation. A ten-year study of creep properties of concrete. Concrete laboratory report No. SP-38, Denver, Colorado; July 1953. 14 pp.

[130] Davis RE, Davis HE, Brown EH. Plastic flow and volume changes of concrete. Proc ASTM 1937;37(Part 2):317—30.

[131] Mamillan MA. A study of creep of concrete. Paris, No. 3 RILEM Bull; July 1959: 15—31.

[132] Illston JM. The creep of concrete under uniaxial tension. Mag Concr Res 1965;17(51): 77—84.

[133] Brooks JJ, Neville AM. A comparison of creep, elasticity and strength of concrete in tension and compression. Mag Concr Res 1977;29(100):131—41.

[134] Li H, Wee TH, Wong SF. Early-age creep and shrinkage of blended cement concrete. ACI Mater J 2002;99(1):3—10.

[135] Whaley CP, Neville AM. Non-elastic deformation of concrete under cyclic compression. Mag Concr Res 1973;25(84):145—54.

[136] Probst E. The influence of rapidly alternating loading on concrete and reinforced concrete. Struct Eng 1931;9:410—29.

[137] Probst E. Plastic flow in plain and reinforced concrete arches. ACI J 1933;30:137—41.

[138] Le Camus B. Recherches sur le comportement du beton et du beton arme soumis a des efforts repetes. Circulaire Serie F, No. 27. Paris: Annales Institut Technique du Batiment et des Travaux Publics; July 1946. 23 pp.

[139] Neville AM, Hirst GA. Mechanism of cyclic creep of concrete, Douglas McHenry international Symposium on Concrete and Concrete structures. ACI Special Publication No. 55; 1978. pp. 83—101.

[140] Brooks JJ, Forsyth P. Influence of frequency of cyclic load on creep of concrete. Mag Concr Res 1986;38(136):139—50.

[141] Brooks JJ, Forsyth P. Relaxation of concrete and its relation to creep under uniaxial compression at various frequencies. Mag Concr Res 1986;38(137):175—82.

[142] Bazant ZP, Panula l. Practical predictions of time-dependent deformations of concrete Part VI: cyclic creep, non-linearity and statistical scatter. Mater Struct 1979;12(69): 169—83. Paris.

11 Methods of Predicting Elasticity, Shrinkage, and Creep of Concrete

Since Chapter 4 has already dealt with methods of predicting elasticity as specified by current standards, they are only mentioned briefly in this chapter in connection with calculating compliance. Methods of predicting creep and shrinkage consist of a set of algebraic equations quantifying the influence of main factors affecting creep and shrinkage discussed in previous chapters. No testing is necessary, and the methods only require specified or characteristic or mean strength, mix composition, and operating conditions. In fact, there are several methods that have been developed over several years and amended at various stages to improve the accuracy of prediction and to account for the effect of recent test findings resulting from the use of new ingredients to make high strength concrete. Nevertheless, long-term experimental data used for development and verification of models have been determined only for normal strength concretes having traditional mix compositions, many of the data sets being collated in the International Union of Laboratories and Experts in Construction Materials, Systems and Structures (RILEM) data bank [1,2]. A comprehensive review of four such models and of statistical indicators of accuracy has been carried out by ACI 209.2R-08 [3]. Most of those methods are presented in this chapter, and their application is demonstrated with worked examples.

The accuracy of standard model predictions is at best around 25−30% and, at worst, much higher. According to ACI 209.2R-08 [3], the main failing of methods is failure to account directly for the influence of type of aggregate and in accounting for type of cement. Another problem is disagreement between researchers on the use of the RILEM test data bank for the source of experimental data to use as verification of models. In consequence, it is recommended that short-term testing be undertaken when a more accurate estimate of long-term creep and shrinkage are required, and when "unknown" aggregates, mineral, and chemical admixture are used in concrete. The second part of this chapter deals with that topic and presents relations between long-term and short-term shrinkage and creep, which may be used to extrapolate measured short-term test data.

Finally, a detailed case study is given that required accurate, ongoing shrinkage and creep data of concrete, which is used to manufacture cable-stayed bridge deck segments in order to calculate the necessary cable forces to be applied during construction, so as to achieve the specified profile of the main deck. All standard prediction methods and prediction from short-term tests are considered in the case study.

Concrete and Masonry Movements. http://dx.doi.org/10.1016/B978-0-12-801525-4.00011-X

Standard Methods of Prediction from Strength, Mix Composition, and Physical Conditions

In the methods presented in this section, creep is quantified by the creep coefficient, which is defined in two ways: First, the creep coefficient, $\phi_{28}(t, t_o)$, defined as the ratio of creep at any age t, after application of load at age t_o, to the elastic strain at the age of 28 days, is:

$$\phi_{28}(t, t_o) = C(t, t_o)E_{c28} \tag{11.1}$$

where $C(t, t_o)$ = specific creep and E_{c28} = secant modulus of elasticity at the age of 28 days.

Second, the creep coefficient $\phi(t, t_o)$, is defined as the ratio of creep at any age t, after application of load at age t_o, to the elastic strain at the same age at application of load, t_o, so that:

$$\phi(t, t_o) = C(t, t_o)E_{cto} \tag{11.2}$$

where E_{cto} = secant modulus of elasticity at age t_o.

Therefore, since creep is the same in both definitions, it follows from Eqs (11.1) and (11.2):

$$\phi(t, t_o) = \frac{E_{cto}}{E_{c28}}\phi_{28}(t, t_o) \tag{11.3}$$

Ross [4] was probably the first person to suggest a creep prediction chart in 1937. Standard curve methods were proposed by Wagner in 1958 [5] and by Jones in 1959 [6], standard values of creep at any time or ultimate creep being modified by factors allowing for cement type, ambient humidity, member size, and mix composition. In 1962, Ulitski [7] expressed shrinkage and creep coefficient for average conditions that were modified algebraically by correction factors for deviations from the averages of member size, relative humidity, and age at loading or age at which shrinkage is reckoned. In 1970, CEB-FIP [8] published their first method based on charts and equations, followed shortly by an ACI Committee 209 method that consisted of equations. Although the former has been developed at various stages into its current form, the ACI method has virtually remained unchanged and is still in current use [9]. In addition, there are two other methods in use that have been considered in the ACI 209.2R-08 review, and they are also included in this chapter: the Bazant—Baweja B3 model [10] and the Gardner—Lockman GL 2000 model [11].

BS 8110: Part 2: 1985

Before describing the four models currently available for estimating creep and shrinkage, it is worthwhile to mention the method of BS 8110: Part 2: 1985 [12], and although it has now been superseded by BS EN 1992-1-1:2004 [4], it is of interest historically because of its simplicity and ease of use. Following an earlier method [13], based on the CEB-FIP 1970 recommendations [8], the UK Concrete Society [14] proposed a simple method for estimating modulus of elasticity and ultimate creep. A modification to that method was made by Parrott [15] in 1979, and a method of estimating shrinkage was included. Values of shrinkage and swelling after periods of exposure of 6 months and 30 years for various relative humidities of storage and volume/surface ratios are given in Figure 11.1. The data apply to concretes made with high quality, dense, nonshrinking aggregates, and having an original water content of 8%, corresponding to approximately 190 L/m^3 of concrete. For concretes with other water contents, shrinkage values of Figure 11.1 are adjusted in proportion to the actual water content.

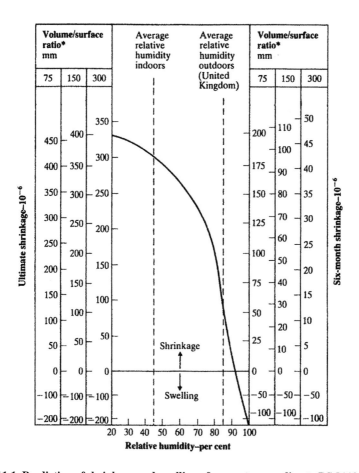

Figure 11.1 Prediction of shrinkage and swelling of concrete according to BS 8110: part 1: 1985 [12]; * volume/surface ratio = 0.5 × effective section thickness.
Source: Concrete Technology, Second Edition, A. M. Neville and J. J. Brooks, Pearson Education Ltd. © Longman Group UK Ltd. 1987.

The equivalent chart for estimating ultimate creep is shown in Figure 11.2. For concrete with an average, high-quality, dense aggregate, at any age t_o, the modulus of elasticity, $E_c(t_o)$, is related to the cube compressive strength, $f_{cu}(t_o)$, as follows:

$$E_c(t_o) = E_{c28}\left[0.4 + 0.6\frac{f_{cu}(t_o)}{f_{cu28}}\right] \qquad (11.4)$$

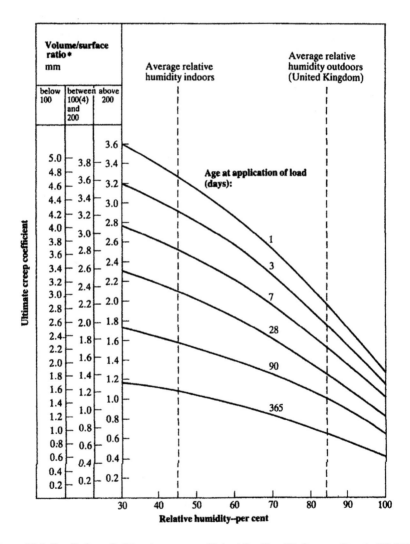

Figure 11.2 Prediction of ultimate creep coefficient for Eq. (11.6) according to BS 8110: part 2: 1985 [12]; * volume/surface ratio = 0.5 × effective section thickness.

Surce: Concrete Technology, Second Edition, A. M. Neville and J. J. Brooks, Pearson Education Ltd. © Longman Group UK Ltd. 1987.

The 28-day modulus of elasticity, E_{c28}, is obtained from the 28-day cube strength, f_{cu28}, by the following expression:

$$E_{c28} = 20 + 0.2f_{cu28} \tag{11.5}$$

The strength ratio term of Eq. (11.4) is best obtained by measurement, but if not possible, the following values may be used with interpolation at other ages:

Age, t_o, days	7	28	90	365
$f_c(t_o)/f_{c28}$	0.70	1.00	1.17	1.25

For lightweight aggregate concrete of density ρ, the modulus of elasticity of normal weight aggregate concrete should be multiplied by $(\rho/2400)^2$.

For a very long time under load, the ultimate creep function, J_∞, is given by:

$$J_\infty = \frac{1}{E_c(t_o)}(1 + \phi_\infty) \tag{11.6}$$

where ϕ_∞ = ultimate creep coefficient, which is obtained from Figure 11.2.

Given the ambient relative humidity, age at application of load and volume/surface ratio, the ultimate total creep function (compliance) can thus be calculated from Eq. (11.6). If there is no moisture exchange, i.e., the concrete is sealed or it is mass concrete, the basic creep coefficient is assumed to be equivalent to that of a concrete member with a volume/surface ratio greater than 200 mm at 100% RH.

For most current prediction models, the range of variables for which they are applicable is appreciably wider than for older methods, particularly for strength and type of cement; ranges are summarized in Table 11.1.

BS EN 1992-1-1: 2004

In 1970, CEB-FIP [8] published their method based on charts and tables, followed by the method of CEB-FIP [19] in 1978, in which creep was divided into irreversible creep (plastic flow) and reversible creep (delayed elastic strain). In addition, plastic flow was divided into a component representing flow for the first day under load (initial flow); coefficients were estimated by a combination of charts and algebraic equations. The division into creep components by the 1978 CEB-FIP method was abandoned in 1993, when it was replaced by CEB Model Code 90 [20], which was then further updated by CEB 99 [17] to include high strength concrete, autogenous shrinkage, and the effects of elevated temperature; that method forms the basis of the current method prescribed by BS EN 1992-1-1: 2004 [16], which now also includes lightweight aggregate concrete.

Table 11.1 Range of Variables for Current Prediction Models [3]

Input Parameter	BS EN 1992-1-1: 2004 [16][a]	ACI 209R-92 [9]	Bazant– Baweja B3 [10]	GL 2000 [11]
Mean 28-day cylinder compressive strength, f_{c28}, MPa	15–120	–	17–70	16–82
Age at loading, t_o, days	>1	≥7	≥t_c	≥ t_c ≥ 1
Period of moist curing, t_c, days	<14	≥1	≥1	≥1
Period of steam curing, t_c, days	–	1–3	–	–
Relative humidity of storage, %	40–100	40–100	40–100	20–100
Type of Portland cement	N, S, R I, II, III	I, III N, R	I, II, III N, S, R	I, II, III N, S, R
Cement content, kg/m^3	–	280–450	160–720	–
Water/cement ratio, w/c	–	–	0.35–0.85	0.40–0.60
Aggregate/cement ratio, a/c	–	–	2.5–13.5	–

For type of cement abbreviations refer to European and American specifications as follows:

Cement	BS EN 1992-1-1: 2004 [16]	ASTM C150-12 [18]
Normal hardening	N	Type I
Slow hardening	S	Type II
Rapid hardening	R	Type III

[a]Method is the same as CEB MC90-99 [17]

Drying Shrinkage

The total shrinkage (ε_{csh}) of normal weight aggregate concrete is comprised of two components: drying shrinkage (ε_{cd}) and autogenous shrinkage (ε_{ca})

$$\varepsilon_{csh} = \varepsilon_{cd} + \varepsilon_{ca} \tag{11.7}$$

Equations for estimating autogenous shrinkage are presented in Chapter 6 (p. 178). Ultimate drying shrinkage, $\varepsilon_{cd\infty}$, is given by:

$$\varepsilon_{cd\infty} = k_h \varepsilon_{cdo} \tag{11.8}$$

where k_h = coefficient depending on the notional size of cross section of the concrete member exposed to drying (see Table 11.2); ε_{cdo} may be calculated from Eq. (11.9) or

Table 11.2 Values of Coefficient k_n for Eq. (11.8)

Notional Size of Cross Section, h_o, mm	Coefficient, k_h
100	1.0
200	0.85
300	0.75
≥ 500	0.70

taken from Table 11.3, which are expected mean values with a coefficient of variation of about 30%.

$$\varepsilon_{cdo} = 0.85[(220 + 100\alpha_{ds1}) \cdot \exp(-0.1\alpha_{ds2}f_{c28})]10^{-6}\beta_{RH} \tag{11.9}$$

in which

$$\beta_{RH} = 1.55\left[1 - 10^{-6}(RH)^3\right] \tag{11.10}$$

where $f_{c28} = 28$-day mean compressive strength (MPa), α_{ds1} and $\alpha_{ds2} = $ coefficients according to type of cement (see Table 11.4), and $RH = $ ambient relative humidity (%).

In Table 11.2, the notional thickness of the concrete member, h_o, is defined as twice the area of cross section divided by the perimeter exposed to drying $= 2 \times$ volume/surface ratio (V/S).

The development of drying shrinkage with time is:

$$\varepsilon_{cd}(t) = \beta_{ds}(t, t_s) \cdot k_h \cdot \varepsilon_{cdo} \tag{11.11}$$

where coefficient $\beta_{ds}(t, t_s)$ is given by:

$$\beta_{ds}(t, t_s) = \frac{t - t_s}{(t - t_s) + 0.04\sqrt{h_o^3}} \tag{11.12}$$

Table 11.3 Nominal Drying Shrinkage, ε_{cdo}, for Concrete Made with Cement CEM Class N Having Different Characteristic Strengths [16]

Characteristic Compressive Strength, MPa		Nominal Drying Shrinkage, 10^{-6}, at Relative Humidity, $RH\%$					
Cylinder	Cube	20	40	60	80	90	100
20	25	620	580	490	300	170	0
40	50	480	460	380	240	130	0
60	75	380	360	300	190	100	0
80	95	300	280	240	150	80	0
90	105	270	250	210	130	70	0

Table 11.4 Coefficients for Eq. (11.9)

Class of Cement—BS EN 1992-1-1: 2004 [16]	Class of Cement—BS EN 197-1: 2011 [21]	Coefficient α_{ds1}	Coefficient α_{ds2}
S-slow hardening	CEM 32.5N	3	0.13
N-normal or rapid hardening	CEM 32.5R, CEM 42.5N	4	0.12
R-rapid hardening	CEM 42.5R, CEM 52.5N and CEM 52.5R	6	0.11

where t = age of concrete, days, and t_s = age at beginning of shrinkage or swelling (days).

BS EN 1992-1-1:2004 [16] does not account for the effect of elevated temperature on drying shrinkage, although CEB MC 90-99 [17] considers the temperature influence during drying, but not during the period of curing (see p. 157).

For drying shrinkage of lightweight aggregate concrete, values for normal weight aggregate concrete should be increased by the following factors:

- 1.5 when 28-day mean strength ≤ 22 MPa, characteristic cylinder strength ≤ 16 MPa, or characteristic cube strength ≤ 18 MPa.
- 1.2 when 28-day mean strength ≥ 28 MPa, characteristic cylinder strength ≥ 20 MPa, or characteristic cube strength ≥ 22 MPa.

Creep

Creep at any age (t) is expressed in terms of the creep coefficient defined in Eq. (11.1). In the case of BS EN 1992-1-1: 2004, the tangent modulus of elasticity at the age of 28 days is used to define the creep coefficient, $\phi_{28}(t, t_o)$ because specified strength classes of concrete are based on 28-day values. The tangent modulus of elasticity can be taken as $1.05 \times$ the secant modulus of elasticity (E_{c28}). When approximate values are sufficient, the ultimate creep coefficient may be estimated from Figure 11.3, provided the concrete is not subjected to a stress greater than $0.45 f_{ck}(t_o)$ at the age of loading t_o. The creep coefficient of concrete made with normal or slow hardening cement and stored at 20 °C may be calculated from:

$$\phi_{28}(t, t_o) = \phi_o \cdot \beta_c(t, t_o) \tag{11.13}$$

where ϕ_o = notional creep coefficient estimated from:

$$\phi_o = \phi_{RH} \cdot \beta(f_{c28}) \cdot \beta(t_o) \tag{11.14}$$

where ϕ_{RH} = factor to allow for relative humidity estimated from:

$$\phi_{RH} = 1 + \frac{1 - 0.01RH}{0.1 \cdot \sqrt[3]{h_o}} \quad \text{for } f_{c28} \leq 35 \text{ MPa} \tag{11.15a}$$

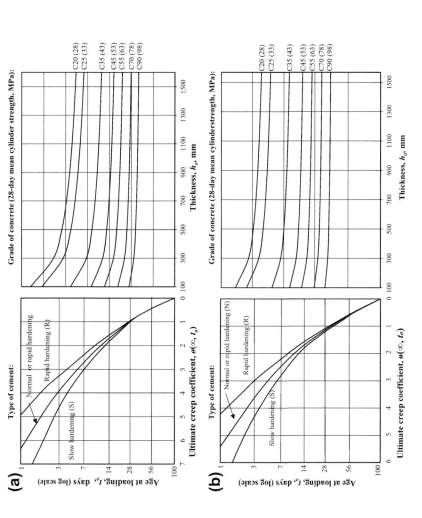

Figure 11.3 Graphical method for estimating 70-year (ultimate) creep coefficient of concrete [16]; grade of concrete is characteristic cylinder strength (See Table 4.1) (a) Relative humidity = 50% (indoors), (b) Relative humidity = 80% (outdoors), (c) Five-stage procedure for estimating 70-year creep coefficient from Figures 11.3 (a) and (b).

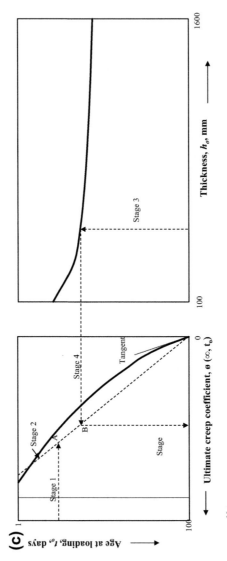

Figure 11.3 cont'd

$$\phi_{RH} = \left[1 + \frac{1 - 0.01RH}{0.1.\sqrt[3]{h_o}}\alpha_1\right]\alpha_2 \quad \text{for } f_{c28} > 35 \text{ MPa} \tag{11.15b}$$

$$\beta(f_{c28}) = \text{factor to allow for strength} = \frac{16.8}{\sqrt{f_{c28}}} \tag{11.16}$$

$$\beta(t_o) = \text{factor to allow for age at loading} = \frac{1}{0.1 + t_o^{0.2}} \tag{11.17}$$

$\beta_c(t, t_o) = $ coefficient to allow for development of creep with time, viz.

$$\beta_c(t, t_o) = \left[\frac{t - t_o}{\beta_H + (t - t_o)}\right]^{0.3} \tag{11.18}$$

where $t = $ age (days); $t_o = $ age at loading (days) and $\beta_H = $ coefficient depending on relative humidity and member size:

$$\beta_H = 1.5\left[1 + (0.012RH)^{18}\right]h_o + 250 \leq 1500 \quad \text{when } f_{c28} = 35 \text{ MPa} \tag{11.19a}$$

$$\beta_H = 1.5\left[1 + (0.012RH)^{18}\right]h_o + 250\alpha_3 \leq 1500\alpha_3 \quad \text{when } f_{c28} > 35 \text{ MPa} \tag{11.19b}$$

$\alpha_{1,2,3} = $ coefficients to allow for influence of concrete strength:

$$\alpha_1 = \left(\frac{35}{f_{c28}}\right)^{0.7} \quad \alpha_2 = \left(\frac{35}{f_{c28}}\right)^{0.2} \quad \alpha_3 = \left(\frac{35}{f_{c28}}\right)^{0.5} \tag{11.20}$$

When the cement used to make concrete differs from normal or slow hardening type, and the storage temperature differs from 20 °C, the age at loading, t_o, may be modified as follows:

$$t_o = t_{oT}\left(\frac{9}{2 + t_{oT}^{1.2}} + 1\right)^{\alpha} \geq 0.5 \text{ days} \tag{11.21}$$

where $t_{oT} = $ temperature adjusted age of concrete given by Eq. (11.22) and $\alpha = $ index depending on class of cement as defined in Table 11.4:

$\alpha = -1$ for cement Class S-slow hardening;
$\alpha = 0$ for cement Class N-normal or rapid hardening;
$\alpha = 1$ for cement Class R-rapid hardening.

The effect of elevated or reduced temperature within the range 0–80 °C during the curing period on creep of concrete is taken into account by adjusting the concrete age (see also p. 316) according to:

$$t_{oT} = \sum_{i=1}^{n} \Delta t_i \cdot \exp - \left(\frac{4000}{[273 + T_i]} - 13.65 \right) \tag{11.22}$$

where t_{oT} = temperature adjusted age that replaces t, T_i = temperature (°C) during time period Δt_i, in which the temperature T_i prevails and n = number of time intervals considered.

BS EN 1992-1-1: 2004 [16] does not consider temperature effects during the creep process, although the CEB MC 90-99 [17] specifies relevant expressions as detailed in Chapter 10.

According to BS EN 1992-1-1 [16], the mean coefficient of variation of $\phi_{28}(t, t_o)$ deduced from the RILEM data bank is of the order of 20%. When a less accurate estimate is considered satisfactory, 70-year values of creep coefficient may be estimated from Figure 11.3 where values are valid for temperatures between -40 °C and $+40$ °C, and a mean relative humidity between $RH = 40\%$ and 100%; the procedure is illustrated in Figure 11.3(c).

For lightweight aggregate concrete, the creep coefficient $\phi_{28}(t, t_o)$ may be assumed equal to the value for normal weight aggregate concrete multiplied by $(\rho/2200)^2$, where ρ = density of lightweight aggregate concrete (kg/m^3).

For the BS EN 1992-1-1: 2004 method, specific creep of normal weight aggregate concrete is given by:

$$C(t, t_o) = \frac{\phi(t, t_o)}{1.05 E_{c28}} \tag{11.23}$$

where E_{c28} = initial tangent modulus of elasticity at the age of 28 days.

Corresponding specific creep of lightweight aggregate concrete is obtained by multiplying normal weight aggregate specific creep by the following factors:

- 1.3 when 28-day mean strength $\leq$22 MPa, characteristic cylinder strength $\leq$16 MPa, or characteristic cube strength $\leq$18 MPa.
- 1.0 when 28-day mean strength $\geq$28 MPa, characteristic cylinder strength $\geq$20 MPa, or characteristic cube strength $\geq$22 MPa.

Compliance

The compliance of concrete, $J(t, t_o)$, represents the total time-dependent strain (elastic strain + creep) at age t under a unit stress applied at age t_o, and is given by:

$$J(t, t_o) = \frac{1}{E_{cto}} + \frac{\phi(t, t_o)}{E_{c28}} \tag{11.24}$$

where E_{cto} = secant modulus of elasticity at age of loading, t_o, and is related to E_{c28} as follows (see also Eq. (4.14)):

$$E_{cto} = \exp\left[S\left(1 - \left(\frac{28}{t}\right)^{0.5}\right)\right]E_{c28} \qquad (11.25)$$

where S = coefficient to allow for type of cement (see also Table 11.3), viz:

$S = 0.20$ for cement Class R—rapid hardening.
$S = 0.25$ for cement of Class N—normal or rapid hardening.
$S = 0.38$ for cement of strength Class S—slow hardening.

The relationship between secant modulus of elasticity and compressive strength is:

$$E_{c28} = 11.03(f_{c28})^{0.3} \qquad (11.26)$$

where f_{c28} = 28-day mean cylinder compressive strength (MPa).

Relationships for modulus of elasticity in terms of characteristic strengths are given in Table 4.1. As stated in Chapter 4, the secant modulus of elasticity applies to normal weight aggregate concretes made with quartzite aggregates. For concrete made with limestone and sandstone aggregates, the modulus of elasticity should be reduced by 10% and 30%, respectively. On the other hand, for basalt aggregates, the value should be increased by 20%.

The estimation of modulus of elasticity applies to concrete cured and stored at a temperature of 20 °C, and there is no provision in BS EN 1992-1-1:2004 [16] for other temperatures. However, CEB MC 90-99 [17] specifies an expression that allows for different temperatures, and this is given in Chapter 4 (Eq. (4.4)).

For lightweight aggregate concrete, the secant modulus of elasticity may be assumed equal to the value for normal weight aggregate concrete multiplied by $(\rho/2200)^2$, where ρ = density of lightweight aggregate concrete between 800 and 2200 kg/m³.

It should be noted that this method does not apply to aerated concrete either autoclaved or normally cured, or lightweight aggregate concrete with an open structure. When more accurate data are need, e.g., where deflections are of importance, tests should be carried out to determine the modulus of elasticity of lightweight aggregate concrete [16].

Total Strain

The sum of compliance times applied stress, σ, plus drying shrinkage plus autogenous shrinkage, measured from the age at loading, is equal to the total time-dependent strain, $\varepsilon_{Total}(t, t_o)$, or:

$$\varepsilon_{Total}(t, t_o) = \sigma \cdot J(t, t_o) + \varepsilon_{cd}(t, t_o) + \varepsilon_{ca}(t, t_o) \qquad (11.27)$$

where σ = applied stress at age, t_o; $J(t, t_o)$ = compliance (Eq. (11.24)); $\varepsilon_{cd}(t, t_o)$ = drying shrinkage from age t_o; and $\varepsilon_{ca}(t, t_o)$ = autogenous shrinkage from age t_o.

Example

It is required to estimate total contraction from the age at loading of 3-m long concrete elements having a volume/surface area ratio of 75 mm, which, after moist curing, are to be stored in air at a temperature of 20 °C, relative humidity of 65%, and subjected to a sustained stress of 10 MPa applied at the age of 14 days for 1 year. Two types of concrete made with Class N (Type I) cement are under consideration, having mean 28-day strengths of (a) 30 MPa and (b) 60 MPa. Details of the composition of the concretes are:

1. Mix proportions, 1:2.28:3.00, $w/c = 0.60$ by mass; air content = 5.0% by volume; cement content = 338 kg/m³; water content = 203 kg/m³; slump = 75 mm; density = 2325 kg/m³.
2. Mix proportions, 1:1.80:2.20, $w/c = 0.50$ by mass; air content = 2.0% by volume; cement content = 425 kg/m³; water content = 213 kg/m³; slump = 50 mm; density = 2340 kg/m³.

Solution
The coefficients of the deformation equations required for the solution are compiled in Table 11.5.

Alternatively, the ultimate creep coefficient may be estimated using Figure 11.3, which yields approximate values as follows:

- when $RH = 50\%$, (a) $\phi_\infty = 3.4$, (b) $\phi_\infty = 2.0$.
- when $RH = 80\%$, (a) $\phi_\infty = 3.0$, (b) $\phi_\infty = 1.8$.

Therefore, taking average values to represent $RH = 65\%$ yields (a) $\phi_\infty = 3.2$, (b) $\phi_\infty = 1.9$, which are slightly greater than calculated values in Table 11.5.

ACI 209R-92

The basis of this model was developed by Meyers et al. [22] and by Branson and Christiason [23] in the 1970s. The latest model [9], published in 1992, follows earlier versions published in 1978 and 1982.

Shrinkage

ACI 209R-92 does not distinguish between autogenous shrinkage and drying shrinkage, so that the method predicts total shrinkage, which, in this section, is simply

Table 11.5 Solution by the Method of BS EN 1992-1-1:2004 [16]

Required Input Data	Coefficient/Component (Location in Text)
Drying Shrinkage, ε_{cd}	
f_{c28}: (a) 30 MPa, (b) 60 MPa.	$k_h = 0.925$ (Table 11.2)
$RH = 65\%$.	$\alpha_{ds1} = 4$; $\alpha_{ds2} = 0.12$ (Table 11.5)
$h_o = 2\ V/S = 150$ mm.	$\beta_{RH} = 1.124$ (Eq. (11.10))
Cement class N.	ε_{cdo}: (a) 413×10^{-6}; (b) 288×10^{-6} (Eq. (11.9))
$t_o = t_s = 14$ days; $t = 365$ days.	$\varepsilon_{cd\infty}$: (a) 382×10^{-6}; (b) 267×10^{-6} (Eq. (11.8))
	$\beta_{ds}(t, t_o) = 0.83$ (Eq. (11.12))
	$\varepsilon_{cd}(t, t_o)$: (a) 316×10^{-6}, (b) 222×10^{-6} (Eq. (11.11))
Autogenous Shrinkage, ε_{ca}	
f_{c28}: (a) 30 MPa, (b) 60 MPa.	$\varepsilon_{ca}(t) = S_c(t)$: (a) 29×10^{-6}, (b) 103×10^{-6} (Eq. (6.30))
$t_o = 14$ days, $t = 365$ days.	$\varepsilon_{ca}(t_o) = S_c(t_o)$: (a) 16×10^{-6}, (b) 55×10^{-6} (Eq. (6.30))
	$\varepsilon_{ca}(t, t_o) = \varepsilon_{ca}(t) - \varepsilon_{ca}(t_o)$: (a) 13×10^{-6}, (b) 48×10^{-6}
Total Shrinkage, $\varepsilon_{cs}(t, t_o)$	
$\varepsilon_{ca}(t, t_o)$, $\varepsilon_{cd}(t, t_o)$.	$\varepsilon_{cs}(t, t_o)$: (a) 329×10^{-6}, (b) 270×10^{-6}
Creep Coefficient, $\phi_{28}(t, t_o)$	
f_{c28}: (a) 30 MPa, (b) 60 MPa.	$\alpha_1 = 0.686$, $\alpha_2 = 0.898$ (Eq. (11.20))
$RH = 65\%$.	ϕ_{RH}: (a) 1.659, (Eq. (11.15a)) (b) 1.304 (Eq. (11.15b))
$h_o = 2\ V/S = 150$ mm.	$\beta(f_{c28})$: (a) 3.067, (b) 2.169 (Eq. (11.16))
Cement class N.	$\beta(t_o) = 0.557$ (Eq. (11.18))
$t_o = 14$ days; $t = 365$ days.	$\alpha_3 = 0.764$ (Eq. (11.17))
$T = 20\ ^\circ$C.	β_H: (a) 477.6, Eq. (11.19a) (b) 418.6 Eq. (11.19b)
	$\beta_c(t, t_o)$: (a) 778, (b) 0.795 (Eq. (11.18))
	ϕ_∞: (a) 2.834, (b) 1.575 (Eq. (11.14))
	$\phi_{28}(t, t_o)$: (a) 2.205, (b) 1.252 (Eq. (11.13))
Compliance (Specific Elastic Strain + Specific Creep), $J(t, t_o)$	
f_{c28}: (a) 30 MPa; (b) 60 MPa.	E_{c28}: (a) 30.60 GPa, (b) 37.67 GPa (Eq. (11.26))
Cement class N.	$S = 0.25$ (Eq. (11.25))
	E_{cto}: (a) 27.59 GPa, (b) 33.96 GPa (Eq. (11.25))
	$J(t, t_o)$: (a) 108.3×10^{-6} per MPa, (b) 62.7×10^{-6} per MPa (Eq. (11.24))
Total Strain (Compliance + Shrinkage), $\varepsilon_{Total}(t, t_o)$	
f_{c28}: (a) 30 MPa, (b) 60 MPa.	(a) $(10 \times 108.3 \times 10^{-6}) + (329 \times 10^{-6}) = 1412 \times 10^{-6}$;
$\sigma = 10$ MPa.	(b) $(10 \times 62.7 \times 10^{-6}) + (252 \times 10^{-6}) = 897 \times 10^{-6}$ (Eq. (11.27))
Total Contraction (Total Strain × Length)	
f_{c28}: (a) 30 MPa, (b) 60 MPa.	(a) $1412 \times 10^{-6} \times 3 \times 10^3 = 4.2$ mm;
Length of member $= 3$ m.	(b) $897 \times 10^{-6} \times 3 \times 10^3 = 2.7$ mm

referred to as shrinkage. Shrinkage, $\varepsilon_{sh}(t, t_c)$, at any time, t, measured from the end of moist curing t_c, is expressed as follows:

$$\varepsilon_{sh}(t, t_c) = \frac{(t - t_c)}{35 + (t - t_c)} \varepsilon_{sh\infty} \tag{11.28a}$$

or from the end of steam curing for 1–3 days:

$$\varepsilon_{sh}(t, t_c) = \frac{(t - t_c)}{55 + (t - t_c)} \varepsilon_{sh\infty} \tag{11.28b}$$

In Eq. (11.28a)) or Eq. (11.28b) the ultimate shrinkage is:

$$\varepsilon_{sh\infty} = = 780 \times 10^{-6} \cdot \gamma_s \tag{11.29}$$

where 780×10^{-6} = ultimate average shrinkage for standard conditions, which is assumed in the absence of specific shrinkage data for local aggregates and conditions at a relative humidity of 40%, and γ_s = product of seven coefficients to allow for influencing factors, viz.

$$\gamma_s = \gamma_{s,tc}\gamma_{s,RH}\gamma_{s,VS}\gamma_{s,s}\gamma_{s,\psi}\gamma_{s,c}\gamma_{s,A} \tag{11.30}$$

For curing periods different from 7 days for moist-cured concrete, coefficient $\gamma_{s,\,tc}$ is given in Table 11.6; for steam-cured concrete, $\gamma_{s,\,tc} = 1$.

The humidity coefficient, $\gamma_{s,\,RH}$, is:

$$\left.\begin{array}{l}\gamma_{S,RH} = 1.40 - 0.010h(40 \leq h \leq 80) \\ \gamma_{S,RH} = 3.00 - 0.030h(80 \leq h \leq 100)\end{array}\right\} \tag{11.31}$$

where h = relative humidity, %.

For relative humidity, h, less than 40%, $\gamma_{s,\,RH} > 1$ should be used, and when $h = 100\%$, there is no swelling [3].

Two methods are recommended for determining the member size coefficient, $\gamma_{s,\,vs}$:

Table 11.6 Curing Period Coefficient for Eq. (11.30)

Period of Moist Curing, t_c, days	Shrinkage Coefficient, $\gamma_{s,\,tc}$
1	1.2
3	1.1
7	1.0
14	0.93
28	0.86
90	0.75

Table 11.7 Shrinkage and Creep Factors for
Member Size when $d < 150$ mm [9]

Member Average Thickness, d, mm	Shrinkage Factor $\gamma_{s, vs}$	Creep factor, $\gamma_{c, vs}$
50	1.35	1.30
75	1.25	1.17
100	1.17	1.11
125	1.08	1.04
150	1.00	1.00

1. Average thickness, d ($=4 \times V/S$)
   For an average thickness less than 150 mm, $\gamma_{s, vs}$ is given in Table 11.7. For an average thickness between 150 and 380 mm:

$$\left.\begin{array}{l} \gamma_{s,vs} = 1.23 - 0.0015d \quad \text{for } (t - t_c) \leq 1 \text{ year} \\ \gamma_{s,vs} = 1.17 - 0.00114d \quad \text{for } (t - t_c) > 1 \text{ year} \end{array}\right\} \tag{11.32}$$

2. Volume/surface ratio, V/S

$$\gamma_{s,vs} = 1.2[\exp(-0.00473 \, V/S)] \tag{11.33}$$

This latter method (2) yields greater coefficients than the average thickness method (1). For either method (1) or (2), the product of all coefficients, γ_s (Eq. (11.30)), should not be less than 0.2. If concrete is subjected to seasonal wetting and drying cycles, use $\gamma_s \cdot \varepsilon_{sh\infty} \geq 100 \times 10^{-6}$, and if concrete is stored under sustained drying conditions, use $\gamma_s \cdot \varepsilon_{sh\infty} \geq 150 \times 10^{-6}$.

The coefficients that allow for composition of the concrete are:

$$\gamma_{s,s} = 0.89 + 0.00161s \tag{11.34}$$

where $\gamma_{s, s}$ = slump coefficient and s = slump of fresh concrete (mm).

$$\left.\begin{array}{l} \gamma_{s,\psi} = 0.30 + 0.014\psi \quad \text{for } \psi = \leq 50\% \\ \gamma_{s,\psi} = 0.90 + 0.002\psi \quad \text{for } \psi > 50\% \end{array}\right\} \tag{11.35}$$

where $\gamma_{s, \psi}$ = fine aggregate coefficient and ψ = fine aggregate/total aggregate ratio by weight (%)

$$\gamma_{s,c} = 0.75 + 0.00061c \tag{11.36}$$

where $\gamma_{s, c}$ = cement content coefficient and c = cement content (kg/m^3 of concrete).

$$\gamma_{s,A} = 0.95 + 0.008A \tag{11.37}$$

where $\gamma_{s, A}$ = air content coefficient and A = air content (%)

Creep

The creep coefficient is that defined in Eq. (11.2) and is expressed as follows:

$$\phi(t, t_o) = \frac{(t - t_o)^{0.6}}{10 + (t - t_o)^{0.6}} \phi_\infty(t_0) \tag{11.38}$$

where $t =$ age of concrete; $t_o =$ age at application of load; $(t-t_o) =$ time since application of load and $\phi_\infty(t_0) =$ ultimate creep coefficient, which is given by $2.35 \times \gamma_c$, where $\gamma_c =$ product of six coefficients representing the influencing factors in creep:

$$\gamma_c = \gamma_{c,to}\gamma_{c,RH}\gamma_{c,vs}\gamma_{c,s}\gamma_{c,\psi}\gamma_{c,A} \tag{11.39}$$

and

$$\phi_\infty(t_o) = 2.35\gamma_{c,to}\gamma_{c,RH}\gamma_{c,vs}\gamma_{c,s}\gamma_{c,\psi}\gamma_{c,A} \tag{11.40}$$

The value of 2.35 in Eq. (11.40) represents the average ultimate creep coefficient that may be assumed in the absence of specific data on creep of concrete made with local aggregates and operating conditions.

For age at application of load greater than 7 days for moist curing, or greater than 1–3 days for steam curing, the coefficient for age at application of load, $\gamma_{c,\,to}$, is estimated from:

$$\begin{aligned}\gamma_{c,to} &= 1.25t_o^{-0.118} \quad \text{(moist curing)}\\ \gamma_{c,to} &= 1.13t_o^{-0.094} \quad \text{(steam curing)}\end{aligned} \Biggr\} \tag{11.41}$$

The humidity coefficient, $\gamma_{c,\,RH}$, is:

$$\gamma_{c,RH} = 1.27 - 0.0067h \quad \text{for } h \geq 40\% \tag{11.42}$$

For relative humidity, h, less than 40%, $\gamma_{c,\,RH} > 1$.

Two methods are recommended for determining member thickness coefficient, $\gamma_{c,\,vs}$:

1. Average thickness, $d\ (=4\ V/S)$
   For average thickness, d, less than 150 mm, $\gamma_{c,\,vs}$ is given in Table 11.7.
   For average thickness between 150 and 380 mm:

$$\begin{aligned}\gamma_{c,vs} &= 1.14 - 0.00092d \quad \text{for } (t - t_o) \leq 1 \text{ year}\\ \gamma_{c,vs} &= 1.10 - 0.00067d \quad \text{for } (t - t_o) > 1 \text{ year}\end{aligned} \Biggr\} \tag{11.43}$$

2. Volume/surface ratio, V/S

When $d \geq 380$ mm or $V/S \geq 95$ mm:

$$\gamma_{c,vs} = \frac{2}{3}[1 + 1.13 \exp(-0.0213(V/S))] \tag{11.44}$$

The coefficients that allow for composition of the concrete are:

$$\gamma_{c,s} = 0.82 + 0.00264s \tag{11.45}$$

where $\gamma_{c,s} = $ slump coefficient and $s = $ slump of fresh concrete (mm).

$$\gamma_{c,\psi} = 0.88 + 0.0024\psi \tag{11.46}$$

where $\gamma_{c,\psi} = $ fine aggregate coefficient and $\psi = $ fine aggregate/total aggregate ratio by weight (%).

$$\gamma_{c,A} = 0.46 + 0.09A \geq 1 \tag{11.47}$$

where $\gamma_{c,A} = $ cement content coefficient and $A = $ air content (%).

Compliance

In the method of ACI 209R-92, the compliance of concrete, $J(t, t_o)$, represents the total time-dependent strain at age t under a unit stress applied at age t_o, and is given by:

$$J(t, t_o) = \frac{1}{E_{cto}}[1 + \phi(t, t_o)] \tag{11.48}$$

where $E_{cto} = $ secant modulus of elasticity at age of loading, t_o, and is related to strength at the age of loading, f_{cto}, as follows (see Eq. (4.14)):

$$E_{cto} = 42.8 \times 10^{-6} \rho^{1.5} \sqrt{f_{cto}} \tag{11.49}$$

and

$$f_{cto} = \frac{t_o}{A + Bt_o} f_{c28} \tag{11.50}$$

where E_c is in GPa, f_c is in MPa, $\rho = $ density of concrete (kg/m^3), and A and B depend on type of cement and curing conditions (see Table 11.8). It may be of interest to note that $f_{c28}/B = $ ultimate compressive strength, $f_{c\infty}$, and $A/B = $ age of concrete (days) when strength is $0.5 f_{c\infty}$.

Table 11.8 Values of Constants A and B for Use in Eq. (11.50)

		constants of Eq. (11.50)	
Type of Cement	**Curing Condition**	**A**	**B**
I	Moist	4.00	0.85
	Steam	1.00	0.95
III	Moist	2.30	0.92
	Steam	0.70	0.98

It may be noted that the required 28-day mean concrete compressive strength, f_{c28}, is required to exceed the specified strength, f'_c, as stipulated by ACI Committee 318R-05 [24].

Total Strain

The expression for total strain, $\varepsilon_{Total}(t, t_o)$, is the same as that for the BS EN 1992-1-1: 2004 method and is given by Eq. (11.24), except that there is no separate contribution from autogenous shrinkage, i.e.,

$$\varepsilon_{Total}(t, t_o) = \sigma \cdot J(t, t_o) + \varepsilon_{sh}(t, t_o) \tag{11.51}$$

where σ = applied stress at the age of loading, t_o, $J(t, t_o)$ = compliance (Eq. (11.45)) and $\varepsilon_{sh}(t, t_o)$ = shrinkage measured from the age at loading.

Example

Using the ACI 209R-92 method, it is required to estimate the total contraction of concrete member given in the example on p. 362. The solution is detailed in Table 11.9.

Bazant—Baweja B3

Following earlier versions [25,26], a simplified model was proposed for estimating creep and shrinkage of concrete in 1978 and 1979 [27], which was extended to include high strength concrete in 1984 [28], followed by further improvements in 1991 [29]. The current model, B3 was published in 1995 [10] and 2000 [30], with a short form of model B3 published in 1996 [31] and 2000 [30]. Details of the scope of application of the full B3 model and its restrictions are discussed in the review by ACI 209.2R-08 [3]. The version presented below is the short form of model B3 [31,32].

Table 11.9 Solution by the Method of ACI 209R-92 [9]

Required Input Data	Coefficient/Component (Location in text)

Shrinkage, ε_{sh}, (t, t_c)

$t_c = t_o = 14$ days, $h = 65\%$.
$\quad d = 4V/S = 300$ mm, $t - t_c \leq 1$ year,
$\quad s$: (a) 75 mm; (b) 50 mm.
$\quad \psi$: (a) $2.28/5.28 \times 100 = 43.2\%$,
$\quad$ (b) $1.8/4.0 \times 100 = 45.0\%$.
$\quad c$: (a) 338 kg/m^3, (b) 425 kg/m^3.
$\quad A$: (a) 5%, (b) 2%.
$\quad t - t_c = 351$ days.

$\gamma_{s,\,tc} = 0.93$ (Table 11.6)
$\gamma_{s,\,RH} = 0.75$ (Eq. (11.31))
$\gamma_{s,\,vs} = 0.78$ (Eq. (11.32))
$\gamma_{s,\,s}$: (a) 1.011, (b) 0.971 (Eq. (11.34))
$\gamma_{s,\,\psi}$: (a) 0.905, (b) 0.930 (Eq. (11.36))
$\gamma_{s,\,c}$: (a) 0.956, (b) 1.009 (Eq. (11.36))
$\gamma_{s,\,A}$: (a) 0.990, (b) 0.966 (Eq. (11.37))
γ_s: (a) 0.471, (b) 0.479 (Eq. (11.30))
$\varepsilon_{sh\infty}$: (a) 367.4×10^{-6}, (b) 373.6×10^{-6}
$\quad$ (Eq. (11.29))
$\varepsilon_{sh}(t, t_c)$: (a) 334×10^{-6}, (b) 340×10^{-6} (Eq. (11.28))

Creep Coefficient, $\phi(t, t_o)$

$t_o = 14$ days, moist curing.
$\quad h = 65\%$.
$\quad d = 4\,V/S = 300$ mm.
$\quad s$: (a) 75 mm, (b) 50 mm.
$\quad \psi$: (a) 43.2%, (b) 45.0%.
$\quad A$: (a) 5%, (b) 2%.
$\quad t - t_o = 351$ days.

$\gamma_{c,\,to} = 0.916$ (Eq. (11.41))
$\gamma_{c,\,RH} = 0.835$ (Eq. (11.42))
$\gamma_{c,\,vs} = 0.864$ (Eq. (11.43))
$\gamma_{c,\,s}$: (a) 1.018, (b) 0.952 (Eq. (11.45))
$\gamma_{c,\,\psi}$: (a) 0.983, (b) 0.988 (Eq. (11.46))
$\gamma_{c,\,A}$: (a) 0.910, (b) 0.64; USE $\gamma_{c,\,A} = 1$ (Eq. (11.47))
γ_c: (a) 0.687, (b) 0.622 (Eq. (11.39))
ϕ_∞: (a) 1.614, (b) 1.462 (Eq. (11.40))
$\phi(t, t_o)$: (a) 1.244, (b) 1.127 (Eq. (11.38))

Compliance (Specific Elastic Strain + Specific Creep), $J(t, t_o)$

f_{c28}: (a) 30 MPa, (b) 60 MPa.
$\quad$ Moist curing.
$\quad$ Type I cement, $t_o = 14$ days.
$\quad \rho$: (a) 2325 kg/m^3, (b) 2340 kg/m^3

$A = 4.00$, $B = 0.80$ (Table 11.8)
f_{c14}: (a) 26.42 MPa, (b) 52.83 MPa (Eq. (11.50))
E_{c14}: (a) 24.66 GPa, (b) 35.21 GPa (Eq. (11.52))
$J(t, t_o)$: (a) 90.85×10^{-6} per MPa, (b) 60.41×10^{-6}
$\quad$ per MPa (Eq. (11.48))

Total Strain (Compliance + Shrinkage), $\varepsilon_{Total}(t, t_o)$

f_{c28}: (a) 30 MPa, (b) 60 MPa.
$\quad \sigma = 10$ MPa

(a) $10 \times 89.09 \times 10^{-6} + 334 \times 10^{-6} = 1243 \times 10^{-6}$;
(b) $10 \times 60.4 \times 10^{-6} + 340 \times 10^{-6} = 944 \times 10^{-6}$
$\quad$ (Eq. (11.51))

Total Contraction (Total Strain × Length)

f_{c28}: (a) 30 MPa, (b) 60 MPa.

(a) $1243 \times 10^{-6} \times 3 \times 1000 = 3.7$ mm
(b) $944 \times 10^{-6} \times 3 \times 1000 = 2.8$ mm

Shrinkage

The mean shrinkage strain, $\varepsilon_{sh}(t, t_c)$, in the cross section at age t, measured from the start of drying t_c, is obtained as follows:

$$\varepsilon_{sh}(t, t_c) = \varepsilon_{sh\infty} k_h S(t) \tag{11.52}$$

Table 11.10 Relative Humidity Coefficient, k_h, of Eq. (11.52)

Relative Humidity, h	Coefficient, k_h
$h < 0.98$	$1 - h^3$
$h = 1$ (swelling)	-0.20
$0.98 \leq h \leq 1$	Linear interpolation

where $\varepsilon_{sh\infty}$ is the ultimate shrinkage, k_h is the humidity dependence coefficient (Table 11.10), and $S(t)$ is the time-dependence coefficient.

The ultimate shrinkage (10^{-6}) is estimated from coefficients to allow for type of cement (α_1) and curing condition (α_2), as well as being dependent on water content $(w, \text{kg/m}^3)$ and 28-day cylinder strength (f_{c28}, MPa), viz.

$$\varepsilon_{sh\infty} = \alpha_1\alpha_2\left[0.019w^{2.1}f_{c28}^{-0.28} + 270\right] \tag{11.53}$$

where α_1 and α_2 are given in Tables 11.11 and 11.12, respectively.

The time-dependence coefficient of Eq. (11.49) is given as follows:

$$S(t) = \tanh\sqrt{\frac{(t - t_c)}{\tau_{sh}}} \tag{11.54}$$

where τ_{sh} = shrinkage half time (time in days to reach half the ultimate shrinkage), which is dependent on member size:

$$\tau_{sh} = 0.196\left(\frac{V}{S}\right)^2 \tag{11.55}$$

where V/S = volume/surface area ratio (mm).

Compared with the full B3 model, the short form does not include the influence of curing condition and specimen size on ultimate shrinkage.

Creep

This method quantifies the effects of influencing factors on specific creep rather than creep coefficient, but specific creep is measured from an elastic strain as determined by a theoretical modulus of elasticity instead of the usual secant modulus of elasticity. The *theoretical or asymptotic modulus of elasticity* is given by the strain resulting from a

Table 11.11 Cement Type Coefficient, α_1, of Eq. (11.50)

Type of Cement	Coefficient α_1
Type I	1.00
Type II	0.85
Type III	1.10

Table 11.12 Curing Condition Coefficient, α_2, of Eq. (11.53)

Type of Curing	Coefficient α_2
Steam	0.75
Water or 100% RH	1.00
Sealed	1.20

unit stress in a creep test of very short duration of 10^{-9} s [3]. The authors claim that the procedure reduces error in the measurement creep due to inaccurate values of secant or static elastic modulus, which are used to isolate elastic strain at starting point for creep. As discussed in Chapter 4, the secant modulus of elasticity is time dependent so that the time taken to apply the load at the start of the creep test is important. Although values of specific creep are different, the definition of compliance, creep function, or total load strain is identical to that given by other methods of prediction.

In the B3 model, basic creep and drying creep are considered separately. Specific basic creep, $C_o(t, t_o)$, is given by:

$$C_o(t, t_o) = q_o \ln\left\{1 + 0.3\left[t_o^{-0.5} + 0.001\right](t - t_o)^{0.1}\right\} \tag{11.56}$$

where q_o is a function of 28-day cylinder strength, f_{c28}:

$$q_o = 2408 f_{c28}^{-0.5} \tag{11.57}$$

Specific drying creep, $C_d(t, t_o, t_c)$, is related to drying shrinkage as follows:

$$C_d(t, t_o, t_c) = q_5[\exp(-3H(t)) - \exp(-3H(t_o))]^{0.5} \quad \text{for } t_o \geq t_c \tag{11.58}$$

where $H(t)$ is a function of relative humidity h (expressed as a decimal):

$$H(t) = 1 - (1 - h)S(t) \tag{11.59}$$

and q_5 is a function of 28-day strength, f_{c28}:

$$q_5 = 6000 f_{c28}^{-1} \tag{11.60}$$

In the foregoing expressions, the units of specific creep, $C_o(t, t_o)$ and $C_d(t, t_o, t_c)$, are 10^{-6} per MPa.

Compliance

The average compliance function, $J(t, t_o)$, at age t caused by a unit uniaxial stress applied at age t_o is the sum of asymptotic elastic strain plus basic creep plus drying creep, i.e.,

$$J(t, t_o) = q_1 + C_o(t, t_o) + C_d\left(t, t_{o,t_c}\right) \tag{11.61}$$

where $q_1 =$ unit strain given by the inverse of the asymptotic modulus of elasticity (E_o), $C_o(t,\ t_o) =$ specific basic creep given by Eq. (11.56) and $C_d(t, t_{o,t_c}) =$ specific drying creep given by Eq. (11.58).

It may be noted that E_o is independent of age and greater than the conventional secant modulus of elasticity normally used to define the start of creep; q_1 is related to the 28-day modulus of elasticity, E_{c28}, and strength, f_{c28}, as follows:

$$q_1 = \frac{0.6}{10^3 E_{c28}} \tag{11.62}$$

and

$$E_{c28} = 4.734\sqrt{f_{c28}} \tag{11.63}$$

In the above expressions, the units of q_1 are 10^{-6} per MPa, with E_{c28} in GPa and f_{c28} in MPa.

Total Strain

When shrinkage is measured together with creep from the age at loading, total strain, $\varepsilon_{Total}(t,t_o)$, is the same as that for the ACI 209R-92 method as given by Eq. (11.54), i.e.,

$$\varepsilon_{Total}(t, t_o) = \sigma \cdot J(t, t_o) + \varepsilon_{sh}(t, t_o) \tag{11.64}$$

where $\sigma =$ applied stress at the age of loading, t_o; $J(t, t_o) =$ compliance (Eq. (11.61)); and $\varepsilon_{sh}(t, t_o) =$ shrinkage measured from the age at loading.

Example

The example given on p. 362 is now used to illustrate the application of the Bazant–Baweja B3 model to estimate the various types of movement leading to the total contraction of the concrete member after 1 year. The solution is presented in Table 11.13.

Gardner and Lockman GL2000

Gardner and Zhao [32] first introduced this model in 1993, which was then modified in 2000 [33] to conform with guidelines prescribed by ACI Committee 209 for evaluating creep and shrinkage models [3]. After further modification [11], the current GL 2000 model was published by Gardner and Lockman in 2001 [34] and modified by Gardner in 2004 [35]. ACI 209.2R−08 [3] points out that, except for compressive

Table 11.13 Solution to Example Using the Bazant−Baweja B3 Model

Required Input Data	Coefficient/Component (Location in Text)
Shrinkage, ε_{sh} (t, t_c)	
f_{c28}: (a) 30 MPa, (b) 60 MPa.	$k_h = 0.725$ (Table 11.10)
$h = 0.65$.	$\alpha_1 = 1.00$ (Table 11.11)
Type I cement.	$\alpha_2 = 1.00$ (Table 11.12)
Moist curing.	$\varepsilon_{sh\infty}$: (a) 784×10^{-6}, (b) 738×10^{-6} (Eq. (11.53))
w: (a) 203 kg/m³, (b) 255 kg/m³.	$\tau_{sh} = 1103$ days (Eq. (11.54))
$V/S = 75$ mm.	$S(t) = 0.511$ (Eq. (11.51))
$t = 365$ days, $t_c = t_o = 14$ days.	$\varepsilon_{sh}(t, t_c)$: (a) 290×10^{-6}, (b) 273×10^{-6} (Eq. (11.51))
Basic Creep, $C_o(t, t_o)$	
f_{c28}: (a) 30 MPa, (b) 60 MPa.	q_o: (a) 482.3, (b) 310.9 (Eq. (11.57))
$t_o = 14$ days, $t - t_o = 351$ days.	$C_o(t, t_o)$: (a) 65.1×10^{-6} per MPa, (b) 42.0×10^{-6} per MPa (Eq. (11.58))
Drying Creep, $C_d(t, t_o, t_c)$	
f_{c28}: (a) 30 MPa, (b) 60 MPa.	q_5: (a) 200.0, (b) 100.0 (Eq. (11.60))
$h = 0.65$.	$H(t) = 0.821$, $H(o) = 1$ (Eq. (11.59))
$t = 365$ days, $t_o = t_c = 14$ days	$C_d(t, t_o, t_c)$: (a) 37.6×10^{-6} per MPa, (b) 18.8×10^{-6} per MPa (Eq. (11.58))
Compliance, $J(t, t_o)$	
f_{c28}: (a) 30 MPa, (b) 60 MPa.	E_{c28}: (a) 25.9 GPa, (b) 36.6 GPa (Eq. (11.63))
	q_1: (a) 23.1×10^{-6} per MPa, (b) 16.4×10^{-6} per MPa (Eq. (11.62))
	$J(t, t_o)$: (a) 125.8×10^{-6} per MPa, (b) 77.2×10^{-6} per MPa (Eq. (11.61))
Total Strain (Compliance + Shrinkage), $\varepsilon_{Total}(t, t_o)$	
f_{c28}: (a) 30 MPa, (b) 60 MPa.	(a) $10 \times 125.8 \times 10^{-6} + 290 \times 10^{-6}$;
$\sigma = 10$ MPa	(b) $10 \times 125.8 \times 10^{-6} + 290 \times 10^{-6}$ (Eq. (11.64))
Contraction (Total Strain × Length)	
f_{c28}: (a) 30 MPa, (b) 60 MPa.	(a) $1550 \times 10^{-6} \times 3 \times 1000 = 4.7$ mm;
Length of member = 3 m	(b) $1043 \times 10^{-6} \times 3 \times 1000 = 3.1$ mm

strength, the model only requires the input data that are available to the engineer at the design stage.

Shrinkage

Shrinkage, $\varepsilon_{sh}(t, t_c)$, at any age t after exposure to drying at age t_c, is determined from the following expression:

$$\varepsilon_{sh}(t, t_c) = \varepsilon_{shu}\beta(h)\beta(t - t_c) \tag{11.65}$$

where ε_{shu} = ultimate shrinkage, $\beta(h)$ = humidity coefficient, and $\beta(t-t_c)$ = coefficient to allow for time of drying.

The ultimate shrinkage is calculated from:

$$\varepsilon_{shu} = 900k\left(\frac{30}{f_{c28}}\right)^{0.5} \times 10^{-6} \tag{11.66}$$

where $k = 1.00$ for Type I cement, $k = 0.75$ for Type II cement, and $k = 1.15$ for Type III cement; f_{c28} = cylinder compressive strength (MPa) at the age of 28 days.

The humidity coefficient of Eq. (11.65) is obtained from:

$$\beta(h) = \left(1 - 1.18h^4\right) \tag{11.67}$$

where h = relative humidity expressed as a fraction and, for $h > 0.96$, $\beta(h)$ is negative, which indicates swelling.

The drying time coefficient of Eq. (11.65) is given by:

$$\beta(t - t_c) = \left[\frac{(t - t_c)}{(t - t_c) + 0.12(V/S)^2}\right]^{0.5} \tag{11.68}$$

where t_c = age of exposure to drying or end of moist curing, and V/S = ratio of volume to exposed surface area (mm).

Creep

In this method, creep is identified by the 28-day creep coefficient $\phi_{28}(t, t_o)$, as defined in Eq. (11.1), and consists of three terms: two terms for basic creep and one term for drying creep. The method allows for the influence of predrying during the curing period, t_c, on creep through a term identified by $\Phi(t_c)$.

The creep coefficient is estimated from:

$$\phi_{28}(t, t_o) = \Phi(t_c) \times \left[2\frac{(t - t_o)^{0.3}}{(t - t_o)^{0.3} + 14} + \left(\frac{7}{t_o}\right)^{0.5}\left(\frac{(t - t_o)}{(t - t_o) + 7}\right)^{0.5}\right.$$
$$\left. + 2.5\left(1 - 1.086h^2\right)\left(\frac{(t - t_o)}{(t - t_o) + 0.12(V/S)^2}\right)^{0.5}\right] \tag{11.69}$$

where t_o = age at loading (days) and the other terms are as identified in the shrinkage expressions.

In Eq. (11.69), when $t_o = t_c$:

$$\Phi(t_c) = 1 \tag{11.70}$$

and when $t_o > t_c$:

$$\Phi(t) = \left[1 - \left(\frac{(t_o - t_c)}{(t_o - t_c) + 0.12(V/S)^2}\right)^{0.5}\right]^{0.5} \tag{11.71}$$

Compliance

In the case of the GL 2000 model, the compliance of concrete, $J(t, t_o)$, is defined as the method of BS EN 1992-1-1: 2004, viz.

$$J(t, t_o) = \frac{1}{E_{cto}} + \frac{\phi(t, t_o)}{E_{c28}} \tag{11.72}$$

where E_{cto} = secant modulus of elasticity at age of loading, t_o, and E_{c28} = modulus of elasticity at the age of 28 days.

The modulus of elasticity E_{ct}(GPa) is related to mean cylinder strength f_{ct}(MPa), as follows:

$$E_{ct} = 3.5 + 4.3\sqrt{f_{ct}} \tag{11.73}$$

and strength development with time is given by:

$$f_{ct} = \beta_e^2 f_{c28} \tag{11.74}$$

in which

$$\beta_e = \exp\left[\frac{s}{2}\left(1 - \sqrt{\frac{28}{t}}\right)\right] \tag{11.75}$$

In Eq. (11.74), the 28-day mean strength, f_{c28}, is related to the specified or characteristic strength, f_c', as follows:

$$f_{c28} = 1.1f_c' + 5.0 \tag{11.76}$$

Also, in Eq. (11.75), t = age of concrete (days) and s depends on type of cement, viz: 0.335 for Type I cement, 0.40 for Type II cement, and 0.13 for Type III cement.

Total Strain

The expression for total strain, $\varepsilon_{Total}(t, t_o)$, is the same as that for the methods of BS EN 1992-1-1: 2004, ACI 209R-92, and Bazant−Baweja B3, i.e.,

$$\varepsilon_{Total}(t, t_o) = \sigma \cdot J(t, t_o) + \varepsilon_{sh}(t, t_o) \tag{11.77}$$

where σ = applied stress at the age of loading, t_o; $J(t, t_o)$ = compliance (Eq. (11.69); and $\varepsilon_{sh}(t, t_o)$ = shrinkage measured from the age at loading (Eq. (11.65)).

Example

The example given on p. 362 is now used to illustrate the application of the Gardner and Lockman GL 2000 model to estimate the various types of movement leading to the total contraction of the concrete member after 1 year. The solution is presented in Table 11.14.

Comparison of Prediction Methods and Recommendations

Comparing the deformations of the foregoing example by the four current methods of prediction suggests that, whatever the choice, the outcome in terms of actual movement is similar and, in fact, for any method, the 1-year total contraction is within $\pm 15\%$ of the

Table 11.14 Solution to Example by the GL 2000 Model

Required Input Data	Coefficient/Component (Location in text)
Shrinkage, $\varepsilon_{sh}(t, t_c)$	
Type I cement.	$k = 1.00$ (Eq. (11.66))
f_{c28}: (a) 30 MPa, (b) 60 MPa.	ε_{shu}: (a) 900×10^{-6}, (b) 636.4×10^{-6} (Eq. (11.66))
$t = 365$ days,	$\beta(h) = 0.789$ (Eq. (11.67))
$t_c = t_o = 14$ days.	$\beta(t - t_c) = 0.585$ (Eq. (11.68))
$V/S = 75$ mm	$\varepsilon_{sh}(t, t_c)$: (a) 415×10^{-6}, (b) 294×10^{-6} (Eq. (11.65))
Creep Coefficient, $\phi_{28}(t, t_o)$	
$t = 365$ days, $t_c = t_o = 14$	$\Phi(t_c) = 1$ (Eq. (11.70))
days, $V/S = 75$ mm	$\phi_{28}(t, t_o)$: (a) 2.077 (Eq. (11.69))
Compliance, $J(t, t_o)$	
f_{c28}: (a) 30 MPa, (b) 60 MPa.	E_{c28}: (a) 27.05 Gpa, (b) 36.81 GPa (Eq. (11.73))
Type I cement	$s = 0.335$, $\beta_e = 0.933$ (Eq. (11.75))
	f_{c14}: (a) 26.11 MPa, (b) 52.23 MPa (Eq. (11.74))
	E_{c14}: (a) 25.47 Gpa, (b) 34.58 GPa (Eq. (11.73))
	$J(t, t_o)$: (a) 115×10^{-6} per MPa, (b) 85.3×10^{-6} per MPa (Eq. (11.72))
Total Strain, $\varepsilon_{Total}(t, t_o)$	
f_{c28}: (a) 30 MPa, (b) 60 MPa.	$\varepsilon_{Total}(t, t_o)$: (a) 1566×10^{-6}, (b) 1147×10^{-6} (Eq.
$\sigma = 10$ MPa	(11.77))
Contraction (Total Strain $\times$ Length)	
f_{c28}: (a) 30 MPa, (b) 60 MPa.	(a) $1566 \times 10^{-6} \times 3 \times 10^3 = 4.7$ mm;
Length of member = 3 m	(b) $1147 \times 10^{-6} \times 3 \times 10^3 = 3.4$ mm

mean estimate of all methods. The most divergence between methods occurred for shrinkage, where estimates were within ±26% of the mean estimate; in the case of compliance, estimates were within ±20% of the mean estimate. In another example, ACI 209.2R-08 [3] compared estimates by different models for 1-year shrinkage and compliance of a member having a volume/surface ratio of 100 mm, moist-cured for 7 days, loaded at the age of 14 days, and stored at 70% *RH*; the specified compressive strength of concrete was 25 MPa. Figure 11.4 shows the deformation—time characteristics and, after 1 year, shrinkage estimates were within ±13% of the mean, whereas

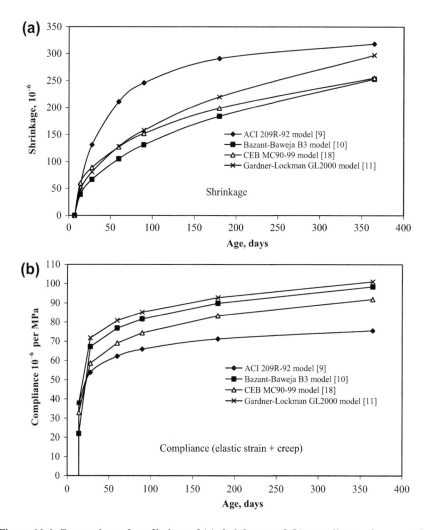

Figure 11.4 Comparison of predictions of (a) shrinkage and (b) compliance of concrete by different models according to ACI 209.2R-08 [3].

compliance estimates were within $\pm 10\%$ of the mean. In the ACI example, the full version of the B3 model was used.

Those comparisons suggest that there is not a great deal of difference between prediction models, but there may be larger differences between predictions for other situations, e.g., concrete members subject to varying stress or strain where stress relaxation and creep recovery occur [3]. However, the real test is a comparison of estimates with actual measured data, which is not readily available and universally acceptable by researchers for the reasons given in the review by ACI 209.2R-08 [3], some of which are given below.

In general, the fact that there are several fundamentally different methods of predicting time-dependent strains suggests a degree of uncertainty in this area of knowledge, which indeed is the case since, in many instances, estimates can be in error by the order of 20—40%. The universal acceptance of any method must be preceded by confirmation by measurements on actual structures but, unfortunately, there are few long-term data of sufficient accuracy available to undertake such an exercise. However, in the first instance, verification using laboratory test data is, of course, necessary, but even here comparison of models is complicated by the lack of agreement on selection of appropriate test data and on the statistical indicator used to assess the accuracy of prediction [3]. A comprehensive review of current methods of prediction and indicators for assessment of accuracy by ACI Committee 209.2R-08 [3] revealed that investigators' findings regarding performance of models are dependent somewhat on test data used in the assessment and the statistical indicator chosen to quantify accuracy. Statistical indicators available are not adequate to uniquely distinguish between models.

An important aspect of this topic is the establishment of the international experimental data collection started in 1978 by Bazant and Panula [26], which was then subsequently extended [1,2,36]. This collection, commonly known as the RILEM data bank, provides an invaluable source of data that can be used as a basis for the comparison of existing prediction models as well as being used to develop and verify new models. Unfortunately, however, there is disagreement by researchers as to the inclusion of some data sets and, in particular, there is inadequate description of the type of cement in order to accurately quantify strength development, especially between different cements used in Europe, the United States, Japan, and the South Pacific [3]. Most data sets are for concretes made with "older type" cements and lack long-term results for concrete made with more modern blended cements containing additives, chemical admixtures, and mineral admixtures. Some investigators have even suggested that separate creep and shrinkage models be developed on a continental basis because of the wide range of cement types [3]. A further problem is that the RILEM data bank lacks test results for the following:

- Creep for drying before loading or loading before drying, which occurs in practice.
- Long term measurements of creep and shrinkage.
- Creep and shrinkage for larger section sizes that are more representative of real structural elements, since smaller specimens may not represent the curing conditions and properties of larger elements.

With regard to concrete containing admixtures, prediction methods assume that any changes in shrinkage and creep of concrete are simply reflected by changes in strength development. This assumption may be an oversimplification. As an alternative approach, the effects of chemical and mineral admixtures can be assessed by the relative deformation method described in Chapter 6 (shrinkage) and Chapter 10 (creep). Here, estimates are made for the plain admixture-free concrete, either by previous experience or a standard prediction model, which are then modified to allow for: (1) changes in mix composition, and (2) the presence of the admixture. Procedures are fully explained in relevant chapters.

In 2005, the accuracy of models for estimating 30-year shrinkage and creep coefficient was compared for one set of laboratory test data in terms of the error coefficient (Eq. (11.78)) [37]. It was found that creep coefficient depended on the type of aggregate and strength of concrete, which most methods failed to take into account (Figure 11.5). The comparison included the CEB MC90 method [17,20] that, although accounting for strength dependency, did not account for aggregate type or, strictly speaking, modulus of elasticity of aggregate. However, the current BS EN 1992-1-1: 2004 method [16], which is based on CEB MC 90, now includes provision for estimating creep and shrinkage of lightweight aggregate concrete (as outlined earlier in this chapter), as well as modulus of elasticity (Chapter 4). For the creep coefficient, the value for normal density concrete is reduced by $(\rho/2200)^2$, where ρ = oven-dry density of lightweight aggregate concrete (less than 2000 kg/m^3). The effect of that provision is shown in Figure 11.5, where it can be seen that, assuming a lightweight aggregate concrete density, $\rho = 1800$ kg/m^3, general trends are in agreement with measured trends for both total and basic creep coefficient, $\phi(t, t_o)$. Furthermore, it seems possible that the influence of the type of normal weight aggregate could be also taken into account through a similar concrete density term for $\rho > 2200$ kg/m^3. For example, in the investigation of Figure 11.5 [38], Stourton aggregate concrete had a lower density than the better quality North Notts Aggregate, and the latter clearly exhibited the greater creep coefficient. Hence, it is recommended that density of concrete or specific gravity of aggregate be should be investigated as a possible indicator of the aggregate stiffness effect on creep coefficient of concrete.

Prediction Using Short-term Test Data

The review by ACI Committee 209.2R-08 [3] points out that shrinkage and creep may vary with local conditions and, since research has shown that short-term measurements improve predictions, the committee recommends the measurement of strength, elastic modulus, shrinkage, and creep, which is similar to the recommendation made by Brooks [37]. This topic is now discussed.

In presenting their GL 2000 model [11], Gardner and Lockman claim that predictions are improved by simply measuring concrete strength development with time and modulus of elasticity. Prediction of shrinkage is improved by using a value of k in Eq. (11.66) determined by interpolation from Table 11.15 with k corresponding to

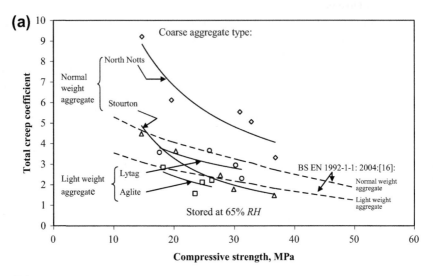

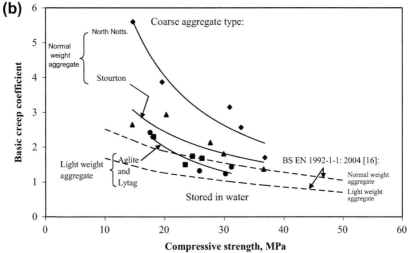

Figure 11.5 Influence of strength of concrete and coarse aggregate type on (a) 30-year total creep coefficient and (b) 30-year basic creep coefficient of concrete loaded at the age of 14 days [37]; creep coefficient is that defined in Eq. (11.2).

coefficient s as calculated from Eqs (11.74) and (11.75) using experimentally determined strength at two ages.

Aggregate stiffness is taken into account in the prediction of shrinkage by using the average of the measured cylinder strength in Eq. (11.66), and in the prediction of compliance by using a strength value back-calculated from the measured modulus of elasticity of the concrete in Eq. (11.73).

Table 11.15 Coefficients *s* and *k* for use in the GL 2000 Model [33]

Type of Cement	Coefficient *s*	Coefficient *k*
Type I	0.335	1.00
Type II	0.400	0.75
Type III	0.130	1.15

According to Bazant and Baweja [31], inaccuracies of the short form of the B3 prediction model are caused by the effect of concrete composition and strength of concrete, and the only way to reduce uncertainty is to conduct short-term tests and use them to amend values of material parameters in the model. The approach is more difficult for shrinkage than for creep and, to improve prediction, the authors recommend the use of a method involving short-term shrinkage tests together with measurements of water weight loss [30].

Most researchers recommend short-term shrinkage and creep tests in order to improve the accuracy of prediction by extrapolation. Clearly, the accuracy of prediction depends on the form of the deformation−time relation used and the degree of fit of experimental points. Hence, accuracy also depends on the duration of the short-term test, so it follows that the longer the test the more accurate is the estimated long-term value. This is illustrated in Figure 11.6, which shows the error coefficient, *M*, after periods under load of 7−180 days for a number of creep tests. The error coefficient has

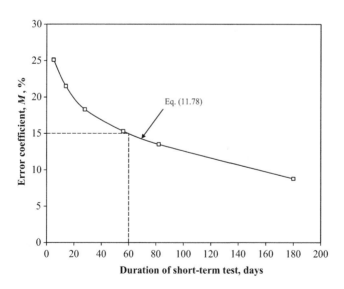

Figure 11.6 Accuracy of predicting shrinkage and creep after 1 year from short-term tests [38].
Source: Creep of Plain and Structural Concrete, A. M. Neville, W. H. Dilger and J. J. Brooks, Pearson Education Ltd. © A. M. Neville 1983.

been used earlier in this book as a statistical indicator of accuracy (Eq. (5.4) and Eq. (12.36), but is repeated here in general form:

$$M = \frac{1}{d_{am}} \sqrt{\frac{\sum (d_p - d_a)^2}{n-1}} \times 100 \tag{11.78}$$

where d_{am} = mean actual long-term value, d_p = predicted value, d_a = actual value, and n = number of data sets.

The value of M in Eq. (11.78) is expressed as a percentage and is analogous to the coefficient of variation, but the deviation is measured from the "true" deformation. Figure 11.6 shows M to decrease with an increase in the short-term test duration and, if a value of $M = 15\%$ is regarded as acceptable, then a minimum test duration may be ascertained, e.g., 60 days to predict 1-year creep in the case of Figure 11 6. Improvements in accuracy can be achieved by increasing the duration of the short-term test, but it should be remembered that the higher cost of testing in terms of the time taken has to be weighed against the amount of improved accuracy of prediction.

Moreover, Neville et al. [38] point out that a corollary of this economic requirement would be the prediction of creep from the elastic strain at the application of load. In fact, several such attempts have been made in terms of the ratio of ultimate creep to the elastic strain, which probably lies between 2 and 5, depending on the environmental storage conditions. Relations between creep and the elastic strain were developed by the U.S. Bureau of Reclamation [39,40], which appears to be a valid approach for give mix and storage conditions. However, the general applicability of that approach has not been established and, furthermore, it has not considered fundamentally correct that the modulus of elasticity of concrete is the only factor affecting the magnitude of creep. Other factors have to be considered as suggested by Kruml [41], who expressed creep coefficient as a function of change in modulus of elasticity with time, ambient conditions, and stress/strength ratio.

The principle of experimental testing to determine a short-term value of creep in order to estimate long-term values has been established historically [38]. Thomas [42] found the ratio of ultimate creep to that occurring in the first year under load to vary little with age at application of load, the ratio increasing from 1.2 to 1.44 as the age at loading increased from 7 to 90 days, respectively. Kruml [41] found that for lightweight aggregate concrete creep after 10 years was approximately 1.3 times the creep after 300 days under load. A linear relation between 2-year creep and 90-day creep of various lightweight aggregate concretes subjected to load at the age of 1 day was demonstrated by Reichard [43].

Neville and Liszka [44] attempted to reduce the duration of short-term tests without loss of accuracy by means of accelerated creep tests in which specimens were stored at a higher temperature. For lightweight aggregate concrete, 7-day accelerated creep at elevated temperatures of 45 and 65 °C was found to be a linear function of 100-day basic creep at 23 °C, viz.

At 45°C : $C_{b100} = 0.96 C_{b7} + 7$ \qquad\qquad (11.79) and

At $65°C$: $\quad C_{b100} = 0.91C_{b7} - 2$ (11.80)

where C_{b100} = specific basic creep (10^{-6} per MPa) after 100 days under load stored in water at 23 °C, and C_{b7} = specific basic creep after 7 days under load stored in water at elevated temperature.

Although the levels of accuracy of Eq. (11.79) and Eq. (11.80) were found to be acceptable, when this approach was applied to concretes made with a range of different aggregates stored in a drying environment, the error coefficient increased and was no better than that based on prediction by short-term tests at normal temperature. In the latter case, Brooks and Neville [45,46] and Brooks [47] developed relations between long-term shrinkage and creep and their short-term values based on earlier work by Neville [48], who had showed that, for concretes with a given cement paste content, the relative increase with time under load is independent of the water/cement ratio. The influence of cement paste content on shrinkage and creep has been quantified in previous chapters, respectively, by Eq. (6.3) and Eq. (10.5), and in general form is:

$$d = d_p(1 - g)^\alpha$$ (11.81)

where d = shrinkage or creep of concrete at a constant stress/strength ratio, d_p = shrinkage or creep of cement paste at a constant stress/strength ratio, g = total aggregate content plus unhydrated cement by volume, and α = parameter reflecting the elastic properties of aggregate and concrete as given by Eq. (6.2), which is sensibly independent of g.

Denoting long-term deformation, i.e., the value to be predicted by d_t, and the short-term deformation, i.e., the value to be determined experimentally by d_{st}, then from Eq. (11.81):

$$d_t = d_{st}\frac{d_{pt}}{d_{pst}}[1 - g]^{(\alpha_t - \alpha_{st})}$$ (11.82)

Now from Eq. (11.81):

$$(1 - g) = \left(\frac{d_{st}}{d_{pst}}\right)^{\frac{1}{\alpha_t}}$$

so that substitution for $(1-g)$ in Eq. (11.82) yields:

$$d_t = d_{pt}\left(\frac{d_{st}}{d_{pst}}\right)^{\frac{\alpha_t}{\alpha_{st}}} = (d_{st})^{\frac{\alpha_t}{\alpha_{st}}} \times d_{pt}(d_{pst})^{\frac{\alpha_{st}}{\alpha_t}}$$ (11.83)

Putting $A = d_{pt} \times (d_{pst})^{-a}$ and $a = \alpha_t/\alpha_{st}$, for creep (C) Eq. (11.83), can be written as:

$$C_t = AC_{st}^a$$ (11.84)

and for shrinkage (S) the corresponding relation can be written as:

$$S_t = A'S_{st}^{a'} \tag{11.85}$$

where A' and a' are analogous to those in Eq. (11.84).

Equations (11.84) and (11.85) have been verified experimentally using concretes made from a range of aggregates and water/cement ratios [45,46]. In fact, in the case of creep, for a constant time under load, coefficient a was approximately equal to unity, thus implying that long-term was directly proportional to short-term creep. Furthermore, although the corresponding shrinkage relation was strictly a power function, analysis of 10-year data [47] revealed it was convenient to use a linear relation for the majority of shrinkage values analyzed without loss of accuracy. This analysis, which included results of other investigators, allowed the coefficients A, A', a, and a' to be expressed as functions of time. For short-term specific basic creep (C_{b28}), total specific creep (C_{T28}), and shrinkage (S_{28}) of 28 days duration, the corresponding long-term deformations $(C_b(t, t_o)$, $C_T(t, t_o)$, $S(t, t_o))$, at any age t after application of load or exposure to drying at age t_o, and for time under load or time of exposure $(t-t_o) \geq 28$ days, are as follows:

$$Basic\ creep : C_b(t, t_o) = C_{b28} \times 0.50(t - t_o)^{0.21} \tag{11.86}$$

$$Total\ creep : C_T(t, t_o) = C_{T28}[-6.19 + 2.156 \log_e(t - t_o)]^{\frac{1}{2.64}} \tag{11.87}$$

$Shrinkage$: When short $-$ term shrinkage $\leq 200 \times 10^{-6}$,

$$S(t, t_o) = B'S_{28}^{b'} \tag{11.88a}$$

in which $B' = (-4.17 + 1.53 \log_e(t - t_o))^2$

and $b' = \dfrac{100}{2.90 + 29.2 \log_e(t - t_o)}$

When short $-$ term shrinkage $\geq 200 \times 10^{-6}$,

$$S(t, t_o) = S_{28} + 100[3.61 \log_e(t - t_o) - 12.05]^{0.5} \tag{11.88b}$$

An analysis of swelling of concrete stored continuously in water was also carried out, but in this case meaningful relations between long-term and short-term values were only possible after short-term test durations of at least 1 year [47]. Swelling relations were of similar form to those of nonlinear shrinkage (Eq. (11.85)), corresponding coefficients being functions of time, $(t-t_o) \geq 1$ year, as follows:

$$\left.\begin{aligned} Swelling : W(t, t_o) &= W_{365}^B \\ B = 0.377(\log{}_e(t - t_o))^{0.55} \end{aligned}\right\} \tag{11.89}$$

where $W(t, t_o)$ = swelling at age t measured from age t_o, and W_{365} = swelling after 365 days.

Examples of the experimental validation of the relations between long-term and short-term deformations are shown in Figures 11.7−11.10. Except for swelling, short-term tests of duration of 28 days are regarded as acceptable for creep and shrinkage since error coefficients are less than 20%. The relations are generally applicable for all types of concrete exposed to any condition since the validation covered a wide range of variables: type of cement, age, aggregate type, mix proportions, and storage environments [46]. If desired, the expressions are readily adaptable for predicting long-term deformation from a short-term test of duration greater than 28 days or 1 year in the case of swelling; for example, the relations between 5-year and 1 year creep and shrinkage are obtained by rearranging Eq. (11.86)−Eq. (11.88). Table 11.16 lists the relevant expressions shown in Figures 11.7−11.10, together with corresponding relations for 50-year deformations.

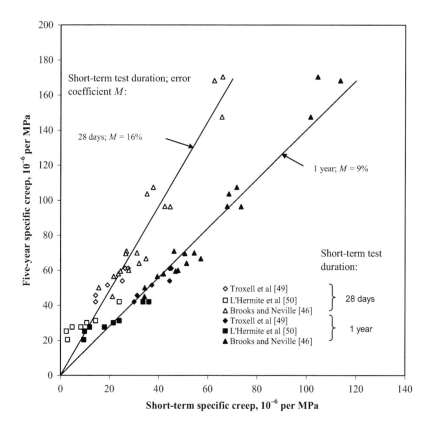

Figure 11.7 **Five year basic creep as a function of (a) 28-day basic creep and (b) 1-year basic creep; relations given in Table 11.16 [46,47].**

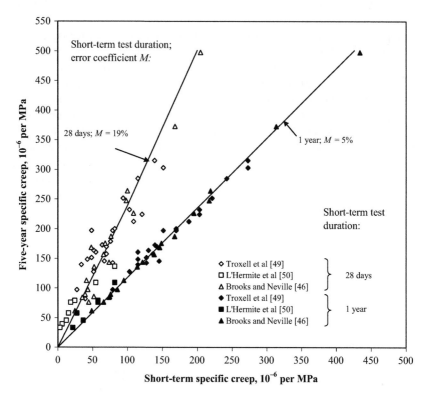

Figure 11.8 Five-year total creep as a function of (a) 28-day total creep and (b) 1-year total creep; relations given in Table 11.16 [46,47].

Table 11.16 Relations between Long-term Deformation and Short-term Deformation [47]

		5- or 10-Year Relation		50-Year Relation	
		Short-term Test Duration		Short-term Test Duration	
Deformation	Eq. No.	28 days	1 year	28 days	1 year
Basic creep	(11.86)	$C_{b5y} = 2.42C_{b28}$	$C_{b5y} = 1.40C_{b1y}$	$C_{b50y} = 3.92C_{b28}$	$C_{b50y} = 2.27C_{b1y}$
Total creep	(11.87)	$C_{t5y} = 2.39C_{t28}$	$C_{t5y} = 1.18C_{t1y}$	$C_{t50} = 2.78C_{t28}$	$C_{t50} = 1.37C_{t1y}$
Shrinkage	(11.88a)	$S_{5y} = 53.57S_{28}^{0.45}$	$S_{5y} = 4.44S_{1y}^{0.79}$	$S_{50y} = 117.55S_{28}^{0.35}$	$S_{50y} = 17.31S_{1y}^{0.61}$
	(11.88b)	$S_{5y} = S_{28} + 388$	$S_{5y} = S_{1y} + 84$	$S_{50y} = S_{28} + 483$	$S_{50y} = S_{1y} + 182$
Swelling	(11.89)	$W_{10y} = W_{28}^{1.64}$ (NA)	$W_{10y} = W_{1y}^{1.20}$	NA	$W_{50y} = W_{1y}^{1.32}$

1y = 1 year, 5y = 5 years, 10y = 10 years, NA = not applicable because of poor correlation (high error coefficient).

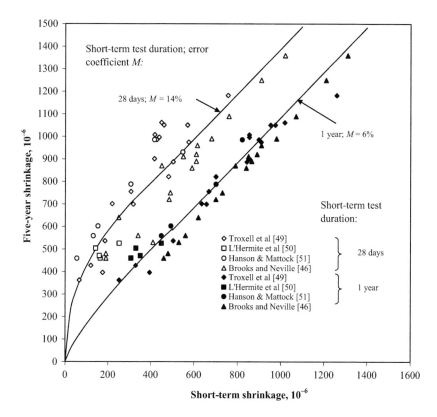

Figure 11.9 Five-year shrinkage as a function of (a) 28-day shrinkage and (b) 1-year shrinkage; relations given in Table 11.16 [46,47].

Case Study

In 1996, a laboratory investigation was undertaken to assess creep and shrinkage of concrete used to construct the cable-stayed highway bridge in the county of Flintshire in North Wales [52]. In order to control the alignment and stresses of the Flintshire Bridge, accurate long-term creep and shrinkage data were required for analysis by a numerical model during the construction stage. Since the concrete mixture included chemical and mineral admixtures, the design model of BS 5400-4: 1990 [53] was deemed to be inappropriate since the creep and shrinkage factors applied for admixture-free concrete. Consequently, laboratory tests were carried out in which specimens were loaded at ages corresponding to the prestressing and tensioning operations of the bridge concrete elements.

Figure 11.11 shows that the asymmetric highway bridge consisted of a single tower, an anchor span constructed from in situ concrete and connected to twin-tension piers. The main span over the navigable river Dee consists of an in situ concrete inverted

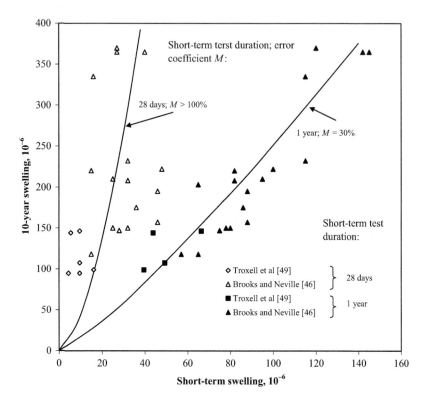

Figure 11.10 10-year swelling as a function of (a) 28-day swelling and (b) 1-year swelling; relations given in Table 11.16 [47].

"trough deck" with both longitudinal and transverse posttensioning. Although the anchor spans were constructed on falsework, the main span was constructed in 8-m concrete segments using the cantilever method. A pair of cables supported each segment with the exception of the first and last segments. As well as limiting stresses during construction and in service, the geometry of the deck had to be controlled to achieve closure and the desired final highway alignment. The primary tools for the latter were the segment-casting profile and cable forces estimated by numerical analysis incorporating creep and shrinkage of concrete. In consequence, the validity of the predicted stresses depended on the accuracy of the model used to predict creep and shrinkage.

The effects of creep and shrinkage were of significance on the performance of the bridge for the following reasons:

• Continuity of deflection and slope at closure of the last 5-m segment number 21 with fixed approach viaduct, where a limited tolerance could be accommodated.
• Increased deflections would cause a change in the reference geometry of the structure and therefore require an increase in cable forces to restore the geometry to the design requirement.
• The effect of long-term creep would be to lower the deck in the main span and increase the tower deflection. This would affect the final road alignment and stay-force distribution.

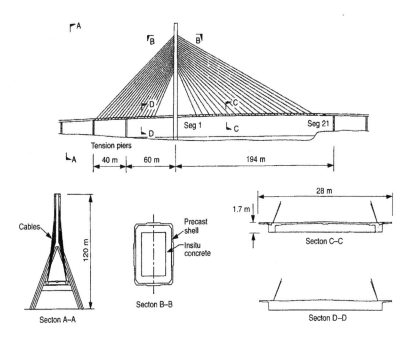

Figure 11.11 The Flintshire cable-stayed bridge [52].

A preliminary sensitivity analysis of the main span deflection was carried out with input data from the CEB 1970 model [8], which revealed that a change in creep coefficient from 1.0 to 1.5 would increase the deflection tip by over 150 mm, and a significant increase in cable installation force would be required to restore the geometry of the deck and tower to the design profile [52]. It was concluded that it was desirable to determine creep and shrinkage characteristics of the type of concrete used (60/30) as precisely as possible. Specific requirements were twofold:

1. One-year creep and shrinkage data for different ages of loading or exposure as the prestressing and tensioning of the bridge segments proceeded, viz. 3, 7 14, 28, 56, and 90 days.
2. Estimates of 50-year creep and shrinkage of bridge segments exposed to an average relative humidity = 80% and having a volume/surface ratio = 200 mm.

The constituents and proportions of the types of concrete used for the construction of the bridge segments are shown in Table 11.17.

Creep and shrinkage laboratory measurements were carried out on 76×255-mm cylinders, two separate casts being made for (1) specimens stored continuously in water at 22 °C to determine basic creep and swelling, and (2) for specimens cured in water until loading and storage in air controlled at 68% *RH* and 22 °C to determine total creep and shrinkage. The 28-day cube compressive strength of concrete for the two casts was 82.0 MPa for the wet-stored tests and 69.1 MPa for the dry stored tests.

Table 11.17 Mix Constituents of the 60/20 Bridge Concrete [52]

Constituent	Quantity, kg/m³ of Concrete	Quantity, kg of Cementitious Material	Quantity, kg of Cement
Ordinary Portland cement	315	0.649	1.000
Ground granulated blast-furnace slag	170	0.351	0.540
Sand	691	1.424	2.193
20-mm limestone	630	1.299	2.000
10-mm limestone	420	0.866	1.333
Plasticizer	1.85	0.00381	0.006
Water	180	0.371	0.571

The applied stress was 15 MPa for creep tests. Shrinkage results are shown in Figure 11.12, elastic strain plus basic creep results in Figure 11.13, and elastic strain plus total creep in Figure 11.14. The 1-year swelling results are given in Table 11.18, along with other pertinent test data.

Table 11.19 compares the measured and estimated modulus of elasticity predicted by standard methods. The measured value is the average of the "wet" and "dry" casts given in Table 11.18, corresponding to the average 28-day cube compressive strength of 75.5 MPa. The equivalent cylinder strength was taken as $0.8 \times 75.5 = 60.4$ MPa, which was used in calculations for the methods of prediction listed in Table 11.19. Compared with the measured secant modulus of elasticity from the creep tests, it can be seen that all methods are reasonably accurate with acceptable deviations from measured values.

Since the laboratory tests were carried out using specimens of different size (volume/surface ratio = 19 mm) to that of the bridge concrete segments (volume/surface ratio = 200 mm), adjustments to the creep and shrinkage results were necessary. Also, because the laboratory storage environments (100% and 68% *RH*) were different from the average annual environmental air *RH* of 80% around the bridge, further adjustments were made to the measured creep and shrinkage data. In the case of the size effect, shrinkage and creep test data were reduced by factors of 0.73 and 0.75, respectively. Those factors are the calculated average of values given by different methods of prediction (see Table 11.20).

In the case of the *RH* correction factor, shrinkage at 80% *RH* (*S*(80)) was calculated from:

$$S(80) = 0.625S(68) \tag{11.90}$$

where $S(68) = $ shrinkage at $RH = 68\%$.

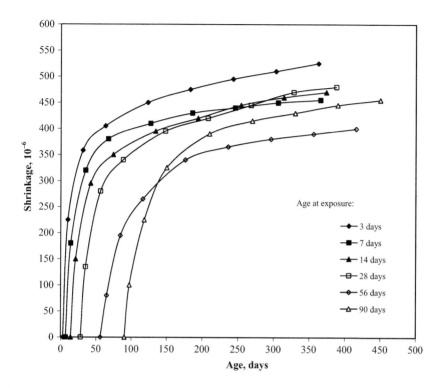

Figure 11.12 Shrinkage of Flintshire bridge concrete exposed to drying after curing in water for different periods [52].

The creep coefficient for 80% RH ($\phi_{28}(80)$) was calculated by linear interpolation between the total creep coefficient at 68% RH ($\phi_{28}(68)$) and the basic creep coefficient at 100% RH, ($\phi_{28}(100)$), i.e.,

$$\phi_{28}(80) = 0.375\phi_{28}(100) + 0.625\phi_{28}(68) \tag{11.91}$$

Tables 11.21 and 11.22, respectively, compare 1-year measured shrinkage and creep coefficient adjusted for a storage environment of 80% RH with estimates by methods of prediction. For the latter, to allow for the effect of chemical and mineral admixtures, it was assumed that the deformation was equivalent to that of plain, admixture-free concrete made with conventional cement having approximately the same strength development to that of the admixture concrete. For the Flintshire Bridge concrete, the measured rate of strength development ratio from 7 to 28 days of the blend of ordinary Portland cement and blast-furnace slag (60/20) concrete was 0.67, which was approximately equivalent to that of ordinary Portland cement [52].

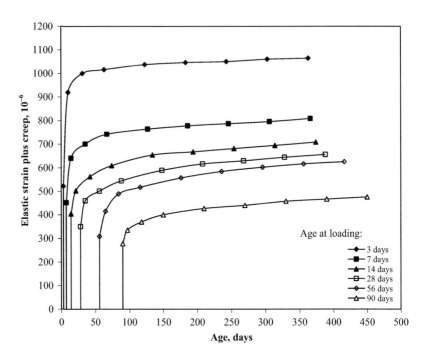

Figure 11.13 Elastic strain and basic creep of Flintshire bridge concrete continuously stored in water subjected to a stress of 15 MPa at different ages [54].

In the case of shrinkage, Table 11.21 indicates that the best accuracy of prediction is within ±30% as achieved by the CEB 90, B3, and GZ methods. The GZ method and the ACI method include an allowance for length of curing before exposure to drying, which has the effect of reducing shrinkage. Although that overall reducing trend of measured shrinkage in Table 11.21 is smaller than those predicted by the ACI 78 and GZ methods, as shown in Figure 11.15, it tends to confirm the observation made by ACI 209 1R-05 [54].

Comparing the predicted creep coefficient with the adjusted measured values of Table 11.22 indicates that the most accurate is the CEB 90 method with deviations within −11% and +17%. The general trend of age at loading on creep of the test concrete generally agrees the trends of the methods of prediction, although most methods tended to overestimate creep for ages at loading of around 7 days (Figure 11.16).

Estimates of 50-year shrinkage and creep are shown in Table 11.23. The estimates were obtained from 1-year measured values of test specimens extrapolated to 50 years by using the relations between 50-year and 1-year deformations given in Table 11.16. Those values include an allowance for the larger size of the bridge segment, viz. the average of the reduction factors given by methods of prediction given in Table 11.20.

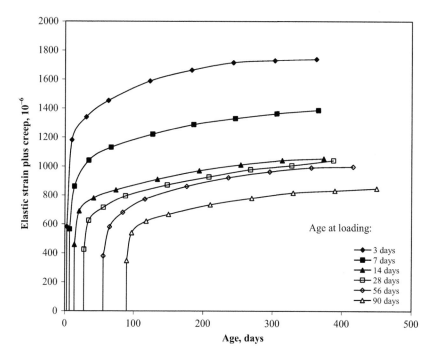

Figure 11.14 Elastic strain and total creep of Flintshire bridge concrete under a stress of 15 MPa applied at different ages after curing in water and then stored in air at *68% RH* [52].

Table 11.18 Measured Elastic Modulus, and 1-Year Swelling, Basic Creep, Total Creep, Shrinkage, and Creep Coefficients of Flintshire Bridge Concrete Test Specimens

Parameter	Age at Loading or Start of Measurement, days					
	3	**7**	**14**	**28**	**56**	**90**
1-year swelling, $10-6$	141	128	92	68	78	72
Secant modulus of elasticity, GPa	28.7w 25.8d	33.2w 26.6d	37.1w 32.8d	42.9w 35.3d	48.5w 39.5d	54.0w 43.1d
1-year specific basic creep, 10^{-6} per MPa	36.2	23.8	20.3	20.4	21.1	13.2
1-year specific total creep, 10^{-6} per MPa	81.1	54.7	39.6	39.6	41.1	33.2
1-year shrinkage, 10^{-6}	525	455	470	480	400	455
1-year basic creep coefficient, $\phi_{28}(t, t)$	1.55	1.02	0.87	0.87	0.90	0.57
1-year total creep coefficient, $\phi_{28}(t, t)$	2.86	1.93	1.40	1.40	1.45	1.17

w = wet cast; d = dry cast.

Table 11.19 Comparison of Measured and Predicted Moduli of Elasticity
of Flintshire Bridge Concrete [52]

	Modulus of Elasticity, GPa, at Age:					
Method	**3 days**	**7 days**	**14 days**	**28 days**	**56 days**	**90 days**
Measured	27.2	29.5	34.9	38.7	43.5	47.9
CEB 70 [8]	29.2 (+7%)	37.2 (+26%)	40.4 (+16%)	46.2 (+1%)	48.2 (+11%)	50.6 (+6%)
ACI 78 [9]	25.3 (−7%)	31.3 (+6%)	35.0 (0%)	37.5 (−3%)	38.9 (−6%)	39.6 (−18%)
CEB 90 [21]	30.3 (+11%)	34.5 (+17%)	37.2 (+7%)	39.2 (+1%)	40.6 (−7%)	41.4 (−14%)
GZ [33]	27.1 (0%)	31.5 (+7%)	34.6 (−1%)	36.9 (−5%)	38.6 (−11%)	39.4 (−18%)

Numbers in parenthesis are deviations from measured value: +ve = overestimate and −ve = underestimate.

Table 11.20 Correction Factors for Adjusting Test Specimen ($V/S = 19$ mm)
Deformation to Bridge Segment ($V/S = 200$ mm) Deformation

	Factor for Method of Prediction:					**Average**
Deformation	**CEB 70**	**ACI 78**	**CEB 90**	**B3**[a]	**GZ**	**Factor**
Shrinkage	0.46	0.42	0.88	0.91	0.97	0.73
Creep coefficient, ϕ_{28}	0.60	0.58	0.79	0.96 for $t_o = 3$ days reducing to 0.88 for $t_o = 90$ days	0.90	0.75

[a]Depends on time under load; t_o = age at loading.

Table 11.21 Comparison of Measured and Predicted 1-Year Shrinkage of Flintshire
Bridge Concrete Test Specimens Adjusted to 80% *RH* of Storage [52]

	Shrinkage, 10^{-6}, for Exposure Age:					
Method	**3 days**	**7 days**	**14 days**	**28 days**	**56 days**	**90 days**
Measured	323	284	294	300	219	284
CEB 70 [8]	199 (−38%)	199 (−30%)	199 (−32%)	199 (−34%)	199 (−9%)	199 (−30%)
ACI 78 [9]	512 (+74%)	466 (+64%)	433 (+47%)	401 (+34%)	368 (+68%)	331 (+17%)
CEB 90 [20]	236 (−27%)	236 (−16%)	236 (−20%)	236 (−21%)	236 (+8%)	236 (−17%)
B3 [10]	288 (−11%)	288 (+1%)	288 (−2%)	288 (−4%)	288 (+31%)	288 (+1%)
GZ [33]	335 (+4%)	282 (0%)	255 (−13%)	237 (−21%)	226 (+3%)	220 (−23%)

Numbers in parenthesis are deviations from measured value: +ve = overestimate and −ve = underestimate.

Table 11.22 Comparison of Measured and Predicted 1-Year Creep Coefficient of Flintshire Bridge Concrete Test Specimens Adjusted to a Storage Environment of 80% *RH* [52]

Method	Creep Coefficient, ϕ_{28}, for Age at Loading:					
	3 days	7 days	14 days	28 days	56 days	90 days
Measured	2.37	1.59	1.40	1.20	1.24	0.95
CEB 70 [8]	2.88 (+22%)	2.52 (+58%)	2.18 (+56%)	1.80 (+29%)	1.59 (+28%)	1.35 (+42%)
ACI 78 [9]	2.59 (+9%)	2.09 (+31%)	1.67 (+19%)	1.39 (+16%)	1.14 (−8%)	0.98 (+3%)
CEB 90 [20]	2.11 (−11%)	1.80 (+13%)	1.58 (+12%)	1.39 (+16%)	1.22 (−2%)	1.11 (+17%)
B3 [10]	3.01 (+27%)	2.31 (+45%)	1.85 (+32%)	1.48 (+23%)	1.19 (−4%)	1.04 (+9%)
GZ [33]	3.11 (+31%)	2.10 (+34%)	1.70 (+21%)	1.45 (+21%)	1.30 (+5%)	1.24 (+31%)

Numbers in parenthesis are deviations from measured value: +ve = overestimate and −ve = underestimate.

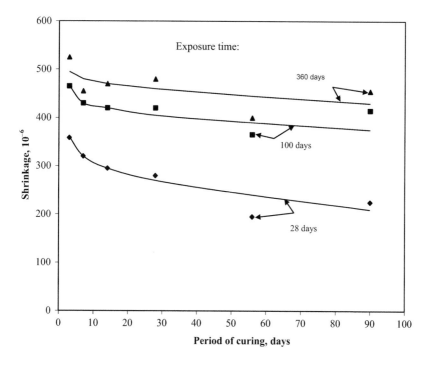

Figure 11.15 Effect of period of curing on shrinkage of Flintshire bridge concrete [52].

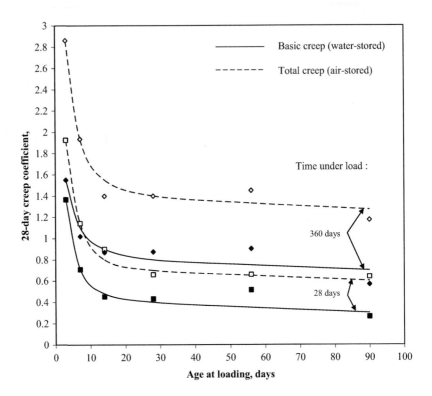

Figure 11.16 Effect of age at loading on 28-day creep coefficient of Flintshire bridge concrete cured in water [52].

Table 11.23 Extrapolated Measured 50-Year Shrinkage and Creep Coefficient of Flintshire Bridge Concrete Test Specimens and Segments [49]

	Shrinkage, 10^{-6}, for Exposure Age:					
Volume/Surface Ratio	**3 days**	**7 days**	**14 days**	**28 days**	**56 days**	**90 days**
19-mm test specimen	505	464	476	482	401	466
200-mm bridge segment	369 (370)	339 (355)	347 (345)	352 (330)	293 (320)	340 (315)

	Creep coefficient, $_{28}$ for age at loading:					
Volume/surface ratio	**3 days**	**7 days**	**14 days**	**28 days**	**56 days**	**90 days**
19-mm test specimen	3.25	2.18	1.92	1.64	1.70	1.30
200-mm bridge segment	2.44 (2.45)	1.64 (1.65)	1.44 (1.45)	1.23 (1.25)	1.28 (1.15)	0.98 (1.05)

Number in parenthesis is "hand-smoothed" value.

The main conclusions of the case study of deformation of concrete bridge segments containing a plasticizer and blast-furnace slag were as follows:

- Standard prediction methods for modulus of elasticity are satisfactory and within $\pm 26\%$ of measured values.
- One-year shrinkage was estimated by standard prediction methods to within $\pm 30\%$ of measured values.
- The CEB 90 method gave the most accurate estimates of 1-year creep coefficient, ϕ_{28}, of the test specimens that were within -11 to $+17\%$.
- An additive factor of $+182 \times 10^{-6}$ is recommended to increase 1-year shrinkage to 50-year shrinkage.
- A multiplying factor of 1.37 is recommended to increase 1-year creep to 50-year creep.
- There was a slight decrease of shrinkage for a longer period of water curing.
- According to standard prediction methods, there is a large variation in size reduction factor required adjust shrinkage and creep of small test specimens to allow for the larger concrete bridge segment, which suggests that further research is warranted. Based on the average values, reduction factors of 0.73 and 0.75, respectively, were chosen for shrinkage and creep.
- For curing periods or ages at loading ranging from 3 to 90 days, 50-year shrinkage varied from 370 to 315×10^{-6}, whereas 50-year creep coefficient varied between 2.45 and 1.05.

Problems

1. For the example given on p. 362, use the BS EN 1992-1-1: 2004 method to calculate the 1-year creep coefficient (ϕ_{28}) if the concretes are made with rapid-hardening cement and, before applying the load, they are cured for 10 days at 40 °C followed by 20 °C for 4 days.
   Answer: (a) 1.880, (b) 1.068.

2. Using the same input data in the example on p. 362, except that lightweight aggregate concrete of density 1500 kg/m³ is specified instead of normal weight aggregate concrete, estimate 1-year values by the BS EN 1992-1-1: 2004 method of: (1) drying shrinkage, (2) creep coefficient (ϕ_{28}), and (3) total contraction of the members. Ignore autogenous shrinkage.
   Answer: (1) (a) 379×10^{-6}, (b) 266×10^{-6}; (2) (a) 1.053, (b) 0.582; (3) (a) 5.7 mm, (b) 3.5 mm.

3. Assuming the same input data for the example on p. 362, except that lightweight aggregate concrete of density 1500 kg/m³ is specified instead of normal weight aggregate concrete, use the ACI 209R-92 method to calculate the contractions of the two members at the age of 70 years due to application of load after moist curing for 28 days.
   Answer: (a) 4.3 mm, (b) 3.1 mm.

4. Use the ACI 209R-92 method to estimate, at the age of 1 year, the contraction of the concrete member having a 28-day mean compressive strength = 30 MPa with the same specification as in the example on p. 362, except that the concrete is steam cured for 7 days before loading.
   Answer: 3.75 mm

5. Compare the following 10-year values as estimated according to the B3 model: (1) shrinkage; (2) basic creep; (3) drying creep; (4) compliance; and (5) contraction for the concrete specified in the example on p. 362 having a 28-day mean compressive strength = 30 MPa, and made with Types I, II, and II cements.

Concrete and Masonry Movements

The solution is summarized in the table below.
Answer:

Deformation	Type I Cement	Type II Cement	Type III Cement
(1) Shrinkage	539×10^{-6}	458×10^{-6}	593×10^{-6}
(2) Basic creep	80.9×10^{-6} per MPa	80.9×10^{-6} per MPa	80.9×10^{-6} per MPa
(3) Drying creep	66.8×10^{-6} per MPa	66.8×10^{-6} per MPa	66.8×10^{-6} per MPa
(4) Compliance	170.9×10^{-6} per MPa	170.9×10^{-6} per MPa	170.9×10^{-6} per MPa
(5) Contraction	6.7 mm	6.5 mm	6.9 mm

6. For the example given on p. 362 and concrete having a 28-day mean compressive strength $= 30$ MPa, use the B3 model to compare 70-year contractions for moist curing and dry curing from the age of 1 day.
   Answer: 7.6 mm (moist), 6.9 mm (dry).

7. For the example given on p. 362, use the GL 2000 method to estimate the contractions of the concrete members after 70 years under load.
   Answer: (a) 5.7 mm, (b) 4.1 mm.

8. In problem 11.7, if the two concrete members are moist cured for 1 day only before loading at 14 days, calculate the contractions after 70 years according to the GL 2000 method.
   Answer: (a) 5.2 mm, (b) 3.8 mm.

9. Discuss the circumstances where you advise undertaking tests to determine shrinkage and creep of concrete rather than use a standard method of prediction.

10. Derive the relation between 70-year shrinkage (S_{70y}) and 60-day shrinkage (S_{60d}) using Eq. (11.88a).
    Answer: $S_{70y} = 71.15(S_{60d})^{0.41}$.

11. Short-term testing of concrete for 60 days revealed: (a) shrinkage $= 150 \times 10^{-6}$ and (b) total creep $= 40 \times 10^{-6}$ per MPa. Use the relations between long-term and short-term deformation to estimate the corresponding deformations after 70 years of drying and sustained load.
    Answer: (a) 550×10^{-6}, (b) 78.63×10^{-6} per MPa.

References

[1] Kutter CH. Creep and shrinkage for windows: the program for the RILEM databank. Weimar Berlin and Karlsruhe: Karlsruhe University; 2007. Version 1.0.

[2] Muller HS, Bazant ZP, Kutter CH. Data base on creep and shrinkage tests. RILEM Subcommittee 5 Report, RILEM TC 107-CSP. Paris: RILEM; 1999. 81 pp.

[3] ACI Committee 209.2R-08. Guide for modeling and calculating shrinkage and creep in hardened concrete. American Concrete Institute; 2008.

[4] Ross AD. Concrete creep data. Struct Eng 1937;15:314–26.

[5] Wagner O. Das kriechen unbewehrten betons, vol. 31. Deutscher Ausschuss fur Stahlbeton; 1958. m74 pp.

[6] Jones TR, Hirsch TJ, Stephenson HK. The physical properties of structural quality lightweight aggregate concrete. College Station: Texas Transportation Institute; August 1959. 46 pp.

[7] Ulitskii I. A method of computing creep and shrinkage deformation of concrete for practical purposes. Beton I Zhelezobet 1962;4:174−80. Translation No. 6030, Commonwealth Scientific and Industrial Research Organisation, Melbourne Australia.

[8] CEB-FIP, International. Recommendations for the design and construction of concrete structures − principles and recommendations. Prague: Comite Europeen du Beton − Federation Internationale de la Precontrainte, FIP Sixth Congress; 1970.

[9] ACI Committee 209. Prediction of creep, shrinkage and temperature effects in concrete structures (ACI 209R-92). American Concrete Institute; 1992. 47 pp.

[10] Bazant ZP, Baweja S. Creep and shrinkage prediction model for analysis and design of concrete structures − model B3. Mater Struct 1995;28:357−65. 415−430, 488−495.

[11] Gardner NJ, Lockman MJ. Design provisions for drying shrinkage and creep of normal strength concrete. Aci Mater J 2001;98(2):159−67.

[12] BS 8110: Part 2: Structural use of concrete: code of practice for special circumstances. British Standards Institution (Withdrawn); 1985.

[13] Concrete Society. A simple design method for estimating the elastic modulus and creep of structural concrete. London; 1978. 1 pp.

[14] Concrete Society. A simplified method for estimating the elastic modulus and creep of normal weight concrete. Training Centre Publication No. TDH-7376. London: Cement and Concrete Association; 1978. 1 pp.

[15] Parrott L,J. Simplified methods of predicting the deformation of structural concrete. Development Report No. 3. London: Cement and Concrete Association; 1979. 11 pp.

[16] BS EN 1992-1-1. Eurocode 2, design of concrete structures. General rules and rules for buildings (see also UK National Annex to Eurocode 2). British Standards Institution 2004.

[17] ASTM C150/C150M-12. Specification for Portland cement. American Society for Testing and Materials; 2012.

[18] CEB Model Code 90-99. Structural concrete − textbook on behaviour, design and performance, updated knowledge of CEB-FIP model code 90. In FIP Bulletin 2, vol. 2. Lausanne: Federal International du Beton; 1999. 37−52.

[19] CEB-FIP. Model code for concrete structures. Paris: Comite Europeen du Beton − Federation Internationale de la Precontrainte; 1978. 348 pp.

[20] CEB-FIP. CEB model code 1990. CEB Bulletin d'information No. 213/214. Lausanne: Comite-Euro International du Beton; 1993. 437 pp.

[21] BS EN 197−1: Cement. Composition, specifications and conformity criteria for common cements. British Standards Institution; 2011.

[22] Meyers BL, Branson DE, Schumann CG, Christiason ML. The prediction of creep and shrinkage properties of concrete. Highways Commission Report No. HR-136. University of Iowa; 1970. 140 pp.

[23] Branson DE, Christiason ML. Time dependent properties of concrete related to design-strength and elastic properties, creep and shrinkage, creep shrinkage and temperature effects. ACI Special Publication, SP-27, American Concrete Institute; 1971.

[24] ACI Committee 318R-05. Building code requirements for structural concrete and commentary. American Concrete Institute; 2005. 430 pp.

[25] Bazant ZP, Osman E, Thonguthai W. Practical formulation of shrinkage and creep of concrete. Paris Mater Struct 1976;9(54):395−406.

[26] Bazant ZP, Panula L. Simplified prediction of concrete creep and shrinkage from strength and mix. Structural Engineering Report No. 78−10/6405. Evanston, Illinois: Department of Civil Engineering, Northwestern University; October 1978. 24 pp.

[27] Bazant ZP, Panula L. Practical predictions of time-depenedent deformations of concrete: part1 — shrinkage, part II — basic creep; part III — drying creep; part IV — temperature effect on basic creep; part V — cyclic creep, non-linearity and statistical matter. Paris Mater Struct 1978;II(65). pp. 301−316, pp. 317−328; II, No. 66, 1978, pp. 415−424, pp. 424−434, No. 69, 1979, pp. 169−174, pp. 175−183.

[28] Bazant ZP, Panula L. Practical prediction of creep and shrinkage of high strength concrete. Mater Struct 1984;17(101):375−8.

[29] Bazant ZP, Kim JK, Panula L, Xi Y. Improved prediction model for time-dependent deformations of concrete: parts 1-6. Mater Struct 1991;24(143):327−45. vol. 24, No. 144, 1991, pp. 409−421; vol. 25, No.134, 1991, pp. 21−28; vol. 25, No. 146, 1991, pp. 84−94; vol. 25, No. 147, 1991, pp. 163−169; vol. 25, No. 148, 1991, pp. 219−223.

[30] Bazant ZP, Baweja S. Creep and shrinkage prediction model for analysis and design of concrete structures: model B3. In: Al-Manaseer A, editor. Proceedings adam neville symposium: creep and shrinkage-structural design effects, SP-194. American Concrete Institute; 2000. pp. 1−100.

[31] Bazant ZP, Baweja S. Short form of creep and shrinkage prediction model B3 for structures of medium sensitivity. Mater Struct 1996;29:587−93.

[32] Gardner NL, Zhao JW. Creep and shrinkage revisited. ACI Mater J 1993;90(3):236−46.

[33] Gardner NJ. Design provisions for shrinkage and creep of concrete. In: Al Manaseer A, editor. The adam neville symposium: creep and shrinkage −structural design effects, SP-194. American Concrete Institute; 2000. pp. 101−34.

[34] Gardner NJ. Comparison of prediction provisions for drying shrinkage and creep of normal strength concretes. Can J Civ Eng 2004;32(5):767−75.

[35] Gardner NJ, Tsuruta H. Is superposition of creep strains valid for concretes subjected to drying creep? ACI Mater J 2004;101(5):409−15.

[36] Hillsdorf HK, Carriera DJ. ACI-CEB conclusions of the Herbert Rusch workshop on creep of concrete. Concr Int 1980;2(11):11.

[37] Brooks JJ. 30-year creep and shrinkage of concrete. Mag Concr Res 2005;57(5):545−56.

[38] Neville AM, Dilger WH, Brooks JJ. Creep of plain and structural concrete. London and New York: Construction Press; 1983. 361 pp.

[39] US Bureau of Reclamation. A ten-year study of creep properties of concrete. Concrete Laboratory Report No. SP-38, Denver, Colorado; 1953. 14 pp.

[40] US Bureau of Reclamation. Creep of concrete predicted from initial modulus values. Concrete Laboratory Report No. C-1242, Denver, Colorado; 1967. 26 pp.

[41] Kruml F. Dihodobe deformacne vlastnosti lahkkch betonov. Stavebnicky Cas 1965;13(3): 137−44.

[42] Thomas FG. A conception of the creep of unreinforced concrete and an estimation of limiting values. Struct Eng 1933;11(2):69−73.

[43] Reichard TW. Creep and drying shrinkage of light-weight and normal-weight concrete. NBS Monograph No. 74. Washington (DC): National Bureau of Standards; 1964. 30 pp.

[44] Neville AM, Liszka WZ. Accelerated determination of creep of lightweight aggregate concrete. Civil Eng Public Works Rev Lond 1973;68(803):515−9.

[45] Brooks JJ, Neville AM. Estimating long-term creep and shrinkage from short-term tests. Mag Concr Res 1975;27(90):3−12.

[46] Brooks JJ, Neville AM. Predicting long-term creep and shrinkage from short-term tests. Mag Concr Res 1978;30(103):51−61.

[47] Brooks JJ. Accuracy of estimating long-term strains in concrete. Mag Concr Res 1984; 36(128):131−44.

[48] Neville AM. Creep of concrete as a function of its cement paste content. Mag Concr Res 1964;16(46):21−30.

[49] Troxell GE, Raphael JM, Davis RE. Long-time creep and shrinkage tests of plain and reinforced concrete. In Proceedings ASTM, vol. 58; 1958. 1101−1120.

[50] L'Hermite RG, Mamillon M, Lefevre C. Nouveaux resultats de recherches sur la deformation et la rupture du beton. Paris Ann Inst Tech Batim Trav Publics 1965;18(207−8): 323−60.

[51] Hanson TC, Mattock AH. The influence of size and shape of member on the shrinkage and creep of concrete. ACI J 1966;63:1017−22.

[52] Brooks JJ, Al-Quarra H. Assessment of creep and shrinkage for the Flintshire bridge. Struct Eng 1999;77(5):21−6.

[53] BS 5400−4:1990, Steel, concrete and composite bridges, code of practice for design of concrete bridges, British Standards Institution. Now withdrawn (replaced by BS EN 1992-2: 2005).

[54] ACI Committee 209.1R-05. Report on factors affecting shrinkage and creep of hardened concrete. American Concrete Institute; 2008. 12 pp.

12 Creep of Masonry

In practice, the overall movement of a structure depends on the net combined effects of elasticity, creep, and thermal movements as well as shrinkage or moisture expansion. Consequently, composite structures involving different materials and restraint can lead to complex movements involving manifestations of creep, such as stress redistribution and relaxation. To avoid cracking and failure, designs for differential movements arising from the use of different materials require the provision of movement joints in composite brickwork clad buildings. Creep causes transfer of stress from brickwork to steel in reinforced brickwork, and loss of prestress in post-tensioned fin and diaphragm walls, but creep can be beneficial in relieving stresses induced by moisture and temperature gradients, and stress concentration in statically indeterminate structures.

Creep of masonry is affected by several factors in a similar manner to elasticity and shrinkage of masonry. It follows that the type of unit, type of mortar, geometry, and time are important factors and, also, the interactive effect of water transfer from mortar to unit just after dry units are laid in the fresh mortar. Additional factors affecting creep are applied external stress, age at loading, and anisotropy.

This chapter opens with a summary of milestones of creep research publications for clay masonry, followed by detailed discussions of the influencing factors. Methods of estimating creep for design are then presented, including Code of Practice guidelines. A data bank of experimental creep results is presented and used to assess the accuracy of the different parameters to quantify creep as well as to compare the accuracy of methods of prediction for clay brickwork.

Historical Background

In 1957, Nylander and Ericsson [1] tested piers constructed from clay, lightweight concrete, and concrete bricks under sustained compressive stresses of 0.8, 0.3 and 0.6 MPa, respectively, and the resulting 400 day specific creep was 25, 80 and 80×10^{-6} per MPa. Lime-rich mortars tended to produce considerably more creep than cement-rich mortar, an observation that was confirmed later by Sahlin [2] who reported tests on clay masonry in which the stress was incrementally increased up to 1 MPa. Poljakov [3] tested brickwork prisms and stated ultimate creep to be between 85% and 135% of the elastic strain for loads of between 0.4 and 0.6 stress/strength ratio. Specimens loaded at the age of 4 days developed 50% more creep than those loaded at 10 days, thus demonstrating that age at loading was an influencing factor. Poljakov also found that compressive loads applied at eccentricity/depth ratios of 0.15−0.35 had no significant effect on creep. An exponential-type expression for creep was developed as a function of age at loading, stress/strength ratio, duration of load, and type of brickwork [3]

Concrete and Masonry Movements. http://dx.doi.org/10.1016/B978-0-12-801525-4.00012-1

After developing apparatus to measure creep [4], the first experiments in the UK were carried out by Lenczner in 1969 [5] using brickwork built with half-scale bricks. However, it was reported that measured values of creep were greater than expected and not thought to be representative of clay brickwork built with full-scale bricks. Subsequently, Lenczner [6,7] used full-scale clay bricks to assess the effect of stress level, mortar type, and damp-proof course (dpc) on creep of hollow brickwork piers. It was found that creep was generally proportional to stress and was greater when built with a weaker mortar. Brickwork built from stronger clay bricks exhibited less creep, the presence of a dpc increased creep of the whole brickwork as well as that above the dpc, and creep increased with a reduction in the relative humidity of storage [7]. Tatsa et al. [8] tested prestressed hollow concrete block prisms and aerated concrete prisms with and without mortar joints after storing in two different environments: (1) dry for at least a week before prestressing, and (2) 24 h under water before prestressing; subsequent storage was 20 °C and 50% relative humidity (RH). For the two curing conditions, the reported creep was very small ranging from 11 to 16×10^{-6} after 210 days, and the ratio of mortar creep to block creep was 4.4 for dry-stored masonry and 16.8 for pre-soaked masonry. For aerated concrete masonry, the corresponding values were 2.3 and 8.9.

The effect of stress on creep of crushed limestone aggregate concrete blockwork hollow piers was investigated by Lenczner in 1974 [9]. In the following year, the effect of stress was investigated on brickwork piers by Lenczner [10] (see Fig. 12.1). He found that creep approximately proportional to the applied stress but, for the same stress/strength ratio, creep in blockwork was much higher than in brickwork. There was little effect on creep of story-high single-leaf clay brick walls resulting from changing the age of loading from 14 to 28 days [11]. Creep of single-leaf brickwork walls was greater than creep of brick piers [11−13], and the earlier finding that creep was less for brickwork built from a stronger clay brick was confirmed [11]. For the same type of mortar and for a given geometry, creep was approximately inversely proportional to the square root of clay brick strength [14], Schubert [15] reported values of ultimate creep coefficient (ratio of ultimate creep to elastic strain) for single-leaf masonry: 0.8−2.3 for lightweight aggregate concrete, 1.0−4.0 for aerated concrete, 0.4 to 1.3 for clay bricks and 1.1−1.9 for calcium silicate bricks.

After finding that creep of cavity walls was greater than creep of piers [16], Lenczner [17] issued design guidelines for creep of clay brickwork in 1980. A hyperbolic equation was found to satisfactorily describe creep as a function of time, and walls built from clay bricks soaked prior to laying showed less creep than for walls built from dry bricks [18].

Creep of lightweight aggregate, square hollow-concrete block, 5-stack bonded prisms was determined by Ameny et al. [19]. It was found that creep was proportional to the stress/strength ratio when the latter varied between 0.17 and 0.4 and the relationship was not affected by an eccentricity of loading of 0.17 of the prism thickness (200 mm). Using a hyperbolic creep-time function, ultimate creep coefficient ranged from 1.09 to 1.64. Creep in the masonry was 18−43% greater than creep in the lightweight concrete blocks themselves. The same authors proposed a composite model, which was followed by further models by Ameny et al. [20]. Those and other composite models for representing creep of masonry are dealt with Chapter 3.

Calcium silicate brick walls exhibited an extrapolated ultimate creep of 122×10^{-6} per MPa, which was 1.45 times the elastic strain [21]. The influence of geometry of cross-section of masonry on creep, as first reported by Lenczner [16], was quantified in terms of the ratio of volume/exposed surface area (V/S) to drying [22−25]. Research using a wide range of unit types [26] confirmed that creep of clay brickwork was significantly affected; in general terms, creep decreased as the unit strength increased. The particular cases of stress relaxation [26] and creep under varying stress [27] were investigated in 1994, which was followed by the measurement of loss of prestress due to creep in post-tensioned masonry [28]. In a similar manner to the effect of unit strength, it was found that creep of masonry decreased as the strength of mortar increased [29]. Creep also decreased as the age at loading increased, but only for ages less than 14 days [30−33], a finding that supplemented earlier findings [3,10]. In 1997, Shrive et al. [34] reported 7-year creep for a range of stress, loading age, and storage condition. Anisotropy of creep of masonry was investigated in 1998 [35], and the important role played by water transfer from freshly laid mortar to dry units on creep of bonded units and mortar was identified in 2000 [36]. Subsequent publications were concerned with measurement of creep of hydraulic lime mortar [37], prediction of creep of mortar [38], the problem of creep testing of clay masonry exhibiting cryptoflorescence [39], and estimation of creep of clay brickwork [40].

Influencing Factors

Stress

Figure 12.1 shows three creep−stress characteristics obtained by Lenczner et al. [10] and Lenczner [12] from tests on hollow brick piers. In the case of the Fletton clay brick piers, stresses up to 55% of the pier strength were applied but in the case of high-strength Butterly brick piers, creep under lower stresses of up to 20−25% of the strength was measured. For the latter brick piers, the creep−stress curves are slightly non-linear but, for practical purposes, creep may be assumed to be proportional to stress. In the case of the Fletton brick piers, it can be seen that throughout the complete range of stress, creep varies in a non-linear way, with creep increasing at a faster rate than stress. The shape of the creep−stress curves implies an increasing contribution of cracking to creep at high levels of stress. At low levels of stress and within the range of usual working stresses (0−3 MPa or 0 to 0.33 stress-strength ratio), creep of both the Fletton brick and Butterly brick piers can be assumed to be approximately linear with respect to stress. Hence, the term specific creep can be justified, i.e., creep per unit of stress (10^{-6} per MPa), to quantify creep for lower levels of stress.

Geometry of Cross-Section

Lenczner's results are plotted in Figure 12.2, where it can be seen that creep of piers is much less than creep of single-leaf walls. His original data were given in terms of the strain ratio (S_r), which was defined as the total load strain (elastic plus creep) divided

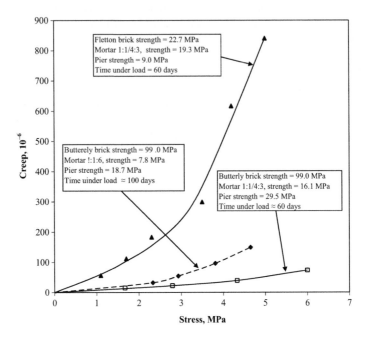

Figure 12.1 Creep hollow brick piers subjected to different stresses at the age of 28 days [10,12].

Figure 12.2 Influence of cross-section geometry on creep of clay brickwork [17].

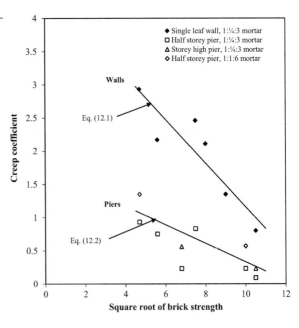

by the elastic strain; the creep coefficient, which is plotted in Figure 12.2, is given by S_r-1. Since the modulus of elasticity and, hence, elastic strain are unaffected by cross-section size and shape of masonry (see Chapter 5), the influence of geometry on creep coefficient is the same as that on creep. Lenczner found the average creep coefficient (ϕ_w) for walls was a function of brick strength as follows:

$$\phi_w = 4.46 - 0.33\sqrt{f_{by}} \tag{12.1}$$

while, for solid brick piers, the average creep coefficient (ϕ_p) was:

$$\phi_p = 1.73 - 0.14\sqrt{f_{by}} \tag{12.2}$$

More detailed investigations into the effect of geometry of clay, calcium silicate, and concrete masonry on creep were reported later [22–25]. Similar patterns of behaviour occurred for the three types of unit, the results for Class B Engineering clay brickwork versus time under load being shown in Figure 12.3. It can be seen that there is a consistent difference in creep at any time for masonry having different cross-sections or volume/surface area ratios. The geometry effect on creep is similar to that on shrinkage as discussed in Chapter 7, the explanation being that creep is less for larger cross-sections because moisture diffusion from the inside to the environment is slower and more difficult due to longer drying path lengths. The 150 day creep shown in Figure 12.3 is plotted as a function of volume/surface ratio in Figure 12.4 together with other types of masonry, where it can be seen that the patterns of behaviour are similar.

According to Figure 12.4, the average creep of a solid pier relative to a single-leaf wall is approximately 0.7, which is greater than the average value of 0.33 as deduced

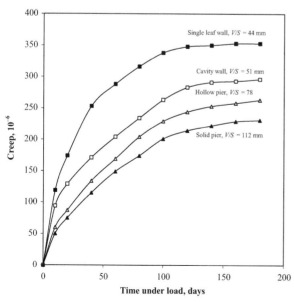

Figure 12.3 Creep - time characteristics of clay brickwork having different geometries [22]; V/S = volume/surface ratio.

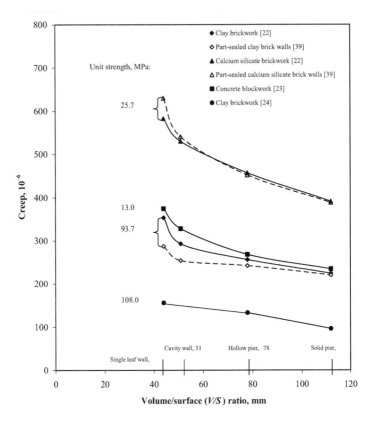

Figure 12.4 Creep as a function of volume/surface area ratio for different types of masonry after 150 days under load.

from Figure 12.2 [17]. The explanation for the difference could be the pre-storage conditions of the brickwork. In Figure 12.2, the drying environment was 50% RH before application of load compared with a mostly sealed environment ($\approx 100\%$ RH) in the case of Figure 12.4. A lower relative humidity for pre-storage would increase pre-drying and have the effect of reducing creep potential, possibly to a greater extent in piers than in single-leaf walls [23].

Compared with vertical creep, *lateral creep* of clay and calcium silicate brickwork was found to be small (-20 to 60×10^{-6}) with no influence of geometry [23] and of a similar level to that reported in an earlier investigation [21]. For practical purposes, *creep Poisson's ratio* can be assumed to be 0.1 and equal to the elastic Poisson's ratio.

Type of Mortar

Lenczner [6] found that creep of Butterly Class B Engineering brick piers was greater when built with a weaker mortar. The average specific creep after 120 days was

49.6×10^{-6} per MPa for a 1:¼:3 cement-lime-sand mortar of 28 day strength 16.1 Ma, compared with an average specific creep of 61.6×10^{-6} per MPa for a 1:1:6 cement-lime-sand mortar having a 28 day strength of 7.8 MPa. For concrete masonry built with different mortars of approximately the same 7 day strength of 4.3 MPa, Ameny et al. [18] reported a 110 day specific creep of 176×10^{-6} per MPa for a 1:3 masonry cement-sand mortar compared with a corresponding value of 136×10^{-6} per MPa for a 1:1:6 cement-lime-sand mortar.

Forth and Brooks [30] investigated the influence of seven different types of mortar (see Table 5.2) on creep of single-leaf brick walls and found that creep generally decreased as the mortar strength increased, as indicated in Figure 12.5. Corresponding creep tests of unbonded mortar prisms whose surfaces were part-sealed to the volume/surface area ratio of the walls (44 mm) followed the same pattern of behaviour (Figure 12.6) and, generally, the creep of mortar prisms was six times greater than creep of the walls. Analysis of the mortar creep-time characteristics was carried out to quantify the creep-strength relationship by the procedure described below.

Like the time dependency of masonry shrinkage discussed in Chapter 7, the development creep of masonry and its component phases is readily described by the Ross

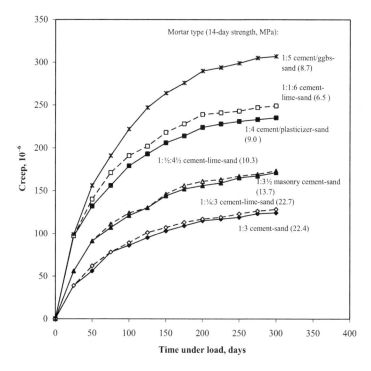

Figure 12.5 Creep of single-leaf walls built with a Class B engineering clay brick and different types of mortar subjected to a stress of 1.5 MPa [30].

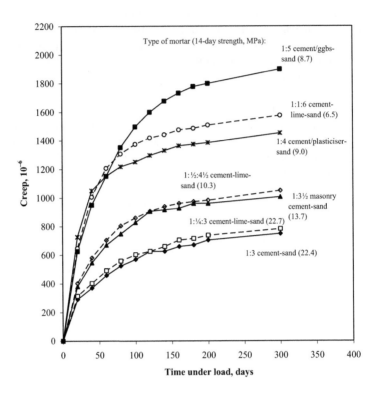

Figure 12.6 Effect of different types of mortar on creep of unbonded prisms part-sealed to a $V/S = 44$ mm; applied stress = 1.5 MPa [30].

hyperbolic equation (7.9). In the case of mortar, if C_m = specific creep (creep per unit of stress) at time t, then:

$$C_m = \frac{C_{m\infty} t}{a_{mc} C_{m\infty} + t} \qquad (12.3)$$

where $C_{m\infty}$ = ultimate specific creep and $a^{-1}{}_{mc}$ = initial rate of specific creep.

In rectified (linear) form, Eq. (12.3) becomes:

$$\frac{t}{C_m} = a_{mc} + \frac{1}{C_{m\infty}} t \qquad (12.4)$$

so that $C^{-1}{}_{m\infty}$ and a_{mc} can be determined, respectively, by the slope and intercept of the linear plot of t/C_{mt} versus t.

Using the creep-time characteristics of unbonded mortar prisms in Figure 12.6, together with data from other investigations [30,42–47], values of $C_{m\infty}$, and a_{mc}, were obtained for different V/S ratios from part-sealed prisms simulating a range of

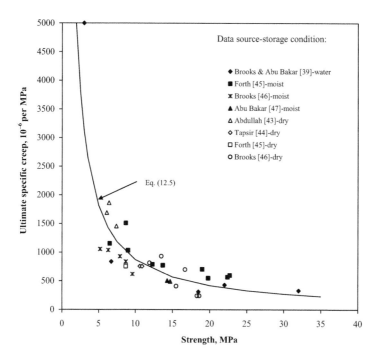

Figure 12.7 Ultimate specific creep of unbonded mortar prisms part-sealed to a $V/S = 44$ mm, as a function of strength at the age of loading; stored under different conditions prior to loading.

geometries of clay brickwork. The tests were carried out at the same time as those to assess the influence of V/S on shrinkage of mortar as described in Chapter 7. In the methods of predicting creep of concrete (Chapter 11), ultimate creep is expressed as independent functions of 28 day strength and age at loading. However, for masonry mortar, it was found to be more convenient to combine the two influences, i.e., to express ultimate specific creep as a function of strength at the age of loading. The results for a $V/S = 44$ mm ($\equiv$ single-leaf wall) are shown in Figure 12.7 for continuously water-stored or moist-cured, and for moist-cured followed by dry-cured mortar. The analysis revealed that, while $C_{m\infty}$ generally varied inversely with strength, the time term, $a_{mc}C_{m\infty}$, remained fairly constant with an average value of 22.0 days. The general relationship of Figure 12.7 is as follows:

$$C_{m\infty} = \frac{10000}{f_m^{1.06}}$$ (12.5)

It should be emphasized that Eq. (12.5) applies to creep of mortar bed joints of single-leaf walls having a $V/S = 44$ mm and, for other geometries of brickwork, the

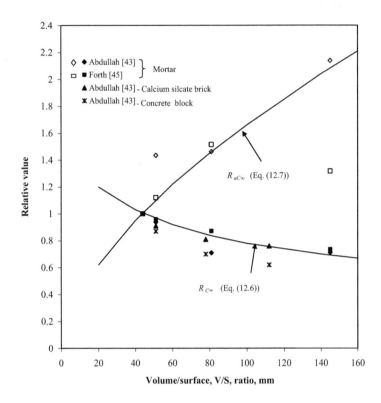

Figure 12.8 Relative coefficients, $R_{C\infty}$ and $R_{aC\infty}$, as a function of volume/surface ratio; equal to 1 when $V/S = 44$ mm.

relative values of ultimate specific creep ($R_{C\infty}$) and the time term ($R_{aC\infty}$) are functions of V/S according to Figure 12.8 and the following expressions:

$$R_{C\infty} = \frac{1.49 + 0.007\left(\frac{V}{S}\right)}{1 + 0.018\left(\frac{V}{S}\right)} \tag{12.6}$$

$$R_{aC\infty} = 0.15\left[\frac{V}{S}\right]^{0.5} \tag{12.7}$$

In fact, these expressions have already been presented in Chapter 7 to account for the influence of V/S on shrinkage of mortar, calcium silicate, and concrete units (Eq. (7.14) and (7.15)). It is apparent that there is a high degree of scatter for $R_{aC\infty}$, which is also the case for the equivalent shrinkage-time term $R_{aS\infty}$ (see Figure 7.17). However, the uncertainty in the level of $R_{aC\infty}$ does not affect long-term estimates of shrinkage and creep appreciably and only has a significant influence for short periods of exposure to drying or time under load.

It may be recalled that the volume/surface (V/S) influence applies to creep of drying masonry where moisture is lost to the environment, i.e., total creep, which probably applies to most practical situations. However, in other situations where moisture loss is prevented, such as in massive masonry elements or foundations covered by earth, creep is less, i.e., *basic creep* occurs. In those instances, there is no V/S influence and relative values of ultimate creep ($R_{c\infty}$) and time term ($R_{aC\infty}$) may be assumed to be those for a large V/S (200 mm), namely, $R_{c\infty} = 0.63$ and $R_{aC\infty} = 2.12$. In addition, for basic creep, the RH of storage is assumed to be 100% (see Eq. (12.8)).

The effect of $R_{c\infty}$ and $R_{aC\infty}$ on creep of mortar may be illustrated by comparing the mortar bed joints in a solid pier ($V/S = 112$ mm) and in a single-leaf wall ($V/S = 44$ mm) both types of masonry sealed for one day and then cured in a dry environment. Assuming a mortar strength, $f_m = 15$ MPa, the ultimate specific creep of the wall is given by Eq. (12.5), which is then multiplied by $R_{c\infty}$ from Eq. (12.6) to obtain the ultimate specific creep of the solid pier. For the single-leaf wall, $a_{cm}C_{m\infty} = 22$, which is multiplied by $R_{aC\infty}$ from Eq. (12.7) to obtain the time constant for the solid pier, The resulting values are listed here:

	Single-Leaf Wall	Solid Pier
$C_{m\infty}$	567×10^{-6} per MPa	425×10^{-6} per MPa
$a_{cm}C_{m\infty}$	22 days	35 days

Therefore, compared with the mortar bed joint of a single-leaf wall, the mortar bed joint of a solid pier has a lower ultimate specific creep and slower rate of development of creep as indicated by the greater time constant $a_{cm}C_{m\infty}$ of Eq. (12.3).

Like shrinkage of mortar, creep of mortar is affected by the *relative humidity of storage* after application of load, creep being greater the lower the relative humidity of storage. An example of that influence is the greater total creep under drying conditions compared with basic creep of sealed masonry, which is illustrated in Figure 12.9. All experimental mortar creep data used in the foregoing analysis to derive Eq. (12.5)–(12.7) were obtained under storage controlled conditions of 21 °C and 65% RH. For other humidity conditions, the following equation may be used, which is based on that given by the CEB-FIP 1990 Model Code for concrete [48]:

$$R_{RH} = 1.33 - 0.005RH \qquad (12.8)$$

where RH = relative humidity (%) and R_{RH} = relative humidity factor (=1 when RH = 65%).

When mortar is initially sealed or moist-cured or dry-cured before loading, substitution of Eq. (12.6)–(12.8) in Eq. (12.3) gives the specific creep at any time of any

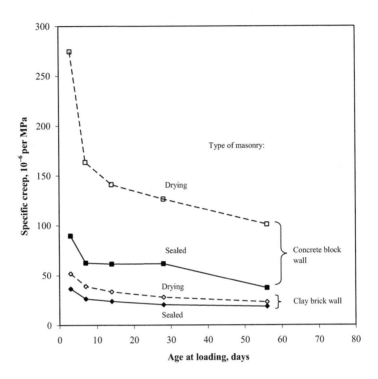

Figure 12.9 Influence of age at loading on creep of masonry after 160 days under load [31,33].

type of mortar subjected to load at any age, for any geometry of cross section and any relative humidity of storage, namely:

$$C_m = \frac{\left[\frac{10000}{f_m^{1.06}}\right] R_{C\infty} R_{RH} t}{22 R_{aC\infty} + t} \tag{12.9}$$

When only the ultimate specific creep of mortar is required:

$$C_{m\infty} = \left[\frac{10000}{f_m^{1.06}}\right] R_{C\infty} R_{RH} \tag{12.10}$$

The above expressions are dependent on the strength of mortar at the age of loading of the brickwork. If 28 day strength is available but the brickwork is to be loaded at a different age, then Table 5.12 may be used to estimate the strength of mortar at the required age of loading.

When assessed using test data [39,43−47], the accuracy of estimating creep of mortar in terms of the error coefficient defined by Eq. (12.36) is approximately 25% [39].

So far in this chapter, the discussion has concentrated on creep of Portland cement based mortar, there being little information available regarding movements of masonry built with lime mortar. However, based on extrapolation of 600 day creep of walls built from hydraulic lime-sand mortar (1:3), the estimated ultimate vertical specific creep was 150×10^{-6} per MPa [38]; the walls were two-brick wide $\times$ 1 m high single-leaf walls built with (undocked) handmade clay bricks of strength 18.6 MPa and suction rate of 2.25 kg/mm^2/min. After storage under polythene sheets, walls were subjected to load at a delayed age of 150 days because of the slow development of strength of the hydraulic lime-sand mortar; storage was at 21 °C and 65% RH. In fact, it was difficult to determine strength of the hydraulic mortar using cubes in the traditional way. The level of creep was less than that of companion walls built with a low strength ordinary Portland cement-lime-sand mortar (1:2:9) of strength 1.2 MPa, which had ultimate vertical specific creep of 306×10^{-6} per MPa. The secant modulus of elasticity was similar for both walls at approximately 3.9 GPa and, when compared in terms of ultimate creep coefficient, the values were 0.57 and 1.2, respectively, for the hydraulic lime-sand mortar wall and cement-lime-sand mortar wall.

Age at Loading

Lenczner and Salahuddin [11] found less creep occurred when age at loading increased. They tested single-leaf walls four bricks wide $\times$ 2.2 m high built from Staffordshire Blue Engineering bricks of strength 60.0 MPa and bonded with a 1:¼:3 cement-lime-sand mortar; after construction, the walls were stored in a controlled environment of 20 °C and 50% RH. After approximately one year under load, the levels of specific creep were 136 and 108×10^{-6} per MPa, when walls were loaded at the ages of 14 and 28 days, respectively.

The trend of less creep as age of loading increases is demonstrated in Figure 12.9 for clay brickwork and concrete blockwork. In those experiments, masonry was either covered with polythene sheets immediately after construction until the age at loading and then allowed to dry in a controlled environment of 21 °C and 65% RH [31] or sealed immediately after construction for the whole period of testing [32]. The materials used were a class B Engineering brick of strength 92.7 MPa and a dense aggregate concrete block of approximate strength 7.0 MPa, bonded with a 1:½:3½ cement-lime-sand mortar of 28-day strength 12.6 MPa. The sealed masonry corresponds to a "basic creep" condition, as defined in Chapter 2, where there is no loss of moisture to the environment, and the drying condition corresponds to "total creep" also as defined in Chapter 2. It can be seen that the decrease of creep is mostly affected at early ages of loading, particularly for concrete blockwork where possibly there was an age effect arising from creep of the block units as well as of mortar bed joints; the former is likely to occur in concrete blockwork constructed with newly made blocks.

Similar trends of a reduction in specific creep as the age at loading increased were observed with masonry walls and solid piers that were allowed to dry before application of load [31], which represents a more realistic situation in civil engineering construction. The creep coefficient also became less as the age at loading increased, but at a lower rate because the elastic modulus tended to increase with age.

Based upon long-term tests of five-stack masonry prisms built with three types of clay unit and three types of mortar, Sayed-Ahmed et al. [34] found that the effect of age at loading on creep reduces significantly after two weeks. Masonry loaded at the age of 7 days yielded 1.2 to 1.7 times the specific creep of masonry loaded at the ages of 14 and 28 days.

The age at loading influence on creep of masonry mainly arises from the effect of age on creep of mortar bed joints and, as stated on page 411, is taken into account in composite model analysis by adopting the strength of mortar at the age of loading to quantify ultimate creep of mortar.

Anisotropy

Creep measured in the vertical and horizontal directions of single-leaf walls is compared in Figure 12.10 [36]; full details of the experiments have already been described in Chapter 5. After curing under polythene sheet for 14 days, the walls were exposed to drying at 21 °C and 50% RH and subjected to a stress of 1.5 MPa. Figure 12.10 demonstrates the anisotropy of three types of wall, and the difference in creep in the vertical and horizontal directions is appreciable ranging from 20% to 65%. For the clay and calcium silicate brickwork, the trend of more vertical creep is as expected according to composite model theory (see Figure 3.10) but, for concrete blockwork, the behaviour is reversed in that horizontal creep is greater than vertical creep. That latter behaviour is, in fact, predicted by composite model theory when masonry is built with a low modulus and high creep unit.

Hydraulic lime-sand mortar appears to behave in a different manner to cement-lime-sand mortar with regard to anisotropy of creep [38]. Tests using very low strength mortars (≈ 1 MPa) indicated that while the wall built with cement-lime -sand mortar exhibited more creep in the vertical direction, the wall built with hydraulic lime-sand mortar underwent more creep in the horizontal direction.

Type of Unit

The vast majority of research into creep of masonry has been reported for clay brickwork built from different types of clay unit, and this section now deals with the influence of clay unit itself and, in particular, its strength before discussing the corresponding situation for calcium silicate and concrete units.

Clay Bricks

As shown in Figure 12.2, Lenczner's experimental results indicate that, for the same type of mortar, creep of clay brickwork decreases as the unit strength increases, his general relationships between creep coefficient of clay brickwork and strength of unit being given by Eq. (12.1) and (12.2). In a later paper, Warren and Lenczner [18] also demonstrated that creep of story-high brick walls was less when built with bricks of greater strength, but the moisture state of clay bricks at the time of

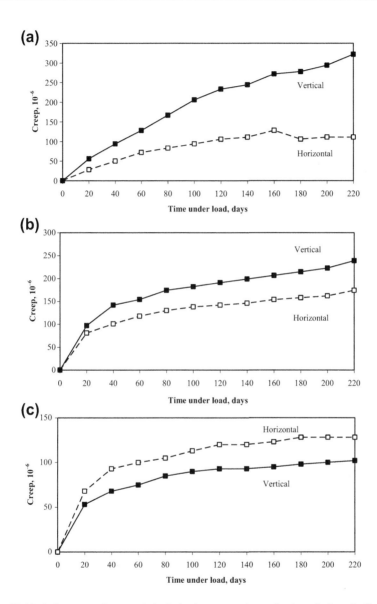

Figure 12.10 Anisotropy of creep of single-leaf masonry due to load applied vertically (bed face direction) and horizontally (header face direction) [36]. (a) Clay brick wall (b) Calcium silicate brick wall (c) Concrete block wall.

construction was found to be an important influence on the creep-strength relationship (Figure 12.11); the moisture state effect is discussed in the next section.

A comprehensive program of tests to investigate the influence of clay unit type on creep of two brick wide × 1 m high single-leaf walls revealed the creep-time

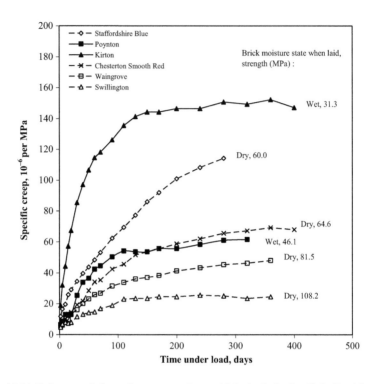

Figure 12.11 Influence of clay unit on creep of story-high single-leaf walls built with a 1:¼: 3 cement-lime-sand mortar, cured and stored at 50% RH and loaded at the age of 28 days [18].

characteristics shown in Figure 12.12(a), creep being measured in the vertical direction. The same pattern of behaviour is reflected in corresponding creep of unbonded units, but measured between header faces (C_{bx}) as shown in Figure 12.13. Like the trend of Figure 12.2, with the exception of the Birtley Old English clay unit, the general trend is one of less creep the higher the unit strength. The creep-time curves of Figure 12.13, together with other experimental data [11,12,23,41,44−46,49], were analyzed in the same way as the mortar creep-time curves described earlier in this chapter using Eq. (12.3) to obtain values of the ultimate specific creep, $C_{b\infty}$, and the rate parameter, a_{cb}. To allow for anisotropy and to obtain creep between bed faces (C_{by}), C_{bx} was multiplied by the elastic modulus ratio, E_{bx}/E_{by} (Table 5.2), the assumption being that anisotropy of creep is proportional to the anisotropy of elastic strain at loading or inversely proportional to the anisotropy of elastic modulus.

Whereas it was reported earlier in this chapter that the ultimate specific creep and creep rate parameters of mortar were dependent on volume/surface area (V/S) ratio (as simulated by part-sealing of specimens), in the case of clay units it was apparent that V/S ratio was not a factor, and the rate parameter, a_{cb}, was reasonably constant

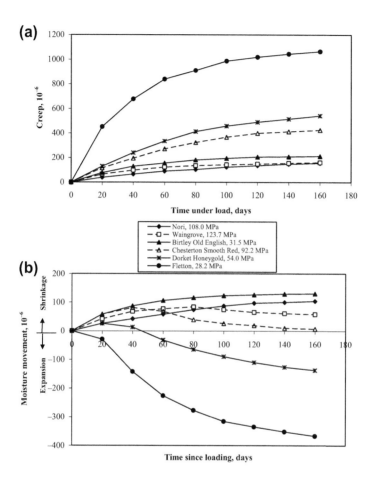

Figure 12.12 (a) Creep of 13-course-high single-leaf walls built from different types of clay unit subjected to a stress of 1.5 MPa and (b) moisture movement of corresponding control walls; unit strengths are given in the legend [26].

with an average of 3.27. Figure 12.14 shows the ultimate specific creep of clay bricks between bed faces could be generally expressed as a function of unit strength:

$$C_{by\infty} = \frac{1700}{f_{by}} - 7 \tag{12.11}$$

and, hence, from Eq. (12.3) the development of specific creep with time is:

$$C_{by} = \frac{\left[\frac{1700}{f_{by}} - 7\right]t}{3.27\left[\frac{1700}{f_{by}} - 7\right] + t} = \frac{(1700 - 7f_{by})t}{3.27(1700 - 7f_{by}) + f_{by}t} \tag{12.12}$$

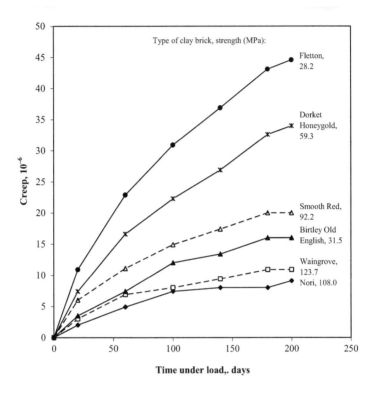

Figure 12.13 Creep different types of clay brick measured between header faces under a stress of 1.5 MPa [45].

It should be reiterated that, unlike creep of mortar, there are no other significant influencing factors in creep of clay units other than strength, stress, and time under load.

Calcium Silicate and Concrete Units

Creep of calcium silicate bricks and concrete blocks is influenced by their strength in a similar manner to creep of clay units and mortar. However, unlike clay units, creep of calcium silicate bricks and concrete block units is influenced by geometry of cross-section and relative humidity of storage in a similar manner to creep of mortar.

Along with clay units and mortar, the development of creep with time of calcium silicate bricks and concrete blocks has been found to be adequately described by Eq. (12.3) and analysis of experimental data [42−44,46] has revealed the average values of $a_{cb}C_\infty$ as listed in Table 12.1. However, it should be mentioned, that compared with mortar, the values are based on very limited test data. In fact, no experimental creep-time results appear to be available for lightweight aggregate or

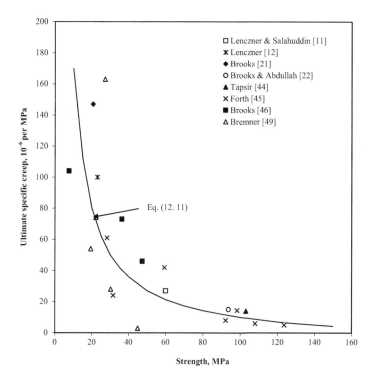

Figure 12.14 Ultimate specific creep of clay units between bed faces as a function of strength [41].

Table 12.1 Average Values of the Time Term, $a_{cb}C_{by\infty}$, for Eq. (12.3) Applied to Units

	Coefficient, $a_{cb}C_{by\infty}$, Days	
Component of Masonry	**Part-Sealed (V/S of Bonded Unit/ Mortar Joint in Single-Leaf Wall)**	**Part-Sealed (V/S of Unbonded Unit/Specimen in Creep Test)**
Concrete block	29 (V/S = 44 mm)	21 (V/S = 25 mm[a])
Calcium silicate brick	44 (V/S = 44 mm)	27 (V/S = 20 mm[b])
Mortar	22 (V/S = 44 mm)	13 (V/S = 19 mm[c])

[a]100 × 100 × 215 mm cut specimen with sealed ends.
[b]with sealed header faces.
[c]75 × 75 × 200 mm specimen with sealed ends.

autoclaved aerated blocks and, in the absence of such data, it is recommended that the concrete block value be used. For the same reason, it was not possible to deduce ultimate specific creep - strength functions for all types of concrete block, and the proposed common curve shown in Figure 12.15 should be regarded as tentative as it is

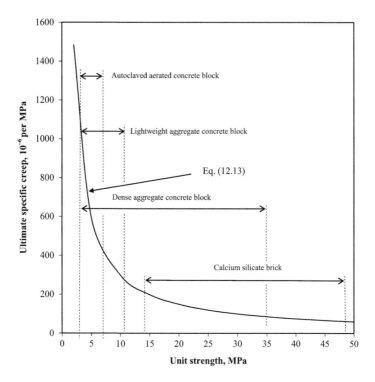

Figure 12.15 Proposed ultimate specific creep of calcium silicate and concrete block units between bed faces as a function of unit strength for a $V/S = 44$ mm ($\equiv$ bonded in a single-leaf wall).

based only on limited creep results for calcium silicate bricks dense aggregate concrete blocks [9,42,44]. The proposed relationship between ultimate specific creep (10^{-6} per MPa) and unit strength (MPa) is as follows:

$$C_{by\infty} = \frac{2966}{f_{by}}$$
(12.13)

The values given in Table 12.1 and Figure 12.14 are based on the V/S ratios stated, and for other geometries of masonry, the relative values of Figure 12.8 or Eq. (12.6) and (12.7) may be used to adjust ultimate specific creep, $C_{by\infty}$, and the time constant, $a_{cb}C_{by\infty}$. Again, although based on limited test results, analysis indicated that the general trends of Figure 12.8 for creep of mortar also apply to creep of calcium silicate and dense aggregate concrete blocks. It is emphasized that relative ultimate creep and relative rate constant apply to bonded masonry materials having a $V/S = 44$ mm, i.e., corresponding to embedment in a single-leaf wall. For example, suppose one is required to develop a creep-time expression for concrete blocks to be embedded in a

Table 12.2 Effect of Header Platen Restraint of Moisture Movement of Control Walls on Assessment of Creep under a Stress of 1.5 MPa [40]

	Type of Clay Unit			
Property	Waingroves, Coal Measure Shale (extruded)	Capel, Weald (Pressed)	MH 1st, Brickearth (Pressed)	Heather, Keuper Marl, (Pressed)
Compressive strength, MPa	84.0	19.4	22.1	30.2
Water absorption, %	7.2	16.9	24.2	20.2
Initial suction rate, kg/m^2/min	0.4	1.4	3.1	2.1
Mortar 28 day strength, MPa	15.7	7.0	14.9	11.6
Moisture movement (header face), 10^{-6}	-138	-77	-38	-85

	Type of Brickwork							
	Waingroves		Capel		MH 1st		Heather	
Property	No Plate	With Plate	No Plate	With Plate	No Plate	With Plate	No Plate	With Plate
Control wall vertical moisture movement, 10^{-6}	-206	-8	-324	92	-2012	-706	-773	-220
Creep, 10^{-6}	600	402	900	484	2740	1430	940	387

−ve sign = expansion.

blockwork solid pier. The required $V/S = 127$ mm (see Figure 7.17) and, from Table 12.2, $a_{cb}C_{by\infty} = 21$ days when $V/S = 25$ mm, so that using the relative values given by Eq. (12.8), when $V/S = 127$ mm, $a_{cb}C_{by\infty} = 21 \div 0.75 \times 1.68 = 47$ days. From Eq. (12.13), the ultimate specific creep for a block strength of 10 MPa ($V/S = 25$ mm) is 297×10^{-6} per MPa, so that using the relative values given by Eq. (12.7), for a $V/S = 127$ mm, the ultimate specific creep $= 297 \times 10^{-6} \div 1.148 \times 0.72 = 186 \times 10^{-6}$ per MPa. Hence, the required specific creep (10^{-6} per MPa) expressed as a function of time is:

$$C_{by} = \frac{186t}{47 + t}$$

Unit/Mortar Bond Interaction

Warren and Lenczner [18] found that *moisture state* of clay bricks at the time of construction is important influence on the creep-strength relationship. They compared single-leaf walls built with different types of docked (wetted) bricks with walls built with different types of dry bricks. For either moisture condition, Figure 12.11 confirms the earlier observation that creep of walls is less when built with greater brick strengths, but the creep-strength relationship depends on the moisture condition of units.

Even when laid dry, the unit moisture state of the unit is determined by its water absorption properties at the time of laying and the subsequent curing conditions of the masonry before application of stress. As was demonstrated with elasticity and shrinkage, those effects change the deformation characteristics of the mortar joints and in particular the bed joints but, except in the case of shrinkage of concrete and calcium silicate units, have little or no effect on unit properties. In the case of creep of mortar joints, the effects of unit absorption and loss of moisture due to pre-drying before loading have been investigated [41] and found not to be appreciably affected by unit absorption when masonry is cured under polythene for some appreciable time before application of load. However, when masonry is exposed to a drying environment soon after constructing masonry with dry units, there is a rapid loss of moisture from bed mortar joints and a reduction of mortar creep resulting from subsequent application of load, which can be quantified by the *mortar creep reduction factor*, γ_c, in terms of the unit absorption, W_a, namely:

$$\gamma_c = \frac{1 - 0.03W_a}{1 + 0.195W_a} \tag{12.14}$$

The mortar creep reduction factor (Eq. (12.14)) is identical to one of the two cases proposed for the elasticity reduction factor (Eq. (5.16)). In the case of masonry prevented from significant pre-drying before application of load or built with docked units, $\gamma_c = 1$.

Creep and Cryptoflorescence

While investigating the influence of unit type on creep of 13-course-high single-leaf clay brickwork walls, Brooks and Forth [26] found that the moisture movements of the control walls (see Figure 12.12(b)) were unusually large vertical expansions in the cases of Dorket Honeygold brickwork (140×10^{-6}) and Fletton brickwork (367×10^{-6}). Those expansions were, in fact, much greater than the irreversible expansions of the corresponding unbonded bricks of 53 and 95×10^{-6}, respectively. Thus, the control walls had a brickwork/brick expansion ratio of much greater than unity, which indicated the presence of enlarged moisture expansion due to cryptoflorescence as discussed in Chapter 9. Consequently, the deduced creep was exaggerated as illustrated schematically in Figure 9.11. In fact, it is likely that the "true" moisture movement of the Dorket Honeygold and Fletton brickwork (as would be indicated, say, by control walls of much larger height and section (see p. 278) was a small shrinkage and this would have a significant effect by lowering the deduced creep.

For example, in the case of Fletton brickwork, the total time-dependent load strain after 160 days (measured on the stressed wall and therefore not affected by cryptoflorescence is equal to the sum of creep and moisture expansion in Figure 12.12, namely, $1065 + (-367) \times 10^{-6} = 698 \times 10^{-6}$. If the "true" moisture strain is shrinkage $= 50 \times 10^{-6}$, then "true" creep $= 698 - 50 \times 10^{-6} = 648 \times 10^{-6}$, which is almost 40% less than the creep as deduced from the actual measured moisture strain of Figure 12.12(b).

It should be noted that in the above-mentioned tests [26], the control walls were not capped with steel header plates unlike the loaded walls used for creep determination and, at the time, it was not realized that the lack of in-plane restraint from header plates may have influenced the extent of enlarged expansion and, hence, deduced creep. A subsequent investigation [40] showed this to be the case after comparing creep deduced from moisture movement of 13-course-high single-leaf control walls with and without steel capping header plates. In those tests, four types of clay unit were chosen to build walls that were known from earlier tests to undergo enlarged moisture expansions due to cryptoflorescence [50]. Enlarged expansions did reoccur, and Table 12.2 confirms that, for all four walls without header plates, they were well in excess of the irreversible moisture expansions of their respective unbonded bricks. Thus the consequence of incorporating steel header plates on the control walls was a significant reduction in deduced creep in all four instances.

Table 12.2 also shows that three of the units used to build the walls featured high water absorptions, high initial suction rates, and low strengths, which are typical properties that tend to be associated with enlarged expansion of brickwork used in small-scale laboratory tests. However, as discussed in Chapter 9, that is not always true as can be seen in the case of the Waingroves unit, which had a much higher strength and much lower water absorption and initial suction rate.

In the above experiments [40], the 13-course-high single-leaf walls were cured for one day under polythene and then exposed to a drying environment of 65% RH before the creep walls were subject to a stress of 1.5 MPa at the age of 7 days. Without steel header capping plates, brickwork/brick expansion ratios could be extremely high, for example, by a factor of 53 in the case of the MH 1st control wall. Although the 50-mm-deep capping plates imposed only a small compressive stress of 0.02 MPa (compared with an applied external stress in the creep test of 1.5 MPa), they significantly suppressed the enlarged moisture expansion and, in consequence, reduced creep by as much as 60%, as can be seen from Table 12.2. However, in the cases of MH 1st and Heather brickwork, enlarged expansions were not totally eliminated. Recalling the importance of good site practice requiring units that have an initial suction rate in excess of $1.5 \, kg/m^2/min$ to be wetted prior to being laid, the two types of unit mentioned would be involved. Depending on the degree of wetting (time of soaking), enlarged expansion would be reduced, but not necessarily eliminated and, in consequence, the deduced creep would also be affected by the degree of wetting. The recommended procedure for the determination of creep of clay brickwork that may be prone to cryptoflorescence are discussed in Chapters 9 and 16.

Prediction of Creep

In this section, the current methods for predicting creep of masonry, including Codes of Practice recommendations, are presented, and expressions for practical use are developed based on the composite model approach. The latter method together with that of Lenczner, BS 5628-5:2005, BS EN 1996-1-1:2005 and ACI 530-05 are assessed for accuracy. In addition to specific creep, two alternative terms used to quantify creep are assessed for accuracy: (1) total load strain (elastic strain plus creep) per unit of stress, also known as the *creep function* or *compliance*, and (2) *creep coefficient* also known as the *creep ratio*, which is the ratio of creep to the elastic strain. The creep function is included in the assessment because of possible inaccuracies in the measurement of the starting point for creep arising from the uncertainty of defining the precise demarcation of creep and the initial elastic strain. The same situation occurs with concrete elements and is why use of creep function is sometimes preferred to quantify creep in concrete technology [51].

Lenczner's Method

As stated earlier in this chapter, Lenczner [17] related creep of masonry to unit strength, f_{by}, and proposed two separate relationships for walls and solid piers, which are given in terms of the creep coefficient by Eq. (12.1) and (12.2), respectively. In terms of ultimate specific creep, $C_{wy\infty}$, the expressions are:

$$\underline{Walls:}\ C_{wy\infty} = \frac{4.46 - 0.33\sqrt{f_{by}}}{10^3 E_{wy}} = \frac{4.46 - 0.33\sqrt{f_{by}}}{3750\sqrt{f_{by}} - 10000} \tag{12.15}$$

$$\underline{Piers:}\ C_{wy\infty} = \frac{1.73 - 0.14\sqrt{f_{by}}}{10^3 E_{wy}} = \frac{1.73 - 0.14\sqrt{f_{by}}}{3750\sqrt{f_{by}} - 10000} \tag{12.16}$$

where E_{wy} is the modulus of elasticity of masonry (GPa) as given by Eq. (5.18).

When expressed as the ultimate creep function, $J_{wy\infty}$, the above expressions become:

$$\underline{Walls:}\ J_{wy\infty} = \frac{5.46 - 0.33\sqrt{f_{by}}}{3750\sqrt{f_{by}} - 10000} \tag{12.17}$$

$$\underline{Piers:}\ J_{wy\infty} = \frac{2.73 - 0.14\sqrt{f_{by}}}{3750\sqrt{f_{by}} - 10000} \tag{12.18}$$

Lenczner's method was developed from tests on clay brickwork built from cement-lime-sand mortar mixes in the volumetric proportions, 1:¼:3 and 1:1:6 [17].

BS 5628-2:2005

This British Standard method [52] gives the creep coefficient as 1.5 and 3.0, respectively, for fired-clay and calcium silicate brick masonry, and dense aggregate concrete block masonry, so that the corresponding values of ultimate specific creep are $1.5 \times$ the elastic strain, and $3.0 \times$ the elastic strain. The specific elastic strain (strain per unit of stress) is the reciprocal of the modulus of elasticity, which is calculated from the characteristic strength of masonry, f_k. Hence, specific creep is:

Clay and Calcium Silicate Masonry

$$C_{wy\infty} = 1.5 \frac{1}{E_{wy}10^3} = 1.5 \left[\frac{1}{0.9f_k 10^3} \right] = \frac{1.67 \times 10^{-3}}{f_k} \tag{12.19}$$

Dense Aggregate Concrete Block Masonry

$$C_{wy\infty} = 3 \frac{1}{E_{wy}10^3} = \frac{3.33 \times 10^3}{f_k} \tag{12.20}$$

The corresponding creep functions are:

Clay and Calcium Silicate Brick Masonry

$$J_{wy\infty} = \frac{2.5}{0.9f_k 10^3} = \frac{2.78 \times 10^{-3}}{f_k} \tag{12.21}$$

Dense Aggregate Concrete Block Masonry

$$J_{wy\infty} = \frac{4}{0.9f_k 10^3} = \frac{4.44 \times 10^{-3}}{f_k} \tag{12.22}$$

The characteristic strength is dependent on unit strength and mortar designation, and is obtained from Tables 5.9 to 5.11.

BS EN 1996-1-1 (Eurocode 6):2005

BS 5628-2:2005 has been superseded by BS EN 1996-1-1; 2005 (Eurocode 6), which includes the UK National Annex [53]. It is recommended that final creep coefficient should be evaluated from tests performed for the project or available from a database. However, in the absence of those sources, BS EN 1996-1-1 lists ranges of creep coefficient for different types of masonry, and these are shown in Table 12.3.

Table 12.3 Final Creep Coefficients of Masonry According to BS 1996-1-1:2005

Type of Masonry	Final Creep Coefficient, $\varphi_{wy\infty}$	
	Range	UK National Annex
Clay	0.5 to 1.5	⎱
Calcium silicate	1.0 to 2.0	
Dense aggregate concrete	1.0 to 2.0	
Manufactured stone and lightweight aggregate concrete	1.0 to 3.0	⎬ 1.5
Autoclave aerated concrete	0.5 to 1.5	⎰
Natural stone	Very low	Normally very low

Since $\varphi_{wy\infty} = C_{wy\infty}\, E_{wy}10^3$, then in general terms the ultimate specific creep is:

$$C_{wy\infty} = \varphi_{wy\infty}\left[\frac{1}{E_{wy}10^3}\right] = \frac{\varphi_{wy\infty}}{f_k 10^3} = \frac{\varphi_{wy\infty}}{Kf_{by}^{\alpha}f_m^{\beta}10^3} \tag{12.23}$$

and the creep function is:

$$J_{wy\infty} = \left[\frac{1}{E_{wy}10^3}\right]\left[\varphi_{wy\infty} + 1\right] \tag{12.24}$$

It can be seen that ultimate creep and creep function depend on the characteristic strength (Eq. (5.20)) which, in turn, depends on type and thickness of mortar joints, mortar type, unit type, and geometrical requirements of the unit. Full details of the various parameters are given in Chapter 5 but, for example, the ultimate specific creep for any type of masonry (except stone) built with general purpose mortar in the UK is:

$$C_{wy\infty} = \frac{1.5}{0.5f_{by}^{0.7}f_m^{0.3}10^3} = \frac{3 \times 10^{-3}}{f_{by}^{0.7}f_m^{0.3}} \tag{12.25}$$

and the corresponding creep function is:

$$J_{wy\infty} = \frac{5 \times 10^{-3}}{f_{by}^{0.7}f_m^{0.3}} \tag{12.26}$$

ACI 530.1R-05

The ACI 530.1R-05 [54] method simply specifies a single value of ultimate specific creep (termed the *coefficient of creep* by ACI) for different types of masonry. The values are given in Table 12.4 together with creep functions.

Table 12.4 Ultimate Creep and Creep Function of Masonry According to ACI 530.1R-05

Type of Masonry	Ultimate Specific Creep, $C_{wy\infty}$, 10^{-6} per MPa	Ultimate Creep Function, $J_{wy\infty}$, 10^{-6} per MPa
Clay	10	$10 + \dfrac{10^{-3}}{0.7f'_m}$ (see Eq. (5.25))
Concrete	36	$36 + \dfrac{10^{-3}}{0.9f'_m}$ (see Eq. (5.25))
Autoclave aerated concrete	73	$73 + \dfrac{10^{-3}}{6.5(f'_{aac})^{0.6}}$ (see Eq. (5.26)

NB. f'_m = specified strength of clay or concrete masonry; f'_{aac} = specified strength of AAC unit.

The ultimate creep function is dependent on the specified compressive strength of the masonry or unit (AAC), the specified compressive strength of clay and concrete masonry depending on the strength of unit, and type of mortar as detailed in Table 5.15.

Composite Model

The vertical creep of brickwork and blockwork masonry is given, respectively, by Eq. (3.51) and (3.53) (see Tables 3.1 and 3.2). Allowing for the development of creep with time, cross-section geometry, relative humidity of storage, and the mortar creep reduction factor, the vertical specific creep of masonry at any time is:

Clay Brickwork

After any time :

$$C_{wy} = 0.86\left[\frac{C_{by\infty}t}{3.27C_{by\infty}+t}\right] + 0.14\gamma_c\left[\frac{C_{m\infty}t}{22R_{aC\infty}+t}\right]R_{C\infty}R_{RH}$$

Ultimate :

$$C_{wy\infty} = 0.86C_{by\infty} + 0.14\gamma_c C_{m\infty}R_{C\infty}R_{RH}$$

$$(12.27)$$

Calcium Silicate Brickwork

After any time :

$$C_{wy} = 0.86\left[\frac{C_{b\infty}t}{44R_{aC\infty}+t}\right]R_{C\infty}R_{RH} + 0.14\gamma_c\left[\frac{C_{m\infty}t}{22R_{aC\infty}+t}\right]R_{C\infty}R_{RH}$$

Ultimate :

$$C_{wy\infty} = [0.86C_{b\infty} + 0.14\gamma_c C_{m\infty}]R_{C\infty}R_{RH}$$

$$(12.28)$$

Concrete Blockwork (Solid Blocks)

After any time :

$$C_{wy} = 0.952\left[\frac{C_{b\infty}t}{29R_{aC\infty}+t}\right]R_{C\infty}R_{RH} + 0.048\gamma_c\left[\frac{C_{m\infty}t}{22R_{aC\infty}+t}\right]R_{C\infty}R_{RH} \Bigg\}$$

Ultimate :

$$C_{wy\infty} = [0.952C_{b\infty} + 0.048\gamma_c C_{m\infty}]R_{C\infty}R_{RH}$$

(12.29)

In the above equations, expressions for the terms $C_{by\infty}$, $C_{m\infty}$, $R_{c\infty}$, $R_{aC\infty}$, R_{RH}, and γ_c have been given earlier in the text but, for convenience, are also summarized in Table 12.5.

Expressions for the creep function can now be readily developed by adding the elastic strains presented in Chapter 5 (Eq. (5.30) to (5.32)) to specific creep given by the above equations. For example, the equations for ultimate creep function are:

Extruded Wire-Cut Perforated Clay Brickwork

$$J_{wy\infty} = \frac{2.15}{f_{by}} + \frac{0.175}{\gamma_e f_m} + 0.86C_{b\infty} + 0.14\gamma_c C_{m\infty}R_{C\infty}R_{RH}$$

(12.30)

Pressed and Slop Molded Clay Brickwork

$$J_{wy\infty} = \frac{3.44}{f_{by}} + \frac{0.175}{\gamma_e f_m} + 0.86C_{b\infty} + 0.14\gamma_c C_m R_{C\infty}R_{RH}$$

(12.31)

Calcium Silicate Brickwork

$$J_{wy\infty} = \frac{1.87}{f_{by}} + \frac{0.175}{\gamma_e f_m} + [0.86C_{b\infty} + 0.14\gamma_c C_{m\infty}]R_{C\infty}R_{RH}$$

(12.32)

Concrete Blockwork (Solid Blocks)

$$J_{wy\infty} = \frac{1.058}{f_{by}} + \frac{A_m}{A'_m}\frac{0.06}{\gamma_e f_m} + \left[0.952C_b + 0.048\frac{A'_m}{A_m}\gamma_c C_{m\infty}\right]R_{C\infty}R_{RH}$$

(12.33)

Corresponding expressions for creep coefficient may also be derived, which are given by:

$$\phi_{wy} = C_{wy}E_{wy} = \frac{C_{wy}}{e_{wy}}$$

(12.34)

Table 12.5 Relationships of Terms in Eq. (12.27) and (12.33)

Term	Relationship	Eq. (no.) in Text
Ultimate specific creep of clay brick, $C_{by\infty}$, as a function of unit strength	$C_{by\infty} = \frac{1700}{f_{by}} - 7$	(12.11)
Ultimate specific creep of calcium silicate brick and concrete block, $C_{b\infty}$, as a function of unit strength	$C_{b\infty} = \frac{2966}{f_{by}}$	(12.13)
Ultimate specific creep of mortar, $C_{m\infty}$, as a function of strength	$C_{m\infty} = \frac{10000}{f_m^{1.06}}$	(12.5)
Relative ultimate creep as a function of V/S, $R_{C\infty}$ (= 1 when V/S = 44 mm) For basic creep, $R_{c\infty} = 0.63$ when V/S = 200 mm	$R_{C\infty} = \frac{1.49 + 0.007\left(\frac{V}{S}\right)}{1 + 0.018\left(\frac{V}{S}\right)}$	(12.6)
Relative time constant as a function of V/S, $R_{aC\infty}$ (= 1 when V/S = 44 mm) For basic creep, $R_{aC\infty} = 2.12$ when V/S = 200 mm	$R_{aC\infty} = 0.15\left[\frac{V}{S}\right]^{0.5}$	(12.7)
Change of relative humidity, R_{RH} (=1 when RH = 65%) For basic creep, $R_{RH} = 0.83$ when RH = 100%	$R_{RH} = 1.33 - 0.005 RH$	(12.8)
Mortar creep reduction factor, γ_c, as a function of unit water absorption, W_a	1. Masonry built with docked units or stored under polythene, $\gamma_c = 1$. 2. Masonry built with dry units, sealed for 1 day and then allowed to dry: $\gamma_c = \frac{1 - 0.03 W_a}{1 + 0.195 W_a}$	(12.14)
Mortar elasticity reduction factor, γ_e, as a function of unit water absorption, W_a	1. Masonry built with wetted or docked bricks: $\gamma_e = 1$. 2. Masonry built with dry units sealed until loading or unsealed masonry built with docked units: $\gamma_e = \frac{1 - 0.016 W_a}{1 + 0.029 W_a}$ 3. Masonry built with dry units, sealed for 1 day and then unsealed: $\gamma_e = -\frac{1 - 0.03 W_a}{1 + 0.195 W_a}$	(5.15) (5.16)

For example, in the case of extruded wire-cut perforated clay brickwork, the ultimate creep coefficient is given by Eq. (5.29) and Eq. (12.27):

$$\phi_{wy\infty} = C_{wy\infty} E_{wy} = \left[\frac{2.15}{f_{by}} + \frac{0.175}{\gamma_e f_m}\right]^{-1} \left[0.86 C_{by\infty} + 0.14 C_m \gamma_c R_{c\infty} R_{RH}\right]$$

(12.35)

Data Banks and Accuracy of Prediction

Whereas Lenczner's method, BS5628-1:2005, BS EN 1996-1-1:2005, and ACI 530-05 methods yield estimates of final values of creep, the composite model allows estimates of creep for any time after application of load. For the assessment of accuracy, the experimental data of Table 12.6 were used, the creep values after the test durations quoted being assumed to be final values of creep for the existing methods, and the actual test durations were used for the composite model method. Theoretically, this assumption should overestimate creep using existing methods; however, the effect is small [41].

Forty-nine sets of published experimental data for clay brickwork were used in the assessment, full details required for the application of all the five methods being given in the creep data bank in Table 12.6. To avoid the possibility of enlarged moisture expansion from cryptoflorescence influencing creep, only experimental masonry creep data have been selected for brick strengths in excess of 45 MPa and water absorptions of less than 8%. Those limits correspond approximately with Class B Engineering clay bricks of BS 3921:1985 (now superseded by BS EN 771-1:2003). In some cases, details are lacking so that estimates (e) have been made. For example, the modulus of elasticity in the tests of Shrive et al. [35] was estimated from equations given in their paper for creep coefficient and specific creep. As stated in Chapter 5 for estimation modulus of elasticity, in most cases of masonry built with extruded or pressed clay bricks, the precise volumes of perforations or frogs were unknown and so, for the BS EN 1996-1-1: 2005 method, it was assumed that perforated clay bricks conformed to Group 2 category of Table 5.14. In the analysis of this chapter, the same assumption was made for assessing elastic strain by that standard. Also, the assessment includes cases for mortar strengths, f_m, >12 MPa, although, strictly speaking, they are outside the specified range of BS EN 1996-1-1: 2005. Also, some of the mortar strengths (5.30) exceeded the maximum allowable limit of Eq. (5.21); however, analysis of those particular data sets revealed no perceptible difference in error coefficient compared with that of the total data sets. For the same reason of lack of details of perforations and frogs, analysis by the ACI 530.1R method of clay bricks was based on gross area rather than net area.

The accuracy of each method was assessed by three different parameters, viz. the error coefficient (M), the average percentage error (E_r), and the percentage of estimates falling within 30% of the actual values (P). The error coefficient given by Eq. (5.4) for assessing the accuracy of modulus of elasticity of mortar, is defined in general terms as:

$$M = \frac{100}{x_{av}} \sqrt{\frac{\sum (x_e - x_a)^2}{n - 1}}$$

(12.36)

Author (Curing Conditions before Loading)	Strength at Loading, MPa	Brick Type & Strength, MPa	Brick Water Absorption, %	Masonry Type, (Volume/Surface Ratio) mm	Stress MPa	Storage RH after Loading, %	under Load, days	Measured E, GPa	Specific Creep, 10^{-6}/MPa
Lenczner [6]. (dry for 28 days)	1. 16.1 (1:¼:3) 2. 7.8 (1:1:6)	Butterly, class B (e), 99.0	4.3 (24 h soak) 5.7 (5 h boil)	1.5-brick-wide hollow pier, (70)	1.7 to 6.0	60e (uncontrolled)	120 120	25.3 25.7	10 23
Lenczner [7]. (dry for 28 days)	1. 17.8 (1:¼:3) 2. 18.4 (1:¼:3)	National star (e), 58.2	4.4(24 soak) 5.5(5 h boil)	1.5-brick-wide hollow pier (70)	1.25 to1.95	60e (uncontrolled) 50	240 240	20.5 19.6	30 62
Lenczner & Salahuddin [11]. (dry for (a) 14 days, (b) 28 days)	1. 14.3e (1:¼:3) 2. 17.2 (1:¼:3)	Staffs. Blue (p), 60.0	1.0 (24 h soak)	4-brick-wide, story high single-leaf wall (44)	3.73 to 4.51	50	365 365	23.8 20.1	136 108
Lenczner [12] (dry for 28 days)	23.3 (1:¼:3)	Butterly, Autumn Brown facing (e), 56.0	3.4e	1. 1.5-brick-wide hollow pier (79) 2. 4-brick-wide, story high cavity wall (80)	0.91	50	420 400	28.3 22.1	39 47
Lenczner [17] (dry for 28 days)	17.5 (1:¼:3) 17.6 (1:¼:3) 17.5 (1:¼:3)	Poynton (p), 46.1 Swillington (e), 108.2	8.0e 5.0e 5.5e	1. 1.5-brick-wide hollow pier (79) 2. 1.5-brick-wide story high hollow pier (79) 1. 1.5-brick-wide hollow pier (79) 2. 1.5-brick-wide story high hollow pier (79)	1.94 1.91 to 1.95 2.5	50	365e 400	14.6 13.4 25.1 28.7 1.8	18 43 5 8 68

Continued

Table 12.6 Previous Investigations Reporting Experimental Data on Creep of Clay Brickwork—cont'd

Author (Curing Conditions before Loading)	Mortar Type Strength at Loading, MPa	Brick Type & Strength, MPa	Brick Water Absorption, %	Masonry Type, (Volume/ Surface Ratio) mm	Stress MPa	Storage RH after Loading, %	Time under Load, days	Measured E, GPa	Measured Specific Creep, 10^{-6}/MPa
Warren & Lenczner [18] (dry for 28 days)		1. Chesterton Smooth red (e), 64.6		4-brick -wide story—high single-leaf wall (44)					
		2. Waingrove (e), 81.5	5.0e		3.0		360	27.8	48
		3. Poynton (p), 46.1	8.0e		1.95		320	16.9	56
		4. Swillington (e), 108.2	5.0e		1.95		360	30.6	24
Brooks & Abdullah [21] (polythene for 10 days. dry for 18 days)	6.6 (1:½:4½)	Armitage B (e), 93.7	4.9	1. 2-brick -wide single-leaf wall × 13 courses high (44)	1.5	65	180	11.8	213
				2. 2-brick -wide cavity wall × 13 courses high (51)				15.8	197
				3. 2-brick Hollow pier × 13 courses high (78)				14	177
				4. 2-brick Solid pier × 13 courses high (112)				15.0	156
Bingel [55] (polythene for 28 days)	13.3 (1:½:4½)	Armitage B (e) 80.6	2.7	2-brick-wide single-leaf wall × 13 courses high	3.0	70	260	17.9	110

Study / Mortar	Unit		Element					
Tapsir et al. [29,44] (polythene for 19 days, dry for 2 days)	Armitage B (e) 103.0	3.7	1. story- high diaphragm wall (80)	1.5	40e (uncontrolled)	120	19.7	71
10.1 (1:½:4½)			2. story-high fin wall (62)				18.8	86
11.5 (1:½:4½)								
Forth et al. [25,26,30]	Nori (e) 108.0	2.5	1. Single-leaf wall × 13 courses high (44)	1.5	65	160	27.8	107
12.6 (1:½:4½)			2. Hollow pier × 13 courses high (78)				27.8	93
Phase 1- (polythene for 14 days, dry for 14 days)	Waingrove (e) 123.7	5.9	Single-leaf wall × 13 courses high (44)				24.4	107
Phase 2a- (polythene for 13 days, dry for 1 day)	Armitage B (e) 98.3	4.5	Single-leaf wall × 13 courses high (44)			300	25.0	83
1. 22.4 (1:0:3)							25.0	87
2. 22.7 (1:½:3)							23.8	117
3. 12.3 (1:½:4½)							23.4	115
4. 13.7 (1M:0:3½)							20.8	155
5. 9.0 (1P:0:4)							20.3	167
6. 6.5 (1:1:6)							19.5	211
7. 8.7 (1S:0:5)								
Phase 2b- (polythene for 14 days, dry for 14 days)	1. Smooth red (e) 92.2	1.7	Solid pier × 13 courses high (112)			160	19.0	100
12.6 (1:½:4½)	2. Nori (e) 108.0	2.5					26.3	75

Continued

Table 12.6 Previous Investigations Reporting Experimental Data on Creep of Clay Brickwork—cont'd

Author (Curing Conditions before Loading)	Mortar Type Strength at Loading, MPa	Brick Type & Strength, MPa	Brick Water Absorption, %	Masonry Type, (Volume/Surface Ratio) mm	Stress MPa	Storage RH after Loading, %	Time under Load, days	Measured E, GPa	Measured Specific Creep, 10⁻⁶/MPa
Shrive et al. [34] Series 1-Moist cured loaded at:	N Mortar (see Table 5.16)	190 × 90 × 57 unit with 2 square & 2 round holes 61.4 (gross area)	9.1	5-stack prism (31)			7 years		
1. 7 days	2.5e				2.43	100e-moist		3.18e	320
2. 28 days	3.5				4.86			3.18e	184
Dry cured loaded at:									
1. 7 days					2.43	20e		3.98e	432
2. 28 days					4.86	uncontrolled		2.48	231
Series 2-	N Mortar	190 × 90 × 57 unit with 10 round holes, 77.8 (gross area)	7.1						
1. Moist cured loaded at 14 days	3.8e				2.43 to 4.86	100e-moist		9.62e	58
2. Dry Cured loaded at 7 days	3.3e				4.86	20e uncontrolled		14.1e	209
3. Dry cured loaded at 28 days	4.6				2.43			11.9e	94
4. Moist cured loaded at 7 days	S Mortar (see Table 5.16)				2.43	100e-moist		5.1e	45
5. Dry cured loaded at 14 days	8.7e				2.43	20e uncontrolled		14.3e	133
6. Dry cured loaded at 28 days	10.9				2.43			14.3e	74

Series 3-	N Mortar	190 × 90 × 57 unit with 10 round holes, 89.2 (gross area)	6.4				
1. Moist cured loaded ay 28 days	4.0		3.64 to 4.86	100e- moist	6.49e	189
2. Dry cured loaded at 28 days	4.0		1.21 to 3.64	20e uncontrolled	6.37e	81
3. Moist cured loaded at 28 days	S Mortar 7.7		1.21 to 3.64	100e-moist	6.90e	93
4. Dry Cured loaded at 28 days	7.7		1.21 to 4.86	20e uncontrolled	5.10e	180

NB. e = estimated value; mortar type: cement-lime-sand; M = masonry cement; P = plasticizer; S = slag replacement of cement; (e) = extruded unit; (p) = pressed unit.

where x_e = estimated value, x_a = actual value, and n = number of data sets.
The average percentage error is:

$$E_r = \frac{100}{n} \sum \frac{(x_e - x_a)}{x_a} \tag{12.37}$$

As stated at the beginning of this section, besides specific creep, the assessment included specific elastic strain and the sum of specific creep and specific elastic strain, i.e., total load strain per unit of stress (otherwise known as the creep function or compliance) and creep coefficient. Although the accuracy of methods of predicting of modulus of elasticity has been dealt with in Chapter 5, the elastic strain referred to in this chapter is that associated with the start of creep measurements, whereas the elastic modulus data of Chapter 5 include creep-independent measurements, such as short-term stress-strain behaviour.

Table 12.7 compares the accuracy of all methods in predicting deformation, and it is apparent that there is some degree of inconsistency between the assessment parameters M and E_r. With large values of deformation, M tends to distorted by small differences between estimates and actual values, whereas E_r is distorted by small differences between estimates and actual values at low levels of deformation. Consequently, the following assessment of accuracy of the methods of prediction is based on more than one of the assessment parameters. An example of inconsistency occurs in the prediction of elastic strain by the BS EN 1996-1-1: 2005 and composite model methods, which have low values of E_r and high P, but the composite model has a much lower M compared to that of the BE EN 1996-1-1: 2005 method.

Overall, although error coefficients were greater, the ranking order of performance of methods for estimating elastic strain was similar to that of modulus of elasticity in Chapter 5. For each method of prediction, Figure 12.16 illustrates the predicted strain versus the measured elastic strain as given by the reciprocal of modulus of elasticity listed in Table 12.6. Deformations using the three terms used to quantify creep of masonry are shown in Figures 12.17−12.19. With the exception of the composite model, when estimating specific creep (Figure 12.17) and creep function (Figure 12.18), it is clear that existing methods reflect the prediction trends of elastic strain (Figure 12.16) in that they underestimate measured large deformations. With regard to predicting the creep coefficient shown in Figure 12.19, Lenczner's method indicates a high variability while the European Codes of Practice methods assume a constant value of 1.5; the ACI method predicts low creep coefficients because the method prescribes only a single-value creep of 10×10^{-6} per MPa.

Of all the four parameters used to assess accuracy, Table 12.7 indicates that the creep function is the most accurate term to predict creep. Although for any particular method, the error coefficient was similar for specific creep, creep function and creep coefficient, the smallest average error occurred for the creep function, the most accurate method being the composite model with a value similar to the accuracy of predicting elastic strain ($E_r \approx 28\%$); the variability was also the smallest (sd $\approx 28\%$) The accuracy of estimating creep coefficient was similar to that of estimating specific creep.

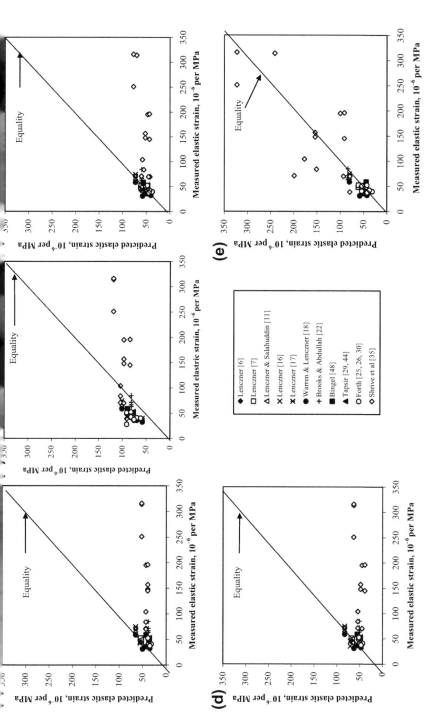

Figure 12.16 Prediction of elastic strain of clay brickwork by different methods (a) Lenczner (b) BS 5628-1: 2005 (c) BS EN 1996-1-1: 2005 (d) ACI 530-05 (e) Composite model.

Table 12.7 Overall Accuracy of Estimating Deformation by Different Methods

Method	Error Coefficient, M %	Average Error, E_r, ± sd %	Percentage within 30% of Actual Value, P
Specific Elastic Strain			
Lenczner	106.0	33.3±26.8	39
BS 5628-1:2005	83.7	61.7±30.4	16
BS EN 1996-1-1:2005	98.0	29.5±26.0	63
ACI 530-05	102.1	38.5±26.9	43
Composite model	50.9	28.1±31.3	69
Specific Creep			
Lenczner	87.4	58.9±35.6	22
BS 5628-1:2005	65.9	127.7±213.3	42
BS EN 1996-1-1:2005	83.4	113.7±195.5	20
ACI 530-05	117.8	84.2±187.5	4
Composite model	44.4	49.4±85.0	53
Creep Function			
Lenczner	87.6	46.1±24.8	29
BS 5628-1:2005	62.9	70.9±69.5	35
BS EN 1996-1-1:2005	84.1	52.2±33.0	27
ACI 530-05	101.7	54.4±26.9	252
Composite model	41.3	26.7±25.3	74
Creep Coefficient			
Lenczner	73.9	74.3±88.9	25
BS 5628-1:2005	62.2	116.6±195.2	22
BS EN 1996-1-1:2005	61.1	116.6±195.2	29
ACI 530-05	108.4	80.3±20.2	6
Composite model	41.0	47.9 ± 75.9	49

To identify possible sources of error, the accuracy of prediction was also analyzed for three groups of data having similar number of observations according to the laboratory where tests were undertaken. Table 12.8 shows the outcome and, for all types of deformation and any method of prediction, lower errors were found for the Cardiff and Leeds data. Using the Cardiff data, similar accuracies were found for the Lenczner method and composite model, which could was partly expected since the Lenczner method was developed from the Cardiff test results. For the Cardiff and Leeds elastic strain data, low errors were also found for the BS EN 1996-1-1: 2005 method. For the Leeds data, the best creep predictions occurred for the composite model, which again could have been expected since the model was developed from associated tests carried out at that location. With all methods of prediction, errors were greater for the Calgary deformations although it must be remembered that input data in Table 12.7 should be regarded as questionable as they were derived from general equations fitted to test data by Shrive et al. [34]; nevertheless, for any group of data, the best predictions were still achieved by the composite model.

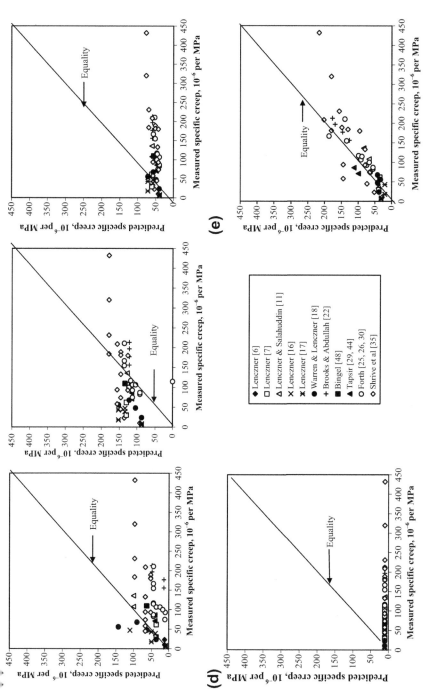

Figure 12.17 Prediction of specific creep of clay brickwork by different methods. (a) Lenczner (b) BS 5628-1: 2005 (c) BS EN 1996-1-1: 2005 (d) ACI 530-05 (e) Composite model.

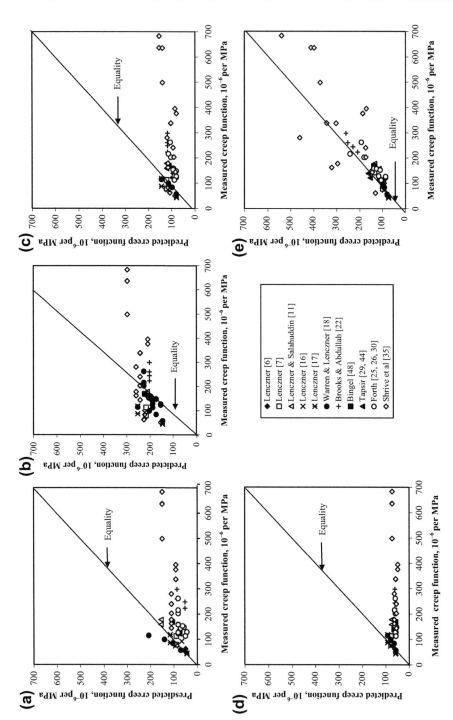

Figure 12.18 Prediction of creep function (elastic strain + creep) of clay brickwork by different methods: (a) BS 5628-1:2005 (c) BS EN

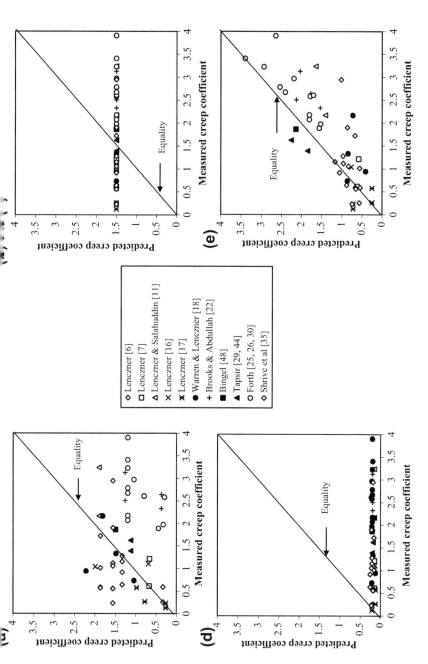

Figure 12.19 Prediction of creep coefficient by different methods. (a) Lenczner (b) BS 5628-1: 2005 (c) BS EN 1996-1-1: 2005 (d) ACI 530-05 (e) Composite model.

Table 12.8 Accuracy of Estimating Deformation of Clay Masonry According to Laboratory Test Group

Method	Error Coeff. M, %	Ave. Error, E, %	Within 30%, P, %
(a) Cardiff [6,7,11,16−18]			
Specific Elastic Strain			
Lenczner	17.4	14.4	94
BS 5628-1	82.5	75.4	0
BS EN 1996	28.7	25.4	69
ACI 530	44.1	43.8	25
C. Model	34.9	27.3	69
Specific Creep			
Lenczner	71.2	52.2	50
BS 5628-1	191.7	311.4	13
BS EN 1996	114.7	255.0	6
ACI 530	112.9	68.4	13
C. Model	55.4	94.6	31
Creep Function			
Lenczner	33.4	20.9	81
BS 5628-1	132.6	151.8	6
BS EN 1996	65.5	73.4	13
ACI 530	40.2	21.7	34
C. Model	32.5	31.8	69
+Creep Coefficient			
Lenczner	58.9	56.3	38
BS 5628-1	94.2	216.0	13
BS EN 1996	94.2	216.0	13
ACI 530	120.4	72.7	13
C. Model	66.5	78.3	31
(b) Leeds [22,24,25,28, 29,40,44]			
Specific Elastic strain			
Lenczner	37.3	24.5	74
BS 5628-1	62.1	63.2	16
BS EN 1996	16.4	11.0	100
ACI 530	23.9	16.9	74
C. Model	18.6	15.0	89
Specific Creep			
Lenczner	82.3	69.5	0
BS 5628-1	32.4	23.6	83
BS EN 1996	52.7	37.1	26
ACI 530	101.5	91.1	0
C. Model	23.5	21.8	68
Creep Function			
Lenczner	68.2	56.3	0

Table 12.8 Accuracy of Estimating Deformation of Clay Masonry According to Laboratory Test Group—cont'd

Method	Error Coeff. M, %	Ave. Error, E, %	Within 30%, P, %
BS 5628-1	27.6	28.0	58
BS EN 1996	42.9	27.7	58
ACI 530	76.4	63.7	0
C. Model	17.2	15.1	95
Creep Coefficient			
Lenczner	69.1	59.5	5
BS 5628-1	48.7	40.4	26
BS EN 1996	42.1	40.4	26
ACI 530	97.9	91.5	0
C. Model	26.2	26.3	58
(c) Calgary [34]			
Specific Elastic Strain			
Lenczner	96.6	67.0	0
BS 5628-1	70.6	44.1	14
BS EN 1996	89.5	59.4	7
ACI 530	92.8	61.9	7
C. Model	45.4	44.0	43
Specific Creep			
Lenczner	82.8	52.3	21
BS 5628-1	56.8	59.1	59
BS EN 1996	89.6	56.1	14
ACI 530	115.5	92.8	0
C. Model	49.9	35.4	50
Creep Function			
Lenczner	84.6	61.1	7
BS 5628-1	55.0	39.4	36
BS EN 1996	84.6	62.5	0
ACI 530	99.2	62.5	0
C. Model	42.2	36.5	50
Creep Coefficient			
Lenczner	78.2	115.0	29
BS 5628-1	71.3	106.4	29
BS EN 1996	63.1	106.4	50
ACI 530	105.8	73.8	7
C. Model	63.4	42.6	57

Whichever parameter is used to assess accuracy, the composite model is consistently the best because more factors influencing creep are taken into account, such as mortar type, absorption of the brick, and geometry of masonry. European Codes of Practice methods relate creep to modulus of elasticity, which, although influenced by type of mortar, is affected in a different manner to creep. Also, Lenczner's method relates creep to brick strength and geometry but does not really account for mortar type.

Published experimental results for deformation of calcium silicate and concrete block masonry are fewer than for clay masonry and in many cases lack test details as can be seen from Table 12.9. Without making several assumptions, an equivalent comprehensive comparison of the methods of predicting deformation to that of clay masonry has not proved to be feasible for calcium silicate and concrete masonry. However, for design purposes, some idea of the order of creep can be ascertained from the information given in Table 12.9. It should be noted that the tests of Abdullah [43] on calcium silicate masonry were carried out on masonry built with units laid "frog down" and, since modulus of elasticity was less for that situation (see p. 107), it is likely that more creep occurred than would be the case with masonry built with units laid "frog-up".

Examples

1. Using all prediction methods, estimate the elastic strain, creep, creep function, and creep coefficient of the clay brickwork reported by Warren and Lenczner as listed in Table 12.8(b) [41].

Solutions

Lenczner method

Elastic Strain When $f_{by} = 81.5$ MPa, from Eq. (5.18), $E_{wy} = 23.9$ GPa and $e_{wy} = 41.9 \times 10^{-6}$ per MPa.

Creep Equation (12.15) gives $C_{wy\infty} = 62.0 \times 10^{-6}$ per MPa.

Total Strain $J_{wy\infty}$ is given by Eq. (12.17) as 103.9×10^{-6} per MPa.

Creep Coefficient φ_{wy} is given by Eq. (12.34) as 1.48.

BS 5628 method

Elastic Strain The modulus of elasticity is given by Eq (5.19). For a brick strength $= 81.5$ MPa and designation (i) mortar, Table 5.9 gives $f_k = 16.0$ MPa. Therefore, from Eq. (5.19), $E_{wy} = 14.4$ GPa and $e_o = 69.4 \times 10^{-6}$ per MPa.

Creep Specific creep is given by Eq. (12.19) as $C_{wy\infty} = 104 \times 10^{-6}$ per MPa.

Creep Function Equation (12.21) gives total strain $J_{wy\infty} = 174 \times 10^{-6}$ per MPa.

Creep Coefficient φ_{wy} is given by Eq. (12.34) as 1.5.

BS EN 1996-1-1 (UK National Annex) method

Elastic Strain For a brick strength $= 81.5$ MPa and mortar strength $= 17.5$ MPa. Equation (5.20) with $K = 0.4$ (Group 2) gives $E_{wy} = 20.5$ GPa and $e_{wy} = 48.7 \times 10^{-6}$ per MPa.

Creep Specific creep $C_{wy\infty} = 73 \times 10^{-6}$ per MPa (Eq. (12.25)).

Examples—cont'd

Creep Function Equation (12.26) gives $J_{wy\infty} = 122 \times 10^{-6}$ per MPa.
Creep Coefficient φ_{wy} as given by Eq. (12.34) is 1.5.

ACI 530.1R-05 method

Elastic Strain Assuming the net area of the unit is equal to the gross area, for a unit strength of 81.5 MPa and type M mortar, from Table 5.15 the specified compressive strength is 26.1 MPa. Therefore, from Eq. (12.25), $E_{wy} = 18.3$ GPa and $e_{wy} = 54.7 \times 10^{-6}$ per MPa.
Creep Specific creep $C_{wy\infty} = 10 \times 10^{-6}$ per MPa.
Creep Function From Table 12.5, $J_{wy\infty} = 64.7 \times 10^{-6}$ per MPa.
Creep Coefficient φ_{wy} as given by Eq. (12.34) is 0.18.

Composite model method

Elastic Strain For an estimated water absorption = 5.0%, $\lambda_e = 0.43$ (Eq. (5.16)) so that for $f_{by} = 81.5$ MPa and $f_m = 17.5$ MPa, Eq. (5.29) gives $E_{wy} = 20.1$ GPa and, hence, $e_{wy} = 42.7 \times 10^{-6}$ per MPa .
Creep (see Table 12.7) The water absorption factor, λ_c, is given by Eq. (12.14) as 0.43. Equation (12.12) gives specific creep of brick, $C_{by} = 12.3 \times 10^{-6}$ per MPa (after time 360 days). Equation (12.9) gives 360-day creep of mortar, $C_m = 414.5 \times 10^{-6}$ per MPa for RH = 50% and V/S = 44 mm (see Table 12.7). Hence substitution in Eq. (12.27) yields specific creep $C_{wy} = 35.5 \times 10^{-6}$ per MPa.
Creep Function After 360 days, Eq. (12.30) gives $J_{wy} = 78.2 \times 10^{-6}$ per MPa.
Creep Coefficient According to Eq. (12.34), $\varphi_{wy} = 0.83$

2. A diaphragm wall is to be constructed from hollow concrete blocks with a 25% cavity by volume and a mortar of 28 day strength 10 MPa. The block strength is 15 MPa based on gross area and the water absorption is 8%. The wall will be post-tensioned at the age of 28 days and, using the composite model method, it is required to estimate the ultimate creep function due to the initial loading and after 60 days under load to decide whether additional post-tensioning is required. After construction, assume the wall will be covered for 1 day and then exposed to the environment having average RH = 65%

Solution

Estimates of the elastic strain and specific creep (C_{wy}) are required. The elastic strain (e_{wy}) is obtained from Eq. (5.28):

$$\frac{1}{E_{wy}} = \frac{1.058}{f_{by}} + \frac{A_m}{A'_m} \frac{0.06}{\gamma_e f_m}$$

where γ_e is given by Eq. (5.16) = 0.30 and $A_m/A'_m = 1/0.75 = 1.33$. Substitution for the block and mortar strengths gives $E_{wy} = 10.3$ GPa and $e_{wy} = \underline{97 \times 10^{-6}}$ per MPa.

Continued

Examples—cont'd

The creep expression for solid concrete blockwork after any time is given by Eq. (12.29), which for hollow blockwork becomes:

$$C_{wy} = 0.952 \left[\frac{C_{b\infty} t}{29 R_{aC\infty} + t} \right] R_{C\infty} R_{RH}$$

$$+ \, 0.048 \frac{A_m}{A'_m} \gamma_c \left[\frac{C_{m\infty} t}{22 R_{aC\infty} + t} \right] R_{C\infty} R_{RH}$$

and the ultimate specific creep is:

$$C_{wy\infty} = \left[0.952 C_{b\infty} + 0.048 \frac{A_m}{A'_m} \gamma_c C_{m\infty} \right] R_{C\infty} R_{RH}$$

The ultimate specific creep of the concrete block, $C_{b\infty}$, is given by Eq. (12.13) as 197.7×10^{-6} per MPa, and the ultimate specific creep of mortar, $C_{m\infty}$, is given by Eq. (12.50) as 871×10^{-6} per MPa.

Now, for a blockwork diaphragm wall, Figure 7.15 gives the $V/S = 85$ mm and the relative ultimate creep, $R_{C\infty}$, as given by Eq. (12.6), is 0.83, while the time relative parameter, $R_{aC\infty}$, as given by Eq. (11.7), is 1.38. The creep water absorption factor, γ_c, is obtained from Eq. (12.14) = 0.3, and, when the RH = 65%, the relative humidity factor $R_{RH} = 1$.

Hence substitution in the above expressions gives $C_{wy} = 103 \times 10^{-6}$ per MPa after 60 days under load, and $C_{wy\infty} = 205 \times 10^{-6}$ per MPa. Consequently, the creep function, after 60 days is:

$$J_{wy} = \underline{200 \times 10^{-6} \text{ per MPa}}$$

and the ultimate creep function is:

$$J_{wy\infty} = \underline{302 \times 10^{-6} \text{ per MPa}}$$

Table 12.9 Previous Investigations Reporting Experimental Data for Creep of Concrete Blockwork and Calcium Silicate Brickwork

Author (Curing Conditions before Loading)	Mortar Type, Strength at Loading, MPa	Unit Type & Strength, MPa	Unit Water Absorption, %	Masonry Type (V/S ratio), mm	Creep Test Storage RH, %	Stress, MPa	Time under Load, days	Measured Modulus of Elasticity, GPa	Measured Specific Creep, 10^{-6} per MPa
Lenczner [9] (dry for 28 days)e	1:1:6, 8.5	Dense aggregate concrete block, 5.6	NA	1¼ wide × 5-course-high hollow pier (82)	50	**1.** 0.78	320	8.1	818
						2. 1.2		8.6	597
						3. 1.62		8.1	470
						4. 1.89		6.8	478
Lenczner [16] (dry for 28 days)e	1:¼:3, 23.3	Lightweight aggregate concrete block, 3.3	23.3	1¼ wide × 5-course-high hollow pier (82)	50	1.06	400	5.4	421
Ameny et al. [19] (dry for 7 days)	**1.** 1M:0:3, 4.7	Hollow lightweight concrete block[a], 12.3	NA	1-block-wide × 5 course high (44)	10 to 60 (un-controlled)	3.1/1.8	110	5.3	191/161
	2. 1:1:6, 4.7					4.1/1.9		7.2	132/140
Ameny et al. [20] dry for 8 days	**1.** 1M:0:2.5, type N. NA	Hollow lightweight concrete block[a]: **1.** Type WF, 6.0e	NA	1-block-wide × 9 high (44)	11 to 81 (un-controlled)	N/WF, 3.5	360	4.8	93
	2. 1:1.5, type M, NA	**2.** Type BF, 8.0	NA			N/BF, 3.56		7.1	176
						M/WF, 3.5		4.8	73
						M/BF, 3.56		8.0	117
Brooks [21] (polythene for 28 days)	1:½:4½, 12.7	Calcium silicate (s), 27.4	NA	2-brick-wide × 13-course-high single-leaf wall (44)	65	3.0	300	11.8	102

Continued

Table 12.9 Previous Investigations Reporting Experimental Data for Creep of Concrete Blockwork and Calcium Silicate Brickwork—cont'd

Author (Curing Conditions before Loading)	Unit Type & Strength, MPa	Unit Water Absorption, %	Masonry Type (V/S ratio), mm	Creep Test Storage RH, %	Stress, MPa	Time under of Load, days	Measured Modulus of Elasticity, GPa	Measured Specific Creep, 10^{-6} per MPa
Brooks and Abdullah [24,43] (polythene for 10 days then dry for 18 days)	Dense aggregate concrete block, 13.0	NA	1. 2-block- wide × 5-course-high single-leaf wall (45)	65	1.5	190	9.9	257
			2. 1¼-block-wide × 5-course- high cavity wall (50)				9.2	224
			3. 1¼-block-wide × 5-course-high hollow pier (82)				9.7	183
			4. 1¼-block-wide × 5-course-high solid pier (110)				9.6	167
	Calcium silicate brick (f); laid 'frog down' 25.7	11.0	1. 2-brick-wide × 13-course-high single-leaf wall (44)			200	5.0	415
			2. 2-brick-wide × 13-course-high cavity wall (51)				5.1	375
	7.3		3. 2-brick-wide × 13-course-high hollow pier (78)				5.5	314
			4. 2-brick-wide × 13-course-high solid pier (112)				5.0	273

Mortar Type, Strength at Loading, MPa: 1:½:4½, 6.1

(polythene for 28 days)	concrete block, 12.6	11.8		high single-leaf wall (45)					
	Calcium silicate brick (f), 26.1	12.3		2-brick-wide × 13-course-high single-leaf wall (44)			100	4.7	187
Tapsir [44]	Dense aggregate concrete block, 14.9	8.8	1:½:4½, 12.3	**1.** 2½-block-wide × 9-course-high fin wall (71)	40e (not controlled)	2.0	120	13.2	150
(Polythene for 19 days then dry for 2 days)		13.0		**2.** 1½-block-wide × 9-course-high diaphragm wall (82)				14.0	133
	Calcium silicate brick (s), 27.1	11.3	1:½:4½, 10.4	**1.** 4-brick-wide × 26-course-high fin wall (65)		1.57		12.8	161
		9.7		**2.** 3-brick-wWide × 29-course-high diaphragm wall (85)				12.1	150
Brooks [46] (polythene for 28 days)	Calcium silicate brick (s)-Malaysia,16.4	12.9	1:½:4½, 7.9	2-brick-wide × 13-course-high solid pier (146)	70	1.5	185	5.9	256
Forth et al. [31,33]	Dense aggregate block, NA (age unknown)	11.4	1:½:4½,	3-stack-high wall: **1.** 44	65	1.5	160		
1. Cured under polythene & tested dry, loaded at:									
3 days		7.5						6.7	275
7 days		9.5						8.0	164
14 days		10.4						8.8	141
28 days		12.6						9.4	127
56 days		13.2						9.7	109

Continued

Table 12.9 Previous Investigations Reporting Experimental Data for Creep of Concrete Blockwork and Calcium Silicate Brickwork—cont'd

Author (Curing Conditions before Loading)	Mortar Type, Strength at Loading, MPa	Unit Type & Strength, MPa	Unit Water Absorption, %	Masonry Type (V/S ratio), mm	Creep Test Storage RH, %	Stress, MPa	Time under Load, days	Measured Modulus of Elasticity, GPa	Measured Specific Creep, 10^{-6} per MPa
2. Cured & tested sealed in polythene, loaded at				2. sealed					
3 days	7.5							6.7	90
7days	9.5							8.0	63
14 days	10.4							8.8	61
28 days	12.6							9.4	62
56 days	13.2							9.7	40

NB. NA = no value available; mortar type: cement-lime-sand; M = masonry cement; P = plasticizer; S = slag replacement of cement; (s) solid unit; (f) frogged unit.
[a]no details of block cavity given.

Summary

The factors influencing creep of masonry have been identified as stress, time under load, geometry, storage relative humidity, type of mortar and type of unit, anisotropy, water absorption of the unit and its moisture content at the time of laying. Creep data banks of clay masonry calcium silicate and concrete masonry have been compiled. However, the data bank does not include creep of clay masonry obtained from laboratory-size test walls and piers that may be prone to cryptoflorescence, which tends to occur in brickwork constructed with clay units having strengths of less than 45 MPa and water absorptions greater than 8%. Because of the lack of creep data for calcium silicate and concrete masonry, accuracy of prediction has been assessed for clay masonry only, and it demonstrates that prediction of masonry movements is by no means an exact science because of the heterogeneous nature of the materials involved and their inherent highly variable properties, not to mention the unit/mortar interfacial effects. Consequently, estimates of creep are similar to those of elasticity and are rather crude, with 20−30% error being regarded as acceptable. The composite model method improves accuracy of prediction simply because more influencing factors are taken into account and the most accurate parameter to quantify creep is the creep function, i.e., the sum of elastic strain plus creep per unit of stress. For a more accurate estimate of creep, assuming previous records are not available for the materials involved, then short-term laboratory testing may be necessary together with extrapolation to obtain longer-term values. This topic is discussed in Chapter 16.

Problems

1. Explain the difference between specific creep, creep function, and creep coefficient.
2. What is the effect of aging on creep of masonry?
3. How can the unit/mortar bond influence creep of masonry?
4. When can enlarged moisture expansion occur?
5. What is the effect of docking or wetting units at the time of laying on creep of masonry?
6. What is the influence of type of mortar on creep of masonry?
7. How does enlarged moisture expansion affect creep of masonry?
8. What would you recommend to avoid the influence of cryptoflorescence when determining creep of masonry in the laboratory?
9. Which is the best parameter for estimating creep by prediction methods and what is a realistic accuracy?
10. What is the effect of geometry on creep of masonry?
11. Compare the rate of development of creep and ultimate creep of single-leaf walls and solid piers.
12. Would you expect the creep of masonry to be the same in the horizontal and vertical directions? Consider clay brickwork, calcium silicate brickwork, and concrete block masonry.
13. Estimate the ultimate specific creep by the BS EN 1996-1-1 method for single-leaf masonry constructed with hollow aggregate concrete blocks with 10 mm mortar joints, and a 1:½:4½ cement-lime-sand mortar having a 28 day strength $= 6$ MPa. The standard-size blocks have 50% cavity volume and strength of 8 MPa based on gross area. After construction, the masonry will be covered for 1 day, and then cured at 65% RH, before being loaded at the age of 28 days and subsequently stored at 65% RH.
   Answer: 393×10^{-6} per MPa (based on gross area).

14. For Question 12.13, calculate the ultimate creep function by the ACI 530.1R method.
Answer: 101×10^{-6} per MPa (based on net area and Type N mortar).
15. For Question 12.13, calculate the ultimate specific creep, creep function and creep co-efficient by the composite model method assuming the blocks have a water absorption = 10%.
Answer: specific creep = 370×10^{-6} per MPa; creep function = 604×10^{-6} per MPa; creep coefficient = 1.78.

References

[1] Nylander H, Ericsson E. Effect of wall perforations on floor slab loads and floor slab deformations in multi-storey houses. Nord Bet 1957;1(Part 4):269—92.
[2] Sahlin S. Structural masonry. Prentice-Hall Inc; 1971. 290 pp.
[3] Poljakov SV. Some problems on creep in ordinary and reinforced masonry. Paris: International Council for Building Research, Studies and Documentation; 1962.
[4] Lenczner D. Design of creep machines for brickwork. Proc Br Ceram Soc 1965;4:1—8.
[5] Lenczner D. Creep in model brickwork. In: Johnson FB, editor. Designing, engineering and constructing with masonry products. Houston, Texas: Gulf Publishing; 1969. pp. 58—67.
[6] Lenczner D. Creep in brickwork. In: Proceedings second international conference on brick masonry, SIBMAC. Stoke-on-Trent: British, Ceramic. Society; 1970. pp. 44—9.
[7] Lenczner D. Creep in brickwork with and without damp proof course. In: Proceedings fourth international symposium on loadbearing brickwork. London: British. Ceramic Society; 1971. pp. 39—49.
[8] Tatsa E, Yishai O, Levy M. Loss of steel prestress in prestressed concrete blockwork walls. Struct Eng 1973;51(5):177—82.
[9] Lenczner D. Creep in concrete blockwork piers. Struct Eng 1974;52(3):97—101.
[10] Lenczner D, Wyatt K, Saluhuddin J. The effect of stress in creep of brickwork piers. In: Proceedings british ceramic. society. loadbearing brickwork, 5; 1975. pp. 1—9.
[11] Lenczner D, Salahuddin J. Creep and moisture movements in brickwork walls. In: Proceedings fourth international brick masonry conference, bruges; 1976. Section 2, p. 2.a.4.0—2.a.4.5.
[12] Lenczner D. Creep and moisture movements in brickwork and blockwork. In: Proceedings international conference on performance of building structures, vol. 1. Glasgow University; 1976. pp. 369—83.
[13] Lenczner D, Salahuddin J. Creep and moisture movements in masonry piers and walls. In: Proceedings first canadian masonry symposium. University of Calgary; 1976. pp. 72—6.
[14] Lenczner D. The effect of strength and geometry on the elastic and creep properties of masonry members. In: Proceedings north american conference. Boulder: University of Colorado; 1978. 23.1—23.15.
[15] Schubert P. Deformation of masonry due to shrinkage and creep. Fifth International Brick masonry Conference, Washington; 1978. pp. 132—138.
[16] Lenczner D. Creep in brickwork and blockwork cavity walls and piers. In: Proceedings fifth international symposium on loadbearing brickwork, 27. London: British Ceramic Society; 1979. pp. 53—66.
[17] Lenczner D. Brickwork: guide to creep. SCP 17. Hertford: Structural Clay Products; 1980. 26 pp.

[18] Warren D, Lenczner D. A creep-time function for single leaf brickwork walls. Int J Mason Constr 1981;2(1):13—20.

[19] Ameny P, Loov RE, Jessop EL. Strength, elastic and creep properties of concrete masonry. Int J Mason Constr 1980;1(Part 1):33—9.

[20] Ameny P, Loov RE, Shrive NG. Models for long-term deformation of brickwork. Mason Int 1984;1:27—36.

[21] Brooks JJ. Time-dependent behaviour of calcium silicate and Fletton clay brickwork walls. In: West WHW, editor. Proceedings british masonry society, stoke-on-trent, 1; 1986. pp. 17—9. presented at *Eighth International Conference on Loadbearing Brickwork* (Building Materials Section), British Ceramic Society, 1983.

[22] Brooks JJ, Abdullah CS. Composite model prediction of the geometry effect on creep and shrinkage of clay brickwork. In: Proceedings of eighth international brick-block masonry conference, vol. 1. Dublin: Elsevier Applied Science; 1988. pp. 316—23.

[23] Brooks JJ, Abdullah CS. Geometry effect on creep and moisture movement of brickwork. Mason Int 1990;3(3):111—4.

[24] Brooks JJ, Abdullah CS. Creep and drying shrinkage of concrete blockwork. Mag Concr Res 1990;42(150):15—22.

[25] Forth JP, Brooks JJ. The effect of cross-section geometry on long-term deformation of clay brickwork. In: Proceedings third international masonry conference. London: British Masonry Society; 1994. pp. 37—43.

[26] Brooks JJ, Forth JP. Influence of unit type on creep and shrinkage of single leaf clay brickwork. In: Proceedings third international masonry conference. London: British Masonry Society; 1994. pp. 31—3.

[27] Bingel PR, Brooks JJ. Prediction of stress relaxation in structural clay masonry from creep. In: Proceedings third international masonry conference. London: British Masonry Society; 1994. pp. 234—8.

[28] Brooks JJ, Bingel PR. Creep of masonry under varying stress. In: Proceedings tenth international brick and block masonry conference, calgary, Canada; 1994. pp. 749—56.

[29] Brooks JJ, Tapsir SH, Parker MP. Prestress loss in post-tensioned masonry: influence of unit type. In: Proceedings ASCE structures congress, Boston, USA; 1995. p. 12.

[30] Forth JP, Brooks JJ. Influence of mortar type on the long-term deformation of single leaf clay brick masonry. In: Proceedings fifth international masonry conference, vol. 1. London: British Masonry Society; 1995. pp. 157—61.

[31] Forth JP, Bingel PR, Brooks JJ. Influence of age at loading on the long-term movements of clay brickwork and concrete block masonry. In: Proceedings seventh north american masonry conference, Vol. 2. Indiana, USA: University of Notre Dame; 1996. pp. 811—21.

[32] Brooks JJ, Abdullah CS, Forth JP, Bingel PR. The effect of age on the deformation of masonry. Mason Int 1997;11(2):51—5.

[33] Forth JP, Bingel PR, Brooks JJ. Effect of loading age on creep of sealed clay and concrete masonry. In: Proceedings fifth international masonry conference, masonry, 8. British Masonry Society; 1998. pp. 52—5.

[34] Sayed-Ahmed EY, Shrive NG, Tilleman D. Creep deformation of clay masonry structures: a parametric study. Can J Civ Eng 1998;25(1):67—80.

[35] Shrive NG, Sayed-Ahmed EY, Tilleman D. Creep analysis of clay masonry assemblages. Can J Civ Eng 1997;24:367—79.

[36] Brooks JJ, Abu Bakar BH. Anisotropy of elasticity and time dependent movement of masonry. In: Proceedings of the fifth international masonry conference, masonry, 8. British Masonry Society; 1998. p. 44.

[37] Forth JP, Brooks JJ, Tapsir SH. The effect of unit water absorption on long-term movements of masonry. Cem Concr Compos 2000;22:273−80.

[38] Bingel PR, Brooks JJ, Forth JP. Creep of hydraulic lime mortar brickwork. In: Proceedings of the sixth international masonry conference, masonry, 9. Stoke-on-Trent: British masonry Society; 1998. pp. 33−5.

[39] Brooks JJ, Abu Bakar BH. Shrinkage and creep of masonry mortar. Mater Struct 2004; 37:177−83.

[40] Forth JP, Brooks JJ. Creep of clay masonry exhibiting cryptoflorescence. Mater Struct RILEM 2008;11(5):909−20.

[41] Brooks JJ. Estimating creep of clay brickwork. Mason Int 2009;22(1):17−22.

[42] Brooks JJ. Composite models for predicting elastic and long-term movements in brickwork walls. In: West WHW, editor. Proceedings british masonry society, stoke-on-trent, 1; 1986. pp. 20−3. presented at *Eighth International Conference on Loadbearing Brickwork* (Building Materials Section), British Ceramic Society, 1983.

[43] Abdullah, CS. Influence of geometry on creep and moisture movement of clay, calcium silicate and concrete masonry [Ph.D. thesis]. Department of Civil Engineering, University of Leeds; 1989, 290 pp.

[44] Tapsir, SH. Time-dependent loss of post-tensioned diaphragm and fin masonry walls [Ph.D. thesis]. Department of Civil Engineering, University of Leeds; 1994, 272 pp.

[45] Forth, JP. Influence of mortar and brick on long-term movements of clay brick masonry [Ph.D. thesis]. Department of Civil Engineering, University of Leeds; 1995, 300 pp.

[46] Brooks JJ. Time-dependent behaviour of masonry and its component phases. EC Science and Technology Programme, Contract No. C11-0925, Brussels; 1996. 106 pp.

[47] Abu Bakar, BH. Influence of anisotropy and curing on deformation of masonry [Ph.D. thesis]. School of Civil Engineering, University of Leeds; 1998.

[48] CEB-FIP Model Code for Concrete Structures. Evaluation of time-dependent behaviour of concrete; 1990 (Bulletin d'information No. 199, Comite Europeen du Beton/Federation Internationale de la Precontrainte, Lausanne, 1991).

[49] Bremner, A. The origins and effects of cryptoflorescence in fired clay masonry [Ph.D. thesis]. School of Civil Engineering, University of Leeds; 2002, 232 pp.

[50] Bremner A, Forth JP, Brooks JJ, Bingel PR. Assessment of potential cryptoflorescence in clay masonry. In: Proceedings of 9th Canadian masonry symposium. Fredericton, Canada: University of New Brunswick; 2001. p. 13.

[51] ACI Committee 209. Prediction of creep, shrinkage and temperature effects in concrete structures. Detroit: American Concrete Institute; 1992.

[52] BS 5628-Part 2. Code of practice for use of masonry: Part 2: structural use of reinforced and prestressed masonry. British Standards Institution; 2005.

[53] BS EN 1996-1-1: 2005, Eurocode No. 6. Design of Masonry Structures. General rules for reinforced and unreinforced masonry structures. See Also: UK National Annex to BS EN 1996-1-1: 2005. British Standards Institution; 2005.

[54] ACI Committee 530-05. Building code requirements for masonry structures. In ACI manual of concrete practice, Part 6. American Concrete Institute; 2007.

[55] Bingel, PR. Stress relaxation, creep and strain under varying stress in masonry [Ph.D. thesis]. Department of Civil Engineering, University of Leed;, 1993, 330 pp.

13 Thermal Movement

The strain due to thermal movement caused by a change in temperature is equal to the product of the coefficient of thermal expansion and the change in temperature. Consequently, thermal movement can be classified as reversible, except in the case of very young hardened cement paste, where the thermal coefficient of expansion changes with hydration during periods of temperature change. In this chapter, the influencing factors relating to thermal movement are presented and discussed in terms of the coefficient of linear thermal expansion, which is expressed in units of $10^{-6}/°C$. The thermal expansion coefficient of concrete or masonry varies depending on the volume and type of aggregate or masonry unit and according to the composition of hydrated cement paste or mortar, its moisture content, and age. In normal circumstances, the thermal expansion coefficient is of interest over the seasonal temperature range (e.g. -20 to $80\,°C$ in the UK) but, clearly, thermal expansion coefficients resulting from very high temperature and in fire conditions are also of importance; however, this topic is beyond the scope of this chapter.

In the first instance, the influence of factors on the thermal expansion coefficient of concrete are discussed, for which more research data have been reported than for masonry, and particularly so for the hydrated cement paste. The latter is, of course, the binding constituent of masonry mortar as well as for concrete. Specific knowledge regarding thermal expansion of masonry units, mortar, and o the composites' brickwork and blockwork is presented later in the chapter.

Concrete

The thermal expansion coefficient of concrete is dependent on both its composition and moisture content at the time of temperature change, the composition influence arising because the two main constituents, aggregate and hydrated cement paste, have dissimilar thermal coefficients as well as different volumetric proportions. According to Neville [1], if there is a large difference in thermal expansion coefficients of aggregate and cement paste, a change in temperature may induce differential movement and rupture of bond at the interface. However, a large difference between the thermal expansion coefficients is not necessarily detrimental when the temperature does not vary outside the range of $4-60\,°C$, but if the two coefficients differ by more than $5.5 \times 10^{-6}/°C$, breakage of bond may occur.

Concrete and Masonry Movements. http://dx.doi.org/10.1016/B978-0-12-801525-4.00013-3

The composite effect of the difference in thermal expansion coefficients and volume contents of paste and aggregate is reflected in the theoretical model expression derived by Hobbs [2], which is given by Eq. (3.1):

$$\alpha_c = \alpha_m - \frac{(\alpha_m - \alpha_a)g2E_a}{E_m + E_a + g(E_a - E_m)} = \alpha_m - \frac{2g(\alpha_m - \alpha_a)}{1 + \frac{E_m}{E_a} + g\left[1 - \frac{E_m}{E_a}\right]} \quad (13.1)$$

where α_c, α_m, and α_a = thermal expansion coefficients of composite (concrete), matrix (hydrated cement paste), and aggregate, respectively; g = aggregate fractional volumetric content; E_m and E_a = elastic moduli of matrix and aggregate, respectively.

The influence of the factors expressed by Eq. (13.1) on the thermal expansion coefficient of concrete is illustrated in Figure 13.1. In theory, it appears that a change in the aggregate volumetric content is a significant factor but, in practice, the thermal coefficient of most types of concrete is only slightly affected since, for a constant workability, the volumetric aggregate content lies between 65% and 80%. Also, according to Eq. (13.1), the thermal expansion coefficient for concrete is affected by the relative

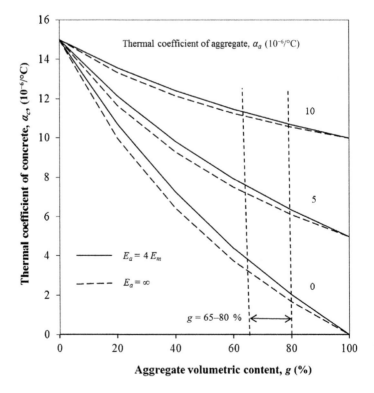

Figure 13.1 Influence of aggregate content and type on thermal coefficient of concrete, with $\alpha_m = 15 \times 10^{-6}/°C$, according to Eq. (13.1).

stiffness of matrix and aggregate, E_m/E_g. However, that influence is negligible, as demonstrated in Figure 13.1. Consequently, it follows that there is little influence of creep of matrix as indicated by a decreasing stiffness ratio, E_m/E_a.

Figure 13.1 is based on an assumed constant thermal expansion coefficient of matrix of $15 \times 10^{-6}/°C$, and it is apparent that thermal expansion coefficient of concrete increases as the thermal expansion coefficient of aggregate (α_g) increases, a feature that is confirmed generally by the experimental test data of Table 13.1, where the dominating influence of type of aggregate can be seen. In fact, a wide range of values can occur within one type of aggregate or rock group, which is also reflected in concrete, albeit to a lesser extent. According to Neville [1], the thermal expansion coefficient for the more common rocks is $0.9-16 \times 10^{-6}/°C$, but for the majority of aggregates it ranges between 5 and $13 \times 10^{-6}/°C$. The thermal expansion coefficients of fine and coarse aggregate can be determined by means of a dilatometer [7].

According to Neville [1], the thermal expansion coefficient of lightweight aggregate concrete is slightly less than for normal weight aggregate concrete, although the range of values in Table 13.2 is similar to those in Table 13.1.

The effect of moisture on the thermal expansion coefficient of concrete is shown in Table 13.3, which compares coefficients for three different types of curing condition. Water curing generally leads to a 45% decrease in coefficients compared with dry curing at 64% RH, with the coefficients resulting from dry/wet curing lying between those resulting from water curing and dry curing. The explanation for the differences is that water has a lower thermal expansion coefficient than that of air (see p. 458).

With regard to the range of thermal expansion coefficients of hydrated cement paste, values vary between 11 and $20 \times 10^{-6}/°C$, depending on the moisture condition [1,9]. The moisture dependence is because the thermal expansion coefficient of cement paste has two components: the *true (kinetic) thermal expansion coefficient*, which is caused by molecular movement of the paste, and the *hygrothermal expansion*

Table 13.1 Ranges of Thermal Expansion Coefficients for Normal Weight Aggregate and Concrete [3,4]

| Aggregate Type | Thermal Expansion Coefficient, $10^{-6}/°C$ | |
	Aggregate[a]	Concrete
Chert	7.4–13.0 (6.6)	11.4–12.2
Quartzite	7.0–13.2 (5.7)	11.7–14.6
Quartz	–(6.2)	9.0–13.2
Sandstone	4.3–12.1 (5.2)	9.2–13.3
Marble	2.2–16.0 (4.6)	4.4–7.4
Siliceous limestone	3.6–9.7 (4.6)	8.1–11.0
Granite	1.8–11.7 (3.8)	8.1–10.3
Dolerite	4.5–8.5 (3.8)	–
Basalt	4.0–9.7 (3.6)	7.9–10.4
Limestone	1.8–11.7 (3.1)	4.3–10.3

[a]The numbers in parentheses are the average values given by ACI 209R [4] and do not necessarily correspond to the average of range of values from Ref [3].

Table 13.2 Thermal Expansion Coefficient of Lightweight
Aggregate Concrete Made with Lightweight Aggregate for a
Temperature Range of −22 to 52 °C [1,6,7]

Aggregate Type	Thermal Expansion Coefficient, 10^{-6}/°C
Pumice	9.4−10.8
Perlite	7.6−11.0
Vermiculite	8.3−14.2
Cinders	Approx. 3.8
Expanded shale	6.5−8.1
Expanded slag	7.0−11.2

Table 13.3 Thermal Expansion Coefficient of Concrete Made with
Different Aggregates and Curing Conditions [1,8]

Aggregate Type	Thermal Expansion Coefficient, 10^{-6}/°C		
	Air Cured at 64% RH	Water Cured	Air Cured then Wetted
Gravel	13.1	7.3	11.7
Granite	9.5	5.3	7.7
Quartzite	12.8	7.1	11.7
Dolerite	9.5	5.3	7.9
Limestone	7.4	4.1	5.9
Sandstone	11.7	6.5	8.6
Portland stone	7.4	4.1	6.5
Blast-furnace slag	10.6	5.9	8.8
Foamed slag	12.1	6.7	8.5

(*swelling*) *coefficient*. The latter arises from an increase in the internal humidity (water vapor pressure) as the temperature increases, with a consequent swelling pressure and expansion of the cement paste. No hygrothermal expansion is possible when the paste is totally dry or when it is saturated since there can be no increase in water vapor pressure. However, at intermediate moisture contents, hygrothermal expansion does occur, and for young hydrated cement paste has a maximum at a relative humidity of 70% (Figure 13.2). As the paste hydrates further, the maximum hygrothermal expansion becomes smaller and occurs at a lower internal relative humidity, because of the increase in crystalline material. In fact, there is no variation in the thermal coefficient in high-pressure steam-cured paste because it contains no gel, just microcrystalline calcium hydroxide (Figure 13.2). Only the values determined on desiccated or saturated cement paste specimens can be considered to represent the true thermal coefficient, although it is the values at intermediate humidities that are applicable in practice. In concrete, the hygrothermal contribution is naturally smaller because of the dominating presence of aggregate [1,10].

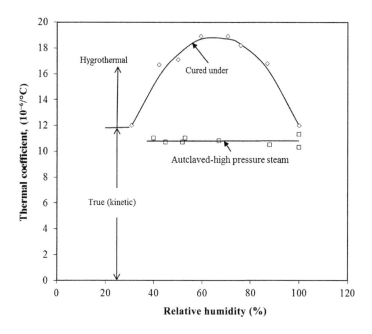

Figure 13.2 Thermal expansion coefficient of young hydrated cement paste as affected by ambient relative humidity of storage after normal curing and high-pressure steam curing [1,8].

The thermal expansion coefficients can be determined according to the methods specified by BS EN 1770: 1998 [11], ASTM [12,13], and the US Army Corps of Engineers [14]. When determined using standard test methods in which the thermal expansion coefficient of concrete corresponds to either the saturated condition or oven-dry condition, an allowance is required for the additional hygrothermal coefficient for air-dried concrete. ACI 209R-92 (Reapproved 2008) [4] specifies different hygrothermal expansion coefficients, according to the expected degree of saturation of the concrete member, as indicated in Table 13.4. On the other hand, in the absence of measured specific data on local materials and the environment, the same standard specifies a method of estimating the thermal expansion coefficient of air-dried concrete using average thermal expansion coefficients for hydrated cement paste and aggregate. For the expected degree of saturation of the concrete member, the hygrothermal expansion coefficient (α_h) is again selected from Table 13.4, but now substituted in the following expression to obtain the thermal expansion coefficient of concrete ($10^{-6}/°C$):

$$\alpha_c = \alpha_h + 3.1 + 0.72\alpha_a \tag{13.2}$$

where α_a = average thermal expansion coefficient of total aggregate ($10^{-6}/°C$) given in Table 13.1, and the number 3.1 is the contribution by the component of thermal expansion coefficient of hydrated cement paste ($10^{-6}/°C$).

Table 13.4 Hygrothermal Expansion Coefficient (α_h) for Different Degrees of Saturation [4]

Environmental Conditions of Concrete Member	Degree of Saturation	Coefficient of Hygrothermal Expansion, α_h 10^{-6}/°C
High humidity, e.g., immersed structures	Saturated	0
Mass concrete pours, thick walls, beams, columns and slabs, where surface is sealed	Between partially saturated and saturated	1.3
Drying external slabs, walls, beams, columns and roofs; unsealed internal walls, columns and slabs (with no mosaic or tiling), and where underfloor heating or central heating exists	Partially saturated decreasing with time to drier conditions	1.5–2.0

If the thermal expansion coefficient of the fine aggregate differs markedly from that of the coarse aggregate, the weighted average by volume of the thermal expansion coefficients of fine and coarse aggregates should be used in Eq. (13.2). ACI 209R-92 [4] also recommends that, for ordinary stress calculations where the type of aggregate and degree of saturation are unknown, an average thermal expansion coefficient of concrete of 10×10^{-6}/°C may be assumed. However, for estimating the range of thermal movements in highways and bridges, the lower and upper bound values of 8.5 and 11.9×10^{-6}/°C are recommended.

Although acknowledging that the thermal expansion coefficient of concrete depends on the type of aggregate and the moisture state of the concrete, and it varies between 6 and 15×10^{-6}/°C, the CEB Model Code 90 [15] specifies that, for the purpose of structural analysis, the coefficient may be taken as 10×10^{-6}/°C. When thermal expansion of concrete is not of great influence in design, BS EN 1992-1-1: 2004 [16] reiterates that the thermal expansion coefficient of concrete may be taken as 10×10^{-6}/°C.

When concrete is steam cured at atmospheric pressure, fresh concrete has a higher thermal expansion coefficient of almost three times that of the thermal coefficient after 4 h due to swelling pressure from expanding air bubbles and water [1]. Air and water have large *thermal coefficients of volume expansion*: $>3000 \times 10^{-6}$/°C for air and $>200 \times 10^{-6}$/°C for water. The thermal expansion coefficient of concrete decreases with age owing to a reduction in the potential swelling pressure due to the formation of crystalline material in the hydrating cement paste.

Concrete thermal expansion coefficient data of Table 13.1 apply to normal temperature. According to Neville [1], above a temperature of 200–500 °C, the thermal expansion coefficient of neat cement paste can become negative due to the loss of water and possible internal collapse. However, this effect is not reflected in the thermal expansion coefficient of concrete, which continues to increase as the temperature increases due to the dominating positive thermal expansion coefficient of aggregate. For example, the thermal coefficient of expansion of concrete generally increases by a factor of two for temperatures above 430 °C compared with temperatures below 260 °C [17]. The trend of increasing thermal expansion coefficient with temperature is confirmed by the thermal strain-temperature characteristics of siliceous and calcareous aggregate concretes over the temperature range of 20 to 1200 °C as specified by [18]. Here, the strain-temperature increases in an exponential-type manner up to maxima of 14000×10^{-6} and 12000×10^{-6} at 670 °C and 800 °C, respectively, for concrete with siliceous and calcareous aggregates. For example, at 300 °C, the chord gradient (which may be considered as a thermal coefficient) of siliceous aggregate concrete is 10.7×10^{-6} per °C compared with the gradient of 17.2×10^{-6} per °C at 600 °C, but in the case of calcareous aggregate, the corresponding values are lower, viz. 7.1×10^{-6} per °C at 300 °C and 11.2×10^{-6} per °C at 600 °C.

Wittman and Lukas [19] investigated the thermal expansion coefficient of hydrated cement paste having a water/cement ratio of 0.4 at temperatures below 0 °C. The strain of 12-mm diameter-specimens in a sealed condition or in water was determined using a quartz-glass dilatometer at incremental temperatures from −25 to 20 °C for specimens aged 5−56 days. The cooling rate was slow so that equilibrium was reached by the time of strain determination, thermal expansion coefficients being calculated at temperature intervals from the strain-temperature plots. Figure 13.3 shows the thermal expansion coefficient of saturated paste as a function of age at different temperatures. As hydration proceeds, the coefficient appears to increase at room temperature, but for temperatures below −10 °C, the reverse trend occurs, while around zero temperature, the coefficient is always small and reaches a minimum after about 10 days.

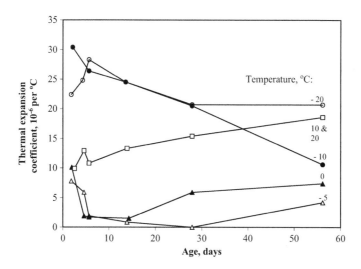

Figure 13.3 Thermal expansion coefficient of saturated hydrated cement paste subjected cooling at low temperature from +20 to −25 °C as a function of age; the individual curves represent different ranges of temperature [19].

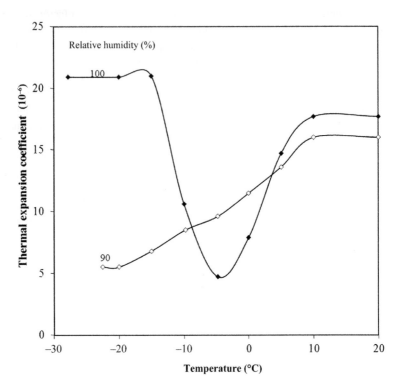

Figure 13.4 Thermal expansion coefficient of hydrated cement paste at the age of 55 days as a function of low temperature, stored at 90% and 100% relative humidity [19].

Figure 13.4 shows the thermal expansion coefficient of the saturated hydrated cement paste at the age of 55 days plotted against temperature. As the temperature is lowered below ambient, the thermal expansion coefficient decreases to a minimum value at about −5 °C before increasing again for even lower temperatures, possibly due to internal swelling pressure of the pore water on the advancing ice front. The initial decrease in thermal expansion coefficient as the temperature is lowered may be explained by the decrease of linear thermal expansion coefficient as water changes into ice. Compared with the thermal <u>volume</u> expansion coefficient of water of $210 \times 10^{-6}/°C$, the thermal linear expansion coefficient of ice at 0 °C is around $50 \times 10^{-6}/°C$. Hence, in terms of linear change of coefficient, as water changes into ice there is an apparent reduction of thermal expansion coefficient of around $20 \times 10^{-6}/°C$.

Figure 13.4 also indicates that the minimum thermal coefficient feature is absent in the case of hydrated cement paste that is slightly dried after a period of initial curing and stored at 90% RH, since there is a continuous decrease in the thermal expansion coefficient as the temperature is lowered [1,19]. At this lower storage humidity, it seems likely that there is room within the paste structure to relieve pore water pressure induced by the advancing ice front [10]. In conclusion, Wittman and Lukas' findings

[19] suggest that the characterization of the thermal expansion coefficient of hydrated cement paste at low temperatures depends on the pore size distribution, degree of hydration, and degree of saturation of the hydrated cement paste.

Example

It is required to estimate the thermal expansion coefficient of a sand-lightweight aggregate concrete having the following mix proportions by mass per m^3 of concrete:

Cement	350 kg/m^3
Lightweight weight fine aggregate	162
Normal weight fine aggregate	550
Lightweight coarse aggregate	424
Added water	180
Air	6% (by vol.)

Assuming the specific gravity of cement $= 3.15$, lightweight fine aggregate $= 1.78$, lightweight coarse aggregate $= 1.35$, and normal weight aggregate $= 2.65$, calculate the volumetric mix proportions of the concrete. Given the thermal expansion coefficients are 4.0 and $8.0 \times 10^{-6}/°C$ for lightweight and normal weight aggregate, respectively, use the ACI 209R method to estimate the thermal expansion coefficient of the concrete, which is to be used to manufacture internal walls of a centrally heated building.

Solution

For 1 m^3 of concrete, the volume proportions corresponding to the mass proportions are:

$$\underset{\text{(cement)}}{\frac{150}{3.15 \times 1000}} + \underset{\text{(lightweight fine)}}{\frac{168}{1.78 \times 1000}} + \underset{\text{(normal weight fine)}}{\frac{550}{2.65 \times 1000}} + \underset{\text{(lightweight coarse)}}{\frac{473}{1.35 \times 1000}} + \underset{\text{(water)}}{\frac{180}{1000}} + \underset{\text{(air)}}{0.06}$$

$$\underset{\text{(concrete)}}{= 1.00 \text{ m}^3}$$

where 1000 kg/m^3 = density of water.

The total volume of aggregate is 0.652 m^3, of which 0.444 m^3 is lightweight coarse-plus-fine aggregate, and 0.208 m^3 is normal weight fine aggregate. Hence, the weighted average of the total aggregate thermal expansion coefficient is:

$$\alpha_a = \left[\frac{0.444}{0.652} \times 4\right] + \left[\frac{0.208}{0.652} \times 8\right] = 5.27 \times 10^{-6} \text{ per } °C$$

From Table 13.3, the average hygrothermal expansion coefficient (α_h) for the environmental condition of the concrete beam is $1.75 \times 10^{-6}/°C$, and substituting in Eq. (13.2) gives the required solution:

$$\alpha_C = 1.75 + 3.1 + 0.72 \times 5.27 = 8.6 \times 10^{-6} \text{ per } °C$$

Masonry

In the previous section, it has been demonstrated that the dominating influence on the thermal expansion coefficient of concrete is the type and volumetric quantity of total aggregate, the latter being 65–80%. In masonry, the same domination exists except even more so because masonry contains more "aggregate" in that the volumetric content of the units is approximately 85% for brickwork and 95% for blockwork.

The theoretical analysis of Chapter 3 of thermal movement of masonry by composite modeling revealed that the expressions derived for moisture movement were also applicable to thermal movement of masonry. Thus, from knowing the thermal expansion coefficients of mortar, brick, or block, the models allow solutions for thermal movement in the vertical and horizontal directions of masonry. Like moisture movement, thermal movements are approximately independent of height and geometry of the masonry, the full solutions for brickwork being given by Eqs (3.84) and (3.85), respectively, viz:

$$\alpha_{wy} = 0.862\alpha_b + 0.138\alpha_m + \frac{0.862(\alpha_m - \alpha_b)}{\left[1 + 24.43\frac{E_{by}}{E_m}\right]} \tag{13.3}$$

in the vertical direction, and in the horizontal direction:

$$\alpha_{wx} = \frac{0.955\alpha_b + \left(0.046 + 0.152\frac{E_m}{E_{bx}}\right)\alpha_m}{\left[1 + 0.152\frac{E_m}{E_{bx}}\right]} \tag{13.4}$$

The corresponding expressions for blockwork are given by Eqs (3.86) and (3.87), respectively, viz:

$$\alpha_{wy} = 0.952\alpha_b + 0.048\alpha_m + \frac{0.952(\alpha_m - \alpha_b)}{\left[1 + 43.75\frac{E_{by}}{E_m}\right]} \tag{13.5}$$

and

$$\alpha_{wx} = \frac{0.979\alpha_b + \left[0.006 + 0.049\frac{E_m}{E_{bx}}\right]\alpha_m}{\left[1 + 0.049\frac{E_m}{E_{bx}}\right]} \tag{13.6}$$

In the above expressions, α_{wy} = vertical thermal expansion coefficient of masonry, α_{wx} = horizontal thermal expansion coefficient of masonry; α_b = thermal expansion coefficient of brick or block, which is assumed to be isotropic; and $\alpha_m =$ thermal expansion coefficient of mortar.

The composite model analysis also revealed that the vertical and horizontal thermal movements of brickwork are similar and mainly dependent on the thermal expansion coefficient of the brick, since there is a negligible change when the mortar thermal expansion coefficient triples from 5 to $15 \times 10^{-6}/^{\circ}$C. It was stated in the previous section that in the case of concrete there is a negligible influence of creep of hydrated cement paste in relieving induced stress in the aggregate. Similarly, in the case of masonry, there is little influence of creep of mortar in relieving induced stress in the brick, as quantified in terms of the modulus ratio, E_b/E_m. For example, assuming α_m and $\alpha_b = 12$ and $6 \times 10^{-6}/^{\circ}$C, respectively, and if E_b/E_m increases by a factor of 4, then α_{wx} only reduces by 4%. The behaviour of thermal movement of blockwork is almost identical to that of brickwork.

Ignoring the contribution of the vertical mortar joints, the vertical and horizontal expressions for thermal expansion coefficients of brickwork become, respectively:

$$\alpha_{wy} = 0.86\alpha_b + 0.14\alpha_m \tag{13.7}$$

and

$$\alpha_{wx} = \alpha_b + \frac{(\alpha_m - \alpha_b)}{\left[1 + 6.26\frac{E_{bx}}{E_m}\right]} \tag{13.8}$$

For blockwork, the approximate equivalent expression for thermal expansion coefficient in the vertical direction is:

$$\alpha_{wy} = 0.952\alpha_b + 0.048\alpha_m \tag{13.9}$$

and the corresponding approximate expression for thermal expansion coefficient in the horizontal direction is:

$$\alpha_{wx} = \alpha_b + \frac{(\alpha_m - \alpha_b)}{\left[1 + 19.85\frac{E_{bx}}{E_m}\right]} \tag{13.10}$$

The similarity of thermal expansion coefficients of brickwork and blockwork, both vertically and horizontally, as given by Eqs (13.5)–(13.11), is demonstrated in Figure 13.5 for mortar thermal expansion coefficients of 10 and $15 \times 10^{-6}/^{\circ}$C. In fact, the following common relationship can be taken to apply to any type of masonry in any direction for any combination of mortar and unit:

$$\alpha_w = 0.90 + 0.93\alpha_b \tag{13.11}$$

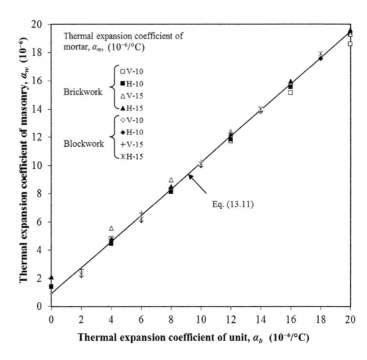

Figure 13.5 Thermal expansion coefficient of masonry as a function of thermal expansion coefficients of unit for mortar thermal expansion coefficients of 10 and 15 × 10^{-6}/°C; V = vertical direction; H = horizontal direction. The horizontal thermal expansion coefficient was calculated assuming $E_{bx}/E_m = 2$ (Eqs (13.8) and (13.10)).

It should be emphasized that Eq. (13.11) is based on theoretical relationships developed from composite modeling of masonry.

Many of the foregoing assumptions and observations concur with the statement by the withdrawn standard, BS 5628-3: 2005 [20], that "the horizontal coefficient of thermal expansion for masonry may be taken as being the same as for units and that thermal movement in the vertical direction may be assessed using the relevant coefficient and the height of units and joints." The current standard, BS EN 1745: 2002 [21] recommends that thermal expansion coefficients should be determined by tests carried out for the project under consideration or by results available from a database, with final values being determined by evaluation of test data.

Some experimental values of the thermal expansion coefficient of units and masonry are shown in Table 13.5. British Standard BS 5628-3: 2005 [20] quotes a range of thermal expansion coefficients for masonry mortars of 11–15 × 10^{-6}/°C. Further information of thermal expansion coefficients of lightweight aggregate concrete used in the manufacture of block units is given in Table 13.2.

British Standard BS 5628-3: 2005 [20] also points out that the type of clay and type of stone influence the thermal expansion coefficients of fired-clay masonry and natural

Table 13.5 Thermal Expansion Coefficients of Units and Mortar According to BS 5628- 3: 2005 [20], and Masonry According to BS EN 1996-1-1: 2005 [22]

| Material | Coefficient of Linear Thermal Movement, $10^{-6}/^{\circ}C$ | |
	BS 5628-3: 2005 (Units)	BS EN 1996-1-1: 2005 (Masonry)
Fired clay	4–8	4–8 (6)
Concrete	7–14	–
Calcium silicate	11–15	7–11 (10)
Dense aggregate concrete and manufactured stone	–	6–12 (10)
Lightweight aggregate concrete	–	8–12 (10)
Autoclaved aerated concrete	–	7–9 (10)
Natural stone	–	3–12 (10)
Natural limestone masonry	3–10	–
Natural sandstone masonry	5–12	–
Natural granite masonry	5–10	–

NB. The numbers in parenthesis are value given in UK National Annex.

stone masonry, respectively, while the thermal expansion coefficient of concrete masonry depends on the type of cement and mix proportions. It follows, therefore, that the thermal expansion coefficient of mortar could be expected to depend on the type of cementitious materials and the mix proportions. Furthermore, it is stated [20] that the magnitude of movement in the horizontal and vertical directions will differ where the thermal expansion coefficients of masonry units and mortar are not equal, and the height and length of masonry units are unequal. The latter statements are borne out by the composite model theory.

According to Ceram Building Technology [23], the range of the thermal expansion coefficients given in Table 13.5 is fairly restrictive, and the range of temperature that facing clay brickwork could be exposed to is of equal interest to the range of thermal expansion coefficients. Their review suggests ranges of temperature are slightly higher for darker colored walls than for lighter colored walls.

Moreover, the ranges of temperature are greater for lightweight materials than for heavyweight materials [24,25], which are shown in Table 13.6 together with general maximum and minimum service temperatures, and corresponding ranges of temperatures for external and internal exposure conditions of the UK.

Beard et al. [26] found a difference of 25 °C in the surface temperature when the temperature in the middle of a brickwork wall was 55 °C. For the purposes of design, Foster and Johnston [27] suggested a more reasonable range of 45 °C, which leads to a commonly adopted thermal linear strain of 300×10^{-6}. Studies carried out in the United States and Canada suggested even greater ranges of temperature [23]. The performance of masonry at very high temperature is covered by BS EN 1996-1-2: 2005 [28].

From their own tests, Ceram Building Technology [23] confirmed a small anisotropy of thermal expansion coefficient for unbonded London Stock clay brickwork. Strains on

Table 13.6 UK Service Temperature Ranges of Materials [23]*

| | Temperature, °C | | |
Location/Material	Min.	Max.	Range
External			
Cladding, walling, roofing			
Heavy weight:			
• Light color	−20	50	70
• Dark color	−20	65	85
Lightweight, over insulation:			
• Light color	−25	60	85
• Dark color	−25	80	105
Freestanding concrete structures or fully exposed structural members:			
• Light color	−20	45	65
• Dark color	−20	60	80
Internal			
Normal use	10	30	20
Empty/out of use	−5	35	40

* © Crown copyright 1979, reproduced with permission from IHS

a wallette (1130 long × 545 high × 215 mm thick) and on a wall (670 long × 930 wide × 230 mm thick) were measured over a temperature range from ambient down to 11 °C and then back again to ambient. An unbonded brick expansion of $7.6 \times 10^{-6}/°C$ was measured between temperatures of 20 and 100 °C according to the method specified by BS 1902: 1990 [29]. The results in Table 13.7 show that although there is some degree of anisotropy for the wallette, there is little anisotropy of thermal coefficient for the wall. The experimental results are in agreement with the theoretical values given by Eqs (13.5) and (13.8) using an assumed $\alpha_m = 12 \times 10^{-6}/°C$.

In the United States, ACI 530-05 [30] recommends a thermal expansion coefficient of $7.2 \times 10^{-6}/°C$ for clay masonry, and the same coefficient of $8.1 \times 10^{-6}/°C$ for concrete masonry and autoclaved aerated concrete masonry.

Table 13.7 Comparison of Thermal Coefficients Determined by CERAM Building Technology [23]

| | Thermal Coefficient $10^{-6}/°C$ | |
Test Specimen	Horizontal, α_{wx}	Vertical, α_{wy}
Wallette	10.4	7.2
Wall	7.1	7.6
Calculated	7.6	8.2

Example

Compare the seasonal thermal movements of a 12 m long × 3 m high clay brick-work in the vertical and horizontal directions undergoing an average temperature change of 30 °C, given that the thermal coefficients of the clay brick are $\alpha_{bx} = 8 \times 10^{-6}/°C$ and $\alpha_{by} = 12 \times 10^{-6}/°C$, and thermal expansion coefficient of the mortar, $\alpha_m = 15 \times 10^{-6}/°C$. The elastic modulus of the mortar is 15 GPa and the elastic modulus of the brick between headers is 30 GPa.

Solution

The thermal expansion coefficient of brickwork in the vertical direction is given by Eq. (13.5) using $\alpha_b = \alpha_{by} = 12 \times 10^{-6}/°C$:

$$\alpha_{wy} = 0.86 \times 12 + 0.14 \times 15 = 12.42 \times 10^{-6} \text{ per } °C$$

In the horizontal direction, the thermal expansion coefficient of brickwork is given by Eq. (13.8) using $\alpha_b = \alpha_{bx} = 8 \times 10^{-6}/°C$:

$$\alpha_{wx} = 8 + \frac{(15 - 8)}{\left[1 + 6.26 \frac{30}{15}\right]} = 8.52 \times 10^{-6} \text{ per } °C$$

Hence, for a temperature change of 30 °C, the seasonal vertical movement is $12.42 \times 10^{-6} \times 30 \times 3 \times 1000 = \pm 1.12$ mm.

The corresponding horizontal movement is $8.52 \times 10^{-6} \times 30 \times 12 \times 1000 = \pm 3.07$ mm.

Problems

1. Assuming the thermal expansion of hydrated cement paste is $15 \times 10^{-6}/°C$ and neglecting the modulus ratio, E_m/E_a, calculate the thermal expansion coefficient of the concrete specified in the example on p. 465 using Eq. (13.1).
   Answer: $9.0 \times 10^{-6}/°C$.

2. Explain the terms true kinetic thermal expansion coefficient and hygrothermal expansion coefficient.

3. Discuss the influence of low temperature on the thermal expansion coefficient of concrete.

4. Does creep of hydrated cement paste affect the thermal expansion coefficient of concrete?

5. If concrete to be used in a bridge undergoes a seasonal increase in temperature of 40 °C, suggest an approximate average thermal expansion, and lower and upper bound values.
   Answer: 400×10^{-6}; $340-476 \times 10^{-6}$.

6. For concrete containing aggregates of different thermal expansion coefficients, how would you estimate the thermal expansion coefficient of concrete?

7. Why is hygrothermal expansion coefficient zero in high-pressure steam-cured concrete?

8. Why is the unit more influential on the thermal expansion coefficient of masonry than aggregate on the thermal expansion coefficient of concrete?

9. Does creep of mortar affect thermal expansion coefficient of masonry?

10. The thermal expansion coefficient of masonry is isotropic and independent of mortar type. Discuss this statement highlighting any limitations.

11. The thermal movement of masonry is reversible. Is this always true?

12. How does the color of clay bricks affect the thermal movement of clay brickwork?

13. For the example on p. 468, calculate the vertical and horizontal movements if the brick thermal expansion coefficient is isotropic and equal to $10 \times 10^{-6}/^\circ C$.
Answer: Vertical $= \pm 0.96$ mm; horizontal $= \pm 3.73$ mm.

References

[1]   Neville AM. Properties of concrete. 4th ed. Pearson Prentice Hall; 2006. 844 pp.

[2]   Hobbs DW. The dependence of the bulk modulus, Young's modulus, creep shrinkage and thermal expansion of concrete upon aggregate volume concentration. Mater Constr 1971;4(20):107−14.

[3]   Browne RD. Thermal movement of concrete, Concrete. J Concr Soc London 1972;6(11): 51−3.

[4]   ACI Committee 209R-92 (Reapproved 2008). Prediction of creep, shrinkage and temperature effects in concrete structures. Detroit: American Concrete Institute; 1992.

[5]   Verbeck GJ, Hass WE. Dilatometer method for determination of thermal coefficient of expansion of fine and coarse aggregate. In: Proceedings of the highways research board, vol. 30; 1951. pp. 187−93.

[6]   Carlson CC. Lightweight aggregate for concrete masonry units. J Am Concr Inst November 1956;53:491−508.

[7]   Valore RC. Insulating concretes. J Am Concr Inst November 1956;53:509−32.

[8]   Bonnell GGR, Harper FC. The thermal expansion of concrete, National Building Studies. Technical Paper No. 7. London: HMSO; 1951.

[9]   Meyers SL. How temperature and moisture changes may affect the durability of concrete. Rock Prod Chicago; 1951:153−7.

[10]   Neville AM, Brooks JJ. Concrete technology. 2nd ed. Pearson Prentice Hall; 2010. 422 pp.

[11]   BS EN 1770. Products and systems for the protection and repair of concrete structures. Test methods. Determination of the coefficient of thermal expansion. British Standards Institution; 1998.

[12]   ASTM C531-00. Standard test method for linear shrinkage and coefficient of thermal expansion of chemical-resistant mortars, grouts, monolithic surfacings, and polymer concretes. American Society for Testing and Materials; 2005.

[13]   ASTM C490/C490M-10. Standard practice for use of apparatus for the determination of length change of hardened cement paste, mortar, and concrete. American Society for Testing and Materials.

[14]   CRD-C 39-81. Test method for coefficient of linear thermal expansion of concrete. US Army Corps of Engineers.

[15]   CEB-FIP Model Code 90. Design code, Comite Euro-International du Beton. Thomas Telford; 1993. 437 pp.

[16] BS EN 1992-1-1. Eurocode 2: Design of concrete structures: general rules and rules for buildings. British Standards Institution; 2004.

[17] Phillio R. Some physical properties of concrete at high temperatures. J Am Concr Inst 1958;54:857−64.

[18] BS EN 1992-1-2. Eurocode 2: Design of concrete structures. General rules. Structural fire design. British Standards Institution; 2004.

[19] Wittman F, Lukas J. Experimental study of thermal expansion of hardened cement paste. Mater Struct 1974;7(40):247−52.

[20] BS 5628−3. (Withdrawn), Code of practice for the use of masonry. Materials and components, design and workmanship. British Standards Institution; 2005.

[21] BS EN 1745. Masonry and masonry products. Methods of determining design thermal values. British Standards Institution; 2002.

[22] BS EN 1996-1-1. Eurocode No. 6: Design of masonry structures. General rules for reinforced and unreinforced masonry structures. See also: UK National Annex to BS EN 1996-1-1: 2005. British Standards Institution; 2005.

[23] CERAM Building Technology. Movement in masonry. Partners Technol Rep; 1996:119.

[24] Morton J. Designing for movement in brickwork. BDA Design Note 10. Brick Development Association; July 1985. 12 pp.

[25] Building Research Establishment. Estimation of thermal and moisture movements and stresses: Part 2. Digest 228, 1979. 8 pp.

[26] Beard R, Dinnie A, Sharples AB. Movement of brickwork-a review of 21 years of experience. Trans J Br Ceram Soc 1983;82.

[27] Foster D, Johnston CD. Design for movement in clay brickwork in the UK. In: Proceedings of the British Ceramic Society, vol. 30; September 1982.

[28] BS EN 1996-1-2. Eurocode 6. Design of masonry structures. General rules. Structural fire design. See also UK National Annex to BS EN 1996-1-2:2005. British Standards Institution; 2005.

[29] BS 1902-5.3. Methods of testing refractory materials. Refractory and thermal properties. Determination of thermal expansion (horizontal method to 1100 °C) (method 1902-503). British Standards Institution; 1990.

[30] ACI Committee 530-05. Building code requirements for masonry structures, ACI manual of concrete practice, Part 6. American Concrete Institute; 2007.

14 Effects of Movements, Restraint and Movement Joints

In the previous chapters, elasticity, creep, shrinkage/moisture movement, thermal movement have been discussed in detail as individual deformation properties of concrete and masonry. In practice, very rarely do those deformations occur independently but instead combine and are often restrained in a complicated manner to influence the movement of the whole structure, or influence the deflection of individual elements especially when reinforcing or prestressing steel is present. For example, restraint of drying shrinkage induces tensile stress which is relieved to some extent by creep, but cracking will occur if the stress exceeds the tensile strength. Furthermore, the risk of cracking is increased where openings, changes in height, thickness or direction of walls exist because of stress concentration.

To place those movements in context, there are other types of movement may contribute to the performance of a structure in terms of cracking or material disruption. These include foundation movement in the form of settlement, subsidence or heave [1], plastic shrinkage of concrete and mortar [2], and chemical movement due to cryptoflorescence (Chapter 9), alkali–aggregate reaction, sulfate attack and carbonation [1]. However, most of these movements are considered to be beyond the scope of this book because, for example, it is considered that plastic shrinkage, foundation movement and some chemical movement can be avoided by good design and site practices. In consequence, the problems and solutions discussed in this chapter relate specifically to those movements discussed in earlier chapters.

The complexity of estimating the likelihood of cracking in masonry and concrete buildings due to restraint of movements is such that there are no ready-made analytical solutions available at the design stage. Instead, from observations of long-term building performance and practical experience, the effects of movement can be accommodated successfully in the structure by the incorporation of gaps or movement joints and the inclusion of mortar bed joint reinforcement in masonry with openings. This chapter deals with joints for both masonry and concrete buildings that are required to accommodate movements, the topics including types of joint, properties and types of fillers and sealants, estimation of total movement, accuracy of construction, and minimum and target widths of joint. Worked examples are given for estimating widths of horizontal and vertical joints. Finally, location and spacing of joints in buildings are discussed. In the first instance, however, problems caused by the effects of movements and restraint are highlighted.

Concrete and Masonry Movements. http://dx.doi.org/10.1016/B978-0-12-801525-4.00014-5

Effects of Movements

Creep at normal levels of stress/strength ratio is not considered to affect strength, although there is likely to be a small increase in the case of concrete and mortar due to a decrease in porosity under prolonged compressive loading. However, when stress exceeds about 60–80% of the short-term strength, creep can lead to a time-dependent failure of concrete known as creep rupture or static fatigue (see p. 281). Hence, creep rupture is possible if the stress is too high relative to the design strength or if the stress is satisfactory but the strength is less than expected. Figure 14.1 illustrates the strain response to compressive stress expressed as fraction of short-term strength. Below an approximate stress/strength ratio of 0.5, creep in linear, i.e., primary creep is proportional to stress but, at higher ratios, creep is non-linear, i.e., creep increases at faster rate than stress due to secondary creep induced by microcracking at the cement paste/ aggregate bond interface. Above a threshold stress/strength ratio of 0.8, a time-dependent failure occurs due to tertiary creep induced by enlargement and linking of the microcracks. In compression, large strains can develop just before failure. Creep rupture also occurs in tension, but at a lower threshold of approximately 0.6 of the short-term strength [4] and, of course, the limiting strains are much less.

The effects of creep under normal stresses are considered to be generally disadvantageous in concrete and masonry structures, although there are exceptions. Long-term deflections of reinforced beams, walls, piers and columns are increased by several times the elastic deformation, while prestress losses occur in prestressed and post-

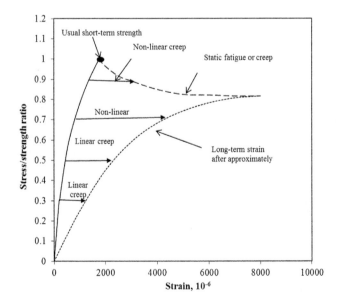

Figure 14.1 Schematic influence of sustained stress on strength and creep of concrete in compression [3].

tensioned elements. In reinforced concrete columns, creep causes a gradual transfer of load from concrete to steel, so that once the steel yields, the concrete takes any increase in load, and the full strength of concrete and steel is developed before failure occurs. The distribution of stresses in composite members is governed by strains in the components and, because of creep, there is a continuous redistribution of stress even in a simply supported reinforced concrete beam under a constant load: the neutral axis is lowered with a consequent decrease in the stress in concrete and an increase of stress in the steel [5].

In post-tensioned masonry walls, prestress losses are greater in concrete block walls than in clay brick walls due mainly to the fact that drying shrinkage is greater in concrete block walls. In clay brickwork, the irreversible moisture expansion of the clay brick opposes the mortar shrinkage whereas, in concrete blockwork, shrinkage of concrete block augments shrinkage of mortar. Table 14.1 compares the ultimate losses due to creep, shrinkage and relaxation of prestressing steel as reported by Tapsir [6]; details of the experiments have been presented earlier in Table 12.9. It can be seen that the total loss of prestress is less for the diaphragm and fin clay brick walls than for either the concrete block walls or calcium silicate walls, mainly due to the much lower contribution from shrinkage of the clay brick walls. In the case of clay brickwork, the major contribution to total loss of prestress is from creep ($\approx 41\%$) compared with the contribution from shrinkage (33%). On the other hand, in the case of calcium silicate and concrete walls, the largest contribution to total prestress loss is from shrinkage (49%), with creep contributing 34%. The smaller contribution to total loss of prestress from shrinkage in clay brickwork is a consequence of the moisture expansion or zero shrinkage of the clay unit. Nevertheless, it seems the effect of moisture expansion of clay brickwork reducing the loss of prestress is not taken into account in the design process [7].

The loss of prestress in calcium silicate masonry walls is similar to that in concrete block masonry walls and, in general, is slightly greater than the general loss of prestress in concrete elements. Of course, the actual losses in any post-tensioned material depend on the same factors that determine the levels of creep and shrinkage of concrete, such as strength, type of mix ingredients and their proportions.

Besides contributing to loss of prestress in prestressed concrete, drying shrinkage increases deflections of asymmetrically reinforced concrete, and differential shrinkage of causes warping and curling of concrete slabs. In high strength or high performance

Table 14.1 Prestress Loss of Post-Tensioned Diaphragm (dia.) and Fin Walls Built from Different Types of Unit [6]

Type of Unit, Strength	Loss of Prestress, %						
	Creep		Shrinkage		Relaxation of Steel	Total	
	Dia.	Fin	Dia.	Fin		Dia.	Fin
Clay, 103 MPa	6.0	6.8	5.0	5.2	3.5 to 4.5	14.5	16.5
Calcium silicate, 27 MPa	7.2	8.0	10.5	11.5	3.5 to 4.5	21.2	24.0
Concrete block, 14 MPa	7.5	8.7	11.3	11.6	3.5 to 4.5	22.3	24.8

concrete, autogenous shrinkage is likely to be greater than drying shrinkage, but since the majority occurs early in the life of concrete it may be possible to delay construction operations until after autogenous shrinkage is complete so as to avoid undesirable effects.

Creep increases the deflection of eccentrically loaded slender columns and can contribute to buckling. In high-rise buildings, creep may cause excessive deflections and other problems as a result of the larger creep deformation of the highly-stressed exterior columns relative to the smaller creep deformation of the concrete core [5]. In long span bridges there may be serviceability problems with excessive deflection due to creep and shrinkage; with prestressed concrete girders, creep may introduce excessive upward deflection if the sustained compressive stress in the bottom fiber is considerably higher than in the top fiber [5].

Cracking of internal partitions and failure of external cladding fixed rigidly to concrete frames can occur due to differential movement. The vulnerability to cracking is demonstrated in the case of differential movement of clay brickwork cladding built tightly in a reinforced concrete frame as shown in Figure 14.2. The outward bowing or buckling of the cladding occurs because there is no room for its expansion due thermal and moisture movements, the outward deflection being exaggerated by the opposing contraction of the concrete columns due to shrinkage and creep. Cracking and spalling can also occur in the cladding and brick slips when they are rigidly fixed

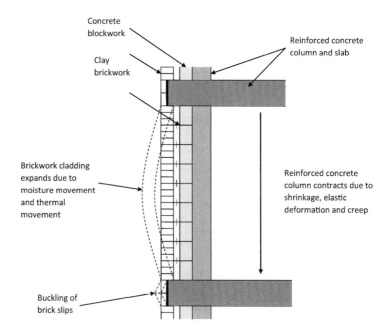

Figure 14.2 Effect of differential movements in brickwork cladding and reinforced concrete frame.

to the end of the concrete slab. Furthermore, an eccentric load is probably induced in the cladding so that creep may enhance the lateral deflection. According to Morton [8], several years may elapse before such signs of distress become apparent.

The solution to the problem in Figure 14.2 is to allow for the differential movement by incorporating a "gap" in the form of movement joint between the underside of the reinforced concrete slab and the top of the clay brickwork cladding; additional mechanical fixing of the brick slips may also be necessary [8]. In a slightly different case where the floor slab does not extend to the outer leaf (Figure 14.3), the steel shelf angle fixed to the ends of the concrete floor slabs supports the panels of clay brickwork cladding, the load being taken by the concrete blockwork inner leaf; flexible wall ties allow for differential movement between the inner and outer leaves. Underneath the

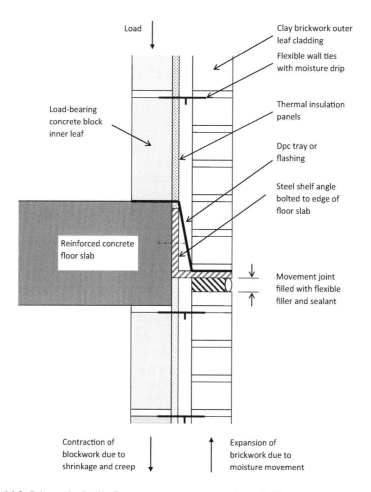

Figure 14.3 Schematic detail of movement joint in a cavity wall [9].

shelf angle is the movement joint comprising a flexible filler and sealant to prevent ingress of moisture.

Distortion and cracking may occur in thin manufactured sheets, such as fiber cement, due to carbonation shrinkage as a consequence of the reaction of atmospheric carbon dioxide with alkali liberated by the hydration of cement [1]. If one side of the sheet is sealed, say, by paint, damage may occur because of differential shrinkage caused by the unsealed side undergoing more carbonation shrinkage.

It would appear from the above examples of the effects of movements that creep is an undesirable property. However, that is not always the case since creep has an advantage of relieving stress concentrations induced by some degree of restraint to free shrinkage and thermal movement or induced by movement of supports. In many instances, free movements do not occur and restraint is often present in both concrete and masonry structures, so that that tensile or compressive forces are induced that can lead to failure by spalling or cracking. Even if localised and not likely to cause structural breakdown, such failures increase the risk of loss of serviceability through lack of durability. Methods of analysis to design for creep and shrinkage in structural concrete and masonry elements can be found in Refs. [5,10−13].

Restraint and Cracking

Restraint to movement is either in the form of *external restraint*, e.g., a surrounding steel frame or foundation, or in the form of *internal restraint* due to moisture or temperature gradients, e.g., the latter occurs in a large un-insulated concrete section due heat of hydration. Remedies to avoid the likelihood of cracking due to internal restraint include the careful selection of ingredients in the concrete mix design, e.g., limiting the temperature rise in the case of mass concrete due to heat of hydration. Sometimes, it is not possible to avoid external restraint to movement in concrete structures, say, by the provision of movement joints, and it is necessary to cast subsequent pours against hardened concrete in order to satisfy requirements of continuity in the structural design. There is a risk of thermal cracking in cantilevered retaining walls for reservoirs, basements, bridge abutments, etc. Restraint from the base of a concrete wall and adjoining sections can be considerable when casting a vertical section of the wall.

When movements of masonry and concrete are restrained, tensile and/or shear stresses are induced that usually lead to the formation of cracks. Figure 14.4 illustrates typical cracking patterns when long masonry walls with openings are restrained by the foundation. Free horizontal movement can occur at the top of the wall but it is restrained at ground level. In the case of expanding clay brickwork (Figure 14.4(a)), even with the presence of the damp proof course (dpc), restraint from ground level affects the main body of the wall so that shear stresses are induced with the result that cracks emanate from corners of the openings. Thermal and moisture expansion of clay brickwork may result in *oversailing* of the dpc (Figure 14.4(b)) and when subsequent contraction is not fully reversible tensile cracks can occur; for example, the "ratchet effect" prevents seasonal movement from being fully reversible [14].

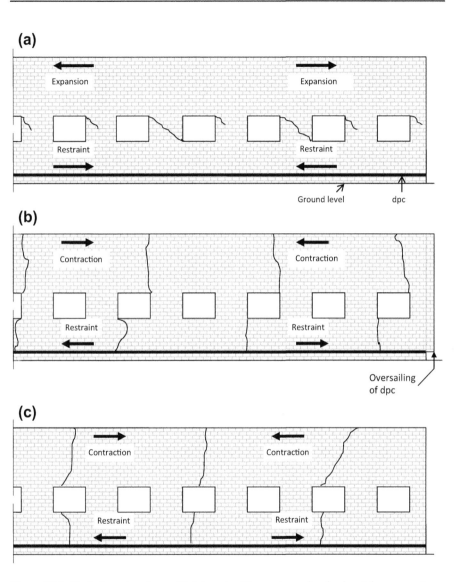

Figure 14.4 Typical crack patterns of brickwork with openings when horizontal movements are restrained. (a) Restraint of expanding clay brickwork (b) restraint of contracting clay brickwork after oversailing of dpc (c) restraint of drying shrinkage of calcium silicate brickwork.

A similar pattern of cracks is likely in calcium silicate brickwork through restraint of drying shrinkage, but here cracks tend to emanate from anywhere from the openings (Figure 14.4(c)) and not just at the corners as in the case of restrained contracting clay brickwork. To avoid the occurrence of cracks, movement joints can be positioned as shown in Figure 14.5.

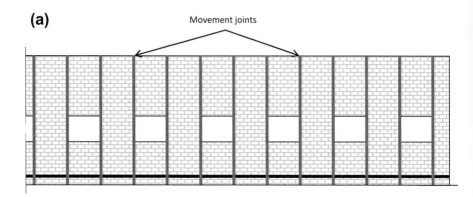

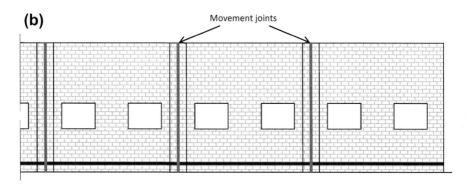

Figure 14.5 Location of movement joints in restrained walls to avoid the development of cracks shown in Figure 14.4. (a) Clay brickwork (b) calcium silicate brickwork.

Another example where damage due to local failure of masonry is likely is illustrated in Figure 14.6. According to BS 5628-3: 2001 [15], short returns of clay masonry of less than 675 mm, are likely to suffer from cracking as shown in Figure 14.6(a), whereas longer returns have sufficient inherent flexibility to accommodate opposing movements of the parallel walls without cracking (Figure 14.6(b)). To avoid cracking in short returns, movement points as indicated in Figure 14.6(c) and (d) should be incorporated with the latter being the preferred option by PD 6697: 2010 [16].

A national survey of the UK carried out by CIRIA [14] in 1986 revealed masonry configurations where cracking from in-plane movement may be expected. Table 14.2 lists structural configurations which are categorised as vulnerable and less vulnerable.

To illustrate the factors involved in the process leading to cracking, consider a concrete or masonry element restrained externally when undergoing drying shrinkage, as shown schematically in Figure 14.7. In normal concrete structures, shrinkage can be as high as 600×10^{-6}, which is approximately six times higher than the failure strain in

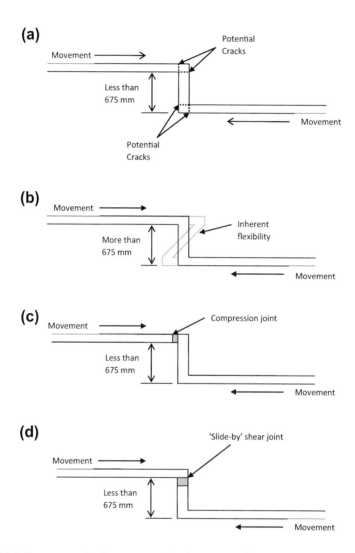

Figure 14.6 Short returns in clay masonry. (a) short return of less than 675 mm with cracking potential (b) short return greater than 675 mm having inherent flexibility (c) short return of less than 675 mm with compression movement joint (d) short return of less than 675 mm with slide-by movement joint (recommended by BS 5628-3: 2001 [15]).

tension. Consequently, if shrinkage is restrained the induced tensile stress can be expected to exceed the tensile strength and a single crack would form as shown in Figure 14.7(b). The process is time-dependent and the development of tensile stress is represented schematically in Figure 14.8. It can be seen that the role of creep is to reduce or relieve the induced stress, which is clearly an advantage, and it is only when the relieved tensile stress exceeds the current tensile strength that cracks are

Table 14.2 Masonry Configurations where Cracking is Likely [14]

Vulnerable	Less Vulnerable
Short return in long straight wall	Long straight walls
Spandrel walls	Stepped or corrugated facades
Link bridges	Long returns (greater than 900 mm)
Long parapets	Simple unbroken shapes
Stronger mortars	Weaker mortars
Discontinuous movement joints	Movement joints in walls
Discontinuous dpc	Restrained walls
Brick slips	Walls under vertical load
Incompressible joint fillers	Bed joint reinforcement in walls
Abrupt curtailment of bed joint reinforcement	
Changes of vertical load	
Slender panels between large walls	
Changes in shape, thickness and height of wall	
Eccentrically confined walls	
Bonding to dissimilar material, e.g., concrete	

able to form as indicated in Figure 14.7. However, cracking may be reduced or avoided by the following measures: incorporating steel reinforcement in the slab; ensuring adequate curing; using low water content in the concrete mix; and including a shrinkage reducing admixture in the concrete mix. For the general case of restrained drying shrinkage of concrete or masonry leading to cracking, the process can be expressed analytically as follows:

$$\sigma_t = S \times R \times E_e \times 10^{-3} \geq f_t' \tag{14.1}$$

where σ_t = induced tensile stress (MPa), S = shrinkage (10^{-6}), R = restraint factor, E_e = effective modulus of elasticity (GPa) and f_t' = tensile creep rupture strength (MPa).

For the case of restrained thermal contraction in a concrete or masonry element, the induced tensile stress is given by:

$$\sigma_t = \alpha \Delta t \times R \times E_e \geq f_t' \tag{14.2}$$

where α = coefficient of thermal expansion (10^{-6} per °C); and Δt = decrease in temperature (°C).

In Eqs. (14.1) and (14.2), the restraint factor, R, is defined as:

$$R = \frac{\varepsilon_f - \varepsilon_a}{\varepsilon_f} \tag{14.3}$$

where ε_f = free strain and ε_a = actual or restrained strain.

Thus, when fully restrained, $\varepsilon_a = 0$ and $R = 1$, and when unrestrained, $\varepsilon_a = \varepsilon_f$ and $R = 0$, so that Eqs. (14.1) and (14.2) confirm that the greater the value of R the greater is the induced tensile stress.

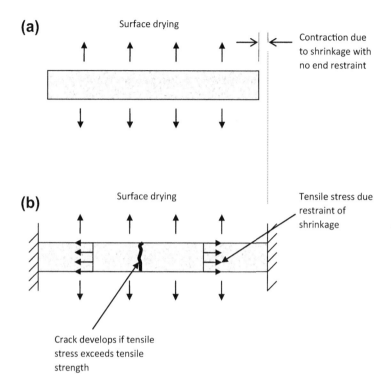

Figure 14.7 Schematic representation of cracking due to restraint of drying shrinkage in concrete or masonry element. (a) No restraint (b) Full restraint.

The effective modulus of elasticity term, E_e, of Eqs. (14.1) and (14.2) accounts for creep in the following way:

$$E_e = \frac{10^{-3}\sigma_t}{e_0 + C} = \frac{10^{-3}\sigma_t}{\frac{\sigma_t}{10^3 E} + \sigma_t C_s} = \frac{E}{1 + 10^3 E C_s} \tag{14.4}$$

where e_o = elastic strain at loading (10^{-6}), C = creep (10^{-6}), E = modulus of elasticity (GPa) and C_s = specific creep (10^{-6} per MPa).

Thus, it can be seen from Eq. (14.4) that creep effectively reduces the modulus of elasticity and, correspondingly, the induced stress given by Eqs. (14.1) and (14.2). It should be noted that the *tensile creep rupture strength*, f_t', in Eqs. (14.1) and (14.2) is <u>not</u> the normal strength, f_t, as determined in a short-term test in the laboratory, but a lower value of approximately $0.6 f_t$ due to the influence of a slow rate of incremental loading on tensile strength as discussed earlier [17].

Equations (14.1) and (14.2) imply that to minimise the risk of cracking, drying shrinkage or thermal contraction should be small, the external restraint should be small, creep should be high and strength should be high.

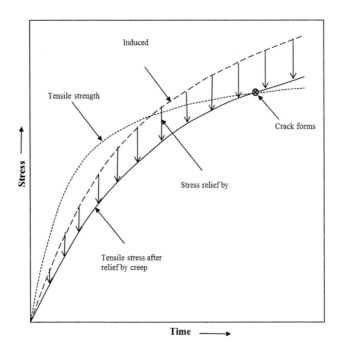

Figure 14.8 Schematic pattern of tensile stress development leading to crack formation in concrete and masonry elements prevented from contracting due to moisture or thermal movement.

In the case of the above example of an externally restrained element, creep is seen to be a desirable property, as it is in most cases involving restraint to movement. However, it should be mentioned that there is a special case of internal restraint where creep is thought to contribute to the possibility of cracking. That case is one of an uninsulated mass concrete having a temperature gradient across its section while undergoing expansion and contraction cycle due the heat generated from hydration of the cementitious materials. It seems that there is too much relief of stress by creep during the heating cycle, which causes greater stress being induced in the cooling cycle [17]. Table 14.3 outlines the stages of stress development across the section of concrete due to internal restraint from the temperature gradient. After placing the large mass concrete, the temperature rise caused by the heat of hydration causes the core to become hotter than the surface layers where heat is lost to the atmosphere unless the mass is completely insulated. The consequent differential expansion results in the core being restrained by the outer section generating compressive stress in the core and tensile stress in the surface layers. These stresses are partly relieved by creep which, at this stage, is large because the concrete is young and of low maturity. At peak temperature, the interior section is subjected to "relieved" compression, while the surface layers are subjected to "relieved" tension, which could crack if there is insufficient tensile strength.

As the concrete cools, the core now tends to contract more than the surface layers so the effect of restraint is now to induce tension and compression in the core and outer surface layers, respectively, which oppose existing stresses induced in the heating stage. The net effect is a lower compression in the core and lower tension and closure of any cracks in the surface layers. Furthermore, creep will again relieve the stress induced in the cooling stage, but to a lesser degree than in the heating stage because the concrete is more mature. Eventually, as the temperature of the concrete approaches the ambient temperature, the stress in the core could change from compression to tension with a risk of cracking, while the surface layers remain in compression.

Considering the situation where the mass concrete is insulated, the temperature gradient is reduced or eliminated and there is no internal restraint and, hence, no cracking. However, if external restraint is present, there is a possibility of cracking during the cooling stage, the process being represented by Eq. (14.2) with Δt is replaced by $(T_p - T_a)$, i.e., the difference between peak (T_p) and ambient temperatures (T_a) of the mass concrete. In other words:

$$\sigma_t = \alpha(T_p - T_a) \times R \times E_e \geq f_t' \tag{14.5}$$

Table 14.3 Development of Induced Stresses in Un-Insulated Mass Concrete Undergoing Early-Age Temperature Cycle due to Heat of Hydration [17]

Temperature	Inner Concrete Core	Outer Concrete Surface Layers
Placement temperature	Uniform temperature-no induced stress	
Heating stage	Warmer inner section wants to expand more than cooler surface layers, but is prevented by stresses induced by restraint:	
	Compression	Tension
	High stress relief by high creep	
Maximum temperature	Creep reduced compression	Creep reduced tension (surface cracks possible if greater than current strength)
Cooling stage	Core wants to contract more than surface layers, but is prevented by stresses induced by restraint:	
	Reduced compression	Reduced tension/crack closure
	Lower stress relief by lower creep	
	Low tension	Low compression
Ambient temperature	Increased tension (cracking possible if greater than current tensile strength)	Increased compression

Source: Concrete Technology, Second Edition, A. M. Neville and J. J. Brooks, Pearson Education Ltd. © Longman Group UK Ltd. 1987.

To minimise the induced tensile stress and, hence, the risk of thermal cracking, the terms of Eq. (14.5) suggests the following actions:

- Reduce the thermal expansion coefficient by careful selection of the concrete mix ingredients, e.g., use lightweight aggregate (see Chapter 13).
- Cool the mix ingredients prior to mixing by cooling the mix water with the addition of ice, which uses the heat from the other ingredients to provide the latent heat of fusion. This will reduce $(T_p - T_a)$, e.g., if $T_a = 20\,°C$, cooling the ingredients to $7\,°C$ will reduce T_p by a corresponding amount.
- T_p can also be minimised by using low-heat cement, e.g., fly ash or ground granulated blast-furnace slag blended with ordinary Portland cement. Blended cements will also reduce the rate of temperature rise as well as the peak temperature.

Internal restraint also occurs when a moisture gradient is present in large sections of concrete drying from the outer surface, and the process induces tensile stress that can lead to surface cracking. The process is similar to that discussed earlier regarding early-age thermal cracking. Figure 14.9 illustrates the mechanics involved for internal restraint of shrinkage in a long cylinder, where surface drying results in a moisture gradient as shown in Figure 14.9(a); in a large section of concrete or masonry, there is little or no loss of moisture at the central section. Assuming that true shrinkage is proportional to the moisture gradient profile, proceeding inwards from the outer surface in a radial direction, the true shrinkage at any section is restrained by a lower true shrinkage at the next adjacent section. Thus, profiles of induced tensile elastic strain and stress develop in the surface layers. Since there is no external force acting on the cylinder, the tensile stress is balanced by compressive stress acting in the central section, and the corresponding compressive elastic strain profile is indicated in Figure 14.9(b). However, because of creep, the induced elastic strain profiles are reduced as shown in Figure 14.9(c) but surface cracking can occur as a result of the maximum tensile stress associated with the maximum tensile strain. Finally, the actual observed shrinkage (S), shown in Figure 14.9(d), is the combination of "true" shrinkage (S_t) and restrained elastic strain after relief by creep (e_c) strain due to induced stress.

Thus, it is important to realise that when referring to shrinkage of concrete or masonry elements, that it is not true shrinkage but a combination of true shrinkage and strain due to internal restraint. This overall shrinkage may be visualised as shrinking of surface layers restrained by a non-shrinking core, which sometimes is referred to as *differential or restrained shrinkage*. In consequence, no laboratory test measures true shrinkage as an intrinsic property of concrete or masonry, so that specimen size should always be reported [17].

Movement Joints

According to BS 5606: 1990 [19], the movement joint is the medium where the changes in dimensions due to both induced and inherent deviations can be absorbed. British Standard 6093: 2006 [20] defines *induced deviation* as a measure of *accuracy*

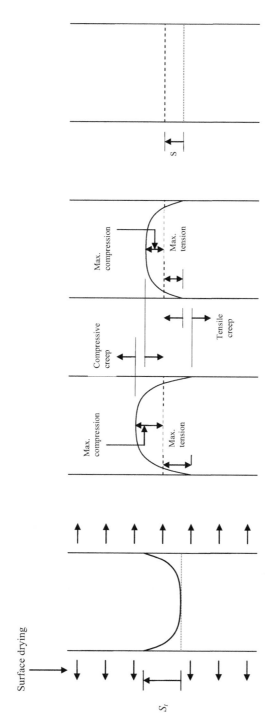

Figure 14.9 Internal restraint of drying shrinkage in a long concrete cylinder drying from its surface [18]. (a) Distribution of true shrinkage S_t (b) Elastic strain induced by restraint of true shrinkage (c) Reduced elastic strain due to creep, e_c (d) Observed shrinkage, $S = S_t + e_c$.

and is the dimensional deviation that occurs as a consequence of operations performed such as setting out, manufacture, assembly, erection etc. *Inherent deviation* is defined as the dimensional variation as a consequence of changes in temperature, humidity, stress etc. [20]. Examples of the latter are: plastic shrinkage, irreversible drying shrinkage, creep, moisture expansion of clay units and early age thermal contraction of mass concrete. Besides accommodating movement, the most relevant non-structural functional requirements of a movement joint in building construction are [21]:

- Durability.
- Resistance to water passage (as ice, liquid or vapor).
- Sound insulation.
- Thermal insulation.
- Resistance to fire.
- Appearance.
- Accessibility for inspection and maintenance.

According to ACI 224.3R-95 (Reapproved 2008) [22], many engineers view joints as artificial cracks, or as a means to either avoid or control cracking in concrete structures. It is possible to create weakened planes in a structure so cracking occurs in a location where it may be of little importance or have little visual impact. *Contraction joints* can be defined as designed planes of weakness that promote cracking of a wall at desired locations, whereas *expansion joints* can be are defined as separations between adjacent sections of a concrete wall to allow free movement caused by thermal expansion, lateral loads or differential movement of foundation elements [23]. Movement joints should not be confused with *construction joints* that allow no provision for movement, but allow construction to be resumed after a period of time since not all concrete in a structure can be placed at the same time [21,22]. For monolithic concrete, construction joints provide continuity of flexure and shear and are watertight through the bonded joint interface. Without that continuity, a weakened region occurs that may serve as a contraction or expansion joint [22]. A contraction joint may be formed by creating a plane of weakness by terminating some or all of the reinforcement either side of that plane. An expansion joint is formed by leaving a gap in the structure of sufficient width to remain open under extreme high temperature conditions [22].

In the UK, construction joints are sometimes referred to as *assembly joints* [1]. Construction joints occur commonly with in situ concrete and may be needed for: (1) joints between elements, (2) dry joints and (3) where depth or extent of concrete pour needs to be limited, e.g., in mass concrete to limit the heat generation. [20]

Despite the above definitions, joint nomenclature is somewhat confusing. In the US, expansion joints are usually dowelled such that movement can be accommodated in one direction, but there is shear transfer in other directions; sometimes, structural joints without any restraint are described as expansion joints. A joint which isolates movement between members and there are no dowels or steel crossing the joint is called an *isolation or free joint* [22]. In the UK [21], as well as *free contraction joints*, other contraction joints for concrete structures containing some reinforcement are termed: *tied partial contraction joints* and *debonded zpartial contraction joints.* Likewise, as well as a *free expansion joints*, there are *reinforced expansion joints.* Furthermore, there are *hinged joints*, *sliding/bearing joints* and *seismic/open joints* [21]. In general, movement joints in the UK tend to be called expansion joints [1], which must be able to accommodate both expansion and contraction due the reversible movements. When needed to accommodate contraction of the containing construction, expansion/contraction joints can be referred to as *soft joints* [1]. Soft joints appear to be specifically used to describe horizontal joints at non-loadbearing wall/floor junctions [20].

For concrete buildings, BS 6093: 2006 [20] deals with structural and major movement joints to cope with all possible sources of movement, and minor movement joints (construction, contraction and soft joints) to accommodate inherent movements such as drying shrinkage. In the US, ACI 224.2R-95 (Reapproved 2008) [22] covers all types of joints (construction, contraction, and expansion or isolation) in different concrete structures: buildings bridges, slabs, pavements, tunnels, walls, liquid retaining structures and mass concrete. In the UK, CIRIA Technical Note 107 deals with the design of movement joints in masonry and concrete buildings [24] and CIRIA Report 146 covers movement joints as part of the design and construction of joints in concrete structures [21]. In addition, movement joints as a safeguard against cracking in buildings are the subject of a Building Research Establishment report [1].

Fillers

Usually, movement joints are formed by incorporating fillers or filler strips during the construction, which may be replaced later by a more suitable material [1]. The filler acts as a spacer to ensure the required width of the joint. Fillers are required to have a compressive strength less than that of the adjoining construction in order to avoid damage to the latter, and they should have adequate recovery if the adjoining construction contracts. Table 14.4 lists properties of typical fillers for movement joints as specified by BS 6093: 2006 [20].

A movement joint that has no other function than to permit size changes due to movements does not actually need to be filled or sealed. In practice, however, it will be important that debris cannot enter and block the joint and thus prevent its intended operation. Moreover, in most cases, the joint will need to prevent ingress of moisture. These considerations dictate the need for a sealant as discussed below.

Sealants

ACI 224.3R-95 (Reapproved 2008) [22] gives a detailed description of the various types of sealants that are commercially available. For many years, oil-based mastics, bituminous compounds, and metallic materials were the only sealants available but, nowadays, *elastomeric* materials are used which have improved performance and longer life. [22]. Sealants are classified as elastic or plastic according to their ability to recover displacement after being stretched or compressed [25]. Type E elastic

Table 14.4 Properties of Fillers for Movement Joints [20]

Filler Type	Typical Uses	Form	Density Range, kg/m³	Tolerance to Water Immersion	Pressure for 50% Compression, MPa	Resilience, % Recovery after Compression
Wood fibre/bitumen	General purpose expansion joints	Sheet, strip	200 to 400	Suitable if immersion is infrequent	0.7 to 5.2	70 to 85
Bitumen/cork	General purpose expansion joints	Sheet	500 to 600	Suitable	0.7 to 5.2	70 to 80
Cork/resin	Expansion joints in water retaining structures where bitumen not acceptable	Sheet, strip	200 to 300	Suitable	0.5 to 3.4	85 to 95
Cellular plastics and rubber	Expansion joints	Sheet, strip	40 to 60	Suitable if immersion infrequent	0.07 to 0.34	85 to 90
Mineral or ceramic fibres or intumescent strips	Fire resistant joints: low movement	Loose fibre or braided and strip	Dependent on degree of compaction	Not suitable	Dependent on degree of compaction	Slight, 36

sealants are described as those able to recover to a specified width and profile and are suitable for joints accommodating reversible movement [26]. On the other hand, plastic sealants, type P, are unable to meet the specification and are suitable for movements that tend to be irreversible [25]. A further sub-classification is based on the modulus of the sealant: HM = high modulus and LM = low modulus, the latter being preferred for minimising stress at the sealant/substrate interface in the case of weak or friable substrates.

At the time of installation, conditions of ambient temperature extremes and high moisture levels are detrimental to the satisfactory application and performance of the sealant. The best time to apply sealants is when temperature and moisture conditions yield a joint width between the mean and maximum values, so that the tensile strain of elastic sealants is reduced [20].

The shape of the movement joint is of less importance with plastic sealants subject to small movements than for elastic sealants and, to a lesser extent, elasto-plastic sealants in joints subject to significant movements. It has been established that for elastic sealants, optimum performance is obtained at a *width to depth ratio* of approximately 2:1 and that, subject to a minimum depth of 5 mm, the width to depth ratio should not be less than 1:1. However, for the practical application, irrespective of movement accommodation, the recommended minimum joint gap width of sealants is 6 mm [20].

The full range of movement between the maximum and minimum joint widths that the sealant can accommodate is termed the *movement accommodation factor* (*MAF*) [25]. For elastic and elasto-plastic sealants applied to *butt joints*, the *MAF* quantifies its ability to accommodate tensile strain [1] and when strained in shear only (*lap joints*), the sealant can usually be expected to perform satisfactorily at twice the stated *MAF* [20]. Sealants are categorized into movement classes: 7.5, 12.5, 20 or 25 in accordance with BS EN ISO 11600: 2003 [26] which should be used as the *MAF*. Features of the movement classes are:

- Class 7.5 sealants have no distinction between plastic and elastic sealant behaviour, although most are plastic in nature.
- Class 12.5 sealants are sub-divided into elastic or plastic types.
- Classes 20 and 25 are considered to be elastic and are sub-divided into high and low modulus classes. Low modulus sealants are suitable for joints exposed to long periods of extension or compression and/or where the substrate material is weak or friable.

Test methods for the determination of properties of sealants are specified by BS EN ISO 11600: 2003 [26].

The specification of joint gap widths should be derived from all deviations to which the joint is subjected. However, minimum joint gap widths can be calculated if the following data are known:

1. The *MAF* of the sealant as a percentage of the minimum joint gap width taken from the material specification or manufacturer's literature.
2. The *total relevant movement* (*TRM*) of components at the joint estimated using BRE Digest 228 [27].

The calculation is based on the ability of the sealant to accommodate the imposed range of tensile strain. The *TRM* excludes irreversible (inherent) movements that close-up the joint and induced deviations which are allowed for separately (see p. 19).

It is emphasised by BS 6093: 2006 [20] that joint gap widths at the time of construction of components and the time of sealing joints might have changed because of conditions giving rise to movements, and it is impossible to ensure that a joint gap no smaller than the calculated minimum will be achieved irrespective of the conditions at the time the joint was formed. The following expression should be used to estimate the minimum joint gap width (W_{min}):

$$W_{min} = \frac{TRM \times 100}{MAF} + TRM \tag{14.6}$$

For example, if a sealant has to accommodate a $TRM = 3$ mm and has a $MAF = 25\%$, then Eq. (14.6) gives $W_{min} = 15$ mm. Only tensile strain imposed on the sealant determines the *MAF*, compressive strain does not contribute. For example, if after assessment of all other deviations $W_{min} = 15$ mm is specified as the minimum joint gap width in cold conditions, the joint can subsequently close to 12 mm due to expansion of the components and still lie within the *MAF* limit of 25% [20]. Further examples are given on p. 499.

Besides the sealant itself, important components of the movement joint are back-up materials, bond breakers and fillers. These components control the sealant joint design giving the correct joint width to depth ratio, prevent wastage, provide a firm surface and ensure that the sealant bonds and adheres only to the substrates [20]. Combinations of these materials are used to give the required depth of sealant and support as shown in Figure 14.10. The use of bond breaker tape prevents adhesion of the sealant to the filler. During installation, it is important not to damage the closed cell foam to avoid bubbles of gas escaping and becoming trapped within the sealant, which will affect its performance.

Total movement and width of joint

In the US, joint width in concrete construction is determined from the thermal movement caused by maximum seasonal temperature rise. Joints vary in width from 25 to 150 mm or more, with 50 mm being typical. Wider joints are used to accommodate additional differential building movement that may be caused by settlement or seismic loading [23].

In the UK, BS 5628-3: 2005 (superseded) [15] specified that the maximum movement in masonry should be no greater than the recommended movement in the joint sealant. It follows that the product of the length of the masonry and the effective strain in a wall should be less than the product of the width of the joint and the permitted strain in the sealant. Publication Document, DP 6697: 2010 [16] recommends that the width of the joint should be sufficient to accommodate both reversible and

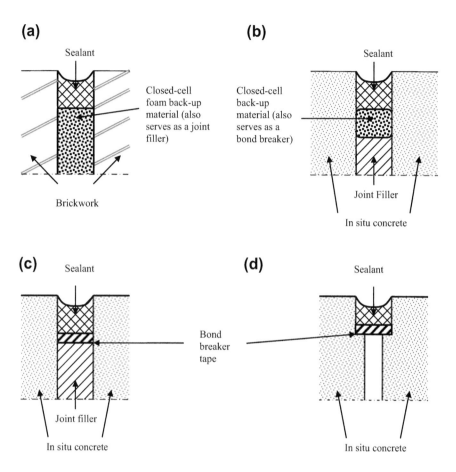

Figure 14.10 Sections through depth of sealed movement joints for masonry and concrete showing use of back-up materials and bond breakers [20]. (a) Masonry joint (b) Concrete joint with filler and back-up material (c) Concrete joint with filler and bond breaker tape (d) Concrete joint with bond breaker tape.

irreversible movements. Clay masonry walls, which are unrestrained or lightly restrained and unreinforced, may expand by 1000×10^{-6} (1 mm/m) during the life of a building due to combined thermal and moisture movements. As a general guide and to allow for compressibility of the filler, the width of the joint (mm) should be about 30% more than the joint spacing (m). For example, movement joints at 12 m centers should be about 16 mm wide. Where a manufacturer can show evidence from experience that the product expands more than 1000×10^{-6} during the life of a building due to combined thermal and moisture movement changes, the foregoing guidance may be modified at the designer's discretion [16]. In calcium silicate masonry undergoing contraction, generally, vertical joints not less than 10 mm wide are required to accommodate horizontal movements [16]. Similarly, in the absence of specific calculations, to accommodate horizontal movement in natural stone masonry, vertical joints not less than 10 mm wide should be provided [16].

In order to assess the width of the horizontal movement joint, Shrive [9] analyzed the movements in a floor/exterior cavity wall where the cavity wall consists of a clay brick outer leaf tied by wall ties to a loadbearing concrete block wall, as shown in Figure 14.3. The analysis assumes that the brickwork is subjected to reversible moisture movement, irreversible moisture expansion and thermal movement due to the temperature difference between the inside and outside walls. On the other hand, if the building is air conditioned, the blockwork will undergo contraction due to elastic strain, creep and shrinkage, there being little contribution from reversible moisture and thermal movements because of the constant humidity and temperature of the internal environment. Seasonal climate effects will change the moisture and temperature of the brickwork relative to the insulated blockwork: expansion of brickwork due to an increase in temperature will close the "gap" or reduce the joint width, while contraction due to a fall in temperature will increase the "gap." Assuming the blockwork is loaded axially but not fully loaded until after the outer leaf is built, the maximum long-term width closure (ΔW) due to *differential vertical movement* is given by:

$$\Delta W = H\left[\alpha_{wbr}(T_{br} - T_{bl}) + M_{ei} + M_{er} + \sigma_w\left(\frac{1}{E_{bl}} + C_{sbl}\right) + S_{wbl}\right] \quad (14.7)$$

where $H =$ height between shelf angles, $\alpha_{wbr} =$ thermal expansion coefficient of brickwork; $T_{br} =$ maximum seasonal temperature of brickwork; $T_{bl} =$ temperature of blockwork; $M_{ei} =$ irreversible moisture expansion of brickwork; $M_{er} =$ reversible moisture expansion of brickwork; $\sigma_w =$ stress on blockwork; $E_{bl} =$ modulus of elasticity of blockwork, $C_{sbl} =$ ultimate specific creep of blockwork and $S_{wbl} =$ shrinkage of blockwork.

For example, assuming $H = 3$ m $= 3000$ mm, $\alpha_{wbr} = 6 \times 10^{-6}$, $T_{br}-T_{bl} = 75\,°C$, $M_{ei} = 300 \times 10^{-6}$, $M_{er} = 100 \times 10^{-6}$; $\sigma_{br} = 1.5$ MPa, $E_{bl} = 16$ GPa, $C_{sbl} = 400 \times 10^{-6}$ per MPa and $S_{wbl} = 400 \times 10^{-6}$, Eq. (14.7) yields $\Delta W = 5.83$ mm. In practice, the *specified width* or *target width* of the movement joint would be greater to allow for properties of the filler and sealant, as well as to allow for inaccuracy of construction as demonstrated in the following section (see Example 2 on p. 500).

With regard to the practice of estimating total movement of masonry and concrete by summing individual movements or inherent deviations as in the case, say, of Eq. (14.7), BRE [27] comments that many naturally exposed materials will be subject to concurrent or interdependent changes of temperature or moisture content with the result that the net movement may be overestimated when thermal movement and moisture movement at normal temperature are treated as fully additive. For example, raising the temperature will cause thermal expansion, but there will be an increase in moisture loss that will accelerate shrinkage of concrete materials compared with shrinkage at normal temperature. On the other hand, irreversible moisture expansion of clay brickwork will increase at high temperature. Furthermore, the effect of a higher temperature on moisture content of concrete is to increase water vapor pressure, which increases the thermal expansion coefficient (see p. 405). It is apparent that the true interactive effect of temperature and moisture on movement is complex and currently cannot be estimated accurately.

Target Width of Joint

In addition to estimating the maximum possible total movement of structural elements due to the inherent deviations of elastic deformation, thermal movement, creep and moisture movements, the width of the joints has to take into account the compressibility of the filler, the *MAF* of the sealant and induced deviations, i.e., the accuracy of construction in terms of variability of dimensions of components separated by the joints. The procedure specified by BS 5606: 1990 [19] for assessing the latter is described below.

The combined deviation (DL_t) of separate elements or components whose deviations are known, say, DL_1 and DL_2 etc. is calculated from the square root of the sum of the squares of the individual deviations, i.e.:

$$DL_t = \sqrt{(DL_1)^2 + (DL_2)^2} \qquad (14.8)$$

From detailed surveys of dimensions of buildings and individual elements constructed from different materials in the UK, ranges of deviations have been compiled by BS 5606: 1990 [19]. The data from the surveys were analyzed to obtain the mean size ($\bar{x}$) and the standard deviation (SD) of n samples of individual size (x_i) of normal distribution curves as shown in Figure 14.11, the SD being given by:

$$SD = \sqrt{\frac{\sum (x_i - \bar{x})^2}{n - 1}} \qquad (14.9)$$

Figure 14.11 indicates that for a size range of ± one single SD, approximately 68% of sizes is included, with the chance that 1 in 3 cases will fall outside the range. For greater multiples of standard deviation, the risk of sizes falling outside the range becomes less. In the cases of deviations of space dimensions between elements and deviations of construction elements, as compiled by BS 5606: 1990 [19], the risk is one in 22 chances, i.e., approximately equivalent to $2 \times SD$. The risk for manufactured components is less, viz. 1 in 80 cases, which is approximately equivalent to $2.5 \times SD$. Examples of deviation data are shown in Tables 14.5 and 14.6, which may be used with Eq. (14.9) for estimating combined deviations or, alternatively, deviations may be obtained from other sources having previous design experience.

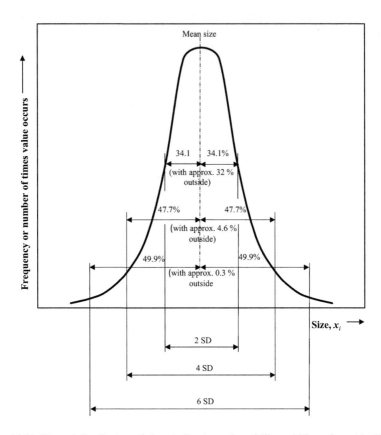

Figure 14.11 Normal distribution of sizes indicating values falling within and outside different ranges of standard deviation (*SD*) [19].

Table 14.5 Average Range of Induced Deviations of Horizontal Space between Walls and between Columns Measured at Floor and Soffit Levels [19]

	Deviation, mm	
Construction Material	**Walls**	**Columns**
Space up to 7 m apart		
Brickwork	±18	NA
Blockwork	±19	NA
In situ concrete	±24	±18
Precast concrete	±17	±13
Steel	NA	±11
Timber	±30	±12
Height between Floor and Soffit of Beams and Slabs		
In situ concrete	±23	
Precast concrete	±19	

Table 14.6 Range of Induced Deviations for Elements in Construction and Manufactured Components [19]

Construction Material/Item	Dimension	Deviation, mm
Masonry (Item T. 1.3 from Table 1-BS 5606)		
Brickwork	Height up to 3 m	±26
Blockwork		±28
Building Plan (Item 1.4 from Table 1-BS 5606)		
Brickwork	Length or width of	±29
In situ concrete	building up to 40 m	±26
Precast reinforced concrete		±38
Steel		±14
Manufactured Components (Item 2.1 from Table 1-BS 5606)		
Precast reinforced concrete	Up to 2 m	±6
	2−6 m	±9
	6−10 m	±12
Fabricated steel	Up to 30 m	±5
Timber - frames	Up to 6 m	±4
- panels		±5
- doors		±3

Examples of applying the procedure to calculate the target width and range of joint sizes are given below.

Examples

1. This example is taken from BS 5606: 1990 clause B 2.2.2 [19]. It is required to calculate the range of joint sizes for separating three timber window frames to be fitted into the prepared spacing between two in situ concrete columns shown in Figure 14.12. The indicated target sizes of the opening and window frames have been previously been calculated by the procedure described in BS 6954-3: 1988 [28].

Solution

The deviation for the space between the in situ concrete columns is ±18 mm (Table 14.5) and the deviation for the component timber window frame is ±4 mm (Table 14.6). Hence, the total deviation as given by Eq. (14.9) is:

$$DL_t = \pm\sqrt{(18)^2 + 4^2 + 4^2 + 4^2} = \pm 19.3 \text{ mm}$$

Now the target size of a single joint $= \frac{2430-2391}{4} = 9.8$ mm and, therefore, the maximum joint size $= 9.8 + \frac{19.3}{4} = 14.6$ mm while the minimum joint width $= 9.8 - \frac{19.3}{4} = 5.0$ mm. Consequently, the jointing technique (sealant and filler) should be capable of accommodating joint sizes in the range of 15

Continued

Examples—cont'd

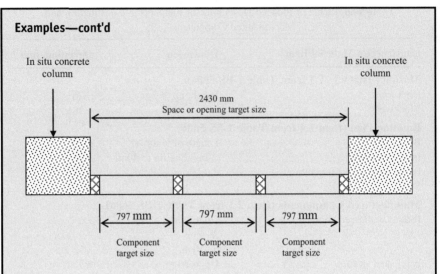

Figure 14.12 Target sizes for prepared spacing between concrete columns fitted with three timber frame windows (components) and four joints [19].

to 5 mm, but for practical application the sealant requires a minimum joint width of 6 mm [20].

2. For the second example, it is required to estimate the target width for the horizontal joint to accommodate the calculated differential vertical movement of 5.83 mm in the movement joint shown in Figure 14.3 (see p. 479).

Solution

a. To avoid complete closure of the joint, the minimum width required is, say, 6 mm, but to allow for the compressibility of the joint filler with, say, a resilience = 70% (see Table 14.3), the minimum width now becomes $6 \times 1/0.7 = 8.6$ mm.

b. Considering the joint sealant, the movement accommodation factor (*MAF*) is needed, which is based on tensile strain capacity as determined by movements that open the joint due to contraction of the brickwork (outer leaf) and expansion of the blockwork (inner leaf). However, there is little expansion by the blockwork since the environmental temperature and humidity are controlled and therefore unchanged, but the clay brickwork contracts due to thermal movement when the outside temperature falls, and also contracts due to reversible moisture movement when drying, namely, $(\alpha_{wbr}(T_{br} - T_{bl}) + M_{er})$ (see Eq. (14.7)). The sum of those two components is 1.65 mm, which is termed the total relevant movement (*TRM*) [27]. The *MAF* of the sealant can now be calculated. On the basis of the total contraction movement of 8.6 mm, approximately 1.65 mm is reversible movement that will impose

Examples—cont'd

tensile strain on the sealant due to seasonal climate changes. Therefore, from Eq. (14.6), the *MAF* is:

$$MAF = \frac{TRM}{W_{min} - TRM} \times 100 = 23.7\%$$

and a sealant of movement class 25% would be appropriate.

c. Consider now the induced deviation due to inaccuracy of construction according to the procedure described by BS 5606: 1990 [19]. To avoid contact of the expanding brickwork and the underside of the shelf angle during service life, the minimum joint width is required to be at least 8.6 mm. However, to minimise the risk of widths being less than 8.6 mm, and in the absence of previous experience, the target joint width needs to be assessed from the complied deviation data of BS 5606: 1990 [19]. From Table 14.5, for 3 m high brickwork, the element deviation $= \pm26$ mm, and the space deviation is ±19 mm for the floor to soffit height between precast reinforced concrete beams (assumed equal to the distance between shelf angles). Therefore, from Eq. (14.8), the total deviation for a single movement joint $= \pm\sqrt{(26^2 + 19^2)} = \pm32.2$ mm. Consequently, to minimise the risk of a lower minimum width than 8.6 mm, a target width of $8.6 + 32.2 \approx 40$ mm is required. That target size means that there is a small risk of one chance in 22 that the joint width will be less than, say, 9 mm or more than a maximum joint width of $(40 + 32) = 72$ mm.

d. For larger joint widths up to the maximum of 72 mm, the *MAF* becomes less and the chosen sealant is therefore satisfactory, but the suppliers of the filler and sealant should be consulted with regard to their installation and performance. In addition, the designer and contractor should be consulted regarding a more accurate construction specification to reduce induced deviation and, hence, the target joint width.

3. In this example, it is required to estimate the target joint width of vertical movement joints separating 8 m lengths of a lightweight concrete blockwork walls exposed to the outside environment. It is estimated that the long-term drying shrinkage of lightweight concrete blockwork $= 300 \times 10^{-6}$ and the reversible moisture movement due to seasonal climate change is $\pm300 \times 10^{-6}$. It is assumed that the thermal expansion coefficient of lightweight concrete blockwork $= 10 \times 10^{-6}$ per °C and the maximum annual temperature change $= 80$ °C. From previous construction experience, the induced deviation of movement joint widths may be taken as ± 10 mm.

Solution

a. The maximum seasonal thermal movement $= \pm(80 \times 10) \times 10^{-6} = \pm800 \times 10^{-6}$ and taking the long-term shrinkage and moisture movement into account, the maximum estimated total long-term contraction of the blockwork wall is $8 \times 1000 \, (300 + 300 + 800) \times 10^{-6} = 11.2$ mm.

Continued

Examples—cont'd

b. To avoid complete joint closure when the wall expands, it is assumed that thermal expansion and reversible moisture (wetting) movement of the wall could occur soon after construction when little irreversible drying shrinkage has taken place. The maximum expansion is $8000 \times (800 + 300) \times 10^{-6} = 8.8$ mm, which would be the smallest width just to prevent complete closure of the joint. However, considering the compressibility of the filler with a resilience of 70%, the minimum width becomes $1/0.7 \times 8.8 = 12.6$ mm, say 13 mm.

c. To select an appropriate sealant, the total relevant movement (TRM) is needed, which is determined by the extension of the sealant or the irreversible shrinkage contraction of the wall, viz. $8000 \times 300 \times 10^{-6} = 2.4$ mm. Substitution in rearranged Eq. (14.6) yields:

$$MAF = \frac{2.4}{13 - 2.4} \times 100 = 22.6\%$$

Therefore, a sealant of movement class 25% is required.

4. For an induced deviation of ± 10 mm, the target width=23 mm and the maximum width is 33 mm. The chosen sealant is satisfactory for the maximum width since the *MAF* becomes less.

Depending on the assumed level of accuracy of construction or induced deviation, from the above examples, it is apparent that target widths of movement joints can be considerably greater than the estimated widths calculated from unrestrained movements, compressibility of filler and the movement accommodation factor of the sealant. The estimated widths are of similar order to those recommended in Codes of Practice, e.g., PD 6697: 2010 [16], and specifying the greater target width ensures that there is only a small risk of any joint width being less than the required minimum width.

Joint Spacing and Location

Concrete Buildings

Depending on the type of the wall, use and service conditions, spacing of contraction joints may vary from 4.5 to 9 m [23]. According to ACI 224.3R-95 [22,23] contraction joint spacing is recommended to be equal to the height of the wall when the latter exceeds 3.6 m, or three times the height of a wall for short walls less than 2.4 m, while the Portland Cement Association (PCA) [23,29] recommends a maximum contraction joint spacing of 6 m and within 3 or 4.5 m of a corner. For walls with openings, the PCA prescribes that the joint spacing should be less than 6.1 m. On the other hand, contraction joints are not necessary if window openings are very wide and separated by small piers (when the wall becomes a frame). However, contraction joints should be located at the center of isolated window openings if the remaining ligament above the opening is at least one-fourth of the width of the opening, and contraction joints should be in line with the jamb below

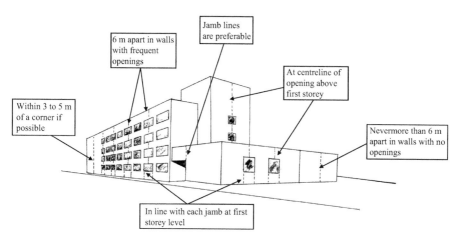

Figure 14.13 Location of contraction joints in a concrete building [22,23].

the first floor openings (see Figure 14.13). That particular requirement may not be aesthetically acceptable, but it is necessary to avoid cracking [23].

According to the PCA [23,29], most buildings of simple rectangular shape and relatively short (60–90 m) do not require expansion joints. However, if they are required, expansion joints should be placed at directional changes along a wall. Spacing of expansion joints can range from 9 to 60 m and is rarely less than 30 m in straight depending on the type of structure.

For determining the need for expansion joints, a report by the National Academy of Science [23,30] recommends an empirical approach. Beam-and-column type buildings, hinged at the base, or slab-and-beam structural frame buildings with heated interiors, the maximum allowable length without expansion joints should be determined according to the change of design temperature as indicated below:

(a) Rectangular multiframed configuration with symmetrical stiffness.

Design temperature change, °C	0 to 15	15 to 39	39 to 50
Allowable building length, m	170	170 to 100	100

(b) Non-rectangular configuration.

Design temperature change, °C	0 to 15	15 to 39	39 to 50
Allowable building length, m	85	85 to 45	45

For all other cases, the rules in Table 14.7 apply and, if more than one of those design features is applicable, the adjustment should be based on the algebraic sum of the individual factors.

In the UK, the design and construction of joints in concrete structures are dealt with by a CIRIA Report 146 [21]. It is generally considered that spacing is a matter of engineering

Table 14.7 Adjustment Factors for Allowable Length of Building [23,30]

Design Condition of Building	Percentage Adjustment to Allowable Length
Air conditioned and heated, operating continuously	+15
Unheated	−33
Fixed column bases	−15
Substantially greater stiffness at one end of the plan dimension	−25

judgment. In large buildings movement joints should be provided at a spacing of 60−70 m, it being recognized that spacing may be increased if the structure is relatively flexible in the horizontal direction or the spacing decreased for smaller exposed structures with stiff columns. To react more rapidly to daily and seasonal temperature fluctuations, un-insulated or lightly-insulated roof slabs, parapets and exposed floor slabs may have closer spacings of 20−40 m. Spacing is influenced by the range of temperature experienced in service, for instance, in Mediterranean regions spacings as close as 25 m are recommended for structures in hot and dry areas compared with 50 m for cooler and wetter areas.

CIRIA [21] recommends that joints are desirable where concrete elements or building profiles change section abruptly, because the smaller section is vulnerable to cracking from temperature variations. Joints are commonly used for a new construction built against an existing construction especially where its form is different or not built from concrete. Except for multi-storey basements, joints provided to accommodate thermal movement and irreversible drying shrinkage are not usually continued below ground level. When needed to accommodate just irreversible shrinkage, it may be worth considering leaving short bays un-concreted until later and omitting movement joints, provided the structure is sufficiently stable and the contractor is in agreement.

Masonry Buildings

With regard to the spacing of movement joints in masonry buildings, BS EN 1996-2: 2006 [31] specifies the following for different types of masonry:

- The horizontal spacing of horizontal movement joints in masonry walls should take into account the type of wall.
- The horizontal distance between vertical movement joints in external non-loadbearing unreinforced masonry walls should not exceed l_m in Table 14.8.
- The distance of the first vertical joint from a restrained vertical edge of a wall should not exceed half the value of l_m in Table 14.8.
- The need for vertical movement joints in unreinforced loadbearing walls should be considered. No recommended values for the spacing are given [31] as they depend on local building traditions, for example, type of floors used and other construction details.
- The position of movement joints should take into account the need to maintain structural integrity of loadbearing internal walls.
- Where horizontal joints are required to accommodate vertical movement in an unreinforced veneer wall or in an unreinforced non-loadbearing outer leaf of a cavity wall, the spacing of the horizontal movement joints should take into account the type and positioning of the support system.

Table 14.8 Maximum Horizontal Distance between Vertical Movement Joints in External Unreinforced Non-Loadbearing Walls [31]

Type of Masonry	Maximum Length, l_m, (UK NA Value) m	Comments
Clay	12 (15)	May be increased with bed joint reinforcement conforming to BS 845-3: 2003 [32] And subject to expert advice
Calcium silicate	8 (9)	Applies for UK value when length
Aggregate concrete & manufactured stone	6 (9)	to height ratio of panel is ≤3:1, and should be reduced when
Autoclaved aerated concrete	6 (9)	length to height ratio is >3
Natural stone	12 (20)	In UK, joint should be located ≤8 m from corner

Additional requirements are prescribed for the UK by PD 6697: 2010 [16], which complements parts of BS EN 1992-2: 2006 [31]:

- With regard to location of movement joints, empirical rules are given that are applicable to the majority of situations. It is not necessary to provide movement joints where the length of internal walls in dwelling is relatively short.
- If a return in the length of clay masonry is less than 675 mm and either adjoining length of masonry exceeds 6 m, the masonry should be interrupted at the return to prevent the development of a mechanical couple and a risk of cracking. This can be affected by the introduction of a vertical compressible joint or a "slide-by" detail. Returns of 675 mm or more should be regarded as having sufficient inherent flexibility to accommodate the stress caused by opposition forces (see Figure 14.6).
- The ratio of length to height of calcium silicate masonry panels should generally not exceed 3:1. As a rule, vertical joints not less than 10 mm wide to accommodate horizontal movements should be provided at intervals between 7.5 and 9 m.
- In internal walls containing openings, movement joints may be needed at more frequent intervals or the masonry above and below the openings may need to be reinforced in order to restrain movement. The design should pay particular attention to long low horizontal panels of masonry, e.g., those under windows.
- To accommodate horizontal movement in natural stone masonry, and in the absence of specific calculations, vertical joints not less than 10 mm wide should be provided at interval no greater than 15−20 mm, and located no more than 7.5 m from an external corner.

Particular features of the building which should be considered when determining location of movement joint positions in masonry are [16]:

- Intersecting walls, piers, floors etc.
- Windows and door openings.
- Changes in height, thickness of walls or type of foundation.
- Chases in walls.
- Movement joints in the building as a whole or in floor slabs.

Advice on the positioning of movements joints is given in BDA Design Note 10 [8]. For example, Figure 14.14 shows the optimum location for a movement joint is at a change in height or thickness of a wall, and the importance of extending movement joints through the tile creasings and cappings. As a general rule, all movement joints should pass through the full thickness of a wall or the outer leaf of a cavity wall and through any finishes that are sufficiently flexible to accommodate the movement [31].

Bed Joint Reinforcement

To avoid the type of cracking in brickwork with openings, such as that shown in Figure 14.4, areas above doors and above or below window may benefit from mortar bed joint reinforcement to distribute tensile stresses [16]. The bed joint steel

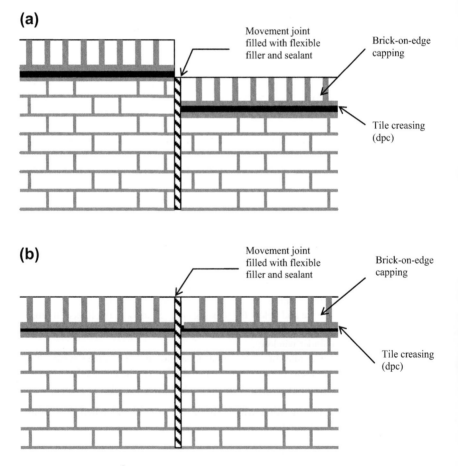

Figure 14.14 Correct location and detail of movement joints in walls [8]. (a) Optimum location for a movement joint at a change in wall height or thickness (b) Movement joint extended through the tile creasing and capping.

reinforcement actually reduces and controls the width of cracks so that they are aesthetically acceptable. Generally, the reinforcement consists of a light steel mesh placed in the bed joints at a vertical spacing of 225 or 450 mm [14]. As with reinforced concrete, the theoretical minimum percentage of reinforcement needed to control cracking is given by the ratio of tensile strength of the brickwork to the yield strength of the reinforcement, typically around 0.1% of the wall cross-sectional area. According to CIRIA [14], to be fully effective reinforcement should be used continuously and not just around openings so that it may be possible to increase joint spacing. To ensure adequate durability, reinforcement should be protected against corrosion in accordance with BS EN 1996-1-1: 2005 [33]; Table 14.9 lists types of steel suitable for exposures to different climates.

In the UK, PD 6697: 2010 [16] specifies reinforcement to be used in masonry walls built of calcium silicate units aggregate concrete masonry units, autoclaved aerated

Table 14.9 Selection of Masonry Reinforcing Steel for Durability [33]

Exposure Class (Local Climatic Conditions)	Minimum Level of Protection, Excluding Cover, for Reinforcement in Location:	
	Bed Joints or Special Clay Units	Grouted Cavity or Quetta Bond Construction
MX1 (dry environment)	Carbon steel galvanized according to BS EN ISO 1461 [34]. Minimum mass of zinc coating 940 g/m^2 or for reinforcement material/ coating reference R1 or R3 [a]	Carbon steel
MX2 (exposure to moisture or wetting)		Carbon steel or, where mortar is used to fill the voids, carbon steel galvanized according to BS EN ISO 1461 [34] to give a minimum mass of zinc coating of 940 g/m^2
MX3 (exposure to moisture or wetting plus freeze/thaw cycling)	Austenitic stainless steel according to BS EN 10088 [35] or carbon steel coated with at least 1 mm of stainless steel or for bed joint reinforcement material/ coating reference R1 or R3[c]	Carbon steel galvanized according to BS EN ISO 1461 [34] to give a minimum mass of zinc coating 940 g/m^2
MX4 (exposure to saturated salt air or seawater) & MX5 (exposure to saturated salt air in an aggressive environment)	Austenitic stainless steel[b] according to BS EN 10088 [35] or carbon steel coated with at least 1 mm of stainless steel or for bed joint reinforcement material/ coating reference R1 or R3[b]	Austenitic stainless steel[b] according to BS EN 10088 [35] or carbon steel coated with at least 1 mm of stainless steel[c]

[a]For internal masonry other than inner leaf of cavity walls, carbon steel reinforcement or bed joint reinforcement with any material/coating reference may be used.
[b]Austenitic stainless steel grades should be selected according to the applicable exposure and environmental aggression. Nor all grades will necessarily be suitable for the most aggressive environments, particularly those where de-icing salts are regularly used, e.g., highways.
[c]See BS EN 845-3: 2003 [32].

masonry units and manufacture stone masonry units to minimise cracking above and below openings. The reinforcement should be long enough to distribute the stress to a position where the vertical cross-sectional area of the wall is able to accommodate it.

Cladding to Framed Structures

The potential problem of differential movement between clay brickwork cladding and concrete blockwork in a reinforced concrete frame has already been discussed in the section dealing with total movement. In general, any type of masonry cladding to framed structures should be designed to prevent cracking as a result of stress generated by differential movement between the masonry cladding and the frame as well as being provided with adequate lateral edge restraint [16]. In the case of calcium silicate or concrete masonry cladding, the differential movement is less than in the case clay brickwork cladding. This is because no irreversible moisture expansion is involved and the long-term moisture movement of both cladding and structure involves shrinkage, thermal expansion of the cladding being the only opposing movement. Steel frame structures are not subjected to shrinkage and so vertical differential movement is due only the thermal and moisture changes in the masonry cladding, so that differential movement between the cladding and steel frame is generally less than with a reinforced concrete frame.

With regard to masonry cladding to timber framed structures, the cladding is normally supported on the same foundations as the framed structure and not on the frame itself, although it is generally tied to it to enhance lateral stability [16]. Since movement of the timber frame and movements of the masonry in response to thermal and moisture changes are dissimilar, there is significant differential movement and building details should accommodate the vertical movement between the timber frame and brickwork cladding as follows [16]:

- 3 mm between sill and brickwork at ground floor level.
- 9 mm between sill and brickwork at first floor level.
- 15 mm between sill and brickwork at second floor level.
- 6 mm allowance at eaves and verge for one storey building.
- 12 mm allowance at eaves and verge for two storey building.
- 18 mm allowance at eaves and verge for three storey building.

Problems

1. Above what threshold level of stress/strength ratio can failure due to creep rupture occur in concrete and masonry structural elements for (a) compression loading, and (b) tensile loading?
2. Is loss of prestress more in post-tensioned calcium silicate brickwork than in post-tensioned clay brickwork? Give approximate prestress loss percentages in both cases.
3. Is creep an advantage or disadvantage in concrete and masonry structures? Would zero creep be desirable?
4. Give examples of external restraint and internal restraint to movement.
5. Sketch typical crack patterns that can occur in restrained walls with openings for (a) clay brickwork, and (b) concrete blockwork.
6. Define restraint factor.

7. Suggest ways of minimising the risk of thermal cracking in mass concrete.
8. Dimensional variation can be caused by (a) induced deviation, and (b) inherent deviation. Define what is meant by those deviations and give examples.
9. Explain the difference between contraction joints, expansion joints and construction joints.
10. What is the main purpose of joint fillers? Give a typical percentage recovery of fillers after applying compression.
11. How are joint sealants classified?
12. When is the best time to apply sealants?
13. Quote the recommended minimum joint width for sealant practical application, and recommended width to depth ratio of sealants for optimum performance.
14. Define the *MAF* of a sealant and state the movement classes of BS EN ISO 11,600?
15. What is *TRM*? Explain its significance.
16. What is the purpose of specifying a target width of a movement joint?
17. Hoe does the ambient temperature affect the allowable length of a concrete building without the use of expansion joints?
18. What is the purpose of bed joint reinforcement in masonry walls?
19. Why is differential movement greater with (a) clay brickwork cladding than with calcium silicate or concrete masonry cladding, and (b) a reinforced concrete frame than with a steel frame.

References

[1] Bonsor RB, Bonsor LL. Cracking in buildings, construction research communications ltd. Build Res Establishment (BRE); 1996:102.
[2] Page CL, Page MM, editors. Durability of concrete and cement composites. Cambridge: Woodhead Publishing; 2007. p. 404.
[3] Rusch H. Researches towards a general flexural theory for structural concrete. ACI J 1960;57(1):1−28.
[4] Brooks JJ, Jiang X. Cracking resistance of plasticised fly ash concrete. In: Cabrera JG, Rivera-Villareal R, editors. Proceedings of the international RILEM conference on the role of admixtures in high performance concrete; 1999. pp. 493−506. Monterrey, Mexico.
[5] Neville AM, Dilger WH, Brooks JJ. creep of plain and structural concrete. London and New York: Construction Press; 1983. 361 pp.
[6] Tapsir SH. Time-dependent loss of post-tensioned diaphragm and fin masonry walls [Ph.D. thesis]. Dept. Civil Engineering, University of Leeds; 1994, 272 pp.
[7] Sinha BP. Prestressed masonry beams. In: Hendry AW, editor. Reinforced and prestressed masonry. Longman; 1991. pp. 78−98.
[8] Morton JA. Designing for movement in brickwork. BDA Design Note 10. Brick Development Association; July 1986. 12 pp.
[9] Shrive NG. Effects of time-dependent movements in composite masonry. In Proceedings of a one-day symposium on the influence of time-dependent strain on structural performance. London: King's College; Dec. 1987. 14 pp.
[10] Arnold AW. Reinforced and prestressed masonry. Longman; 1991. 213 pp.
[11] Rusch H, Jungwirth D, Hillsdorf HK. Creep and shrinkage. Their effect on the behaviour of concrete structures. New York: Springer-Verlag; 1983. 284 pp.
[12] Gilbert RI. Time effects in Concrete structures. In Developments in civil engineering, vol. 23. Elsevier; 1988. 321 pp.

[13]   ACI Committee 209R-92 (Reapproved 2008). Prediction of creep, shrinkage and temperature effects in concrete structures. Detroit: American Concrete Institute; 1992.

[14]   CIRIA. Movement and cracking in long masonry walls. CIRIA Practice Note Special Publication 44, Construction Industry Research & Information Association; 1986. 8 pp.

[15]   BS 5628−3: 2005 (Superseded), Code of practice for the use of masonry. Materials and components, design and workmanship, British Standards Institution.

[16]   PD 6697. Recommendations for the design of masonry structures to BS EN 1996-1-1 and BS EN 1996-2 (UK). British Standards Publication; 2010.

[17]   Neville AM, Brooks JJ. Concrete technology. 2nd ed. Pearson Prentice Hall; 2010. 422 pp.

[18]   Illston JM, Dinwoodie JM, Smith AA. Concrete, timber and metals. The nature and behaviour of structural materials. Von Rostrand Reinhold; 1979. 663 pp.

[19]   BS 5606:1990, Guide to accuracy in building, British Standards Institution.

[20]   BS 6093: 2006, Design of joints and jointing in building construction. Guide, British Standards Institution.

[21]   Bussell M, Cather R. Design and construction of joints in concrete structures. CIRIA Report 146; 1995. 79 pp.

[22]   ACI Committee 224. 3R-95 (Reapproved 2008). Joints in concrete construction. American Concrete Institute; 1995. 42 pp.

[23]   Concrete Q & A, joint spacing in concrete walls. Concr Int Sept. 2011;33(9):87−8.

[24]   Alexander SJ, Lawson RM. Design for movement in buildings. CIRIA Technical Note 107; 1981. 54 pp.

[25]   BS 6213: 2000 + A1: 2010, Selection of construction sealants. Guide, British Standards Institution.

[26]   BS EN ISO 11600: 2003+A1: 2011, Building construction. Jointing products. Classification and requirements for sealants, British Standards Institution.

[27]   Building Research Establishment. Estimation of thermal and moisture movements and stresses: Part 2; 1979. Digest 228, 8 pp.

[28]   BS 6954-3:1988, ISO 3443-3:1987, Tolerances for building. Recommendations for selecting target size and predicting fit, British Standards Institution.

[29]   Portland Cement Association. Building movements and joints. EB086; 1982. 64 pp.

[30]   National Academy of Science. Expansion joints in buildings. Technical Report No. 65, Washington DC; 1974. 31 pp.

[31]   BS EN 1996-2: 2006, Eurocode 6. Design of masonry structures. Design considerations, selection of materials and execution of masonry (see also UK NA to BS EN 1996-2: 2006), British Standards Institution.

[32]   BS EN 845-3:2003 + A1:2008, Specification for ancillary components for masonry. Bed joint reinforcement of steel meshwork, British Standards Institution.

[33]   BS EN 1996-1-1: 2005, Eurocode 6, Design of masonry structures. General rules for reinforced and unreinforced masonry structures (see also UK NA to BS EN 1996-1-1: 2005), British Standards Institution.

[34]   BS EN ISO 1461: 2009, Hot dip galvanized coatings on fabricated iron and steel articles. Specifications and test methods, British Standards Institution.

[35]   BS EN 10088−1: 2005, Stainless steels. List of stainless steels, British Standards Institution.

15 Theoretical Aspects of Creep and Shrinkage of Mortar and Concrete

Before summarizing existing theories of creep and shrinkage of mortar and concrete, this chapter outlines the structure of cement paste. A basic understanding of the cement gel and pore system is necessary because, as is the case with many other properties, it plays a key role in determining shrinkage and creep of mortar and concrete. In particular, the number of pores and spacing, the amount of pore water and its rate of removal are very important features.

The state of water within the cement paste is then described in detail, and experimental evidence presented to demonstrate that the gel pore water is structural in nature, having significant stiffness as opposed to free water held in larger capillaries, which has a low stiffness. That property is then used to develop a new theory to quantify internal stress acting on the solid gel of hydrated cement paste in terms of the pore size/spacing ratio or porosity and pore water content. Essentially, compared with the externally applied stress, the analysis predicts lower stress on the solid gel in the case of sealed cement paste (basic creep) but a higher stress on the solid gel in the case of drying cement paste (total creep). Hence, there is an increase of creep, which is known as drying creep.

Finally, the analysis is used to simulate creep under various scenarios and experimental evidence is presented to confirm theoretical predictions. Particular cases highlighted are drying creep under tensile loading, the water/cement ratio effect on creep, and temperature influences on creep.

Structure of Hydrated Cement Paste

Table 15.1 outlines the constituents of concrete when observed at the engineering level, and then at the microscopic and submicroscopic levels. At the engineering level, concrete is a multiphase composite material, consisting of coarse aggregate embedded in a matrix of mortar, which itself consists of fine aggregate particles embedded in a matrix of hardened cement paste. Depending on the degree of hydration, the macroscopic level reveals that cement paste comprises the products of hydration and unhydrated cement grains. The hydration products consist of cement gel or C-S-H (calcium silicate hydrates) and crystals of $Ca(OH)_2$ (calcium hydroxide) with a semi-continuous system of water-filled or empty capillary pores. The cement gel is also known as tobermorite, after a naturally occurring mineral.

Concrete and Masonry Movements. http://dx.doi.org/10.1016/B978-0-12-801525-4.00015-7

Table 15.1 Structure of Concrete at Different Levels of Observation [1]

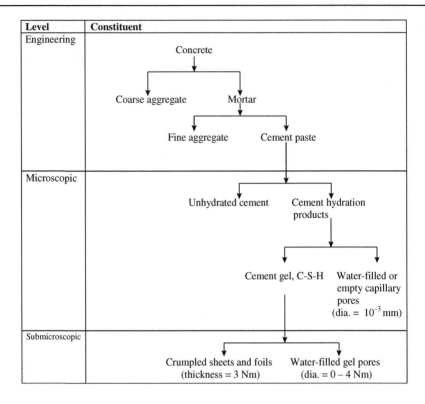

Figure 15.1 shows a simplified model of the arrangement of cement gel and the relatively large capillary pores (dia. $= 10^{-3}$ mm). Also shown are much smaller gel pores between the gel particles, which can only be observed at the submicroscopic level (Table 15.1). In the hydrated cement paste, the largest component of the hydration products is the C-S-H, which exceeds the Ca(OH)$_2$ by a factor of approximately seven by mass, and is of colloidal size and properties (about 10 nm in cross-section) [1].

The resulting mass of C-S-H is very porous (about 28% by volume), the average diameter of the gel pores being 2 nm or 2×10^{-9} mm, which implies that only a few molecules of water can be adsorbed on a solid surface. The C-S-H is a mixture of ill-formed intertwined particles, some fibrous or needle–shaped, but mostly crumpled sheets and foils, which form a continuous system of water-filled gel pores. The C-S-H, interwoven with crystals of Ca(OH)$_2$, adheres to the unhydrated cement particles and fills some of the space that existed between the particles prior to hydration [1].

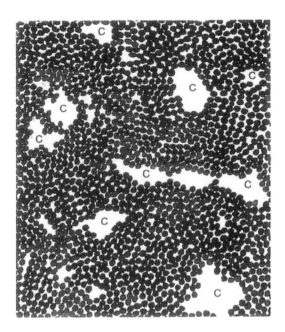

Figure 15.1 Simplified model of the structure of hydrated cement paste: solid dots are gel particles, spaces are gel pores, and larger spaces marked C are capillary pores [2].

According to Bazant [3], a more preferable classification of pores is that of the International Union of Pure and Applied Chemistry given in Table 15.2. Gel pores consist of micropores and mesopores, but micropores cannot support menisci and therefore do not show the capillary effects thought to be involved in shrinkage and creep mechanisms (see "Shrinkage" later in the chapter). The division between mesopores and macropores approximates to the size where capillary effects become negligible, and these pores do not exist in well-hydrated pastes made with water/cement ratios of 0.4 or less. Also, when water is removed by vacuum drying, mesopores and macropores collapse to form micropores. The larger capillary pores are classified as voids or microcracks.

Table 15.2 Classification of Pores in Hardened Cement Paste [3]

Pore Designation	Diameter	Remarks
Micropores	Less than 2.5 nm (25 A)	Part of C-S-H
Mesopores	2.5–50 nm (25–500 A)	Capillary pores
Macropores	50 nm to 10 μm	
Entrained air voids	10 μm to 0.1 mm	Not directly linked to
Entrapped air voids		shrinkage mechanisms
Preexisting microcracks		

The water in excess of that required for hydration fills the remainder of space between the original grains of cement, namely, the larger capillary pores. With moist curing and sufficiently low water/cement ratios, some of the capillary pores become segmented by the products of hydration, but in the absence of such curing the pores become emptied [4]. If no water movement to or from the cement paste is permitted, the reactions of hydration use up the water until too little is left to saturate the solid surfaces, and the relative humidity within the paste decreases. This is known as *self-desiccation*. Since C-S-H can form only in water-filled space, self-desiccation leads to a lower hydration compared with moist-cured paste. However, in self-desiccated pastes, with water/cement ratios in excess of 0.5, the amount of mixing water is sufficient for hydration to proceed at the same rate as when moist-cured [1].

Figure 15.2 illustrates the probable structure of the C-S-H in which the solid phase encloses a larger gel pore, with the interparticle bonds being likely chemical in nature since the gel is not a true gel but is of limited-swelling type [4]. However, owing to a large specific surface area and the close proximity of the sold surfaces separating the gel pores, forces of attraction exist that are usually referred to as van der Waals forces, which can be considerable because of the high specific surface of the C-S-H. Thus, the nature of the interparticle bonds may be either physical or chemical. It is estimated that there are 7×10^{13} C-S-H particles per mm^3 and, because of their size, the shape of the individual C-S-H particles cannot be determined by observation. There are, however, strong indications that they are in the form of thin, rolled, or crumpled sheets, averaging 1 μm in length, 3 nm thick, and 10 nm wide in the rolled state. Each

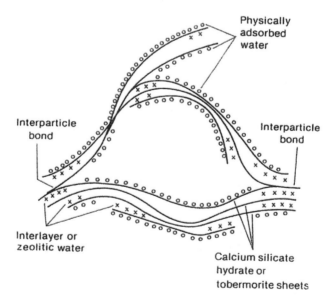

Figure 15.2 Feldman and Sereda's representation of the structure of hydrated cement paste [5].

hydrated C-S-H sheet is composed of two or three solid monomolecular layers with a monomolecular layer of zeolitic water in between [1].

The State of Water

In addition to interlayer or zeolitic water between layers of C-S-H sheets, Figure 15.2 shows that there is adsorbed water of the surface of sheets and pore water between particles, and there is also water between agglomerations of particles or in capillaries (Figure 15.1). As a consequence, water in the hydrated cement is held with various degrees of firmness. At one extreme, there is free water, which is beyond the surface forces of the solid phase and is located in the larger capillary pores. At the other extreme, there is chemically combined water or water of hydration, forming a definite part of the hydrated compounds. Between these two categories, there is gel water, which consists of adsorbed water held by the surface forces (van der Waals forces) and interlayer water held between the C-S-H or tobermorite sheets. In addition, gel water includes lattice water, which is defined as that part of the water of crystallization not chemically associated with the principal constituents of the lattice [4].

There is no technique for determining the distribution of water between these different states, nor is it easy to predict these divisions from theoretical considerations, as the energy of binding of combined water in the hydrate is of the same order of magnitude as the energy of binding of the adsorbed water [4].

A convenient division of water in the hydrated cement paste, necessary for investigation purposes, though rather arbitrary, is to divide it into two categories: evaporable and non-evaporable. Several methods are used, all of which essentially divide water according to whether or not it can be removed at a certain vapor pressure, such a division being actually arbitrary because the relation between vapor pressure and water content of hydrated paste is continuous. However, in general terms, the non-evaporable water contains nearly all chemically combined water and also some water not held by chemical bonds. As hydration proceeds, the amount of non-evaporable water increases, and since the amount of non-evaporable water is proportional to the solid volume, the volume of non-evaporable water can be used as a measure of cement paste present, i.e., the degree of hydration [1].

Evaporable water includes the free water and some of the more loosely held adsorbed water, and can be determined by the loss in mass on heating to 105 °C. The non-evaporable water is then deduced from the original water content, but if that is unknown, it can be measured as the mass loss on heating to 1000 °C.

The manner in which water is held in the cement paste determines the energy of binding. For instance, the energy of the non-evaporable water is 1.7 kJ/g, while the energy of the water of crystallization of calcium hydroxide is 3.6 kJ/g. Likewise, the density of water varies from approximately 1.2 for non-evaporable water to 1.1 for gel water, and of course 1.0 for free water [1]. It is suggested that the increase in density of the adsorbed water at low surface concentrations is not the result of compression, but is caused by the orientation of the molecules due to the action of surface forces

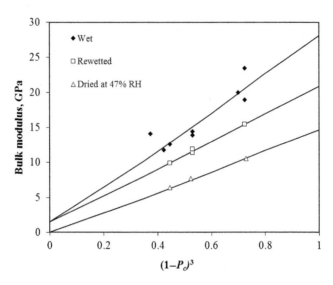

Figure 15.3 Effect of drying on bulk modulus of cement paste of different porosity, P_c [6].

resulting in a so-called disjoining pressure. The disjoining pressure is the pressure expected to maintain the film of adsorbed water molecules against external action. Confirmation that the properties of adsorbed water are different from those of free water is afforded by measurements of the adsorption of microwaves by hardened cement paste [1].

Verbeck and Helmuth [6] reported evidence concerning the state of water in cement pastes as obtained from nuclear magnetic resonance (NMR) studies. In those techniques, one of the parameters measured is the transverse relaxation time, which is a measure of the strength of the local fields surrounding the protons. In strong fields, the relaxation time is short, because resonating protons react quickly to local forces. Conversely, in weak fields, relaxation times are long. For example, for liquid water, the relaxation time is about 2.5 s compared a time of about 7×10^{-6} s for ice, indicating that water molecules in ice have a strongly fixed orientation. Results of pulsed NMR measurements on C-S-H gave relaxation times of $300–500 \times 10^{-6}$ s, indicating that water in these systems was under the influence of local forces. Based on those results and the fact that a decrease of elastic modulus occurred after drying, the authors concluded that much of the evaporable water in cement pastes seems to be part of the solid structure.

The results of bulk modulus of elasticity of cement pastes referred to by Verbeck and Helmuth [6] are shown in Figure 15.3, which clearly indicates that, for a given capillary porosity (P_c), wet pastes had a much higher bulk modulus than pastes dried to 47% RH. Re-wetting also caused an increase in modulus but not complete recovery. The intercept for the wet curve at $P_c = 0$ corresponds to the bulk modulus of free water $= 2$ GPa. Bulk modulus was determined from the elastic and shear moduli calculated from measurements of the fundamental and torsional resonant frequencies of thin slab specimens, while porosity was determined by calculation from measurements of evaporable water content [7].

It is interesting to note that, in Figure 15.3, the estimated bulk modulus of pore water (K_w) removed by drying, using a composite hard two-phase model developed for concrete (p. 18), is much greater than the bulk modulus of free water. Assuming pore water to be in place of aggregate and the dry cement paste is the matrix, then according to the model, the ratio of the bulk modulus of water to the bulk modulus of the dry cement paste is given by:

$$\frac{K_w}{K_{dp}} = \frac{1}{V_w}\left[\frac{K_{sp}}{K_{dp}} - (1 - V_w)\right] \tag{15.1}$$

where K_{sp} = bulk modulus of saturated paste, K_{dp} = bulk modulus of dry paste at 47% RH, and V_w = fractional volume of water removed.

From Figure 15.3, for a known capillary porosity, the bulk moduli of the paste are given and, assuming that V_w is equal to the capillary porosity, P_c, the bulk modulus of water can be found from Eq. (15.1). The foregoing assumption may not be precise, but is thought to be reasonable because any removal of low-modulus free water, which does not contribute to the modulus of the saturated paste, will be compensated by removal of some gel water that is not taken into account in capillary porosity. For example, from Figure 15.3, when $P_c = 0.26$, $K_{sp} = 12.4$ GPa and $K_{dp} = 5.6$ GPa, substitution in Eq. (15.1) gives $K_w = 31$ GPa. In other words, the bulk modulus of capillary water is over 15 times that of free water. If less water is removed by drying, Eq. (15.1) would yield a greater bulk modulus of water. Furthermore, at lower water/cement ratios and lower porosities, it seems that the estimated bulk modulus of pore water in cement paste is greater.

Parrott's [8] explanation of the lower modulus for dry cement paste was a dehydration of the calcium silicate hydrate rather than removal of gel water, because the latter would not be removed until the RH was well below 50% RH. Also, in larger capillaries, it could be expected that reversible behaviour would occur, and Figure 15.3 shows this not to be the case. However, dehydration of C-S-H might be expected to cause a decrease of strength, which is not the case since the reverse is true.

In the same paper [8], Parrott's own results demonstrated an increase in elastic strain of cement paste as the moisture content decreased. There was a large increase in compliance (approximately equal to the elastic strain per unit stress) with drying shrinkage, which was taken as a measure of increasing moisture loss. The increase in compliance indicated that the elastic modulus was less. Elastic stress–strain curves were measured on cement paste specimens of at least 28 days old, strain being that recorded after a time of 85 s after an incremental load change.

Sereda and Feldman [9] also showed that the elastic modulus of cement paste was significantly greater at saturation than when dried. They determined the elastic modulus from the load-deflection results of centrally-loaded discs of cement pastes made with different water/cement ratios and conditioned at various relative humidity from saturation to 0% RH. Figure 15.4 shows that, for any water/cement ratio, the modulus decreased on first drying from saturation to 0% RH. On subsequent

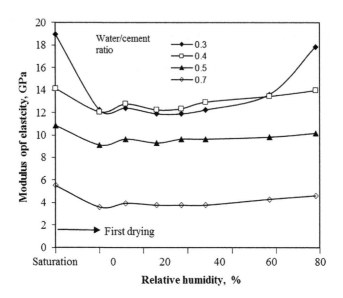

Figure 15.4 Influence of drying on modulus of elasticity of cement paste [9].

adsorption, the modulus remained unchanged up to 50% RH, and then increased but not to the original value at saturation. Corresponding strength results (Figure 15.5) generally showed the reverse trend, namely, the strength at saturation was less than at 0% RH and then slightly decreased on adsorption; the strength at resaturation was similar to the original strength.

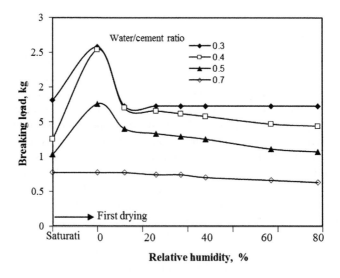

Figure 15.5 Influence of drying on strength of cement paste [9].

Hobbs [10] suggests that water in cement paste takes the form of three types:

- Capillary water that is not under the influence of any solid surface.
- Adsorbed surface water under the influence of one solid surface. This can be treated as a liquid of high viscosity and is partly load-bearing.
- Water that is under the influence of two solid surfaces or interlayer water having only a few molecules in thickness. This water does not behave like a liquid and is best considered as part of the solid structure.

Other sources [11,12] agree that the modulus of wet concrete is greater than for dry concrete, while strength varies in the opposite sense. According to Illston et al. [11], drying removes water from the larger pores, which maybe load-bearing, with a consequent reduction of elastic modulus with a decreasing moisture content of the hydrated cement paste. More significantly, water is also removed by drying from the finer pores and from the layers of the solid material. This water is bound solidly, and it can be regarded as part of the solid material, contributing to its stiffness, so that this loss of water is a second cause of a reduction in modulus with drying.

The above behaviour is supported by the results of an investigation into 30-year creep and shrinkage of concrete [13]. Here, the modulus of elasticity was compared in terms of the secant modulus on unloading saturated specimens and specimens stored dry at 65% RH. Strength was compared in the same way, and also with specimens not previously subjected to load. It was shown that strength of dry specimens was 20–30% greater than saturated specimens, while strength of previously loaded specimens was approximately 10% greater than that of the companion (load-free) control specimens. Table 15.3 shows the actual results. The opposite tend occurred for modulus of elasticity of wet-stored and dry- stored specimens. In 15 out of 18 cases, Table 15.3 indicates that the unloading secant modulus was greater for wet specimens, generally by 25%.

From the review, it may be concluded that the properties of pore water held within the hydrated cement paste are different from normal free water in that it is capable of taking significant load and maybe more so at lower water/cement ratios and porosities. That conclusion is used as the basis of a new theory for drying creep, which is presented later in this chapter.

Existing Mechanisms of Drying Shrinkage and Creep

It is not proposed to discuss in detail all existing theories and mechanisms of drying shrinkage and creep of concrete, since they are well-documented elsewhere: for example, in Neville et al., [4], Illston et al., [11], Mindess and Young [12], and Bazant [3], and more recently, Ulm et al., [14]. Only very brief summaries are given here [15], which are followed by the presentation of a new theory of drying creep that was developed to account for the experimentally observed effects of influencing factors described in earlier chapters. The theory was developed for hardened cement paste (C-S-H), which is the source of creep, and therefore applies to both concrete and mortar. Coarse and fine aggregate or sand act mainly as important restraints to creep

Table 15.3 Long-Term Strength and Modulus of Elasticity Results [13]

Water/Cement Ratio	30-year strength, MPa				15- or 20-year Unloading Elastic Modulus, GPa	
	Stored Wet		Stored Dry		Stored Wet	Stored Dry
	Loaded	Load-Free	Loaded	Load-Free		
0.80	21.5	20.2	30.8	25.3	20.9	18.9
0.67	36.2	26.9	59.3	51.9	29.1	31.1
0.58	45.4	36.7	75.6	62.7	40.8	37.8
0.54	50.8	46.6	78.6	64.5	52.1	38.1
0.50	61.6	50.6	78.4	53.3	54.4	43.8
0.80	24.6	24.3	31.4	25.7	23.0	17.4
0.67	32.6	34.3	50.7	41.3	27.8	24.2
0.56	42.4	42.0	53.9	50.0	33.4	26.1
0.48	46.4	51.3	51.5	–	39.9	28.4
0.40	56.8	55.4	55.2	54.2	57.9	31.1
0.67	23.0	24.8	31.1	–	12.9	16.4
0.62	28.5	32.3	36.4	30.2	17.6	14.7
0.55	29.4	28.8	36.6	34.5	19.3	15.9
0.45	22.4	25.3	36.0	35.5	21.1	18.5
0.86	27.0	35.8	39.2	41.1	12.8	16.7
0.75	31.2	41.0	48.4	45.1	34.8	20.4
0.63	41.3	50.8	62.4	57.1	33.0	24.2
0.55	50.0	44.0	60.1	61.6	34.5	21.9

"Loaded" refers to average of two specimens previously loaded for 30 years; "load-free" refers to the accompanying control shrinkage or swelling specimens. Creep specimens were unloaded for 15 years and then stored for 15 years. Underlined values are modulus after 15 years.

and shrinkage of the hardened cement paste, although other effects may play a role though the transition zone at the paste/aggregate.

Shrinkage

When C-S-H is exposed to a drying environment, a relative humidity gradient is created between the C-S-H and the surrounding air. Initially, moisture in the form of free water is lost from the larger capillaries, the result being little or no change in volume. However, the initial loss of moisture creates an internal humidity gradient within the C-S-H, and to maintain hygral equilibrium, adsorbed water is transferred from the gel pores, the process of which results in a reduction in volume of the C-S-H. Subsequently, depending on the degree of drying, interlayer water may also be transferred to the empty gel pores and larger capillaries.

The general consensus is that the reduction in volume of the cement paste is caused by compression in the solid framework of the C-S-H to balance the capillary tension of the increasing curvature of the menisci of adsorbed water as the gel and capillary pores empty. This mechanism is known as the *capillary tension theory*, which is believed to apply to relative humidities between 100% and approximately 40%, at which point the menisci become unstable [11,12]. At lower relative humidities, one school of thought proposes the cause of shrinkage to be the change in *surface energy* of the C-S-H as firmly held adsorbed water molecules are removed. Alternatively, drying causes the *disjoining pressure*, which exists in the interlayer water located within the areas of hindered adsorption, to be relieved as water molecules are removed, and consequently a reduction in volume occurs [11,12].

The theories apply to reversible behaviour, and shrinkage is not fully reversible, probably because additional chemical and physical bonds are formed during the process of drying. Moreover, carbonation occurs, which prevents ingress of moisture on re-wetting.

An alternative viewpoint is that shrinkage may be considered as load strain comprising elastic-plus-basic creep of the solid C-S-H induced by the capillary stress (see p. 521).

Creep

Although there have been numerous proposed theories [3,4,11,12,14], the exact mechanism of creep is still uncertain, and it could be that several mechanisms are required to explain all the phenomena. It is generally agreed that creep is related to the internal movement of adsorbed or interlayer water since C-S-H, from which all evaporable water has been removed, exhibits little creep. Movement of water to the outside environment is essential for drying and total creep, and it will be shown later that internal movement of water can also contribute to basic creep, because all pores do not remain full of water in mass or sealed concrete due to hydration. There is a strong dependency of basic creep on porosity or strength, which is indirect evidence that empty or part-empty pores govern much of creep. The creep of the solid skeleton of C-S-H and very-long-term creep after all the water has disappeared may be due to viscous flow of sliding between particles [15].

Drying Creep Theory

The new theory for drying creep now presented [16] was developed to account for the features mentioned in the previous paragraph and the influencing factors referred to in earlier chapters. First, it is appropriate to present a historical review of drying creep.

In 1931, Davis and Davis [17] were probably the first researchers to report the effect of relative humidity of storage on creep of concrete. However, it was Pickett [18] in 1942 who emphasized that "the amount and rate of plastic flow (creep) in concrete has been found to depend upon rate of drying," and that "shrinkage cannot account for additional creep unless inelastic strain, not proportional to stress, is produced." Nowadays, those original observations are generally referred to as the *Pickett effect* or drying creep, which is defined in Chapter 2 as the creep in excess of basic creep of sealed concrete after taking into account shrinkage as measured on a separate unloaded specimen.

In his paper, Pickett [18] quotes Lynam [19] as saying "shrinkage and creep are two aspects of the same phenomenon; it is impossible to separate them and no good end is served by trying to do so." Lorman [20] was also reported as saying "shrinkage or swelling due to loss or gain of moisture and creep due to seepage are interrelated phenomena." Nevertheless, the separation of shrinkage and creep has been adopted by all researchers and is the common approach for the predictive methods given in international Codes of Practice.

Pickett [18] attempted to explain the phenomena of drying creep in terms of a nonuniform shrinkage and a nonlinear stress–creep relationship. However, in a later paper [21], he admitted that the foregoing only explained a small part of drying creep, which is now accepted as the microcracking effect. Wittmann and Roelfstra [22,23] suggested that drying creep was the result of the suppression of microcracking that occurs in the load-free control shrinkage specimen, and stresses due to nonuniform drying. They concluded that drying creep was an apparent mechanism related to the "shrinkage-induced stress." On the other hand, Bazant and Chern [24] labeled the majority of drying creep as a "stress-induced shrinkage," which is the real mechanism caused by breakage of bonds during microdiffusion of water between capillary pores and gel pores. By measuring the curvature of eccentrically loaded specimens to eliminate the influence of shrinkage, Reid [25] and, independently, and Bazant and Xi [26] prevented microcracking to demonstrate the existence of the "stress-induced shrinkage." In the small eccentrically loaded tests, the stress was compressive so that microcracking was suppressed, but there was still an additional curvature compared to basic creep. In the large eccentrically loaded tests, a tensile stress was induced so that microcracking occurred, which caused even more curvature.

In another approach, Wittmann [27] considered creep and shrinkage to be coordinated phenomena, and used the activation energy approach to estimate the equivalent shrinkage stress (σ_{sh}) of neat cement paste by equating the rate of shrinkage to the rate of basic creep. Hence, for creep under drying conditions and subjected to an external stress (σ), the total equivalent stress ($\sigma + \sigma_{sh}$) can be estimated. He found that the rate of total creep calculated from the equivalent stress and activation energy agreed with the experimental measured rate of total creep. In Wittmann's tests, where the relative humidity was 40%, the average equivalent stress/strength ratio for shrinkage was 0.28. According to Neville et al. [4], in some cases, where large shrinkage and a high water/cement ratio exist, the equivalent stress/strength ratio can exceed unity, so that cracking may occur. Wittmann's approach does not appear to have been developed further, probably because of the complex integration of the rate of creep expression involving the activation energy. However, that approach was probably the first to imply that shrinkage is a "stress-induced creep."

Earlier in 1964, Ruetz [28] quantified creep of hardened cement paste specimens under drying conditions in terms of a "creep enlargement factor,"defined as the ratio of total creep to basic creep. For water/cement ratios of 0.20–0.80, he found that the creep enlargement factor occurred very quickly (measured within 15 min after loading) and reached the same maximum of approximately 2.3, but at earlier times for the lower water/cement ratios. Ruetz deduced that some of the capillary water must be removed before creep enlargement begins. From both tension and torsion tests, he found increases in creep under drying conditions and concluded that "the process of shrinkage will orientate itself in the direction of applied external load, although the orientation effect cannot be the mechanism which motivates the increase of creep with drying."

Nagataki and Yonekura [29] implied that drying shrinkage is basic strain induced by capillary stress, i.e., the sum of elastic strain and basic creep due to capillary stress. They calculated the capillary stress from pore size distribution as measured by mercury intrusion, other pore sizes being deduced from evaporable and non-evaporable water contents. The capillary stress increased with time, and although the calculated basic strain was less than the measured shrinkage at early ages, the agreement was considered satisfactory, bearing in mind the assumption used for estimating the rate of moisture diffusion. After observations that drying creep became small with high-strength concrete, they surmised that drying creep of low-strength concrete was due to the difference in capillary stress between the loaded (creep) and control (shrinkage) specimens.

In the 60 years since Pickett's research, there has been no satisfactory explanation of drying creep of concrete and one that has universal agreement. Most existing theories are based on the modification of drying shrinkage by the external load so as to increase deformation by a shrinkage-induced stress or a stress-induced shrinkage. The approach adopted here is different in the respect that pure total creep is considered independently from drying shrinkage, and is the result of an increase in stress acting on the solid gel when water is first removed from the gel pores at the same time as the external load is applied. Total creep may be considered as basic creep under a greater stress and the extra creep is drying creep. Shrinkage or, strictly, the internal (capillary) stress arising from water removal, is assumed to be the same in both the loaded and load-free specimens and so does not contribute any additional deformation. The exception to this would be concrete subjected to extreme operating conditions, for example, very high temperature when the high total stress could cause nonlinear transient thermal strain [30–32].

The stress concentration at the boundary of holes in metals is well known, the theoretical analyses having been developed in the 1930s to explain why metals under cyclic loading often failed at much lower loads than expected. Timoshenko and Goodier [33] showed that a small circular hole in a plate subjected to a uniform uniaxial stress, σ, produced a stress of 3σ at the edge of the hole, but the stress rapidly reduced as the distance from the hole increased. In the case of a plate of finite width, where the diameter of the hole was half of the plate width, the maximum stress was 4.3σ [34]. The case of a row of holes in a plate at right angles to the applied stress gave a maximum stress of 3.24σ when the spacing of the holes was four times the radius of the hole [35].

The shape of the hole is also a factor. For example, an elliptical hole in an infinite plate with its larger axis at right angles to the nominal external stress increased the maximum stress to 5σ when the major axis was twice the minor axis [33].

The solutions for the stress concentration produced by small spherical and cylindrical inclusions are given by Goodier [36]. For a spherical inclusion that is appreciably more rigid than the surrounding medium, the stresses in the medium are dependent on the Poisson's ratio. Assuming a value of 0.2 for the latter, the vertical or hoop stress is shown in Figure 15.6(a) and is given by:

$$p = \sigma_c - \sigma_c \left[\frac{a_m^3}{4r^3} + \frac{3a_m^5}{4r^5} \right] \tag{15.2}$$

where p = vertical stress, σ_c = nominal external stress, a_m = radius of the sphere, and r = distance from center of sphere.

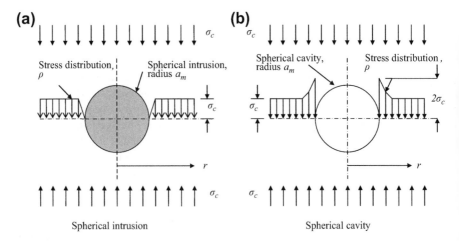

Figure 15.6 Local stress distributions caused by: (a) a small spherical intrusion (Eq. (15.2)) and (b) a spherical cavity (Eq. (15.3)), in a medium subjected to an external compressive stress, σ_c.

For a spherical cavity, the corresponding vertical stress is shown in Figure 15.6(b) and is given by:

$$p = \sigma_c + \sigma_c \left[\frac{a_m^3}{4r^3} + \frac{3a_m^5}{4r^5} \right] \tag{15.3}$$

Comparing the two cases, it can be seen that, when $r = a_m$, the stress is zero for the rigid inclusion, but it is twice the nominal compressive stress for the cavity. It is interesting to note that at the top of the cavity, the theory predicts a lateral tensile stress, which is believed to explain the failure of concrete in compression by vertical splitting [36].

A simple physical model for the cement gel is now assumed, which is that the average radius of the pores is a_m and they are equally spaced throughout the gel at a distance b between centers. Considering the case of sealed gel in which the pores are completely full of water and under an external compressive stress, it is assumed that the entrapped water will act as a rigid inclusion relative to the hydration products (solid gel) and thus the local stress will be given by Eq. (15.2). For two adjacent pores at a spacing $b = 3a_m$, the stress distribution (from Eq. (15.2)) is shown in Figure 15.7(a), but without any allowance being made for the "row of holes effect" derived by Howland [35], who showed that only a small increase in maximum stress occurred for a plate. It can be seen that the effect of the water-filled gel pores is to reduce the stress on the solid gel relative to the nominal stress.

For the corresponding case of empty adjacent pores or cavities, the stress distribution given by Eq. (15.3) is shown in Figure 15.7(b). In contrast to the previous case, there is a general increase in stress on the solid gel for the empty pore compared with

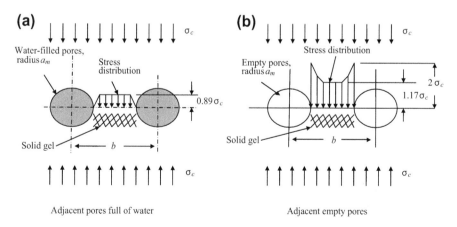

Figure 15.7 Stress distribution on solid gel between adjacent pores: (a) full of water and (b) empty, when hydrated cement paste (gel) is subjected to a nominal compressive stress, σ_c, and $b = 3a_m$.

the nominal stress, and consequently that stress is considerably greater than that for the water-filled pores. Therefore, it may be implied that, for the same external nominal stress, the deformation of the solid gel with empty pores would be greater than when they are full of water.

The average stress (σ) acting on the solid gel is given by:

$$\sigma = \frac{1}{0.5b - a_m} \int_{a_m}^{0.5b} p \cdot dr \tag{15.4}$$

where p is given by (15.2) or (15.3).

When the gel pores are full of water, the average stress, σ_f, is:

$$\sigma_f = \left(\frac{\sigma_c}{0.5b - a_m}\right)\left[(0.5b - a_m) + 0.125\left(a_m^3\left[\frac{4}{b^2} - \frac{1}{a_m^2}\right] + 1.5a_m^5\left[\frac{16}{b^4} - \frac{1}{a_m^4}\right]\right)\right] \tag{15.5}$$

and for the empty gel pores, the average stress, σ_e, is:

$$\sigma_e = \left(\frac{\sigma_c}{0.5b - a_m}\right)\left[(0.5b - a_m) - 0.125\left(a_m^3\left[\frac{4}{b^2} - \frac{1}{a_m^2}\right] + 1.5a_m^5\left[\frac{16}{b^4} - \frac{1}{a_m^4}\right]\right)\right] \tag{15.6}$$

For the case when $b = 3\,a_m$ (Figure 15.7), the average stresses are $0.56\sigma_c$ and $1.44\sigma_c$ for the full pores and empty pores, respectively. In other words, total removal of the adsorbed water from the gel increases the average stress on the solid gel by a factor of 2.57. The range of stresses induced by full and empty pores, relative to the nominal compressive stress for a range of pore size/spacing (a_m/b) ratios, is shown in Figure 15.8. Expressing the ratio of empty pore stress to the full pore stress as the *stress enlargement factor* (SEF), it can be seen that total removal of water results in a gradual increase in the SEF until a pore size/spacing ratio of 0.45 when there is a dramatic increase in the SEF (Figure 15.9). Correspondingly, it can be assumed that creep would increase when the pores are completely emptied of water, say, for total creep in severe drying conditions.

As stated earlier, Ruetz [28] defined the ratio of measured total creep to basic creep as the *creep enlargement factor* (CEF), which in this analysis is not quite equal to the SEF since the latter would induce elastic strain as well as creep. The SEF is actually equal to the ratio of total creep compliance (total creep plus elastic strain) to the basic creep compliance (basic creep plus elastic strain).

It is relevant to note that the average pore size/spacing ratio (a_m/b) is related to the porosity. In a unit cube of cement paste, the number of pores = $1/b^3$ which is equal to the porosity (P) divided by the volume of one pore, i.e., $3P/4\pi a_m^3$. Consequently:

$$\frac{a_m}{b} = 0.62P^{1/3} \tag{15.7}$$

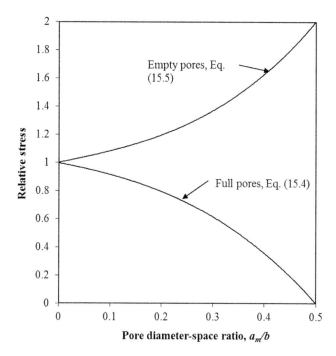

Figure 15.8 Effect of pore diameter–space ratio on average stress on the solid gel for empty pores and for water-filled pores.

For example, considering the gel porosity, which is constant, say, at approximately 28% [1], the average $a_m/b = 0.40$, and from Eqs. (15.5) and (15.6), the corresponding average stresses are $0.38\sigma_c$ and $1.62\sigma_c$ for the full and empty gel pores, respectively. Therefore, total removal of gel pore water results in an SEF of 4.3. If some of the capillaries are included with the gel pores, then the porosity, average a_m/b and SEF will all be greater. In fact, porosity is now the total porosity and is dependent on the water/cement ratio (w/c) and degree of hydration (h) [37]:

$$P = \frac{w/c - 0.17h}{0.317 + w/c} \tag{15.8}$$

Consequently, Figure 15.9 also demonstrates that an increase in porosity or water/cement ratio causes an increase in the SEF and, by implication, an increase in the CEF.

   In reality, the foregoing situation applies only when there is total removal of water from the pores, such as after prolonged drying or when there is a rapid expulsion of water, say, by heating after application of the load. In the latter situation, a large SEF may account for transitional thermal creep [38] and transient thermal creep [30–32]. On the other hand, under normal environmental conditions of temperature and humidity, moisture diffuses slowly from the larger pores to the drier outside

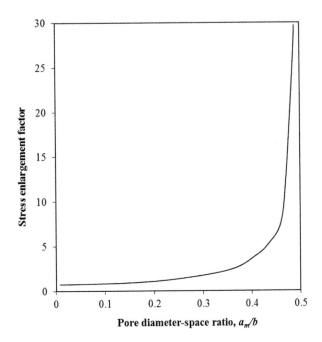

Figure 15.9 Stress enlargement on solid gel with empty pores compared with water-filled pores.

environment, and from the smaller pores to the larger pores. Some of the smaller gel pores may never become completely empty depending on the level of relative humidity. To model the moisture diffusion time-dependency effect, the stress induced by a part-empty pore has been analyzed as depicted in Figure 15.10. Here, an empty pore of radius a'_m, within an original filled pore of radius a_m, is assumed to produce a stress distribution as given by Eq. (15.6), but only the average stress (σ_{pe}) acting on the solid gel (over a distance $= b-2a_m$) is required for the analysis of creep of the solid gel:

$$\sigma_{pe} = \left(\frac{\sigma_c}{0.5b - a_m}\right)[(0.5b - a_m)] + 0.125\left(X^3 a_m^3\left[\frac{4}{b^2} - \frac{1}{a_m^2}\right] + 1.5X^5 a_m^5\left[\frac{16}{b^2} - \frac{1}{a_m^4}\right]\right)$$

(15.9)

where $X = a'_m/a_m$.

Figure 15.11 shows the SEF for gradual water removal, as represented by increasing values of X, from initially full pores of different pore size/spacing ratios. For a given a_m/b, the important feature is that almost immediately on removal of some water, the SEF on the solid gel is much greater than unity. Then as X increases there is a steady period before the SEF increases sharply when the pores are between 50% and 75% empty; the SEF values when $X = 1$ (empty pores) are those shown in Figure 15.7. The SEF for high values of a_m/b is very large, and as will be seen later, they correspond to the normal range of water/cement ratios used in concrete. Consequently, it

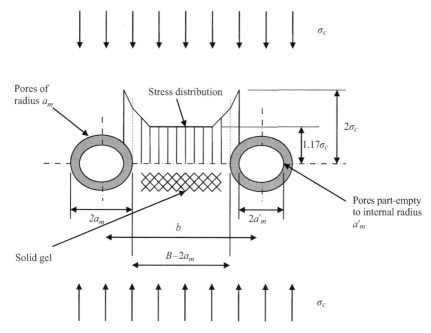

Figure 15.10 Stress distribution on solid gel between adjacent part-empty pores with a spacing of $b = 3a_m$ when hydrated cement paste is subjected to an external compressive stress, σ_c.

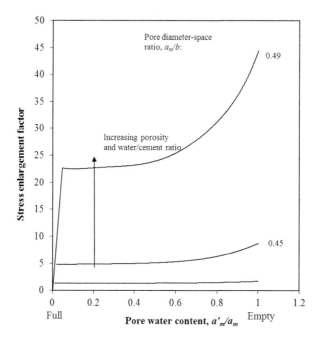

Figure 15.11 Effect of water removal from initially full pores on the stress enlargement factor (full when $a'_m/a_m = 0$ and empty when $a'_m/a_m = 1$).

seems as if the theory would overestimate total creep in predicting correspondingly a very high CEF. However, the assumption that all the pores are initially full of water is unlikely to be true for the reasons outlined next.

It is well known that a higher porosity or a higher water/cement ratio (represented here by the average a_m/b) results in a weaker paste exhibiting more creep [4]. To account for this fact, the theory should predict an increase of stress acting on the solid gel as the water/cement ratio increases. Figures 15.8 and 15.11, respectively, show this to be the case provided the pores are empty and part-empty ($\equiv$ total creep), but Figure 15.8 also shows that is not true not for full pores ($\equiv$ basic creep). However, even in the case of well-cured, sealed cement paste it is thought that some internal transfer of moisture occurs from smaller to larger part-empty capillaries as hydration of cement proceeds [4]. Thus, there will be an SEF, although much smaller than in the case of drying because moisture diffusion is slower and pores would not empty completely. Therefore, some water movement is required to explain why basic creep is approximately proportional to the inverse of strength when the water/cement ratio changes. Furthermore, the capillary porosity must be a factor in creep rather than just the smaller gel porosity, which is approximately constant and independent of the water/cement ratio. Thus, the SEF of Figure 15.11 cannot be assumed to equate with the CEF, i.e., the total creep/basic creep ratio, because the theory in its present form does not model the water/cement ratio influence on basic creep correctly.

Estimated Stress on Solid Gel for Different Water/Cement Ratios

Since realistic cement pastes have different sizes of pores filled with different amounts of water, average stresses acting on the solid gel will vary in a complex manner. However, an overall average stress for this situation can be estimated using the theory by making the following simplifications:

- For a given porosity, the pore is a single size.
- Some pores are full, others are part-full, and some are empty of water.
- Stresses on the solid gel are in proportion to the fractions of full, part-full, and empty pores.
- The water content of part-filled pores is proportional to the degree of hydration, so that pores are full at zero hydration and empty at full hydration.
- Unhydrated cement reduces the stress on the solid gel.

Consider cement pastes having a range of water/cement ratios from 0.2 to 0.6 and subjected to load at the age of 28 days. The degree of hydration (h, %) can be estimated by the following [39]:

$$h = 191.6 - 67\left(w/c^{-0.5}\right) - 55.2t_o^{-0.33} \tag{15.10}$$

where t_o = age, days.

For the specified water/cement ratios, the degree of hydration, total porosity, and pore radius/spacing ratio, calculated from Eqs. (15.10), (15.8), and (15.7), respectively, are shown in Table 15.4. It can be seen that for the large range of water/cement ratios, the range of a_m/b is small and therefore the potential SEF is large (Figure 15.9).

Table 15.4 Estimated Degree of Hydration (h), Total Porosity (P_t), and Pore Radius/Spacing Ratio (a_m/b) for Cement Pastes of Different Water/ Cement Ratio (w/c) at the Age of 28 days

w/c	h	P_t	a_m/b
0.2	0.23	0.31	0.42
0.3	0.51	0.35	0.44
0.4	0.67	0.40	0.46
0.5	0.78	0.45	0.48
0.6	0.87	0.49	0.49

Using the values of a_m/b, the average stress on the solid gel when the pores are full of water can be calculated by Eq. (15.5). To obtain the average stress when the pores are part full, X is required for application of Eq. (15.9). By assuming that $X = h$ as a simple approximation so that, for example, the pores are empty when $h = 1$, the levels of average stress relative to the externally applied stress are given in Table 15.5 together with the average stresses when the pores are completely empty (Eq. (15.6)).

The degree of hydration (h) at low water/cement ratio indicates there is a significant proportion of unhydrated cement, the fractional volume (V_{uc}) of which can be estimated from [37]:

$$V_{uc} = \frac{0.317(1 - h)}{w/c + 0.3178} \tag{15.11}$$

The presence of unhydrated cement reduces the stress on the hydrated products, i.e., the solid gel. To allow for this effect, the relative stiffness of the paste can be calculated using a two-phase composite model [40]:

$$\frac{1}{E_c} = \frac{1 - V_{uc}^{0.5}}{E_p} + \frac{V_{uc}^{0.5}}{E_p\left(1 - V_{uc}^{0.5}\right) + E_{uc}V_{uc}^{0.5}} \tag{15.12}$$

where E_c = elastic modulus of the composite (solid gel + unhydrated cement), E_p = elastic modulus of the solid gel, and E_{uc} = elastic modulus of the unhydrated cement.

Table 15.5 Stresses on Solid Gel When Pores Are Full, Part-Full, and Empty

w/c	($X = h$)	Stress with Full Pores, σ_f	Stress with Part-Full Pores, σ_{pf}	Stress with Empty Pores, σ_e
0.2	0.23	$0.312\sigma_c$	$1.003\sigma_c$	$1.688\sigma_c$
0.3	0.51	$0.269\sigma_c$	$1.045\sigma_c$	$1.74\sigma_c$
0.4	0.67	$0.183\sigma_c$	$1.149\sigma_c$	$1.818\sigma_c$
0.5	0.78	$0.108\sigma_c$	$1.307\sigma_c$	$1.892\sigma_c$
0.6	0.87	$0.044\sigma_c$	$1.508\sigma_c$	$1.956\sigma_c$

Table 15.6 Influence of Unhydrated Cement on the Stiffness of the Solid Gel

w/c	h	V_{uc}	E_p/E_c
0.2	0.23	0.470	0.41
0.3	0.51	0.252	0.59
0.4	0.67	0.145	0.71
0.5	0.78	0.084	0.79
0.6	0.87	0.046	0.86

Assuming $E_{uc} = 10\,E_p$, the relative stiffness, E_p/E_c, is shown in Table 15.6 for the various water/cement ratios calculated using Eq. (15.12). The stiffness ratio implies the unhydrated cement reduces the stress acting on the solid gel, for example, from 100% for a pure gel to 41% when $w/c = 0.2$.

Sealed Cement Paste

Let us now consider various cement pastes that have been sealed and cured for 28 days and then subject to load, i.e., basic creep conditions. It is assumed that some pores will be full of water and some will be part empty due to hydration, but there are no totally empty pores. For an average porosity or pore size/spacing ratio, the average stress acting on the solid gel depends on the proportion of full pores and part-full pores. If the distribution of stress is assumed to be in proportion to the fractions of full pores and part-full pores, the overall average stress, σ_s, is:

$$\sigma_s = \sigma_f F + \sigma_{pf}(1 - F) \tag{15.13}$$

where F = volume fraction of full pores, (1–F) = volume fraction of part-full pores, and σ_f and σ_{pf} are given in Table 15.5.

The overall average stress for different values of F is shown in Figure 15.12, after taking into account the presence of unhydrated cement by multiplying σ_s (Eq. (15.13)) by E_p/E_c (Table 15.6). The theory now predicts the correct pattern of behaviour for basic creep of sealed paste. For the condition when F is less than approximately 0.8, i.e., 80% of pores are full and 20% are part-full of water, basic creep increases (through an increase of stress) with an increase of water/cement ratio. It can be noted that, for most cases, the stress acting on the solid gel is less than the nominal external stress, i.e., the relative stress is less than unity. Also, under prolonged loading with any further water movement, the stress on the solid gel increases, as simulated by a decrease in the fraction of full pores (F).

Drying Cement Paste

Now consider cement paste to have been exposed to drying just after loading so that water is lost to the outside environment. In this case, which is that of total creep, it is

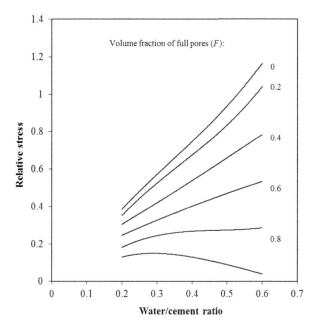

Figure 15.12 Induced stress on sealed solid gel for various fractions of pores filled with water (Eq. (15.11)).

assumed that there are no full pores, but some pores are part-empty and some are empty in various proportions. The average stress on the solid gel, σ_d, is:

$$\sigma_d = \sigma_{pe}(1 - F') + \sigma_{e'}F' \tag{15.14}$$

where $(1-F') = $ volume fraction of part-empty pores, $F' = $ volume fraction of empty pores, σ_e is given by Eq. (6), and $\sigma_{pe} = $ stress due to the part-empty pores.

The stress due to the part-empty pores, σ_{pe}, is assumed equal to the nominal external stress, σ_c, because Figure 15.13 shows this to be the case for all values of a_m/b ($\equiv$ water/cement ratios) for up to approximately $X = 0.5$. Figure 15.14 shows the average stress acting on the solid gel for various values of F', as given by Eq. (15.14) and multiplied by the stiffness ratio, E_p/E_c, of Table 15.6. Generally, the relative stress increases (and therefore total creep) as the water/cement ratio increases and it is greater than that for the sealed paste (Figure 15.12). It also increases with prolonged drying under load, as simulated by an increase in the fraction of empty pores (F').

Test Cases

In the following test cases, it is assumed that hardened cement paste, mortar, or concrete has been moist-cured at atmospheric pressure. For cases of high-pressure steam curing, the theory is not applicable.

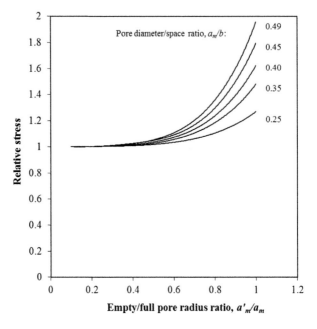

Figure 15.13 Effect of water removal from initially full pores on increasing the stress on solid gel, according to Eq. (15.6).

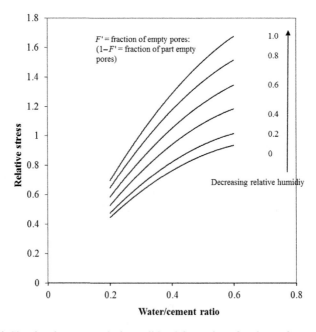

Figure 15.14 Simulated stress on drying solid gel for various fractions of empty pores.

Having developed the theory to account for the influence of water/cement ratio on basic creep as well as on total creep, other important influences on creep are now considered that do not appear to have been explained satisfactorily by previous theories. As has already been stated, the SEF as defined in the new theory is analogous to the CEF defined by Ruetz [28] as the ratio of total creep to basic creep. Assuming that creep is proportional to stress, the theoretical SEF will now be compared with the experimentally determined CEF for different scenarios. Although developed for cement paste, the theory applies also to mortar or concrete, since the influence of the aggregate in restraining creep of paste can be assumed to be the same for basic and total creep, so that the SEF is equal for cement paste, mortar, and concrete.

Under drying conditions, drying shrinkage occurs, and as usual, it is assumed that it is the same in the specimen under load as in the control specimen; but significantly, it is assumed that there is no influence on creep of an interaction of shrinkage with stress arising from the external load. Only in the cases of high stresses, which can cause nonlinear creep, and in predrying before loading would a shrinkage–stress interaction be considered relevant. Also, under normal operating conditions, the microcracking contribution to drying creep is assumed to be small.

Water/Cement Ratio and Relative Humidity of Storage

To simulate the influence of relative humidity of storage, first consider basic creep of the sealed cement paste having $F = 0.6$, i.e., 60% of pores are full of water and 40% of pores are part empty; the relative stress on the solid gel is given by the $F = O. 6$ curve in Figure 15.12. Now consider total creep of the drying paste with relative stresses on the solid gel (Figure 15.14) ranging from $F' = 0.2$ ($\equiv$ high humidity) to $F' = 1.0$ ($\equiv$ very low humidity). The corresponding range of relative stress ratios or SEF is plotted in Figure 15.15 as a function of water/cement ratio and simulated change of relative humidity. The trend of SEF with water/cement ratio indicates total creep increases more than basic creep until the water/cement ratios is around 0.5. The trend is similar to that of CEF for several previously reported experimental results. In the case of cement paste [28], Figure 15.16 indicates the right order of magnitude and so does Figure 15.17 for the 10-year results for concrete [41]. The CEF for concrete appears to be largely unaffected by the time under load (Figure 15.18), but increases when the relative humidity decreases, as predicted by the SEF curves of Figure 15.15.

The particular case of predrying concrete before application of the load should be mentioned, as the theory predicts a relative stress on the solid gel ≥ 1 (see Figure 15.13) immediately after the load is applied. On the other hand, sealed concrete nearly always has a lower relative stress (Figure 15.12), and therefore the SEF is greater than 1. This implies that total creep is greater than basic creep for the predried concrete. However, in Ruetz's tests [28], Figure 15.19 shows the opposite to occur in that the predried concrete had a CEF of less than unity, so that there was a total creep reduction compared with basic creep. The explanation is thought to be that predrying shrinkage is actually the result of capillary stress and creep [29], so that some creep has already taken place and there is less creep potential by the time the external load is applied.

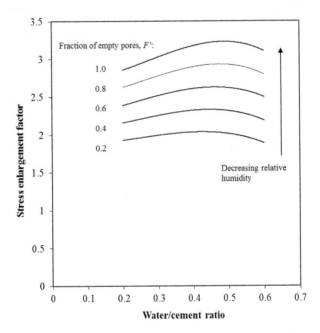

Figure 15.15 Simulated creep enlargement factor (ratio of total creep to basic creep) of cement paste.

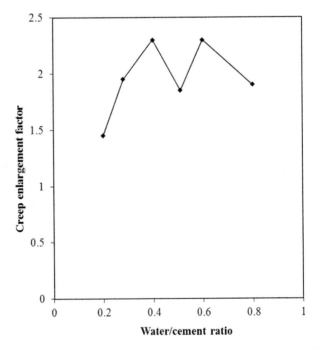

Figure 15.16 Creep enlargement factor after 28 days for cement paste having different water/cement ratios; stored at 40% RH [28].

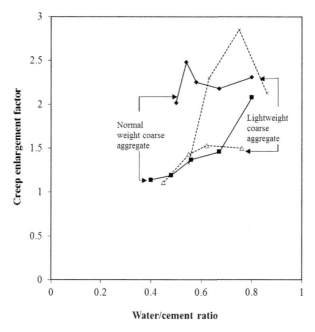

Figure 15.17 Ten-year creep enlargement factor for concrete made from different aggregates; stored at 65% RH 16, 41.

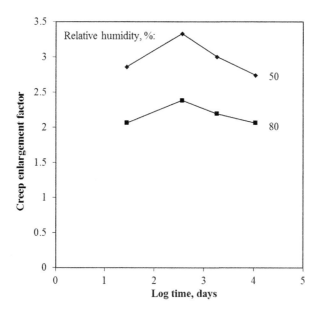

Figure 15.18 Effect of relative humidity of storage on creep enlargement factor [42].

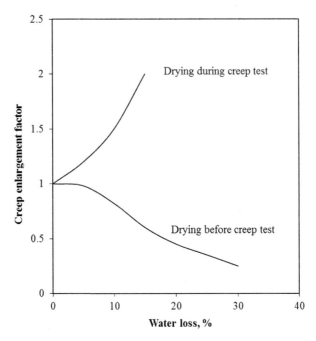

Figure 15.19 Effect on creep of cement paste water loss as a percent of evaporable water at 20 °C. [28]

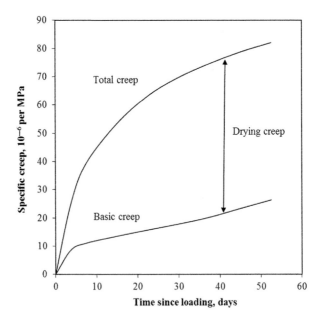

Figure 15.20 Creep in tension [16, 43]. Concrete cured in water for 28 days then loaded (a) in air at 65% RH (total creep), and (b) in water (basic creep).

Tensile Creep

Most existing theories of creep fail to explain why drying creep occurs under tensile loading of concrete exposed to drying conditions, because superposition of shrinkage or shrinkage-induced stress, which is much greater than the strain or stress due to the tensile load, should result in a "contraction creep." However, the proposed theory does predict drying creep in tension, and in fact drying creep under any mode of loading, because the stress acting on the solid gel is dependent on the direction of externally applied stress, the pore size/spacing ratio, and the rate and degree of water removal from the pores. Under tensile loading, water removal from the pores results in a predicted SEF in the same manner as in compression. Theoretical confirmation is given in Figure 15.18, which shows some tensile test data yielding a CEF of between 3 and 4. The existence of drying creep under torsional loading was demonstrated by Ruetz [28].

Transitional Thermal Creep

As discussed earlier, transitional thermal creep relates specifically to sealed concrete when heated after the load is applied. Figure 15.21 shows the SEF based on the theoretical case of sealed paste ($\equiv$ basic creep) with initial state of $F = 0.6$. At elevated temperature, the basic creep of the solid gel is already greater than at 20 °C, but then a sudden increase of temperature would cause a rapid diffusion of water from full to part-full pores. The latter is simulated by $F = 0.2$ and 0 (very high temperature), which results in the predicted SEF shown in Figure 15.21.

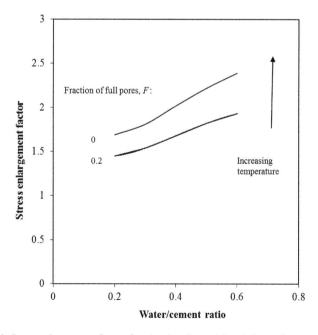

Figure 15.21 Sress enlargement factor for simulated transitional thermal creep.

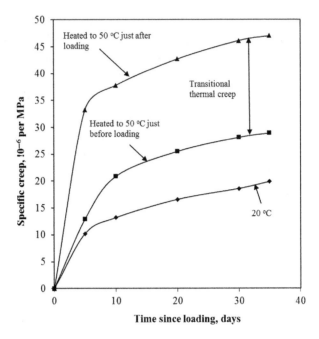

Figure 15.22 Transitional thermal creep of ordinary Portland cement concrete, having a creep enlargement factor of 1.64 after 30 days [44].

Figure 15.22 demonstrates the difference in experimental creep when concrete is subjected to elevated temperature before (accelerated basic creep) and after loading (accelerated plus transitional thermal creep). After 35 days under load, the CEF for transitional thermal creep is approximately 1.7 relative to the basic creep at 50 °C, so it can be seen that there is reasonable agreement with the SEF from the theory.

A similar simulation can be made for transient thermal strain or transient thermal creep [30–32], i.e., the very high strain developed in drying concrete heated to very high temperature after the load is applied. In this case, the basic creep at elevated temperature would be even greater than for transitional thermal creep, and so would the SEF due to rapid emptying of pores by expulsion of gel and interlayer water from the concrete to the outside environment.

Final Remarks

Based on the analysis of stress concentration due to the presence of rigid inclusions and holes in a surrounding medium subjected to an external stress, the new theory explains the mechanism of drying creep of cement paste and concrete. Based upon previous experimenters' results and observations, it is assumed that water within the gel pores is structural and is capable of withstanding significant load. It is demonstrated that, when first exposed to drying, compared with the nominal external stress, an increase in stress on the solid gel occurs as the pores are emptied of water. Compared with sealed paste, there is a stress enlargement for drying paste that mainly arises from the fact that the water-filled pores of the sealed paste reduce the stress acting on the solid gel to below that of the externally applied stress. The analysis implies the following:

- The concept of a stress-induced drying shrinkage or shrinkage-induced stress is not required to account for drying creep.
- The stress on the solid gel, and therefore creep, depend on the pore size/spacing ratio and pore water content.
- Some water movement within sealed paste is required to explain the reported influence of water/cement ratio on basic creep.
- Compared with basic creep, drying and total creep will always increase in the direction of the applied external stress due to subsequent pore water removal. Consequently, the theory also explains drying tensile creep, transitional thermal creep, and transient thermal strain.

Problems

1. Explain the terms "stress enlargement factor" and "creep enlargement factor."
2. What is known as the Pickett effect?
3. Describe the structure of C-S-H.
4. State the different types of water held within hydrated cement paste.
5. Describe the types of pores existing within cement paste.
6. What is capillary tension theory?
7. Does drying creep occur under tensile load?
8. What factors affect the stress concentration at the edge of cavity in a material subject to external load?
9. When cement paste is under an external stress, assuming gel pore water to be load-bearing, is the internal stress on the solid gel greater or smaller than the external stress for:
   a. empty pores?
   b. pores full of water?
10. For the previous question, how is the internal stress on the solid gel affected by:
    a. hydration of sealed cement paste?
    b. water loss by drying to the environment?
    c. an increase in water/cement ratio?

References

[1] Neville AM. Properties of concrete. 4th ed. UK: Pearson Education Ltd; 2006. 844pp.
[2] Powers TC. The physical structure and engineering properties of concrete. Chicago: Research and Development Bulletin 90, Portland Cement Association; 1958. 39pp.
[3] Bazant ZP, editor. Mathematical modeling of creep and shrinkage of concrete. John Wiley & Sons; 1988. p. 459.
[4] Neville AM, Dilger WH, Brooks JJ. Creep of plain and structural concrete. London and New York: Construction Press; 1993. 361pp.

[5]   Feldman RF, Sereda PJ. A model for hydrated cement paste as deduced from sorption-length change and mechanical properties. Mater Struct Paris 1968;1(6):509–19.

[6]   Verbeck GJ, Helmuth FH. Structure and physical properties of cement paste. In: Proceedings of the fifth international symposium on the chemistry of cement, Tokyo, 1968, vol. 3. Cement Association of Japan; 1969. pp. 1–32.

[7]   Helmuth RA, Turk DH. Elastic moduli of hardened Portland cement and tricalcium silicate pastes: effect of porosity. Special report 90. In: Symposium on structure of Portland cement paste and concrete. Washington: Highway Research Board; 1966. pp. 135–44. PCA Research. Dept. Bulletin 210.

[8]   Parrott LJ. The effect of moisture content upon the elasticity of hardened cement paste. Mag Concr Res March 1973;25(82):17.

[9]   Sereda PJ, Feldman RF, Swenson EG. Effect of Sorbed water on some mechanical properties of hydrated Portland cement pastes and Compacts. Special Report 90. In: Symposium on structure of Portland cement paste and concrete. Washington: Highway Research Board; 1966. pp. 58–73. PCA Research Dept. Bulletin 210.

[10] Hobbs DW. Movement: creep and shrinkage. Adv Concr Technol; 2001:10. British, Cement Association, Nottingham University, UK.

[11] Illston JM, Dunwoodie JM, Smith AA. Concrete, timber and metals, the nature and behaviour of structural materials. Van Nostrand Reinhold; 1979. 663pp.

[12] Mindess S, Young JF. Concrete. Prentice-Hall; 1981. 671pp.

[13] Brooks JJ. 30 year creep and shrinkage of concrete. Mag Concr Res 2005;57(9):545–56.

[14] Creep, shrinkage and durability mechanics of concrete and other quasi-brittle materials. In: Ulm EJ, Bazant ZP, Wittmann FH, editors. Proceedings of the sixth international conference, CONCREEP-6. MIT, Cambridge (MA), USA: Elsevier; 2001. p. 811.

[15] Brooks JJ [Chapter 7]. In: Newman J, Choo BS, editors. Advanced concrete technology: concrete properties. Elsevier; 2003.

[16] Brooks JJ. A theory for drying creep of concrete. Mag Concr Res 2001;53(11):51–61.

[17] Davis RE, Davis HE. Flow of concrete under the action of sustained loads. ACI J 1931;27:837.

[18] Pickett G. The effect of a change in moisture content on the creep of concrete under a sustained load. ACI J 1947;38:333–55.

[19] Lyman CG. Growth and movement in Portland cement concrete. Oxford University Press; 1934. p. 60.

[20] Lorman WR. The theory of concrete creep. Proc ASTM 1940;40:1084.

[21] Pickett G. Effect of aggregate on shrinkage of concrete and a hypothesis concerning shrinkage. ACI J 1956;27(5):581–90.

[22] Wittmann FH, Roelfstra P. Total deformation of loaded and drying concrete. Cem Concr Res 1980;10:601–10.

[23] Wittmann FH, Roelfstra P. Time-dependent deformation of a drying composite material. Proc SMIRT 1985;8. Brussels, Paper H5/8.

[24] Bazant ZP, Chern JC. Concrete creep at variable humidity: constitutive law and mechanism. Mater Struct RILEM 1985;18(103):1–20.

[25] Reid SG. Deformation of concrete due to drying creep. In: Bazant ZP, Carol I,E, Spon FN, editors. Proceedings of fifth international RILEM symposium on creep and shrinkage of concrete; 1993. pp. 39–44.

[26] Bazant ZP, Xi Y. New test method to separate microcracking from drying creep: curvature creep at equal bending moments and various axial forces. In: Bazant ZP, Carol I,E, Spon FN, editors. Proceedings of fifth international RILEM symposium on creep and shrinkage of concrete; 1993. pp. 77–82.

[27] Wittmann F. Kriechen bei gleichzeitigem schwinden des zementsiens. Rheol Acta 1966;5:198–204.

[28] Ruetz W. A hypothesis for the creep of hardened cement paste and the influence of simultaneous shrinkage. In: Proceedings of an international conference on the structure of concrete. London: Cement and Concrete Association; 1968. pp. 146–53.

[29] Nagataki S, Yonekura A. The mechanisms of drying shrinkage and creep of concrete. Transactions Jpn Concr Inst 1983;5:127–40.

[30] Khoury GA, Grainger BN, Sullivan PJE. Transient thermal strain of concrete: literature review, conditions within specimen and behaviour of individual constituents. Mag Concr Res 1985;37(132):131–44.

[31] Khoury GA, Grainger BN, Sullivan PJE. Strain of concrete during first heating to 600°C under load. Mag Concr Res 1985;37(133):195–215.

[32] Khoury GA, Grainger BN, Sullivan PJE. Strain of concrete during first cooling from 600 °C under load. Mag Concr Res 1986;38(144):3–12.

[33] Timoshenko S, Goodier JN. Theory of elasticity. 2nd ed. Kogakusha: McGraw-Hill; 1951. 506pp.

[34] Howland RCJ. On the stresses in the neighbourhood of a circular hole in a strip under tension. Transactions Royal Soc Lond Ser a 1930;229:49–86.

[35] Howland RCJ. Stresses in a plate containing an infinite row of holes. Proc Royal Soc Lond 1935;148:471–91.

[36] Goodier JN. Concentration of stress around spherical and cylindrical inclusions and flaws. Trans A.S.M.E 1933;55:39–44.

[37] Neville AM, Brooks JJ. Concrete technology. Harlow: Longman; 2008. 438pp.

[38] Illston JM, Sanders PD. The effect of temperature change upon the creep of mortar under torsional loading. Mag Concr Res 1973;25(84):136–44.

[39] Cabrera JG, Lynsdale CJ. The effect of superplasticizers on the hydration of normal Portland cement. Cemento; 1996:532–41.

[40] Counto UJ. The effect of the elastic modulus of the aggregate on the elastic modulus, creep and creep recovery of concrete. Mag Concr Res 1964;16(48):129–38.

[41] Brooks JJ. Accuracy of estimating long-term strains in concrete. Mag Concr Res 1984;36(128):131–45.

[42] Troxell GE, Raphael JM, Davis RE. Long-time creep and shrinkage tests of plain and reinforced concrete. Proc ASTM 1958;58:1101–20.

[43] Brooks JJ, Neville AM. A comparison of creep, elasticity and strength of concrete in tension and in compression. Mag Concr Res 1977;29(100):131–41.

[44] Boukendakdji M. Mechanical properties and long-term deformation of slag cement concrete [Ph.D. thesis]. Department of Civil Engineering. University of Leeds; 1989, 290pp.

16 Testing and Measurement

This chapter is mainly concerned with methods of subjecting concrete and masonry to different types of sustained constant load and the measurement of strain due to creep and shrinkage. Measurement of strain and parameters associated with other types of movement has already been discussed in previous chapters:

- Standard methods of determining static modulus of elasticity of concrete are described in Chapter 4 (p. 75) and determination of dynamic modulus of elasticity in the same chapter (p. 76). Methods for measuring the elasticity of mortar, brick and block units, and masonry are referred to in Chapters 7 and 8.
- Measurement of autogenous shrinkage of concrete is discussed in Chapter 6 (p. 169).
- The standard methods of determining coefficient of thermal expansion of concrete and masonry, units, and mortar and masonry are described in Chapter 13 (p. 462).
- Measurement of irreversible moisture expansion of clay brick units is covered in Chapters 8 and 9.

Firstly, methods of applying uniaxial compressive and tensile loads are described, and then measurement of movement by different types of strain gauges is covered, including practical guidance on experimental procedures for installing strain measuring equipment for the systems used for concrete and masonry for many years in the Civil Engineering Laboratory at Leeds. Details are given of a prescribed standard method of test for creep of concrete by ASTM C 512-02 [1], which has existed since 1969 together with specifications by European test methods, such as the RILEM method described in CPC 12:1983 [2]. BS EN 1355: 1997 [3] also prescribes a method of determining creep of test specimens taken from prefabricated components of autoclaved aerated concrete (AAC) or lightweight concrete with an open structure. More recently, BS ISO 1920-9: 2009 [4] specifies a procedure for determining creep concrete that is similar to the RILEM method. At present, there is no standard method of determining creep of masonry, but recommendations are proposed in this chapter.

In most instances where creep tests are undertaken and data reported, drying shrinkage is also determined and reported from measurements using control, load-free, specimens cast at the same time as the creep specimens. In those tests, the ends of the control specimens are sealed or covered to replicate drying conditions of creep test specimens whose ends are fitted with steel platens to transmit load, and also shrinkage is measured by the same type of strain gauge as that used for the measurement of total time-dependent strain. In these instances, shrinkage is required in order to isolate creep from the total measured time-dependent strain, as well as being a movement property of interest in its own right. However, there are other situations where shrinkage or moisture expansion is required to be measured independently of creep where prescribed methods are slightly different since they generally require the

Concrete and Masonry Movements. http://dx.doi.org/10.1016/B978-0-12-801525-4.00016-9

change in total length of the specimen determined by comparator as the indicator of shrinkage; those "independent shrinkage test methods" are described at the end of this chapter and include ASTM standards for determining shrinkage of masonry mortar and units.

Methods of Load Application

The ideal requirements for creep testing can be stipulated as follows. The loading system has to be able to maintain a constant known stress with a minimum of maintenance and subsequent manual adjustment. In uniaxial load testing, there should be a uniform stress distribution over the cross-section of the specimen without requiring an unduly heavy frame. Since the demarcation between creep and elastic strain is not easily determined, the apparatus should be capable of applying the load very quickly. It is also desirable that the loading system be reasonably compact to make possible operation in a room with controlled temperature and humidity.

Compression Apparatus

The majority of creep tests have been undertaken on specimens loaded in compression— cylinders and prisms—subjected to uniaxial stress. Generally speaking there are four loading methods: dead load, spring loaded, dynamometer loaded, and hydraulically loaded [5].

The dead load system is normally used in the form of a lever arm to provide a mechanical advantage, such as that shown in Figure 16.1, where the load is applied by a water-filled cylindrical tank or steel weights suspended from the end of a 20:1 lever arm via a steel ball to three concrete specimens in series with a load-cell. The system is also capable of being adapted to perform stress relaxation tests by automatically emptying water according to the reduction in load required in order to maintain a constant strain in the specimens. For masonry tests, a higher load is generally required for the larger cross-sectional area of a representative specimen, which can be achieved by interconnecting two lever arms by a crosshead beam, as shown in Figure 16.2. Here, a 13-course high single-leaf calcium silicate wall is loaded in compression by the crosshead beam via steel spreader beams, steel rollers, and a wall header plate grouted and leveled to the top of the wall; that arrangement ensures that a uniformly distributed load is achieved. An identical companion wall not subjected to load is also shown in Figure 16.2 for the measurement of shrinkage/thermal movement.

Lenczner [8] was the first to design test rigs for determining creep of brickwork built with half-scale bricks. Rather than use a long lever arm system, which requires large dead weights and occupies large spaces, he applied the load by a mechanical system based on screw-worm and gear wheel principle. This system was used later for testing of half-story high hollow piers built with full-scale bricks. The apparatus consisted of a central tension shaft passing through the pier and connected to a steel base, the load being applied manually by a large hand-wheel operating on the gear and worm mechanism. The test rig capacity was 600 kN, the compressive load being

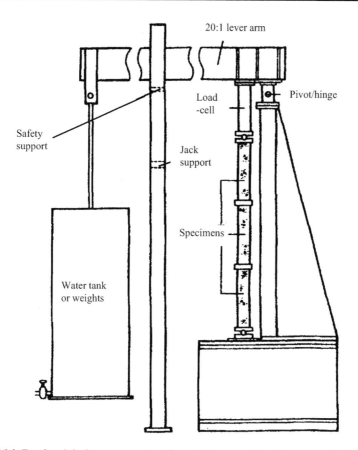

Figure 16.1 Dead-weight lever arm system for applying a sustained compressive load to concrete specimens [6].

measured by four pre-calibrated 150 kN proving rings. The crosshead was operated by an electric motor to bring it to the approximate position on the top of the pier, but the final adjustment as well as actual application of the load was achieved by the large hand-wheel that controlled the movement of the crosshead to very fine limits [9].

In a spring-loaded system (Figure 16.3), one or more heavy coil springs are held in a compressed position between steel plates in a frame, the coils and concrete specimen being in series. Because the stiffness of the spring is less than that of concrete, the energy stored in the spring ensures that the magnitude of the sustained load is only little reduced by the change in length of the specimen due to creep or shrinkage. Without a spring, creep of the specimen would rapidly reduce the applied stress. The tension in the rods can be increased to compensate for the loss of load, and this is sometimes done with a spring system in the early stages after application of load when the rate of creep in highest. A problem with a spring-loaded test frame is the application of load cannot be applied as quickly as with a lever arm system. To

Figure 16.2 Double lever arm system for applying a sustained load to a calcium silicate wall located behind a load-free control wall [7].

measure the applied load, the spring may be calibrated beforehand and the load increased until the spring shows the required displacement. The design of the frame may be modified to accommodate a loading hydraulic jack in series with a load-cell so that the specimen can be loaded rapidly, then the tie rods of the frame tightened and the jack unloaded but left available to be used later to reapply part of the load lost due to creep and shrinkage; alternatively, the jack may be removed and used to repeat the process on another frame.

For higher-stress applications, a combination of the spring-loaded and lever arm systems is possible where a tie rod and spring replace the suspended weight at the end of the lever arm. The load is applied to the specimen by tensioning the tie rod against the spring, which is located on the upper surface of the lever arm, the energy stored in the spring ensuring a near-constant load [5,10].

High loads can be applied more easily and can be maintained to a greater accuracy by the use of a hydraulic system, as shown in Figure 16.4. Such a system is compact and flexible in that it can be used for application of predetermined variable stress.

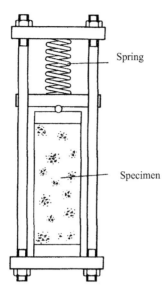

Figure 16.3 Spring-loaded compressive creep frame [5].
Source: Creep of Plain and Structural Concrete, A. M. Neville, W. H. Dilger and J. J. Brooks,
Pearson Education Ltd. © A. M. Neville 1983.

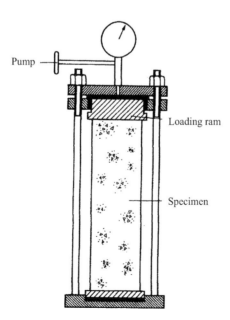

Figure 16.4 Hydraulic creep frame used by L'Hermite [11].

However, the maintenance of sustained load is sensitive to small changes in length of the specimen and also to small leakage of the hydraulic fluid, so that frequent adjustment is necessary. This difficulty is overcome with a stabilized hydraulic system [5,12], which utilizes an auxiliary spring-loaded hydraulic cylinder. The system is capable of accommodating several creep specimens at the same time [13,14] and has been found to operate satisfactorily over long periods of time. An alternative arrangement involves the use of nitrogen through a spring-loaded pressure-reducing valve [15].

For measuring creep of story-high masonry walls and piers, Lenczner [16,17] used a hydraulic system to apply compression which, for a 29-course high × 4-brick wide single-leaf wall, consisted of three hydraulic jacks interconnected through a manifold and three calibrated proving rings of 200 kN capacity. Each test rig consisted of four 63 mm dia. steel tension shafts, a base, a distribution plate, and two crossheads. The shafts were screwed to bosses welded to the base, and at the top they were supported by means of cylindrical units, a distribution plate, and crossheads. The jacks were positioned between the crossheads, and the purpose of the intermediate crosshead was to maintain the load in the event of a drop in hydraulic pressure due to accidental leakage of hydraulic fluid.

Leeds Creep Test Frames

Concrete and Mortar

For simultaneous creep testing of large numbers of concrete and mortar specimens, a relatively inexpensive frame is recommended of the type shown in Figure 16.5. Two specimens are held in series with a steel tube dynamometer by four steel tie rods, strain readings being taken by a demountable hand-held mechanical gauge (see next section). The test frame is suitable for 76 dia. × 255 mm long cylindrical specimens or 50 × 50 × 250 mm prismatic specimens for concrete made with a maximum size of aggregate of 10 mm. The specimens require plane and square ends to ensure satisfactory seating in the steel recessed end-plates so that eccentricity of loading is minimized. Vertically cast cylindrical specimens therefore require the upper surface to be ground prior to assembly, whereas horizontally cast prisms have the advantage that no preparation is required provided square recessed end-caps are available instead of circular caps for cylindrical specimens. Prisms are particularly useful for creep testing of mortar, and it is also possible to test individual masonry bricks between header faces provided suitable rectangular end-caps are available.

Figure 16.6 shows an assembly of creep frames on a tiered wooden rack with slotted supports that accommodate the frame trunnions. The assemblage is stored in a drying environment and the mortar prisms under load are part-sealed together with corresponding part-sealed control prisms. This type of experiment was designed to provide creep and shrinkage data for modeling of mortar in masonry, and in fact the four frames have different amounts of sealing, as shown in Figure 16.7, in order to simulate drying of mortar joints in different types of 13-course-high brickwork, built with standard bricks, viz. 215 × 102.5 × 65 mm, and standard 10 mm mortar joints.

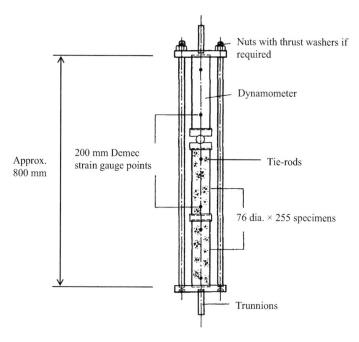

Figure 16.5 Simple creep frame with cylindrical concrete specimens and dynamometer [18].

Figure 16.6 Part-sealed mortar prisms under load for determining creep and corresponding control prisms for determining shrinkage. Part sealing simulates volume/surface ratios of mortar in different types of brickwork indicated in Figure 16.7 [19].

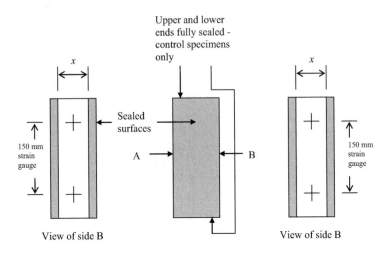

Parameter	Type of Brickwork			
	Single Leaf	Cavity Wall	Hollow Pier	Solid Pier
Volume/surface ratio, mm	44	51	81	146
Dimension x, mm	45	40	30	22

Figure 16.7 Partial sealing of $50 \times 50 \times 230$ mm mortar prisms to simulate drying of mortar joints in different types of brickwork [19].

Since the ends of the specimens under load are effectively sealed by the steel caps and spacers, for identical drying conditions, the ends of control specimens are sealed.

Part-sealing of specimens can be achieved by application of two layers of bitu-mastic paint followed by a third layer and, while tacky, a layer of polyurethane sheet. Silicone sealant at the edges of the polythene sheet is also recommended. A less laborious alternative is the application of self-adhesive aluminum waterproofing tape. In the case of completely sealed specimens to represent mass or large volume con-crete in the determination of basic creep, the process of sealing is carried out after fixing strain gauge points to the concrete, care being taken to fix the sealant around the gauge points. The effectiveness of sealing may be assessed by monitoring weight measurements of a fully sealed specimen over a period of time, and both of the foregoing methods of sealing having been found to show negligible weight loss after one year.

An alternative method of simulating approximate conditions of basic creep of concrete is to store the specimens and creep frames in water, as shown in Figure 16.8. This is also a convenient method of determining creep of concrete at elevated tem-perature since the water tanks can readily be equipped with thermostatically controlled immersion heaters. The frames are supported horizontally by their

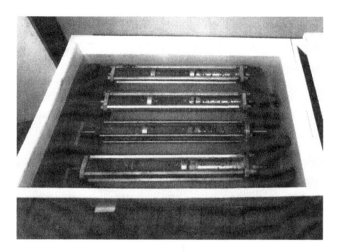

Figure 16.8 Load test frames stored in water to measure basic creep of concrete; control specimens also stored to measure swelling.

trunnions in slots in the top of submerged reinforced concrete stands so that the surfaces of concrete test specimens are completely covered with water. When strain measurements are carried out, a notched wood block is placed underneath each trunnion to just expose the surfaces of the dynamometer and concrete cylinders at the gauge position; this avoids the hand-held demountable mechanical strain gauge from becoming wet. The frame can then be rotated to take readings at the other gauge positions. If basic creep is required using test frames stored in water for lengthy periods, it is advisable to manufacture frames from stainless steel components.

For creep testing of normal-strength concrete, the mild steel dynamometer dimensions are approximately 300 mm long with a 76 mm internal dia. × 2 mm thick wall; four pairs of 200 mm strain gauge points are equally spaced circumferentially. The maximum load capacity is 100 kN, which represents a potential maximum stress of approximately 20 MPa that can be applied to the concrete specimen. With a 200 mm Demec gauge (see Figure 16.5), readings are accurate to within ± 1 division of the dial indicator, which is equivalent to a strain of $\pm 8 \times 10^{-6}$ or a change in load of ± 0.66 kN. Hence, the stress on 76 mm dia. concrete cylinders can be controlled to within an accuracy of ± 0.15 MPa.

After taking zero-strain readings, calibration of dynamometers can be carried out by applying loads between 5 and 100 kN in a standard laboratory test machine in increments and decrements of 20 kN. After ensuring consistency of readings at the same load and proportionality of load and strain, the required datum strain readings should be calculated for each of the four circumferential positions according the precise load to be applied to the concrete specimens.

Several hours before assembly of test frames, components should be placed in their test environment to attain temperature equilibrium. Zero readings at all four circumferential positions are then taken on the dynamometer, and if they differ from those taken at the time of calibration, the datum readings should be adjusted accordingly.

Test frames are assembled in their vertical position by resting the lower flange of the test frame on a wooden stand having a hole to accommodate the trunnion. During assembly, it is important to ensure that dynamometer and concrete cylinder gauge positions are numbered and in line for convenience of strain measurement using a demountable mechanical gauge. The two concrete specimens should be firmly seated in the recess of the lower flanges and steel spacers, the upper one containing a ball joint lubricated with grease. After seating the dynamometer in the upper ball-joint spacer, it is placed on top of the ball joint, and then the upper flange and trunnion are placed over the tie-rod ends and the four nuts hand-tightened. During assembly of the test frame, it is important to check that the specimens and dynamometer remain central within the tie rods throughout the assembly to avoid eccentricity of load.

For testing of high-strength concrete, the same type of test frame may be used, but made with high-strength steel for the tie rods and dynamometer with an increased wall thickness of 3 mm. This permits a maximum load of approximately 200 kN and concrete stress of 40 MPa. In this case, using a 200 mm Demec gauge to monitor strain in the dynamometer means the stress applied to the specimen can be controlled to within ±0.27 MPa.

An alternative method of monitoring load in the dynamometer is by a pair of electrical resistance strain (ers) gauges fixed on opposite sides and suitably protected and waterproofed to ensure reliability of performance, especially if to be used for the determination of basic creep in water storage. After the initial preparation, monitoring of readings and load adjustment are more convenient with this system.

Since there is no spring in the system to store energy, the loss of load due to creep and shrinkage (if drying occurs) of concrete has to be compensated manually by tightening the four nuts, a procedure that has to carried out frequently after first loading but afterward much less so depending on the type of concrete and its creep-time characteristic. Typically, after first application of load, load adjustments and readings should be taken daily for one week, and then once per week for the first month, and then once per month for the first year. The procedure is to take gauge readings on the dynamometer at all four positions and compare them with the original readings on initial application of load. The tie rods are then retensioned by tightening the nuts in small steps and in sequence to reestablish the original datum readings. It is important to carry out this operation carefully to avoid causing undue eccentricity of load. The adjusted load is deemed satisfactory if the average difference between adjusted dynamometer gauge readings and original datum readings is within one division of the gauge. For ease of the manual load adjustment, it is advisable to incorporate thrust washers between the flange and tensioning nuts of the tie rods.

At the same time as taking readings on loaded specimens, strain readings of control specimens are required together with recordings of temperature and, if stored in a drying environment, relative humidity (RH).

Masonry

Figure 16.9 shows the arrangement for testing of masonry first developed by Abdullah [19], the principle of which is basically the same as that for determining creep of

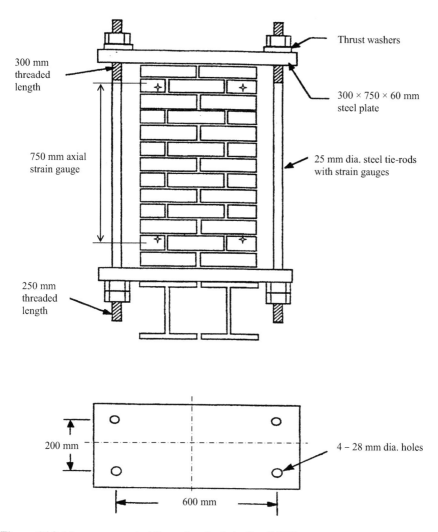

Figure 16.9 Masonry creep test frame for single leaf wall [19].

concrete. Essentially, 13-course high × 2-brick wide brickwork or 5-course high × 1-block wide blockwork is constructed on a 60 mm thick base plate with drilled holes for tie-rods, the base-plate being supported on steel beams of sufficient depth to allow the tie-rods to be locked in position. Care is required to ensure the wall is built symmetrically within the four tie-rod holes in the base plate. An identical control wall is also constructed on an adjacent but thinner steel plate grouted and leveled on the laboratory floor. Immediately after any breaks in construction, the walls are covered with polythene sheet, and on completion of construction, the walls are recovered with polythene sheet for the specified period of curing. Other unbonded control specimens, such as mortar prisms and unbonded units, are also stored and covered with the walls.

The same loading system configuration is used for testing cavity walls but with a wider upper load plate. For larger cross-sections of masonry, such as solid and hollow piers, besides having square plates for the base and upper load plates, 36 mm dia. tie-rods are used to accommodate the higher load. When testing hollow piers and cavity walls, holes may be provided in the upper load plate to simulate ventilation of the cavities.

One day before the age of load application, the polythene sheet is removed, tie-rods are bolted to the lower base plate, and the top of the load wall is capped with mortar to allow the upper load steel plate to be carefully lowered over the tie-rods and leveled on top of the wall. During this process, the polythene sheet is also removed from the control wall, and the upper surface of this wall is coated with a waterproof membrane so that front and rear surfaces are exposed to drying in an identical manner to the load wall. After leveling the upper plate on the load wall, 750 mm gauge points are attached to both sides of the walls for axial strain measurement (see Figure 16.9). At this time, if required, other gauge points are also installed for measurement of lateral stain (400 mm) and also individual bonded and unbonded units can be fitted with strain gauges. On completion of these tasks, walls and specimens should be re-covered by the polythene curing sheets.

Part of the experimental preparation is the calibration of tie-rods that act as load-cells for monitoring the load on the masonry wall. Strain in the four 25 mm dia. steel tie rods is measured by ers gauges, which require calibration in a standard laboratory test machine to obtain their load-strain characteristics. When fixing the ers gauges, it is essential to use a gauge/steel adhesive that is free from "zero-drift" and to coat the gauge with a robust cover for protection and satisfactory long-term performance. The calibration procedure involves applying three or four cycles of load to minimize hysteresis and stabilize the gauge adhesive before recording full load-strain readings for the specified operating range of load using a suitable strain-measuring unit and data logger. As a precaution, a backup mechanical strain gauge system is recommended consisting of predrilled holes at each ers gauge location suitable for a 200 mm Demec strain gauge.

At the stipulated age of the masonry when the load is to be applied, polythene curing sheets are removed, zero readings taken on walls and auxiliary specimens, and the laboratory temperature and humidity readings noted. The tie rods are tensioned in turn and in small increments by manually tightening the upper nuts (Figure 16.9) until the required load is achieved. The wall strains are then remeasured to obtain the elastic strain and to calculate the secant modulus of elasticity. The load on the wall is checked regularly during the first stages of testing and nuts retightened occasionally to compensate for loss of load due to creep and shrinkage of the masonry. At this stage, any creep test rigs for unbonded mortar specimens and units are assembled and subjected to load.

Tension Apparatus

Tensile creep tests are difficult to perform with accuracy because it is not easy to apply a uniformly distributed small tensile stress to the specimen that is free from

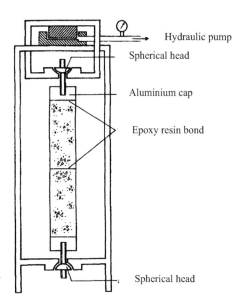

Figure 16.10 Test frame used by Akatsuka [20] for determination of creep of concrete in tension.

eccentric loading. In the past, attempts have been made to use anchorages embedded in the ends off the specimen. Moreover, the use of end-plates glued with epoxy resin, as illustrated in Figure 16.10, has not always proved successful, especially on damp concrete and for strength testing where unrepresentative failure often occurs in the vicinity of the bond [6,21]. However, Bisonnette et al. [22] successfully bonded aluminum attachment plates to the ends of 70 mm square prisms after grinding to remove the weak superficial layer of cement paste.

A bobbin-shaped specimen was used by Elvery and Haroun [21], and later by Brooks and Neville [6], the specimen being fitted with steel end-caps through which the tensile load was transmitted. The bobbin-shaped specimen comprised a central cylindrical section, 76 dia. × 178 mm, and the overall length was 356 mm with a 6° angle at the intersection of the cone -shaped ends and central section. In the original design [6], after casting the specimen, the top and bottom of the mold had to be removed in turn to fit end-caps using a special capping jig (Figure 16.11); the bond between end-cap and specimen was a quick-setting cement paste. Although successful in terms of high rate of strength failures (>90%) in the central section and away from the change in section, the time for capping specimens was laborious and lengthy.

In a later improved design, the lengthy capping procedure was eliminated by adapting the upper and lower sections of the steel casting mold to act as the end attachments to the specimen [23]. The new design further minimized the possibility of eccentric loading as well as overcoming the problem of preparation time. Figure 16.12 shows the specimen having one-half of the central section of the mold

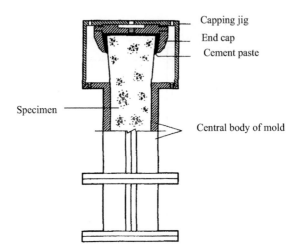

Figure 16.11 Half-section through mold and capping assembly for tensile bobbin-shaped concrete specimen [6].

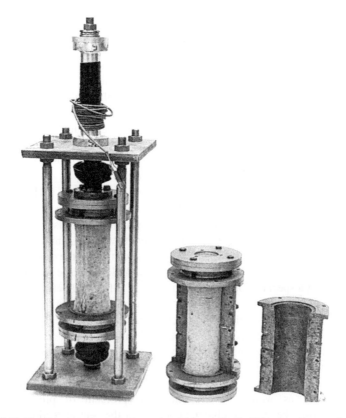

Figure 16.12 Tensile creep rig and improved mold for casting bobbin-shaped concrete specimen; upper and lower components of mold form end-caps when central body of mold is removed [23].

removed with the top and bottom sections of the mold forming the end attachments ready for connecting to semi-universal joints in the creep rig shown in the figure or in a universal test machine for strength determination. The detailed specification for the manufacture of the tensile mold to cast a bobbin-shaped specimen with end attachments is given in Figure 16.13.

A dead-weight lever arm system for determining creep in direct tension is shown in Figure 16.14. The test rig can accommodate three bobbin-shaped specimens, the

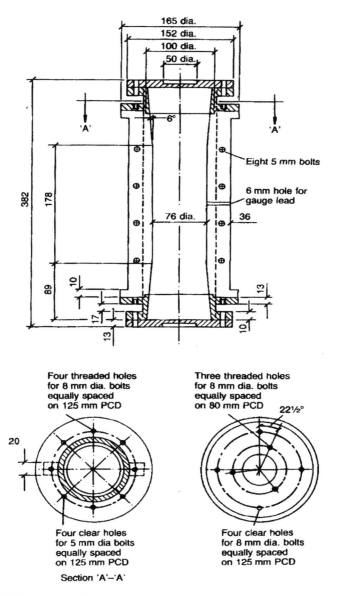

Figure 16.13 Detailed specification for improved tensile bobbin-shaped specimen mold [23].

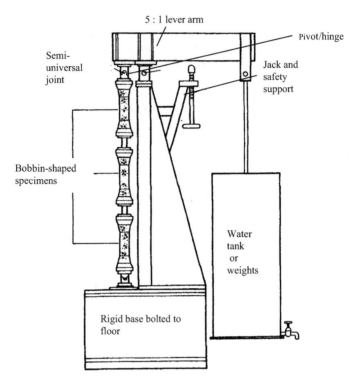

Figure 16.14 Dead load lever arm system for loading concrete specimens in direct tension [6].

load being applied by weights or a water tank suspended at the end of a lever arm pivoted to transmit the load through semi-universal joints fitted at the top and bottom of the specimen assembly. In fact, the test rig is the same as that used for testing in compression, which is fitted with a longer lever arm in the reverse direction (see Figure 16.1). Bisonnette et al., [22] also used a lever arm system to amplify load applied by a pneumatic jack, controlled by a pressure regulating valve, to three prismatic concrete specimens via a load-cell and hinged steel rods.

An alternative method of determining creep in tension was used by Ross [5,24], who applied pressure to the inner face of a thin-walled hollow concrete cylinder with open ends to produce circumferential tension. The pressure was applied by a flexible bag the ends of which had to be supported separately to prevent any axial stress in the walls of the cylinder.

Apparatus for Other Types of Loading

Flexural and torsional creep tests are relatively easier to perform than uniaxial tension, since in the latter shrinkage of drying concrete has to be taken into account; there is no deflection or rotation of load-free drying control specimens [5,25]. Apparatus for multiaxial creep tests has been reported by Hannant [26], Gopalakrishnan et al.,

[27] and Bazant et al., [28]; details of the equipment can also be found in Neville et al., [5].

Measurement of Movement

The change in deformation with time due to load, moisture movement, and thermal movement can be measured by gauges of various types: electrical resistance gauges, mechanical gauges, displacement transducers, and acoustic gauges. When selecting a gauge, expense is obviously important, but careful consideration has to be given to the suitability of the gauge for the following:

- Number of concrete specimens or measuring points on site
- Working environment, e.g., wet or dry conditions
- Convenience of access particularly if on site and if the building/structure is in use
- Gauge preparation time
- Accuracy of measurement
- Time required for measurement

Electrical Resistance Strain (ers) Gauge

These gauges operate on the principal that a change in electrical resistance is directly proportional to strain. They are very sensitive and can measure strain to within $\pm 1 \times 10^{-6}$. Figure 16.15 shows an *external or surface mounted* 30 mm ers gauge on a

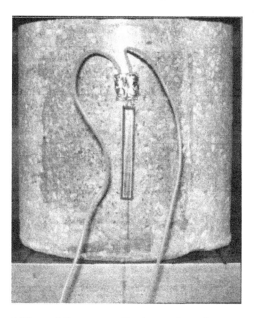

Figure 16.15 External 30 mm foil ers gauge fixed to surface of a concrete core for measurement of short-term axial strain under compressive loading.

cylindrical concrete core. Generally, a second gauge is fixed diametrically opposite, but up a total of four equally spaced gauges are recommended so that the average strain is measured to compensate for any suspected eccentricity of load. It is important to prepare the local surface of the concrete for satisfactory bond and operation of the gauge. At that point, any large pores should be filled, and then the surface abraded and smoothed to remove laitance, and finally cleaned with solvent to remove grease and debris. Discussions with the gauge manufacturer/supplier for a suitable quick-setting adhesive are recommended, particularly if the concrete is damp at the time when gauges require fixing, say, during the period of curing. After fixing, depending on the location, gauges may require a surface coating to protect from moisture and accidental damage.

With the *embedment or internal* type, the gauge is encapsulated with a protection layer of resin/sand surface roughness to assist with keying to the concrete. Strain gauges may be fixed at the center of the mold by making two small holes in the end of the strain gauge through which a thin wire is inserted for supporting the strain gauge. As illustrated in Figure 16.16, the wire is wrapped round one-half of the mold, tightened, and the ends tied firmly prior to being clamped between the flanges on assembly of the other half of the mold. During casting of concrete, this arrangement requires careful filling of the mold in layers and compaction with gentle vibration to avoid displacement of the gauge.

An advantage of ers gauges is their suitability for automatic data logging recording systems; but in general, they are not suitable for long-term creep tests of concrete because of the danger of zero drift, which mainly arises from the creep of the bonding material between the gauge and concrete. However, this problem may be overcome by fixing ers gauges to a metal backing material such as steel or aluminum that is in contact with the concrete. For example, in the tests of Bissonnette et al. [22], movement was

Figure 16.16 Embedment type ers gauges fixed in molds prior to casting specimens for measurement of short-term stress-strain characteristics. Left: 60 mm axial foil gauge in tensile bobbin mold with the upper end attachment removed. Middle: 60 mm axial foil gauge in 150 mm dia. compression cylinder. Right: 30 mm lateral foil gauge in 150 mm dia. compression cylinder.

measured by extensometers comprised of monolithic aluminum half-rings comprised of two stiff legs (in contact with concrete) connected by a flexible part instrumented with ers gauges in a full Wheatstone bridge arrangement; the potential difference was proportional to the diametrical displacement of the half ring.

A particular advantage of surface mounted ers gauges is that they can be readily used to manufacture laboratory load-cells for monitoring load in both short- and long-term tests. For example, ers gauges can replace the 200 mm Demec gauge discs on the steel dynamometers of the creep frame shown in Figure 16.5. This has the advantage of speeding up the time required for checking and adjusting the load before taking strain readings of concrete specimens loaded in this type of test rig.

Mechanical Gauge

Mechanical strain gauges have the advantage of being independent of time and temperature effects. This type of instrument is more universally known as the Demec gauge or Whittemore gauge [29]. Demountable mechanical gauges are portable and have the additional advantage in that one gauge can be used to measure strain in a large number of specimens and therefore the strain equipment is relatively inexpensive, although it is labor intensive in terms of time required for measurement. The gauge consists of an invar bar with two conical locating points, one fixed and the other pivoting on a knife edge, the point locating in predrilled steel discs attached to the concrete. Movement of the pivoting point is measured by a dial gauge attached to the strain gauge. Figure 16.17 shows a set of gauges with different gauge lengths and sensitivities; typically a 200 mm gauge would be used for concrete laboratory specimens (Figure 16.18) and a 750 mm gauge used for masonry (Figure 16.9); gauge lengths up to 2000 mm are available. Each gauge box set has a standard setting bar and standard length bar made from low thermal expansion invar steel. After minimum surface preparation, the predrilled stainless steel discs are easily attached to concrete or masonry using a rapid-setting adhesive and set to an initial length with the standard setting bar. To ensure consistency of readings throughout the test program, before taking a set of readings it is essential to check and note the gauge zero reading using the standard length bar, and adjust the dial gauge if necessary; this is important if different operators are used at different stages of the project.

In addition to dial versions, digital versions of Demec gauges are available from the 100 mm gauge upwards. The digital indicator is connected to a data processor recorder, thus reducing the time required for measurement when there are a large number of gauge positions. The indicator displays the spindle movement digitally and one increment represents 50% of one division on the dial gauge. For example, with the 200 mm gauge, one division of the dial version represents a strain of 8×10^{-6} while one increment of the digital version represents a strain of 4×10^{-6}.

Carlson Strain Meter

A good quality, although fairly expensive gauge is the Carlson strain meter shown in Figure 16.20; like the ers gauge, it operates on the principle that a change in

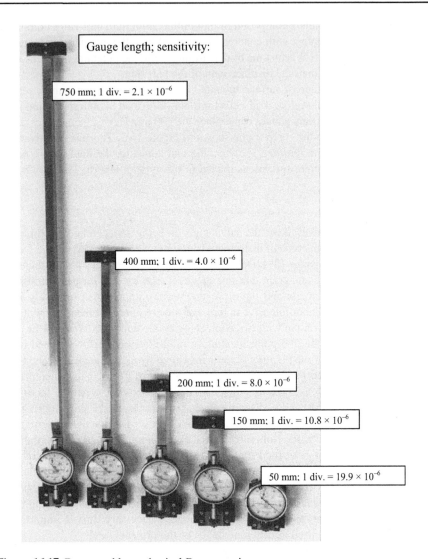

Gauge length; sensitivity:

750 mm; 1 div. = 2.1×10^{-6}

400 mm; 1 div. = 4.0×10^{-6}

200 mm; 1 div. = 8.0×10^{-6}

150 mm; 1 div. = 10.8×10^{-6}

50 mm; 1 div. = 19.9×10^{-6}

Figure 16.17 Demountable mechanical Demec strain gauges.

resistance of a coil of elastic steel wire is proportional to the strain imposed by movement in concrete. The gauge shown is for embedment in concrete but may also be used to measure external strain on concrete when fixed with saddle mounts. It measures strain and temperature by two coils of highly elastic steel wire, one of which increases in length and electrical resistance when a strain occurs, while the other coil decreases in length and electrical resistance. The ratio of the two resistances is independent of temperature (except for thermal expansion), and therefore the change in resistance ratio is a measure of strain. The total

Figure 16.18 Measuring strain of concrete specimens with a 200 mm demountable mechanical Demec strain gauge.

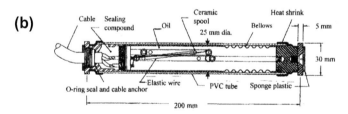

Figure 16.19 Carlson strain meter. (a) Standard strain meter, (b) Cross-section of 200 mm strain meter showing components [29].

resistance is independent of strain since one coil increases and the other coil decreases by the same amount due to the change in length of the meter. Therefore, the total resistance is a measure of temperature [29]. Standard gauge lengths range between 200 and 500 mm, and miniature versions are available ranging from 100 to 250 mm [29].

Acoustic Gauge

Also known as a *vibrating wire gauge*, the acoustic gauge was first reported for use in the UK by Potocki [30] and then by Tyler [31]. Gauges are available as embedment types or surface-mounted types (Figure 16.20). They are reasonably priced with a good long-term stability, and are suitable for measurement of static strain or very slow changes of strain. The sensitivity of a gauge having a 100 mm gauge length is 1×10^{-6}, and if required, several gauges may be monitored automatically by specially designed data logging equipment.

The gauge consists of a pre-tensioned fine steel wire enclosed in a stainless steel or acrylic tube clamped between two end-flanges or end mountings. A change of strain produces a change of tension in the wire. A dual-purpose electromagnetic coil at the center of the gauge housing and a current pulse is energized by a monitoring unit causing the wire to vibrate at a natural frequency determined by the tension in the wire [29]. The vibrating wire induces a voltage in the wire at a frequency corresponding to that of the vibrating wire, which is usually monitored by the measuring unit in terms of the time required to complete a given number of cycles, i.e., the inverse of frequency, f. The change in strain, $d\varepsilon$, is given by:

$$ d\varepsilon = K \left[\frac{1}{t_1^2} - \frac{1}{t_2^2} \right] \qquad (16.1) $$

where t = period of vibration of the wire = f^{-1}, t_1 being the datum or initial period of vibration and t_2 being the period after straining;

K = gauge factor = $\frac{4}{Eg} \rho l^2$;
E = modulus of elasticity of the wire; g = acceleration due to gravity; ρ = density of the wire and l = length of the wire.

(a)

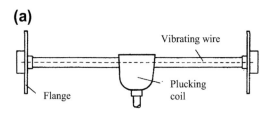

(b)

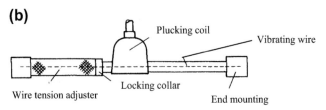

Figure 16.20 Acoustic or vibrating wire strain gauge. (a) Internal gauge, (b) Surface mounted gauge.

Figure 16.21 demonstrates applications of the embedment and external gauges for measurement of strain of laboratory concrete and masonry test specimens. The fixing procedure for internal gauges is the same as that described for internal ers gauges on p. 568, while a quick-setting adhesive similar to that used for Demec points is suitable for the fixing the mounting blocks of the external acoustic gauge. With the surface-mounted gauge, the adjuster permits some wire tensioning during installation of the gauge whereas the embedment gauge requires tensioning prior to installation. Although the embedded acoustic gauge is designed to replace the displaced concrete with a stainless steel tube of similar stiffness and therefore strain

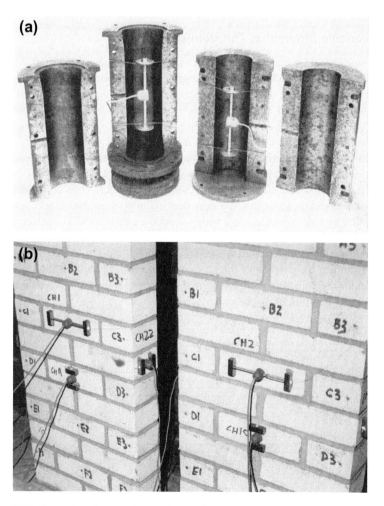

Figure 16.21 Internal and external acoustic or vibrating wire strain gauges. (a) Setting-up embedment acoustic gauges in tensile and compressive molds before casting concrete specimens. (b) External acoustic gauges fixed to calcium silicate brick wall to measure axial (50 mm) and lateral (150 mm) strains of embedded bricks.

should be unaffected, tests carried out to assess the effects of gauge embedment on 76 mm dia. concrete specimens revealed the following differences [6]:

- Strength of gauged concrete specimens was approximately 19% lower than strength of ungauged specimens.
- Strain of gauged specimens under equal short-term and long-term loading was approximately 17% greater than for ungauged specimen for either wet or dry storage.
- Shrinkage was unaffected by gauge embedment.

For less mature concrete of lower stiffness, e.g., testing at early ages, the manufacturer recommends that embedded gauges be fitted with acrylic tubes instead of stainless steel.

Linear Variable Differential Transformer (LVDT)

These gauges are displacement transducers that have good long-term stability and a large range of sizes is available for many engineering applications. They can be used for external measurement of concrete movement and, although rather expensive, they are reusable. Specialized applications are early autogenous shrinkage and cyclic displacement under cyclic loading at high frequency; typically, a gauge with maximum travel of ± 1.0 mm would have a resolution of 1×10^{-3} mm.

The LVDT is an inductive device, its only moveable part being a ferromagnetic core that develops a variable coupling between a primary winding and two identical secondary windings connected in series (see Figure 16.22). The coil assembly is housed in the annulus of a hollow stainless steel cylinder, the center of which contains the close-fitting core connected to a spring-loaded extension rod in contact with the concrete specimen. The primary winding is driven by an AC electrical signal, and as the core moves off center, the secondary coils pick up the signal by magnetic induction. In fact, the position of the core varies the voltage induced into each of the two secondary windings in an opposite manner so that a differential voltage occurs that is a linear function of displacement [29].

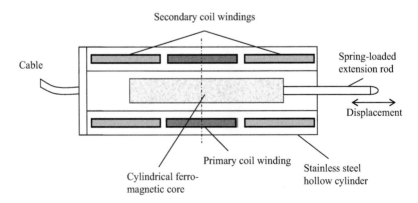

Figure 16.22 Schematic arrangement of Linear variable differential transformer (LVDT) components.

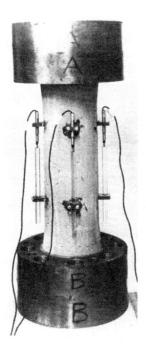

Figure 16.23 Linear variable differential transformer (LVDT) arrangement used for hysteresis loop measurement of concrete by Ashbee et al., [32].

An important aspect of externally mounted LVDTs is the quality of bearing surface at the contact point of the extension rod and the method of the attachment to concrete. To measure hysteresis loops of concrete specimens, Ashbee et al., [32] used LVDTs fixed the between Tufnol blocks glued to the surface of specimens, glass rods being used as bearing surfaces at the point of contact with the LVDT extension rod (see Figure 16.23). However, adopting the same technique in another investigation [33] highlighted the difficulty of zeroing the LVDTs at the start of testing, and as an improvement, micrometer screw gauges were substituted, as illustrated in Figure 16.24. Moreover, an improvement to the method of attaching the LVDT/micrometer gauge support blocks to the concrete was accomplished by the use of brass threaded inserts cast into the concrete to provide a firm anchorage for steel blocks. Figure 16.24 gives the details of the LVDT/micrometer arrangement and the design of support block inserts.

Standard Methods of Test for Creep Determination

The following methods are all applicable to creep of concrete since there are no standard methods prescribed for creep of masonry. For the latter it is recommended that tests be carried out as specified on p. 555 for concrete and calcium silicate

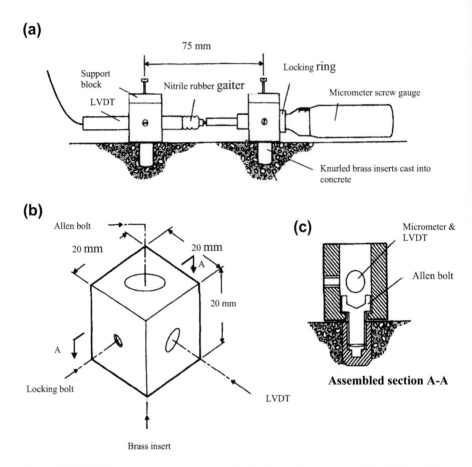

Figure 16.24 Strain gauge arrangement and fixing details for linear variable differential transformer (LVDT). (a) Arrangement of LVDT and micrometer screw guage, (b) Details of guage support block, and (c) Assembled through support block [33].

masonry and brickwork built with high-strength clay units. However, in the case of clay brickwork where it is suspected that that enlarged moisture expansion of the control wall due to cryptoflorescence may be possible, it is recommended that movements of unbonded mortar and brick specimens be measured in order to estimate moisture movement of the control wall by the composite model. By this means, an exaggerated estimate of creep is avoided (see p. 244).

ASTM C512-10

This method is applicable to molded cylinders of concrete with a maximum size of aggregate ≤ 50 mm [1]. The molds have to conform with ASTM C 192 [34], the use of both horizontal and vertical molds being permitted.

A loading frame is prescribed that is capable of applying and maintaining the required load on the specimens, and in its simplest form consists of header plates bearing on the ends of loaded specimens, a load-maintaining element that may be a spring or a hydraulic capsule or ram, and threaded rods to take the reaction of the loaded system. Bearing surfaces of the header plates should be plane within 0.025 mm. Specimens in a single frame may be stacked, but the length of a single frame or single specimen should not exceed 1.8 m. Springs may be used to maintain the load, the initial compression being applied by means of a portable jack or testing machine. If springs are used, a spherical head or a ball joint should be provided to ensure axial loading, and end-plates should be rigid enough to ensure uniform distribution of stress. An acceptable frame is shown in Figure 16.25. The load should be measured to within 2% by a permanently installed hydraulic pressure gauge or by a hydraulic jack and a load-cell inserted in the frame when the load is applied or adjusted.

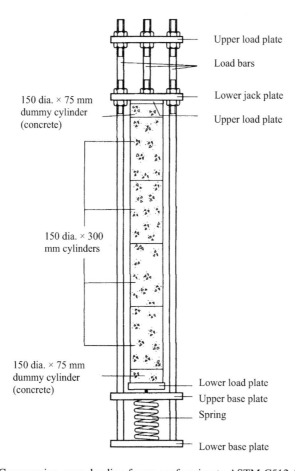

Figure 16.25 Compression creep loading frame conforming to ASTM C512-10 [1].

The axial strain in the specimen should be determined to the nearest 10×10^{-6}, but no direction is given as to whether the strain-measuring device should be embedded, attached, or portable. However, if a portable apparatus is used, the gauge points should be attached in a positive manner; attached gauges relying on friction are not permitted. External strain should be measured on at least two gauge lines spaced uniformly on the periphery of the specimen. Internal strain must be measured axially and the gauge length must be at least three times the maximum size of aggregate. The strain-measuring system must be capable of operating for at least a year without change in calibration. According to the standard [1], "systems in which varying strains are compared with a constant length standard bar are considered most reliable, but unbonded ers gauges are satisfactory."

The specimens should be in the form of cylinders 150 ± 1.6 mm in diameter with a length of at least 292 mm. When the specimen is in contact with the steel bearing plates, its length must exceed the gauge length by at least the diameter of the specimen. If, however, there is contact with another specimen, i.e., in a stack, a length exceeding the gauge length by only 38 mm is adequate but dummy specimens have to be provided at the ends: their length must be at least equal to their diameter of 150 mm. The ASTM method requires that at least six specimens be cast from each batch of concrete for each test condition. Two of these are required to determine compressive strength of concrete, two are loaded in creep frames, and two remain unloaded as control specimens to determine deformations due to causes other than load, i.e., drying shrinkage and thermal movement.

Three curing conditions are recognized: standard, mass, and "other." For the standard condition, immediately after casting the specimens should be stored at 23.0 ± 1.7 °C and covered to prevent evaporation. At the age of between 20 and 48 h, moist curing at the same temperature should start and continue until the age of 7 days. Moist curing is defined as that in which free water is maintained on the surface, but neither storage in water nor exposure to a stream of running water is permitted. Subsequent storage should be in air at a temperature of 23.0 ± 1.1 °C and $50 \pm 4\%$ RH. If mass curing conditions are desired, at the time of casting or demolding, the specimens have to be enclosed and sealed in moisture-proof jackets, e.g., copper or butyl rubber, and have to remain in those jackets throughout the test. "Other" curing conditions are a description of the situation when information is required for specific applications: different test age and ambient conditions are stipulated.

It is important to ensure axial loading when placing the test specimens in the loading frame. When stacked specimens and external gauges are used, it is helpful to apply a small preload (not more than 1.4 MPa) and to note the strain variation around the specimens. If necessary, the specimens should then be realigned for better strain uniformity. The stress/strength ratio at the time of application of load should not exceed 0.4. Before and after application of load, strain readings should be taken immediately, then 2 to 6 h later, and monthly up to the age of one year. At the time of each strain reading, the load should be measured, and if need be, adjusted. The need is defined as a variation of at least 2% from the correct value. Strain readings on the control specimens should be taken at the same time.

The ASTM C512-10 method defines the total load-induced strain as the difference between the strain values of the loaded and control specimens. If the strain immediately after loading is subtracted, creep is obtained. Use of specific creep is recommended and calculation of the rate of creep (with respect to logarithm of time) $F(t_0)$ by the Bureau of Reclamation is suggested; see Eq (16.2).

RILEM TC

This method prescribes prismatic or cylindrical specimens with a slenderness height–width ratio of 3, 4, or 5 and width being at least four times the maximum size of aggregate[2]. The preferred dimensions of the prisms are $100 \times 100 \times 400$ mm or $150 \times 150 \times 500$ mm, which may be cast horizontally or vertically.

After casting, all specimens should be stored during the first 24 h in their molds in a humid room. Subsequently, both prior to and during the creep test, they should be stored at a temperature of 20 ± 1.0 °C under one of the four following conditions: (1) sealed, (2) water, (3) in water for 6 days after demolding and then in dry air at an RH of $50 \pm 5\%$ or $65 \pm 5\%$, and (d) in air at an RH of $50 \pm 5\%$ or $65 \pm 5\%$.

The loading apparatus is required to maintain the required load permanently in time with a precision equal to at least 3%, and in order to ensure a uniform distribution of stress, one of the loading platens should be able to rotate slightly. The strain gauge length should be at least four times the maximum aggregate size and not less than 100 mm, but preferably greater than 150 mm; gauges should be equidistant from the ends of the specimens and at a distance of at least 0.25 of the specimen height. Embedment-type gauges or surface-mounted gauges with an accuracy of not less than 20×10^{-6} are permitted with at least two surface gauge positions (molded sides for prisms and diametrically opposed generators for cylinders). The required specimens are three specimens for the measurement of creep under load, three control specimens for shrinkage, and three specimens for the determination of static modulus of elasticity and compressive strength at the age of application of load for the creep test.

For the measurement of total deformation under load, an initial stress should be applied corresponding approximately to 20% of the final sustained stress. The maximum scatter between the deformations at different points of strain measurement should not exceed 25% of the mean value. If this is not the case, the specimens should be realigned and the initial loading repeated until the scatter is satisfactory. Specimens should then be left at rest for at least 1 h and then initial strain readings taken. The loading should continue with a minimum of at least three intermediate measurements of strain prior to reaching the final sustained stress. The total duration of loading process should be as short as practicable and not greater than 10 min.

Measurements of applied stress and deformation under load as well as of control specimens should be made regularly: daily during the first week, weekly during the 3 months, and then monthly. It is recommended that the age of application of load is 28 days and the sustained stress is equal to one-third of the compressive strength of concrete at the age of commencement of the creep test.

BS EN 1355: 1997

This method applies to creep of AAC or lightweight aggregate concrete (LAC) where samples for test specimens are taken from a production element, which should be representative and have a section of 100×100 mm section and height of 300 mm [3]. From the central part of the component, at least two specimens should be cut not less than 2 days after autoclaving or casting. They should be cut in such a way that their longitudinal axis is perpendicular to the rise of the mass during manufacture (AAC) or in the plane of the compression force acting in the component when used in the structure (LAC).

The standard prescribes no reinforcing bars within the gauge length of the specimen, but if unavoidable, bars that are perpendicular to the longitudinal axis may be accepted in exceptional cases. For AAC, the planeness of load-bearing surfaces should be less than 0.2 mm, and if not, grinding or capping is necessary. In the case of LAC, planeness may be less than 0.5 mm provided that equalizing layers of 12 ± 2 mm soft fiberboard are inserted between the ends of the specimen and loading platen. Deviations of other surfaces should not exceed 1 mm and the angle between the load-bearing surface and longitudinal surface of the creep specimens should not deviate from a right angle by more than 1 mm per 100 mm.

Specimens are required to be conditioned to obtain a uniform moisture distribution by drying at a temperature of $\leq 60\,^{\circ}\text{C}$ until the moisture content is $6 \pm 2\%$. Thermal equilibrium with the laboratory environment is also required by storing specimens (protected against moisture changes) for at least 72 h prior to testing at a temperature of $20 \pm 2\,^{\circ}\text{C}$. During the creep test, specimens should be stored at $20 \pm 2\,^{\circ}\text{C}$ and $60 \pm 5\%$ RH, although other conditions are permitted.

For AAC, the creep test age at loading is optional, but for LAC, the preferred age at loading is 28 days. The test procedure is identical to the RILEM method. The actual moisture contents of creep and control specimens at the beginning and end of testing, together with dry density, should be determined.

BS ISO 1920-9: 2009

The prescribed test apparatus is similar to that of the RILEM method with the exception that cylinders of 100 mm dia. are preferred, but other sizes may be used taking into account the maximum size of aggregate [4]. Planeness and perpendicularity of the specimen within the tolerances specified may be achieved by capping, lapping, or fitting steel bearing end-plates or bonding specimens together with a thin layer of epoxy resin. The stipulated curing and storage conditions are similar to the RILEM method.

The load should be measured to at least 2% using a hydraulic system or a spring-loaded system. For the former, the apparatus consists of a rigid frame in which there are the load-sustaining hydraulic cell, three test cylinders with end-plates, a hemispherical seat, and a load-cell. The bearing surfaces in contact with the specimens should not vary from the plane by more than 0.05 mm. The strain-measuring system should be capable of performing for one year without a change in calibration, strain being measured to the nearest 10×10^{-6} by attached or portable devices with reference points positively fixed; gauges relying on friction contact are not permitted. Deformations should be determined on gauge lines space uniformly around the periphery of the specimen. The gauge reference points should be evenly spaced about the mid-height of the specimen and the number of gauge lines should not be less than two for control specimens and not less than three for loaded specimens. The effective gauge length should not be less than three times the maximum aggregate size and not greater than:

1. 260 mm for large specimens without end-plates, and 160 mm for small specimens without end-plates.
2. 150 mm for large specimens having attached end-plates, and 100 mm for small specimens having attached end-plates.

Small specimens are defined as those having a maximum nominal aggregate size ≤ 25 mm and large specimens are defined as those having a maximum nominal aggregate size >25 mm.

The procedure for measurement of deformation under load is similar to that described for the ASTM C512 and RILEM methods, and like the former method, calculations should include the term $F(t_o)$ derived from Eq (16.2) representing the compliance, ε:

$$\varepsilon = \frac{1}{E} + F(t_o)\log_e (t + 1) \tag{16.2}$$

where $1/E =$ initial elastic strain, $t_o =$ age at loading, $t =$ age of concrete, and $F(t_o) =$ rate of creep with respect to logarithm of time under load $(t - t_o)$.

The above expression was first suggested by the US Bureau of Reclamation [35] for elastic strain plus creep-time for the first year under load, and hence the recommendation that it be used as a parameter for comparing creep behaviour of different concretes.

Independent Shrinkage/Moisture Expansion Tests

When drying shrinkage is required to be determined independently of creep, i.e., not using a creep companion control specimen, the apparatus often prescribed is similar to that shown in Figure 16.26. Here, a *length comparator* is used for tests of drying shrinkage and moisture expansion of clay bricks units. The specimen under test requires stainless steel balls glued to each end, which at the time of measurement locate in conical seats in the base of a steel frame and in the pointer of a dial gauge or LVDT fixed to the frame. The steel frame is usually susceptible to temperature changes and an invar bar is provided so that corrections can be made for thermal expansion [36]. Prior to taking readings of specimen, the length of the invar bar is compared with the original length at the start of testing, and any change is due to thermal movement of the frame and is used to correct the specimen reading.

In the US, ASTM C490/C490M-11 [37] prescribes that the length comparator should have a dial micrometer or other measuring device graduated to read to 0.002 mm or less, accurate within 0.002 mm in any 0.002 mm range and accurate within 0.005 mm in any

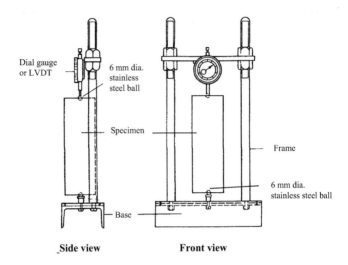

Figure 16.26 Comparator for measuring shrinkage of concrete and moisture expansion of clay bricks [36].

0.2 mm range. In addition, the measuring device should have sufficient range (at least 8.0 mm) to cater for small variations in specimen length. The terminals of the comparator should be plain, polished, and heat-treated and fitted with collars that just allow free rotation of the gauge studs that are cast into the ends of the specimen.

In the method prescribed by BS ISO: 1920 [38], drying shrinkage of sample prepared in the field or laboratory is determined by a horizontal or vertical comparator. The size of prismatic specimen is specified as $75 \times 75 \times 280$ mm or $100 \times 100 \times 400$ mm with maximum size of aggregate of 25 mm. Specimens should be moist-cured for 7 days and then, for the measurement of shrinkage, stored at a temperature of $22 \pm 2\,°C$ and $55 \pm 5\%$ RH.

Earlier in this chapter, the determination shrinkage of AAC by BS EN 1355: 1997 [3] was mentioned in connection with the measurement of creep. However, BS 680: 2005 [39] also prescribes a similar test to determine *shrinkage of autoclaved aerated concrete (AAC)*, which includes specimens of concrete block masonry units in the form of prisms cut from a new production batch. The size of the specimens should be 40×40 mm in cross-section and greater than 160 mm in length, and preferably three samples prepared from the upper third, middle, and lower third of the production unit or masonry unit in the direction of rise of the mass during manufacture. The longitudinal axis of the specimens should be perpendicular to the direction of rise and preferably in the longitudinal direction of any reinforcement. For masonry units, the longitudinal direction should be in the vertical direction corresponding to the height of the unit. Gauge plugs for the length measurement device should be attached to end faces of specimens.

Initially, the prisms should be conditioned by saturation in water at a temperature of $20 \pm 2\,°C$ for at least 72 h, and then stored in sealed plastic for a further 24 ± 2 h.

The prisms should be subsequently dried in air at a temperature of $20 \pm 2\,°C$ and $45 \pm 5\%$ RH until a constant length is recorded. Mass should be recorded as well as length using a comparator. Finally, the test specimens should be dried to a constant mass at a temperature of $105\,°C$ to determine dry density and moisture content.

BS 680: 2005 [40] specifies two parameters to be determined: (1) the *reference value for drying shrinkage*, $\varepsilon_{cs,ref}$, and (2) the *total value of drying shrinkage*. The former is determined from a plot of relative length change versus moisture content and is given by the length change between the moisture contents of 30% and 6% by mass. The latter is given by the length change from the end of conditioning period until a constant length occurs under storage of $20 \pm 2\,°C$ and $45 \pm 5\%$ RH.

ASTM C157/157M-08 [40] prescribes a test method for determining the *length change by comparator of hydraulic-cement mortar and concrete* using four prismatic specimens of size $25 \times 25 \times 285$ mm (mortar) and of size sizes of $100 \times 100 \times 285$ mm (concrete) provided the latter has a maximum aggregate size of less than 50 mm. If the maximum aggregate size is smaller, then the specimen size for concrete should be $75 \times 75 \times 285$ mm. After casting the specimens with end-face gauge studs, they should be cured in their molds for 23.5 ± 0.5 h and then stored in lime-saturated water at a temperature of $23 \pm 0.5\,°C$ for a minimum times of 15 min (50 mm prism) and 30 min (100 mm prism) before measuring their initial length. The specimens are then required to be cured in lime-saturated water at a temperature of $23 \pm 2\,°C$ until the age of 28 days before taking another length reading. Subsequently, readings are required in two storage environments:

1. Lime-saturated water tanks at a temperature of $23 \pm 2.0\,°C$ at ages of 8, 16, 32, and 54 weeks. Prior to taking readings, specimens should be placed in water tanks at a temperature of $23 \pm 0.5\,°C$ for 15 or 30 min according to the size of specimen.
2. From the end of curing, in a drying environment of $50 \pm 4\%$ RH at ages of 8, 16, 32, and 64 weeks.

A similar test is prescribed by ASTM C596-09 for *mortar-containing hydraulic cement* [41]. Specimens should be cured as specified by ASTM C157 except that they should be moist-cured in their molds for 24 ± 0.5 h, or if they have insufficient strength, for 48 ± 0.5 h. After removal from their molds, the former specimens should be cured in lime-saturated water for 48 h, and the latter specimens cured in lime-saturated water for 24 h. At the age of 72 ± 0.5 h, specimens should be removed from water, wiped with a damp cloth, and an initial comparator reading taken before air storage at $50 \pm 4\%$ RH for 25 days. Length measurements are required after 4, 11, 18, and 25 days of air storage. An approximate value of ultimate shrinkage may be obtained by extrapolating shrinkage versus the reciprocal of time (including moist-curing period) plotted on log scales.

Standards ASTM C 426-10 [42] and ASTM C 1148-9a [43] deal with *shrinkage of masonry units and mortar*, respectively. In the case of units, a value of *equilibrium shrinkage* is required after drying under specified conditions of temperature and time. Shrinkage may be measured by strain gauges with a gauge length of 254 mm or by a length comparator for which the specimen ends require gauge plugs. Whole unit or portions of face shells may be used provided they are cut lengthwise from hollow units of length ≥305 mm long, and are 100 mm wide × full length of the face. Shorter concrete bricks may be joined by an epoxy bond and a 254 mm demountable strain gauge used.

The procedure commences by immersing specimens in water at a temperature of 23 ± 1.1 °C for 48 ± 2.0 h and recording initial readings of length or strain at saturation together with temperature. After draining for 1 min ± 5 s over a 9.5 mm or larger mesh and removing visible surface water by blotting with a damp cloth, the saturated and surface-dry specimens are then weighed and the weight recorded. Within a period of 48 h after removal from water, specimens should then be stored for up to 48 h continuously in air at a temperature of 24 ± 8 °C and humidity of less than 80% RH. At least three specimens should be dried by placing in an electric oven controlled to a temperature of 50 ± 0.9 °C and brought to equilibrium with air having an RH of 17 ± 2.0%; the latter is achieved by the air immediately above a saturated solution of calcium chloride at 50 °C.

At the end of 5 days of drying, including any preliminary period of drying in air up to 48 h, specimens should be removed from the oven and cooled to 23 ± 1.1 °C within 8 h. After cooling, the length and weight readings of the specimens should be recorded; the air temperature should be 23 ± 2.8 °C at the time of readings. The specimens should then be returned to the oven for a second period of drying. The duration of the second and any subsequent periods of drying should be 44 ± 4 h. Following any further periods of drying, the cooling stage should be repeated followed by readings of weight and length. The foregoing procedure is repeated until equilibrium conditions are achieved, i.e., when the average length change is 0.002% or less over a span of 6 days of drying, and when the average weight loss is 48 h of drying is 0.2% or less compared to the last previously determined weight. If the drying shrinkage at equilibrium is not apparent, then a value may be obtained graphically from a plot of shrinkage (%) versus drying time from which equilibrium shrinkage can be estimated for a rate of shrinkage of 0.002% in 6 days [42].

To determine the drying *shrinkage of masonry mortar*, ASTM C1148-92a [43] requires five test prismatic specimens of size 25 × 25 × 285 mm, which should be cast with end-face gauge plugs and cured as stipulated by ASTM C 157/C157M-08 [40] or cured in molds for 48 ± 0.5 h, and after removal specimens should be moist cured until the age of 72 h. The length change should be measured by comparator after 4, 11, 18, and 28 days of air storage. If the shrinkage does not stabilize after 28 days of drying, then ultimate shrinkage may be estimated from a plot of shrinkage versus time.

Tests for determining *moisture expansion and shrinkage of aggregate concrete and manufactured stone masonry units* are specified by BS EN 772-14: 2002 [44]. Moisture expansion is that occurring between the initial condition and after soaking in water for four days at a temperature of 20 ± 2 °C, while shrinkage is that occurring between the initial condition and after drying for 21 days in a ventilated oven at a temperature of 33 ± 3 °C. Altogether, at least three specimens are required for each test and should be whole masonry units or cut from masonry units, which should be stored in airtight bags immediately after sampling until testing after a storage time of 28 days unless otherwise specified. After 14 days, specimens should be removed from the airtight bags and stored for a further 14 days in a laboratory at a temperature of at least 15 °C and RH not exceeding 65%. Initial measurements should be carried out after 6 h on specimens conditioned in a laboratory controlled to a temperature of 20 ± 2 °C and RH = 50–65%. Four-day final moisture expansion should be made allowing water to drain for10 min while 21-day final shrinkage should be made after allowing the specimen temperature to stabilize in the laboratory for 6 h.

Problems

1. Discuss the systems of applying the load suitable for the determination of long-term creep.
2. Discuss the different types of strain measurement suitable for short-term and long-term movements in (a) masonry and (b) concrete.
3. For projects involving a large number of specimens, which method of load application and strain measurement would you recommend?
4. It is required to determine creep and moisture movement of masonry built with a new type brick made from recycled materials. What size of masonry, method of load application, and method of strain measurement would you recommend?
5. Discuss the merits or otherwise of measuring basic and total creep of concrete by (a) mechanical gauge, (b) electrical resistance gauge, and (c) LVDT.
6. What is a length comparator used for? Describe the apparatus.
7. Define equilibrium shrinkage and describe how you would measure it.
8. Name standard methods of determining creep of concrete available to the designer.
9. In laboratory creep and shrinkage testing, how would you (a) simulate mass concrete, (b) drying of a concrete I-section beam, and (c) drying of mortar joints in a solid 2-brick wide masonry pier?
10. What specific issues should be considered when selecting a strain gauge to determine creep and shrinkage?

References

[1] ASTM C512/C512M-10. Standard test method for creep of concrete in compression. American Society for Testing and Materials; 2010.
[2] RILEM TC/CPC 12. Measurement of deformation of concrete under compressive load, 1983, RILEM recommendations for the testing and use of construction materials. RILEM; 1994. pp. 38–40.
[3] BS EN 1355: 1997. Determination of creep strains under compression of autoclaved aerated concrete or lightweight aggregate concrete with open structure. British Standards Institution; 1997.
[4] BS ISO 1920-9:2009. Testing of concrete. Determination of creep of concrete cylinders in compression. British Standards Institution; 2009.
[5] Neville AM, Dilger WH, Brooks JJ. Creep of plain and structural concrete. London and New York: Construction Press; 1983. 361 pp.
[6] Brooks JJ, Neville AM. A comparison of creep, elasticity and strength of concrete in tension and in compression. Mag Concr Res 1977;29(100):131–41.
[7] Brooks JJ. Time-dependent behaviour of calcium silicate and Fletton clay brickwork walls. In: West WHW, editor. Proceedings British Masonry Society, stoke-on-trent, no.

1; 1986. pp. 17–9 (presented at eighth international conference on loadbearing brick-work (building materials section), British Ceramic Society, 1983).

[8] Lenczner D. Design of creep machines for brickwork. In: Proceedings British Ceramic Society, No. 4; July 1965. pp. 1–8.

[9] Lenczner D. Creep in concrete blockwork piers. Struct Eng 1974;52(3):87–101.

[10] Neville AM. The measurement of creep of mortar under fully controlled conditions. Mag Concr Res 1957;9(25):9–12.

[11] L'Hermite R. What do we know about the plastic deformation and creep of concrete?, vol. 1. Paris: RILEM Bull; 1959. pp. 21–51.

[12] Ali I, Kesler CE. Rheology of concrete: a review of research, bulletin no. 476, Engineering Experiment Station. Urbana: University of Illinois; 1965. 101 pp.

[13] Meyers BL, Pauw A. Apparatus and instrumentation for creep and shrinkage studies, vol. 34. Highw Res Rec; 1963. pp. 1–18.

[14] Best CH, Pirtz D, Polivka MA. A load-bearing system for creep studies of concrete, vol. 224. ASTM Bull; 1957. pp. 44–47.

[15] Illston JM. Load cells for concrete creep testing. London Engineering 1963;195(5054):318–9.

[16] Structural Clay Products Ltd. compiled by Lenczner D. In: Foster D, editor. Brickwork: guide to creep, SCP17; 1980. 26 pp.

[17] Lenczner D. Creep and moisture movements in brickwork and blockwork. In: Proceedings of an International Conference on Performance of Building Structures, vol. 1. University of Glasgow; 1976. pp. 369–83.

[18] Neville AM, Liszka WZ. Accelerated determination of creep of lightweight concrete. London Civ Eng Public Works Rev 1973;68(803):515–9.

[19] Abdullah, C. S., Influence of geometry on creep and moisture movement of clay, calcium silicate and concrete masonry [Ph.D. thesis]. Department of Civil Engineering, University of Leeds; 1989, 290 pp.

[20] Akatsuka Y. Methods of evaluating tensile creep and stress relaxation of concretes subjected to continuously increasing loads. Trans Jpn Soc Civ Eng September 1963;97:1–12.

[21] Elvery RH, Haroun W. A direct tensile test for concrete under long- or short-term loading. Mag Concr Res 1968;20(63):111–6.

[22] Bisonnette B, Pigeon M, Vaysburd AM. Tensile creep of concrete: Study of its itivity to basic parameters. ACI Mater J 2007;104(4):360–8.

[23] Brooks JJ, Wainwright PJ, Al-Kaisi AF. Compressive and tensile creep of heat-cured ordinary Portland and slag cement concretes. Mag Concr Res 1991;42(154):1–12.

[24] Ross AD. Experiments on the creep of concrete under two dimensional stressing. Mag Concr Res 1954;6(16):3–10.

[25] Glucklich J, Ishai O. Creep mechanism in cement mortar. ACI J 1962;59:923–48.

[26] Hannant DJ. Equipment for the measurement of creep of concrete under multiaxial compressive stress, vol. 33. Paris: RILEM Bull; 1966. pp. 421–422.

[27] Gopalakrishnan KS, Neville AM, Ghali A. Creep Poisson's ratio of concrete under multiaxial compression. ACI J 1969;66:1008–20.

[28] Bazant ZP, Kim SS, Meiri S. Triaxial moisture-controlled creep tests of hardened cement paste at high temperature. Paris Mater Struct 1979;12(72):447–55.

[29] US Army Corps of Engineers. Engineering and design–instrumentation for concrete structures, chapter 2, engineering manuals. Publication, No. EM 1110-2-4300; 2012. 67 pp.

[30] Potocki FP. Vibrating-wire strain gauges for the long-term internal measurement in concrete. Eng 1958;206(5369):964–7.

[31] Tyler RG. Developments in the measurement of strain and stress in concrete bridge structures. Road Research Laboratory; 1968. Report, LR 189.

[32] Ashbee RA, Heritage CAR, Jordan RW. The expanded hysteresis loop method for measuring the damping capacity of concrete. Mag Concr Res 1976;28(96):148–56.

[33] Brooks JJ, Forsyth P. Influence of frequency of cyclic load on creep of concrete. Mag Concr Res 1986;18(136):139–50.

[34] ASTM C192/C192M-13. Standard practice for making and curing concrete test specimens in the laboratory. American Society for Testing and Materials; 2013.

[35] US Bureau of Reclamation. Creep of concrete under high intensity loading. Report No. C-820. Denver, Colorado: Concrete Laboratory; April 1956. 6 pp.

[36] Illston JM, Dinwoodie JM, Smith AA. Concrete, timber and metals. The nature and behaviour of structural materials. Van Nostrand Reinhold, International Student Edition; 1979. 663 pp.

[37] ASTM C 490/490M-11. Standard practice for use of apparatus for determining the length change of hardened cement paste and concrete. American Society for Testing and Materials; 2011.

[38] BS ISO 1920-8, 2009. Testing of concrete. Determination of the drying shrinkage of concrete for samples prepared in the field or in the laboratory. British Standards Institution; 2009.

[39] BS EN 680: 2005. Determination of the drying shrinkage of autoclaved aerated concrete. British Standards Institution; 2005.

[40] ASTM C157/C157M-08. Standard test method for length change of hardened hydraulic-cement mortar and concrete. American Society for Testing and Materials; 2008.

[41] ASTM C596-09. Standard test method for drying shrinkage of mortar containing hydraulic cement. American Society for Testing and Materials; 2009.

[42] ASTM C426-10. Standard test method for linear drying shrinkage of concrete masonry units. American Society for Testing and Materials; 2010.

[43] ASTM C1148-92a. Standard test method for measuring the drying shrinkage of masonry mortar. American Society for Testing and Materials; 2008.

[44] BS EN 772–14: 2002. Methods of test for masonry units. Determination of moisture movement of aggregate concrete and manufactured stone masonry units. British Standards Institution; 2002.

Author Index

Subject Index

Note: Page numbers followed by "f" indicate figures; "t", tables; "b", boxes.

Printed and bound by CPI Group (UK) Ltd, Croydon, CR0 4YY

08/05/2025

01864800-0002